Investigations into Living Systems, Artificial Life, and Real-World Solutions

George D. Magoulas
University of London, UK

Managing Director: Lindsay Johnston
Editorial Director: Joel Gamon
Book Production Manager: Jennifer Yoder
Publishing Systems Analyst: Adrienne Freeland
Assistant Acquisitions Editor: Kayla Wolfe
Typesetter: Christina Henning
Cover Design: Jason Mull

Published in the United States of America by
Information Science Reference (an imprint of IGI Global)
701 E. Chocolate Avenue
Hershey PA 17033
Tel: 717-533-8845
Fax: 717-533-8661
E-mail: cust@igi-global.com
Web site: http://www.igi-global.com

Library of Congress Cataloging-in-Publication Data

Investigations into living systems, artificial life, and real-world solutions / George D. Magoulas, editor.
pages cm
Includes bibliographical references and index.
Summary: "This book provides original research on the theoretical and applied aspects of artificial life, as well as addresses scientific, psychological, and social issues of synthetic life-like behavior and abilities"-- Provided by publisher.
ISBN 978-1-4666-3890-7 (hardcover) -- ISBN 978-1-4666-3891-4 (ebook) -- ISBN 978-1-4666-3892-1 (print & perpetual access) 1. Biological systems--Simulation methods. 2. Artificial life--Mathematics. 3. Neural networks (Computer science) 4. Social sciences--Mathematics. I. Magoulas, George D.
QH324.2.I63 2013
571--dc23

2012051533

British Cataloguing in Publication Data
A Cataloguing in Publication record for this book is available from the British Library.

Editorial Advisory Board

List of Reviewers

Table of Contents

Detailed Table of Contents

Chapter 3

Carlos Gershenson, Universidad Nacional Autónoma de México, México

This paper discusses how concepts developed within artificial life (ALife) can help demystify the notion of death. This is relevant because sooner or later we will all die; death affects us all. Studying the properties of living systems independently of their substrate, ALife describes life as a type of organization. Thus, death entails the loss of that organization. Within this perspective, different notions of death are derived from different notions of life. Also, the relationship between life and mind and the implications of death to the mind are discussed. A criterium is proposed in which the value of life depends on its uniqueness, i.e. a living system is more valuable if it is harder to replace. However, this does not imply that death in replaceable living systems is unproblematic. This is decided on whether there is harm to the system produced by death. The paper concludes with speculations about how the notion of death could be shaped in the future.

Chapter 4

Kenichi Minoya, Nagoya University, Japan
Tatsuo Unemi, Soka University, Japan
Reiji Suzuki, Nagoya University, Japan
Takaya Arita, Nagoya University, Japan

Human beings have behavioral flexibility based on a general faculty of planning for future events. This paper describes the first stage of a study on the evolution of planning abilities. A blocks world problem is used as a task to be solved by the agents, and encode an inherent planning parameter into the genome. The result of computer simulation shows a general tendency that planning ability emerges when the problem is difficult to solve. When taking social relationships, especially in the collective situation, into account, planning ability is difficult to evolve in the case that the problem is complex because there is a conflict between personal and collective interests. Also, the simulation results indicate that sharing information facilitates evolution of the planning ability although the free rider problem tends to be more serious than the situation where agents do not share information. It implies that there is a strong connection between evolution of the planning ability and symbolic communication.

Chapter 5

K. B. Patange, Deogiri College, India
A. R. Khan, Maulana Azad College, India
S. H. Behere, Dr. Babasaheb Ambedkar Marathwada University, India
Y. H. Shaikh, Shivaji Arts, Commerce and Science College, India

Frequency of noise can affect human beings in different ways. The sound of firecrackers is a type of intensive impulsive noise, which is hazardous. In this paper, the noise produced by firecrackers during celebration festivals in Aurangabad (M.S.), India is measured. The noise is analyzed from the study of power spectra for different types of firecrackers. Noise measurements of firecrackers show that they produce high sound pressure peak levels at their characteristics frequencies. Plots of noise power versus frequency for different crackers are presented and the inferences are discussed. Typical firecracker peak noise levels are given.

Substantial research has been dedicated to simulating the spread of infectious diseases. These simulation models have focused on major urban centers. Rural people have drastically different interaction and travel patterns than urban people. This paper describes a generic simulation package that can simulate the spread of an epidemic on a small rural town. This simulation package is then used to test the effectiveness of various mitigation strategies.

By measuring to what extent hospitals meet or exceed patient's expectations, hospital managers can determine the needed service design and delivery improvements that contribute to patient satisfaction and revisit intention. It is necessary to evaluate quality of health care services from patient perspective. This research investigates the factors, including hospital performance, hospital stay, hospital facilities, interaction with patients, service quality, and patient security culture, that affect significantly patient satisfaction and revisit intention in Jordanian hospitals using structural equation modeling. Data were collected from five main hospitals. The results showed that hospital performance has no significant effect on patient satisfaction and revisit intention. This result indicates that the patients are facing troubles in admission, registration, waiting time, and response time for results of medical tests. Also, the hospital stay, hospital facilities, service quality, and patient security culture are found significantly important in achieving patient satisfaction and revisit intention. Further, the interaction with patients' requirements and needs significantly related to service quality and hospital stay. These results shall provide policy and planning manager a great assistance in determining the factors that improve hospital performance, maintain quality medical services, and plan future improvements in the design and development of medical health care services in Jordan.

Generating chaotic attractors from nonlinear dynamical systems is quite important because of their applicability in sciences and engineering. This paper considers a class of 2-D mappings displaying fully bounded chaotic attractors for all bifurcation parameters. It describes in detail the dynamical behavior of this map, along with some other dynamical phenomena. Also presented are some phase portraits and some dynamical properties of the given simple family of 2-D discrete mappings.

Electrooculogram (EOG) signals have been used in designing Human-Computer Interfaces, though not as popularly as electroencephalogram (EEG) or electromyogram (EMG) signals. This paper explores several strategies for improving the analysis of EOG signals. This article explores its utilization for the extraction of features from EOG signals compared with parametric, frequency-based approach using an autoregressive (AR) model as well as template matching as a time based method. The results indicate that parametric AR modeling using the Burg method, which does not retain the phase information, gives poor class separation. Conversely, the projection on the approximation space of the fourth level of Haar wavelet decomposition yields feature sets that enhance the class separation. Furthermore, for this method the number of dimensions in the feature space is much reduced as compared to template matching, which makes it much more efficient in terms of computation. This paper also reports on an example application utilizing wavelet decomposition and the Linear Discriminant Analysis (LDA) for classification, which was implemented and evaluated successfully. In this application, a virtual keyboard acts as the front-end for user interactions.

This research proposes an agent-based simulation model combined with the strength of systemic dynamic mathematical model, providing a new modeling and simulation approach of the pathogenesis of AIR. AIR is the initial stage of a typical sepsis episode, often leading to severe sepsis or septic shocks. The process of AIR has been in the focal point affecting more than 750,000 patients annually in the United State alone. Based on the agent-based model presented herein, clinicians can predict the sepsis pathogenesis for patients using the prognostic indicators from the simulation results, planning the proper therapeutic interventions accordingly. Impressively, the modeling approach presented creates a friendly user-interface allowing physicians to visualize and capture the potential AIR progression patterns. Based on the computational studies, the simulated behavior of the agent–based model conforms to the mechanisms described by the system dynamics mathematical models established in previous research.

This paper revisits the problem of active learning and decision making when the cost of labeling incurs cost and unlabeled data is available in abundance. In many real world applications large amounts of data are available but the cost of correctly labeling it prohibits its use. In such cases, active learning can be employed. In this paper the authors propose rough set based clustering using active learning approach. The authors extend the basic notion of Hamming distance to propose a dissimilarity measure which helps in finding the approximations of clusters in the given data set. The underlying theoretical background for this decision is rough set theory. The authors have investigated our algorithm on the benchmark data sets from UCI machine learning repository which have shown promising results.

Preface

Synthesizing and evolving intelligent behaviour, analyzing natural processes and biological systems, and imitating their behaviour in complex artificial systems are considered major challenges in artificial life and intelligent systems design. Addressing these challenges requires crossing boundaries between various disciplines and thinking across established fields of study. In this context, mathematical and computational methods, and analysis and modelling tools, which are used in innovative ways, are critical to get insights that go beyond conventional knowledge domains.

This book brings together researchers from various disciplines whose work aims to address issues related to the above mentioned challenges. It is a collection of original research and development work in artificial life, analysis, and modelling of living systems and real world applications. It can play the role of a reference for those working in the area as it captures the state-of-the-art and provides insight into the future of artificial life and intelligent systems research. It can also be used as an advanced upper-level course supplement and resource for instructors to help them design activities that would assess the benefits of the various approaches and technologies. High-level undergraduate and postgraduate students can find in this book examples of artificial life and intelligent systems design and implementation that cover key stages in their development, and analyses of how they are performing or perceived by users. It may benefit readers from different disciplines helping them to gain holistic view of the various aspects of analysis, design, and synthesis, and technologies' deployment for complex artificial systems.

The book is organized in three parts. In Section 1, "Analysis and Modelling of Living Systems," contributions address open problems in living systems through new models, frameworks and methods for describing biological and cognitive functions, natural processes and phenomena, and examine their potential to provide real-world solutions. Developing a comprehensive understanding of the operation of living systems, and of their components and ways they communicate with each other play a key role in deciphering life processes. Chapters of this part combine hypothesis-driven experimentation with computer-based modelling and simulation to analyse and model biological functions, natural phenomena, and real life behaviours.

Section 2, titled "Mathematical and Analytical Techniques with Applications," presents mathematical and analytical techniques for modelling living systems and provides insights on how these can be exploited further in real-world applications. The importance of mathematics in the study of living systems and in the development of artificial life applications is well established. Advances in applied mathematics have facilitated the extension and development of analytical approaches and techniques, which are frequently supported by computer simulation, in order to tackle some challenging real problems in this area.

Section 3, titled "Intelligent Information Processing and Applications," focuses on knowledge-based and data-driven computational methods for modelling and exploratory data analysis of biological and cognitive processes. It proposes intelligent information processing and representation models that incorporate Artificial Intelligence (AI) techniques, such as symbolic AI, fuzzy logic and neural systems. Heuristic search and real world test cases are also employed to evaluate the effectiveness of the computations in applications.

In what follows, an overview of the book chapters is presented.

SECTION 1: ANALYSIS AND MODELLING OF LIVING SYSTEMS

In *Resistance of Cell in Fractal Growth in Electrodeposition,* Shaikh, Khan, Patange, Pathan, and Behere present a study of dynamic electrical resistance of the electrodeposition cell. Electrodeposition is a technique commonly used for the growth of metallic dendritic patterns, which often exhibit fractal pattern formations. The authors aim to explore the patterns developed as a result of Diffusion Limited Aggregation (DLA), which is often encountered in various processes in physical and chemical sciences, and engineering. To this end, the electric resistance of the circular electrodeposition cell is measured in real time using a computer-based data acquisition system. Their acquisition system has been designed for this particular study and is implemented based on standard analogue to digital controller ADC interfaced to the computer through the printer port. It allows cell voltage and current through the cell to be sampled at pre-defined intervals and measures the dynamic electrical resistance of the electrodeposition cell during growth. The findings suggest that the study of growth pattern under constant electric field conditions could help improving our understanding of the DLA.

Propagation of Front Waves in Myelinated Nerve Fibres: New Electrical Transmission Lines Constituted of Linear and Nonlinear Portions, by Nkomidio and Woafo, is a study that aims to improve our understanding of the impulse propagation in myelinated nerve fibres. To this end, the authors consider Ranvier nodes as nonlinear portions and myelin sheath as linear portions. Moreover, they extend the results of their analysis to the propagation of front waves in electrical transmission lines with alternated linear and nonlinear portions. The work explores in particular the effects of the components of the linear portion on the profile and velocity of the wave. Through numerical simulations, the authors show that the front wave introduced in the nonlinear portion deforms itself in the linear portion but recovers its initial profile and velocity in the next nonlinear portion. This form of deformation and recovery can be used for the development of new and low cost electrical transmission lines for high amplitude nonlinear signals, and for the research of neuronal prosthesis.

In *What Does Artificial Life Tell us about Death?* Gershenson discusses how concepts developed within artificial life (ALife) can help demystify the notion of death in living systems. The aim is to provide insights on how living information can be maintained and/or reproduced in digital systems. Adopting the general notion of life as a process or organization and describing living systems from a functional perspective, the author examines notions of death from a systematic point of view. Within this perspective, different notions of death can be derived from different notions of life, resulting in loss of organization. In addition, the author explores the relations between notions of life and death and the mind, cognition, awareness, and consciousness. A criterion is proposed in which the value of a living system depends on its uniqueness and replaceability. In general, living systems are difficult to replace because of their epigenesis, i.e. the information acquired in their lifetime through experience. In ALife however, it is easy to maintain and reproduce the organization acquired through development of digital organisms, as epigenetic information can be stored and replicated.

A Constructive Approach to the Evolution of the Planning Ability by Minoya, Unemi, Suzuki, and Arita explores the dynamics inherent in the mechanism of evolutionary acquisition of planning abilities in social environments. They adopt a constructive approach which attempts to create not only a symbolic model of a living system but also a symbolic living object. Their model is elaborated without direct and precise reference to empirical biological reality. A blocks world problem is used as a task to be solved by the agents and encode an inherent planning parameter into the genome. Simulation experiments show there is a general tendency for planning ability to emerge when the problem is difficult to solve. When considering social relationships, especially in the collective situation, the planning ability is difficult to evolve when the problem is complex because there is a conflict between personal and collective interests. Also, the simulation results indicate that sharing information facilitates evolution of the planning ability, although the free rider problem tends to be more serious in this case than in the case where the agents do not share information. According to the authors, the results imply that there is a strong link between evolution of the planning ability and symbolic communication. These findings provide insight into the mechanism of the co-evolutionary dynamics of the planning and symbolic communications when two kinds of groups, sharing and no sharing information, coexist in the same population.

In *Noise Power Spectrum for Firecrackers,* Patange, Khan, Behere, and Shaikh investigate the sound of firecrackers which is a type of intensive impulsive noise that is hazardous. It is well known that noise is often a limiting factor for the performance of a device or system. In the context of living systems, the term "noise" is associated with variance amongst measurements obtained from identical experimental conditions. The authors experiment with noise produced by firecrackers during celebration festivals in Aurangabad, India. The noise is analyzed from the study of power spectra for different types of firecrackers. Noise measurements of firecrackers show that they produce high sound pressure peak levels at their characteristics frequencies. Plots of noise power versus frequency for different crackers are presented and the inferences are discussed. Typical firecracker peak noise levels are also described. As frequency of noise can affect human beings in different ways, the findings of this study may provide further insight into the extent environmental noise affects biological functions and the relation between external noise and noise that depends on intrinsic system properties.

In the same vein, *Traffic Noise: 1/F Characteristics* by Patange, Khan, Behere, and Shaikh explores the characteristics of the so-called pink noise. Samples of Traffic Noise are collected from selected locations from busy roads of Aurangabad city in Maharashtra state (India) and the data are analyzed. It is observed that in many cases the traffic noise possesses pink noise (1/f noise) prevailing over appreciable range of frequency. The log-log plot of noise power versus frequency results in a straight line with a slope approximately equal to unity confirming the presence of pink noise. After certain frequency, the noise power no longer behaves like pink noise (1/f noise) and becomes more or less constant with random fluctuations. Plots of noise power versus frequency on log-log basis for different locations studied are presented and the inferences are discussed. The results could contribute to our understanding of the environmental noise and the extent the natural environment influences the behaviour of living systems.

SECTION 2: MATHEMATICAL AND ANALYTICAL TECHNIQUES WITH APPLICATIONS

Existence of Positive Solutions for Generalized p-Laplacian BVPs, by Lian, Wong, Lo, and Yeh, concerns a mathematical problem which arises when modelling living systems. It is widely acknowledged that nonlinear equations and numerical analysis techniques are important to the description of complex multi-component systems, in general, and living systems in particular. In this work, the authors examine the existence of positive solutions for nonlinear boundary value problems (BVP). In particular, the authors look at multiple solutions for higher order BVPs with p-Laplacian operator. Their theoretical results have implications for a class of BVPs that has received attention because a number of physical, biological, and chemical phenomena are described in this way.

In *Mathematical Model to Assess the Relative Effectiveness of Rift Valley Fever Countermeasures,* Gaff, Burgess, Jackson, Niu, Papelis, and Hartley study mathematical modelling of infectious diseases. This is an area that has attracted attention because it concerns the mechanics of disease propagation, intervention strategies, and control, as well as sensitivity of countermeasures. The authors use a Rift Valley Fever (RVF) model to study the efficacy of countermeasures to disease transmission parameters. RVF is a viral infectious disease that propagates through infected mosquitoes and primarily affects animals but also humans. The RVF model consists of a deterministic ordinary differential equation system that predicts the spread of RVF in mosquitoes and a livestock population. Disease control and countermeasures for RVF include the application of insecticide to vector populations targeting either adult mosquitoes or mosquito larvae, livestock vaccination, or finally culling of exposed and/or infected animals. The authors first extend the mathematical model of RVF to include transmission of RVF between two vector populations and a single livestock population. The new model allows them to experiment with four types of disease intervention strategies: vector adulticide, vector larvicide, livestock vaccination, and livestock culling. They explore each of these approaches through simulation for various degrees of intensity and efficacy of intervention and evaluate the sensitivity of the models to the various disease transmission parameters. Results suggest that, under certain conditions, livestock vaccination, and culling offer the greatest benefit in terms of reducing livestock morbidity and mortality. As there is currently lack of comparative evaluation studies between countermeasures for RVF, this work can provide decision maker insights about disease intervention strategies and control.

Uribe-Sanchez and Savachkin, in *Resource Distribution Strategies for Mitigation of Cross-Regional Influenza Pandemics,* develop new pandemic mitigation models that offer dynamic decision support capabilities. Their aim is to assist public health policy makers in developing effective dynamic predictive distribution strategies of limited resources during influenza pandemics. The authors study the evolution of disease, the population dynamics and employ simulation optimization to generate dynamic resource distribution strategies. While the underlying simulation mimics the disease and the population dynamics of the affected regions, their optimization model generates progressive allocations of mitigation resources, including vaccines, antivirals, healthcare capacities, and social distancing enforcement measures. The proposed approach minimizes the impact of ongoing outbreaks and the expected impact of the potential outbreaks, considering measures of morbidity, mortality, and social distancing, translated into the cost of lost productivity and medical expenses. The model was implemented on a simulated outbreak involving four million inhabitants and compared to pro-rata and myopic strategies. The pro-rata policy allocates the total available resources to all network regions a-priori, in proportion to the regional population, while the myopic policy allocates the available resources from one actual outbreak region to the next, each time trying to cover the entire population at risk of the region. The simulation results provide evidence that on average, the new strategy outperforms both the pro-rata and the myopic policy at all levels, providing efficient resource utilization.

A Computational Model of Mitigating Disease Spread in Spatial Networks, by Kim, Li, Zhang, Sen, and Ramanathan, examines the problem of damage spreading and containment in spatial networks. The authors focus on disease spreading in two-dimensional fixed-radius random networks, and pay particular attention on the kinematics of disease spreading with respect to how damage is controlled by their model. They propose control strategies that are potentially relevant in diverse contexts spanning a range of spatial and temporal scales, such as culling during epidemics in farm animals, fire fighting to wildfires and social bullying in community networks. In addition, they analyze both the sensitivity of disease progression and the effect of the containment process with respect to various parameter settings as well as the correlation of model parameters. The findings of this study suggest that the radius of containment process is the most critical parameter, e.g. the kinematics of the spatial-temporal patterns is particularly sensitive to the radius of the containment process region. Thus finding the best available values in the computational model would allow reducing damages from disease spread of a future pandemic. Insights from this study can be useful to control other virus spread problems in spatial networks, such as disease spread in a geographical network and virus spread in a brain cell network.

Easton, Carlyle, Hunt, Anderson, and James' work, titled *Simulating the Spread of an Epidemic in a Small Rural Kansas Town,* is concerned with the spread of infectious diseases and the impact of mitigation strategies. Predicting whether or not a disease impacts an entire town and becomes a pandemic is extremely sensitive to changes in probabilities or mitigation strategy, and has an immense impact on the number of infected individuals. In this context, mathematical models have been used to derive interesting theoretical results, such as how fast a disease dies out and how a disease spreads from host to host. Also, substantial research has been dedicated to simulating the spread of infectious diseases with a focus on major urban centres. In contrast to existing models, which assume that time is divided into periods and each individual is classified in a particular state for an entire period, the authors generate a contact network to simulate how the disease spreads and consider multiple different states. Their work also exploits the concepts of disease tracks, which enable different groups of individuals to have distinct disease paths. Thus, each disease track could be associated with different parameter values. Furthermore, the contact network enables the diseases to spread according to the individual's habits. Moreover, the authors focus on people living in rural areas, who have drastically different interaction and travel patterns than urban people. They demonstrate the use of their model and simulation package in modelling and predicting the spread of an epidemic on a small rural town and the effectiveness of various mitigation strategies.

A Structural Model to Investigate Factors Affect Patient Satisfaction and Revisit Intention in Jordanian Hospitals, by Al-Refaie, investigates the factors, including hospital performance, hospital stay, hospital facilities, interaction with patients, service quality, and patient security culture, that have an impact on patient satisfaction and revisit intention in Jordanian hospitals. The author combines qualitative and quantitative methods in order to model patient satisfaction and uses structural equation modelling as a means to empirically support the effectiveness of the derived models. The study employs data collected from five main hospitals and shows that hospital performance has no significant effect on patient satisfaction and revisit intention. The findings indicate that patients are facing difficulties with admission and registration services as well as waiting and response time for test results. Also, hospital stay, hospital facilities, service quality, and patient security culture are found to play an important role in achieving patient satisfaction and revisit intention. This form of analysis and modelling based on patient's perceptions can contribute to our understanding of the extent hospitals meet patient's expectations and can be useful to hospital managers to determine service design requirements and delivery improvements that contribute to patient satisfaction and revisit intention. Moreover, the findings of this work can provide assistance to policy makers and planning managers in determining the factors that improve hospital performance, maintain quality medical services, and plan future improvements in the design and development of medical health care services in Jordan.

In *Generating Fully Bounded Chaotic Attractors,* Elhadj investigates how dynamical systems make a transition from regular behaviour to chaos. The issue of generating chaotic attractors has several applications in artificial intelligence and artificial life application where nonlinear dynamics emerge. The author explores a particular class of 2-D nonlinear mappings, which has been widely studied because it is the simplest example of a dissipative map with chaotic solutions. In this case, fully bounded chaotic attractors emerge for all bifurcation parameters. The work describes in detail the dynamical behaviour of this map and discusses other dynamical phenomena. The work presents some phase portraits and discusses dynamical properties of the given simple family of fully bounded 2-D discrete mappings, revealing some new chaotic attractors.

The issue of constructing self similar patterns is investigated by Singh, Mishra, and Jain in the context of fractal geometry. Their work, titled *Orbit of an Image under Iterated System II,* concerns orbital pictures that are expressed in terms of transformations of an Iterated Function System-IFS. An orbital picture of this type is a mathematical structure that is developed by following the path of an object under IFS. This is typically implemented by using one-step feedback process namely, the function iterative procedure. Although, the one-step process works well for contractive transformations, sometimes problems arise when the transformations are non-contractive. The authors extend previous work introducing the role of superior iterative procedure to find the orbital picture under IFS. They use this procedure to generate orbital pictures of different variability for non-contractive and non-expansive transformations using two-step feedback process. Their algorithm mode is experimentally evaluated. The findings show that the superior iterative procedure generally works very well to construct orbital pictures in case of non-contractive transformations. Moreover, it converges smoothly wherein one-step process does not converge. This means that the sequence of objects obtained by superior iterations gives an attractive orbital picture, while the sequence of objects obtained by the function iterative procedure oscillates and does not converge towards a regular pattern. These promising results demonstrate the potential of this approach to compute 2-variable orbital pictures, which is useful in computational mathematics and fractal image processing applications, e.g. biological modelling, computer graphics, image compression, and in other application areas of fractal geometry.

Along the same line the work titled *Superior Koch Curve* by Prasad investigates the design of Superior Koch Curve with different scaling factor. The Koch curve is the limiting curve obtained by applying the self similar divisions to infinite number of times. In the Koch curve, self-similar patterns can be obtained by dividing a line into three equal parts and replacing the middle segment of this straight line by a triangle of the same length as the segment being removed and then applying this construction an infinite number of times on the resulting segments. This process of fractals construction yields an iterated function system. The particular curve proposed in this work has been designed using the technique of superior iteration, i.e. the scaling factor is based on superior iteration. The proposed design approach has implication in a range of real world applications, such as computer graphics and antenna miniaturization.

SECTION 3: INTELLIGENT INFORMATION PROCESSING AND APPLICATIONS

Chowdhury, Scoglio, and Hsu, in *Mitigation Strategies for Foot and Mouth Disease: A Learning-Based Approach,* study models that learn from data to predict the spread of the Foot and Mouth Disease (FMD) in time. Their work is in the field of predictive epidemiology studying disease dynamics in order to predict future outbreaks. This is an area where several analytical spatio-temporal models exist to spatially locate epidemic outbreaks in time mainly employing explicit mathematical models. In contrast, the authors' approach is data-driven, assuming that spatial information regarding disease dynamics is imprecise, and exploits learning-based predictive models as a promising alternative. In their formulation, local information regarding the temporal evolution of infection is hard-coded in geographical regions, and a utility function is defined to assess the cost-effectiveness of mitigation strategies. Their study analyzes temporal predictions from neural network, autoregressive, Bayesian network, and Monte-Carlo simulation models which are validated using FMD incidences reported in Turkey. The findings support their claim that it is possible to generalize local learning-based models to recover the latent spatial parameters. In addition, the authors perform simulations of mitigation strategies based on the predictive models to study the cost-effectiveness of culling, vaccination and movement strategies in order to minimize the total number of infected livestock at the end of a period under study. They conclude that neural networks are better predictors than other models and that vaccination and movement ban strategies are more cost-effective than premise culls only before the onset of an epidemic outbreak.

In *Considerations on Strategies to Improve EOG Signal Analysis,* Wissel and Palaniappan explore on Human-Computer Interfaces that employ Electrooculogram (EOG) signals, e.g. interfaces for eye-based applications, hands-free interfaces and wearable embedded systems. Their approach consists of a feature extraction stage and a classification stage. They explore various strategies for improving the analysis of EOG signals, investigating extraction of features using parametric methods in the frequency and time domains. Their methods range from straightforward time-based techniques that take the whole buffer content without any further operations to sophisticated wavelet decomposition based on Discrete Wavelet Transform-DWT to obtain specific features in a particular time and frequency resolution. For the classification stage, the authors employ Nearest Neighbour (NN) classifiers, Artificial Neural Network (ANN) and Linear Discriminant Analysis (LDA). Considering a virtual keyboard as the real life context for their study, they perform a comparison of the various feature extraction-classification combinations using data recorded from real user interactions. They particularly identify four promising strategies: template matching using NN and three DWT-based classifiers. Their results indicate that parametric Auto Regressive modelling using the Burg method, which does not retain the phase information, gives poor classification. In contrast, they find that the projection on the approximation space of the fourth level of Haar wavelet decomposition yields feature sets that enhance the class discrimination and reduce the feature dimensionality, making the method much more efficient in terms of computation. They also find that the computation cost for ANN and LDA is not significantly different when training is done off-line. Lastly, the authors implement a combination of DWT-based LDA classifier with a virtual keyboard acting as the user front-end and perform real time testing with a small sample of users getting some very promising detection rates.

An Autonomous Multi-Agent Simulation Model for Acute Inflammatory Response, by Wu, Ben-Arieh, and Shi, proposes an agent-supported system to enable computer simulation for acute inflammatory response-AIR. AIR is the initial stage of a typical sepsis episode, often leading to severe sepsis or septic shocks. The process of AIR has attracted the interest of clinicians because it affects more than 750,000 patients annually in the United State alone. Existing approaches employ mathematical models describing system dynamics and offer limited ability of capturing behavioural variations since they use deterministic scalar parameters. In order to model the inherent stochastic nature of the biological system and provide the ability to include the correct boundary conditions, the authors combine agent-based modelling with dynamic mathematical models. The proposed agent-supported system employs multiple agents that exploit knowledge of the variables describing the system dynamics and incorporate various indicators of AIR progression. This approach allows to link system dynamics and real AIR environment-related information. Moreover, agents allow defining autonomous and probabilistic behaviours, capturing the stochastic nature of the AIR progression episode. Unlike the use of strict mathematical model, the agent-supported system simulation allows the designers and the users to simulate and observe the interactions among different agents; thus, it is more intuitive and flexible than the traditional mathematical models. Outcomes of this work can be useful to clinicians interested in predicting the sepsis pathogenesis for patients. Using the prognostic indicators from the simulation results, clinicians can plan appropriate therapeutic interventions, and visualise and capture potential AIR progression patterns.

In *Rough Set based Clustering Using Active Learning Appro*ach, Kandwal, Mahajan, and Vijay revisit the problem of active learning and decision making when a large number of unlabeled data is available but their labelling incurs a high cost. The main aim of a clustering approach that employs active learning would be to generate or sample the unlabeled instances in such a way that they self-organise into small groups with minimal overlapping. In this context, the authors extend the basic notion of Hamming distance to propose a dissimilarity measure, which helps finding the approximations of clusters in the given data set. Moreover, they introduce Rough sets as a tool to deal with inexact, imprecise, or vague knowledge in the real world data and identify rough clusters. Their algorithm partitions the dataset into k clusters and tries to maximise the intracluster similarity whilst minimising the intercluster similarity. Active learning is applied at each iteration with the learner actively selecting a batch of unlabeled samples for training to improve the internal model, i.e. make cluster adjustment as quickly as possible. Their algorithm compares the feature vector for a new instance with feature vectors for clusters centres, when there is a matching the feature vector is labelled, otherwise the distance is calculated and used to label the vector with the clusters where the similarity is maximum. The authors evaluate successfully the proposed algorithm using benchmark data sets from the UCI machine learning repository.

The issue of managing imprecision, inherent in the operation of living and artificial intelligence systems, is also investigated in the work of Danish Lohani, titled *Intuitionistic Fuzzy 2-Metric Space and Some Topological Properties* in the context of fuzzy topology. Fuzzy topology has a wide range of applications in quantum physics, nano technology, and brain research. It is a specialised domain of Fuzzy theory which has found applications in the field of science and engineering, e.g. population dynamics, chaos control, nonlinear dynamical systems and medicine. In this work, the focus is on the notion of intuitionistic fuzzy metric space and the concept of intuitionistic fuzzy normed space. The author studies the intuitionistic fuzzy 2-metric space, which provides a more suitable functional tool to deal with the inexactness of the metric, or 2-metric in some situations. He discusses analogues of precompactness and metrizability and establishes some theoretical results in this space.

Folding Theory for Fantastic Filters in BL-Algebras, by Lele, also contributes to fuzzy logic theory examining the notion of n-fold fantastic and fuzzy n-fold fantastic filters in BL-algebras. The work goes beyond the typical formulation of fuzzy logic as a system of formal deductive systems that support deduction under vagueness. It equips mathematical fuzzy logic with abstract algebraic semantics in a way analogous to the classical (Boolean) logic. The author shows that every n-fold (fuzzy n-fold) fantastic filter is a filter (fuzzy filter), but the converse is not true. Building on the concept of level set of a fuzzy set in a BL-algebra, the author characterises fuzzy n-fold fantastic filters and establishes an extension principle for n-fold and fuzzy n-fold fantastic filters in BL-algebras. These theoretical results are important not only for BL-algebra but also as foundations of methods of fuzzy logic in a broad sense.

Psillakis, Christodoulou, Giotis, and Boutalis' work, titled *An Observer Approach for Deterministic Learning Using Patchy Neural Networks with Applications to Fuzzy Cognitive Networks,* contributes both to theory and applications by combining Fuzzy Cognitive Networks and dynamic systems with new

Neural Network models. The authors build on previous work that established Fuzzy Cognitive Networks (FCNs) as an alternative operational extension of Fuzzy Cognitive Maps-FCM. Their aim is to increase FCMs suitability for control and adaptive decision making applications by better representing interactions with the system they represent and dynamic behavioural patterns. In this work, the authors derive a new identification methodology that consists of designing a suitable observer that provides asymptotic estimation of the respective nonlinear vector field, and then employing a localised neural network to extract and store the information of this estimate. To this end, a new class of localised neural networks, called Patchy Neural Networks (PNNs), are introduced. PNNs employ basis functions that are "patches" of the state space and are used to identify the unknown nonlinearity from the observer's output and the state measurements. The authors propose a scheme that achieves learning in a single pass from the respective patches and does not need standard persistency of excitation conditions. Furthermore, PNNs weights are updated algebraically, reducing the computational load of learning significantly. This identification procedure is applied to the learning problem of the dynamics of FCNs and a Duffing oscillator providing evidence for the effectiveness of the proposed approach.

George D. Magoulas
University of London, UK

Section 1
Analysis and Modelling of Living Systems

Chapter 1
Resistance of Cell in Fractal Growth in Electrodeposition

Y. H. Shaikh
Shivaji Arts, Commerce and Science College, India

K. B. Patange
Deogri College, India

A. R. Khan
Maulana Azad College, India

J. M. Pathan
Maulana Azad College, India

S. H. Behere
Dr. Babasaheb Ambedkar Marathwada University, India

ABSTRACT

This paper presents the study of dynamic electrical resistance of the electrodeposition cell during the growth metallic dendrites showing fractal character. The electric resistance of the circular electrodeposition cell is measured in real time using a computer based data acquisition system. The data acquisition system constructed is capable of measuring the cell voltage and the current through the cell under program control at pre-decided intervals. This allows for the measurement of the dynamic electrical resistance of the electrodeposition cell. The system is based on standard analogue to digital controller ADC interfaced to the computer through the printer port.

1.1 INTRODUCTION

Fractal pattern formation (Yaroslavsky, 2007) is one of the important phenomena in nature. One of the commonly used techniques for the growth of metallic dendritic patterns showing fractal character is elctrodeposition (Atchison, Burford, & Hibbert, 1994). Patterns developed as a result of Diffusion Limited Aggregation (DLA) (Cronemberger & Sampaio, 2006; Chakrabarty et al., 2009; Hibbert, 1991) are often found in various processes in physical sciences, chemical sciences, and engineering.

The electro-depositions obtained in circular cell (Vicsek, 1992; Costa, Sagues, & Vilarrasa, 1991) geometry under constant cell operating voltage conditions indicate that the growth at thc outer part of the depositions is relatively denser

DOI: 10.4018/978-1-4666-3890-7.ch001

(Shaikh, Khan, Pathan, Patil, & Behere, 2009). The increased branching at the later stage of development of the growth was attributed to the increased electric field (Argoul, Huth, Merzeau, Arnrodo, & Swinney, 1993; Fleury, Chazalviel, Rosso, & Sapoval, 1990) due to the reduction in gap between cathode and anode. This change is gradual and continues with the growth and with the evolution of the growth the electrical properties of the cell undergo changes. This motivated us to measure the dynamic electrical resistance of the electrodeposition cell as this will help better understanding of the mechanism.

1.2 Design and Construction of Computer Controlled Data Acquisition System Using ADC 0809 and Printer Port Interface

The data acquisition system was based on an 8 bit, 8 input ADC 0809 (Tocci & Widmer, 1998; Ram, 1991; Anderson, 1996) interfaced to the computer through the printer port (Anderson, 1996; Peacock, 1994) with the eight analogue inputs independently selectable through the address lines on the ADC through the software is as shown in experimental set up in Figure 1. The ADC used has an 8-bit resolution, which was found to be sufficient for the present work. The ADC was operated at full voltage range of 0 – 5 V dc, this resulted the accuracy in measurement of 1 part in 256 over the full range of operation. This corresponds to a resolution of 0.02 volts and was sufficient for the present application. The voltage measurement employed a potential divider to cover the desired range of voltage.

Calibration of the divider was checked and adjusted before use. The cell voltage was measured using a standard digital voltmeter and the values were compared with the data read in the computer and its equivalent cell voltage. Table 1 gives a typical set of readings, the column labeled actual cell voltage is the voltage measured across the cell using a digital voltmeter.

The two columns under ADC measurements gave the corresponding values of digital values read by the program and then converted to value of analogue voltage across the cell. For comparison, the measured voltage across the cell is plotted against actual cell voltage in Figure 2. The resulting plot is a reasonably linear, the straight line joining the points is the least square fit applied to the data points. The least square fit line fairly fits the data as is seen from the value of r^2 = 0.9993. The slope of the line is 1.0017 indicating a linear relationship and the measured values are in agreement with the true values of the cell voltages.

For the measurement of current, a resistance was included in series with the electrodeposition cell and the voltage developed across the resistance was measured. The actual current was obtained in controlling programme using the ratio of potential difference to the resistance $I = \frac{V}{R}$, V is the voltage measured across the series resistance R. Ideally a low series resistance is desirable so that the potential drop across the sensing resistor is small enough and can be neglected. A low series resistance develops low voltage across it for the present working conditions thus the potential developed is to be amplified for recording. Typically a current of 0.5A through a series resistance of 0.2 ohm will develop a potential dif-

Figure 1. Experimental set up for resistance of cell in electrodeposition

Table 1. Comparison of measured and actual voltage across the cell

S.No.	Actual cell voltage	ADC Measurement	
		Digital value	Voltage (V)
1	0.0	0	0.000
2	0.5	9	0.527
3	1.0	17	0.938
4	1.5	26	1.523
5	2.0	34	2.051
6	2.5	43	2.520
7	3.0	51	2.930
8	3.5	60	3.574
9	4.0	68	4.160
10	4.5	77	4.512
11	5.0	85	5.039
12	5.5	94	5.391
13	6.0	102	5.977
14	6.5	111	6.328
15	7.0	119	6.973
16	7.5	128	7.559
17	8.0	137	7.910
18	8.5	145	8.496
19	9.0	154	9.199
20	9.5	162	9.492
21	10.0	171	10.078

ference of 0.1 V. Managing low resistance under the experimental conditions works out to be tricky. Therefore a compromise has to be arrived at, about the selection of the value of the current sensing resistor.

The full scale voltage for the ADC circuit used is 5Vdc and the expected current values are normally within 1 A. To obtain 5V (full scale voltage) at a current of 1 A, a 5 ohm sensing resistor is needed. There are two options for connecting the series resistance, on the positive power supply side of the circuit as shown in Figure 2 and 3 on the negative or ground side of the power supply as in circuit as shown in Figure 4.

The drawback of connecting it on the positive side of the power supply is that the potential difference developed across the sensing resistor is floating (not with respect to common ground). This requires the shifting of the zero by an amount equal to the instantaneous value of the cell voltage for interfacing with the circuit. Another probable solution is to use a floating ground. The advantage of connecting the current sensing resistor on the ground side is that the same can be directly connected to the measuring circuit. The disadvantage of this is that the common ground for the cell voltage measurement is to be shifted or else ac-

Figure 2. The plot of actual cell voltage against measured voltage across the cell

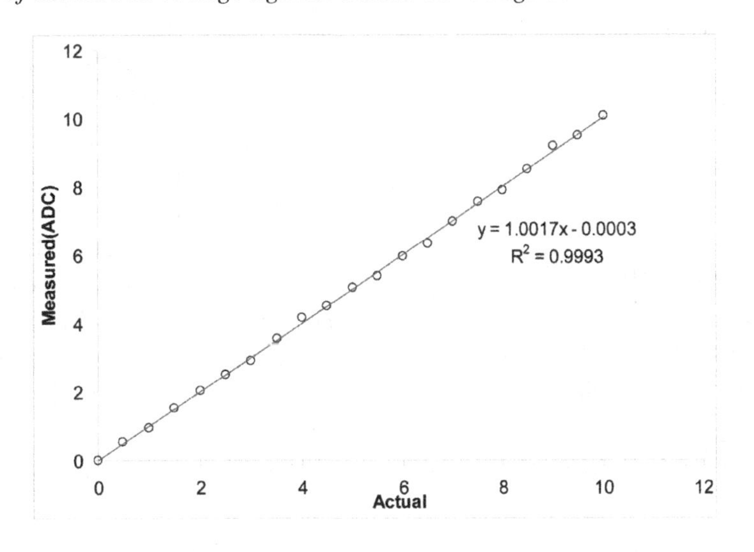

Figure 3. Series resistance on the positive side of power supply

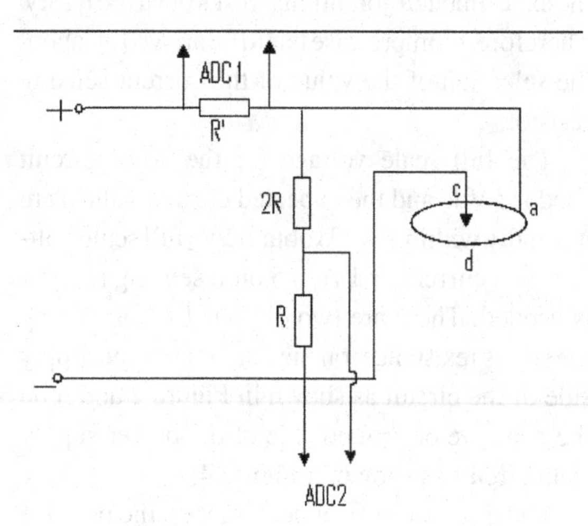

Figure 4. Series resistance on the negative side of power supply

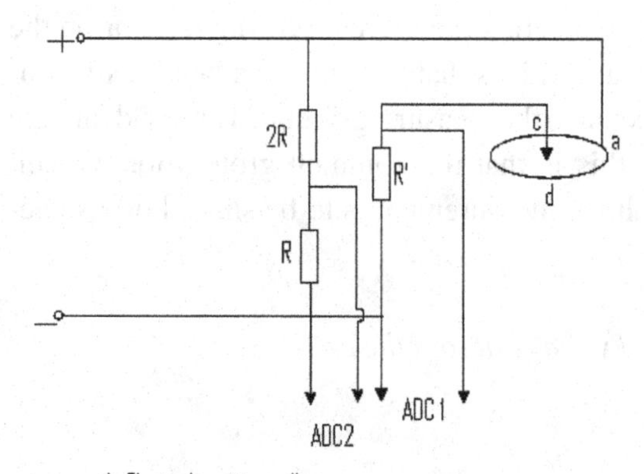

tual cell voltage has to be obtained subtracting the potential drop across the current sensing resistor. In both the cases the cell voltage will not remain constant and the actual cell voltage has to be used. As the change in voltage across the cell is not too large during a set of experimental conditions, this method is acceptable as the main point of concern is the resistance of the electrodeposition cell.

There is one more interesting feature of this type of arrangement that, as the growth proceeds, the current through the cell also increase. This causes a higher drop across the current sensing resistor, which in turn results is a decrease in the effective cell voltage. Effectively, as the growth proceed the cell voltage gradually decrease. This effect we noticed in a series of experiments under changing working conditions. The electrodeposition under constant cell voltage conditions gives rise to a growth which is a result of increasing electric field conditions in the cell. This is so because as the growth develops, the distance between the cathode and the anode is reduced, hence the electric field is increased. This increase in the electric field during the growth can be partly compensated by the potential drop across the current sensing resistor. We observed that when a series resistance is present in the circuit, the growth on the outer part of the pattern is less branched in contrast to the one in the absence of the series resistance. Under constant voltage conditions, as the growth becomes larger in size, the branching trend is found to be more (some what similar to higher voltage growth). This difference in structure of the growth is found to be reduced with the use of a series resistance, which contributes to the decrease in cell voltage (with the increase in size of growth). However, provisions can be made for completely controlled conditions to obtain growth under predetermined conditions.

As we have used only two analogue inputs, the two MS bits of address lines were permanently hard wired to zero (ground) and the two inputs were selected controlling the LSB of the address lines (pin 25 of ADC). When this line is held low, input IN 0 is selected (Pin 26 of ADC 0809) and when this pin is held high (1), IN 1 is selected (Pin 27 of ADC 0809). To select an analogue input (say IN 0 or IN 2), valid address of the analogue input (0 for IN 0 or 1 for IN 1) is made available at pin 25 of ADC. The Address Latch Enable (ALE line, Pin 22 of ADC 0809) is pulsed through the program and the address is latched when the pin undergoes a low to high transition. This was implemented using one of the output lines of the printer port. The conversion of analogue input to digital output is initiated by giving a short pulse (duration of more than 100 ns typically) to the start of conversion pin (START line, Pin 6 of ADC 0809). Actual conversion begins at the negative going edge of the pulse applied to the START input. The end of conversion is indicated by the changing status, from low to high, of the EOC line (Pin 7 of ADC 0809). For taking a measurement the program keeps watching this pin, and as soon as thin pin goes high, the procedure for reading in the digital value is initiated. The 8 bit digital output, split up into two nibbles by the multiplexer in the circuit, is then read in by the program in succession. The program then combines the two nibbles to obtain the value of the digital input, which is then converted into the voltage or current using the corresponding multiplying factor.

1.3 Controlling Programme Logic for Data Acquisition System

Figures 5 and 6 are flowcharts for programme and control logic of Data acquisition system used for measurement of resistance of electrodeposition cell.

The basic programme logic is shown in Figure 5. For the measuring the dynamic resistance of the cell with time, voltage across the cell and the current through the cell is measured at regular intervals of time. For measuring the voltage across the cell a potential divider is sued and for measuring the current through the cell, a series resistance is used as a sensor.

For measuring the resistance, first the voltage is read by the ADC and the programme converts the reading to its equivalent voltage. Through the controlling programme, the first analogue input of the ADC (In-0) is selected controlling the address lines of the ADC and the address latch enables. First the address of the input to be selected in made available at the three address lines and the ALE line is pulsed to latch the address. To initiate conversion of the analogue input into digital output, a start conversion (SC) pulse is given by the controlling programme to the ADC. The conversion commences at the positive going edge of the SC pulse. Now the end of conversion (EOC) is indicated by the EOC output of the ADC. The programme keeps watching the EOC signal. If the EOC signal is not found within expected time, it announces failure and programme stops. Once the EOC signal is detected, the reading of the digital signal begins.

As the circuit used is capable of reading one nibble (4 bits) at a time, the programme first selects lower nibble through the multiplexer and reads it. Then the second nibble is selected and read in. After reading both the nibbles, the programme combines the two nibbles and makes the full eight bit number. This number is then converted into the actual measured quantity using the multiplying factor. After measuring the voltage, following exactly the same procedure as is shown in the Figure 6 the current is also measured. The resistance of the cell is then calculated dividing the potential difference across the cell 'V' by the current through it 'I'.

1.4 Measurement of Resistance of the Electro-Deposition Cell at Regular Interval of Time During the Growth Process

We measured the electrical resistance of the electrochemical cell (Center for Polymer Studies, 2010) at regular intervals of time during the growth process. A typical growth of copper dendrite (Sander, 2000; Ben-Jacob, 1993) under normal working condition takes about 10 to 30 minutes. We recorded the voltage across the electrodeposition cell and the current through the cell at regu-

Figure 5. Flowchart 1

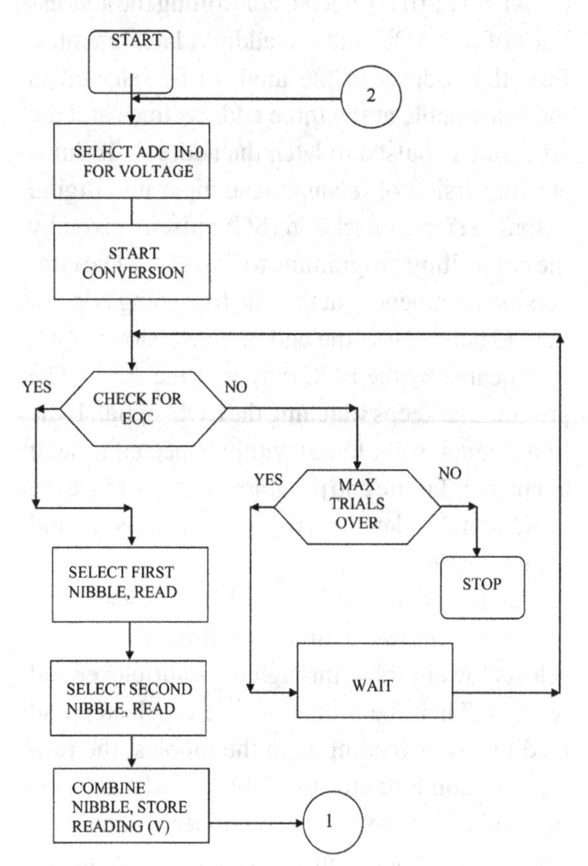

Figure 6. Flowchart 2

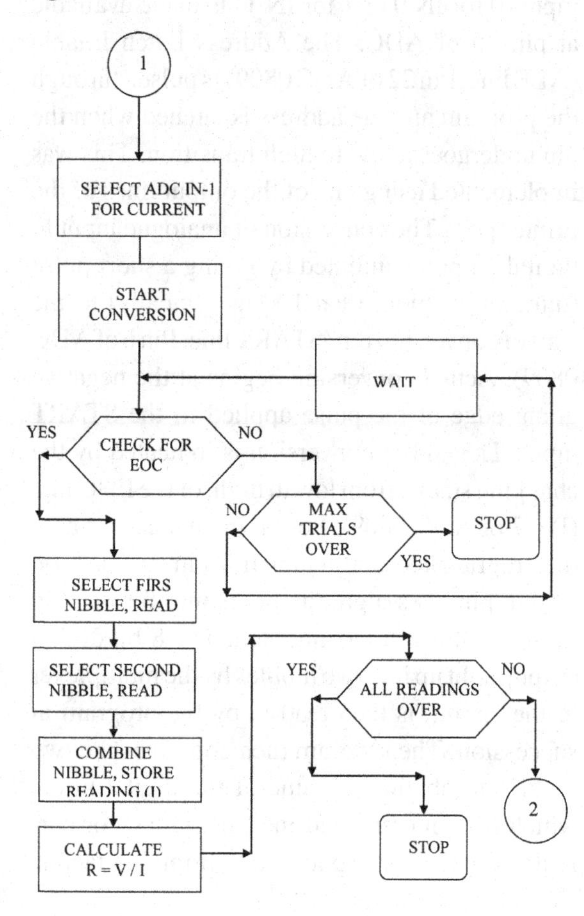

lar intervals of 10 seconds. As a regulated power supply is used, the voltage across the cell is expected to be constant throughout. As a series resistance is used to measure current the drop of voltage across this resistance gets subtracted from the actual supply voltage and the cell voltage is lowered by this amount. Thus during the growth studied, the voltage across the cell remains more or less constant except for the drop across the current sensing resistor. However as the growth proceeds, the current through the cell is also found to increase. This increase in current results in a corresponding decrease in cell voltage. A typical set of measurements recorded at a regular interval of 0.2 min (12 sec), for the growth of copper dendrite is presented. The voltages and currents are the actually measured values at the corresponding time point and the resistance is the resistance of the cell calculated as $R = \frac{V}{I}$. Figure 7 is a plot of current versus time and Figure 8 shows the change in resistance with time as the growth proceeds.

1.5 Result and Discussion

Results from series of experiments showed that as time proceeds the current through the cell was found to increase and there was a slight decrease in voltage across the cell. The value of resistance of the cell is calculated. It is observed that during the initial phase of growth the current is low and thus the electrical resistance of the cell is high, also the current increases faster with growth. As the time proceeds and growth develops, the increase in current becomes more or less linear, in

Figure 7. Showing the plot current with time

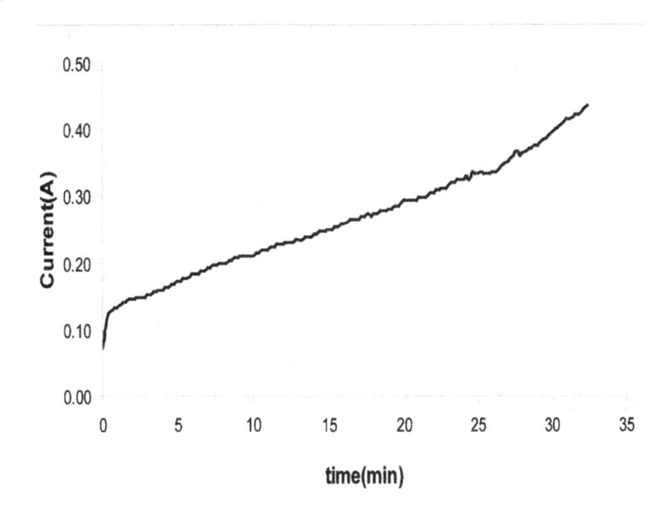

Figure 8. Shows plot of time versus resistance of cell

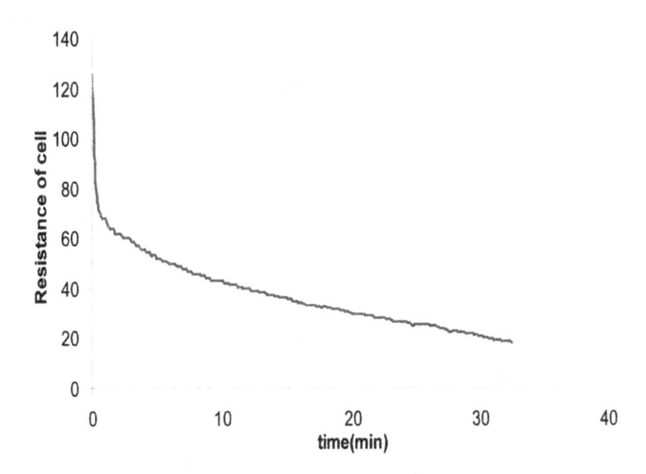

contrast to the initial phase where it is faster rising and nonlinear. This phase lasts for a minute or so depending on the working conditions prevalent in the cell. Later the current increase is more or less linear and so is the electric resistance. In the present experiment the initial resistance measured was 125 Ω which decreased rapidly to about 60 Ω in about 2 seconds. Then the resistance continued to fall slowly with time up to about 18 Ω after 32 min. The cell resistance has dropped to about $1/7^{th}$ of its original value. Careful inspection of the current versus time plots shows that the rate of increase of current with time is very fast in the beginning, which settles down to a value as the growth takes shape. Also initially the rate of increase of current with respect to time is higher as compared to the later stages where the electrodeposit has grown appreciably. This is due to the decrease in resistance of the cell, increase in the area of the electrodes and increase in electric field (Patil, Chisty, Khan, Basit, & Behere, 2001) itself during the developed stages of the growth. The findings suggest the study of growth pattern under constant electric field conditions would help understanding the DLA.

REFERENCES

Anderson, P. (1996). *Use of a PC printer port for control and data acquisition*. The Electronic Journal for Engineering Technology.

Argoul, F., Huth, J., Merzeau, P., Arnrodo, A., & Swinney, H. (1993). Experimental evidence for homoclinic chaos in an electrochemical growth process. *Physica D. Nonlinear Phenomena*, *62*, 170. doi:10.1016/0167-2789(93)90279-A

Atchison, S. N., Burford, R. P., & Hibbert, D. B. (1994). Chemical effects on the morphology of supported electrodeposited metals. *International Journal of Electroanalytical Chemistry*, *371*, 137. doi:10.1016/0022-0728(94)03245-9

Ben-Jacob, E. (1993). From snowflake formation to the growth of bacterial colonies, Part I: Diffusive patterning in non-living systems. *Contemporary Physics*, *34*, 247–273. doi:10.1080/00107519308222085

Center for Polymer Studies. (2010). *Growing rough patterns: Electrodeposition*. Retrieved from http://polymer.bu.edu/ogaf/html/chp41.htm

Chakrabarty, R. K., Moosmüller, H., Arnott, W. P., Garro, M. A., Tian, G., & Slowik, J. G. (2009). Low fractal dimension cluster-dilute soot aggregates from a premixed flame. *Physical Review Letters*, *102*(23). doi:10.1103/PhysRevLett.102.235504

Costa, J. M., Sagues, F., & Vilarrasa, M. (1991). Growth rate of copper electrodeposits: Potential and Concentration effects. *Physical Review Letters*, *43*(12), 7057–7060.

Cronemberger, C. M., & Sampaio, L. C. (2006). Growth of fractal electrodeposited aggregates under action of electric and magnetic fields using a modified diffusion-limited aggregation algorithm. *Physical Review Letters*, *73*(4).

Fleury, V., Chazalviel, J.-N., Rosso, M., & Sapoval, B. (1990). The growth speed of electrochemical deposits. *International Journal of Electroanalytical Chemistry*, *290*, 249. doi:10.1016/0022-0728(90)87434-L

Hibbert, D. B. (1991). Fractals in chemistry. *Chemometrics and Intelligent Laboratory Systems*, *11*, 1–11. doi:10.1016/0169-7439(91)80001-7

Patil, A. G., Chisty, S. Q., Khan, A. R., Basit, M. A., & Behere, S. H. (2001). Fractal growth in copper sulphate solution. In *Proceedings of the Indian Science Congress Association*, Delhi, India.

Peacock, C. (1994). *Interfacing the standard parallel port*. Retrieved from http://www.beyondlogic.org/spp/parallel.pdf

Ram, B. (1991). *Fundamentals of microprocessors and microcomputers* (5th ed.). Retrieved from http://www.national.com

Shaikh, Y. H., Khan, A. R., Pathan, J. M., Patil, A., & Behere, S. H. (2009). Fractal pattern growth simulation in electrodeposition and study of the shifting of center of mass. *Chaos, Solitons, and Fractals*, *42*(5), 2796–2803. doi:10.1016/j.chaos.2009.03.192

Tocci, R., & Widmer, N. (1998). *Digital systems: Principles and applications* (8th ed.). Upper Saddle River, NJ: Prentice Hall.

Vicsek, T. (1992). *Fractal growth phenomena*. Singapore: World Scientific.

Witten Jr., T. A. & Sander, L. M. (2000). Diffusion-limited aggregation, a kinetic critical phenomenon. *Critical Review Letters, 47*(19).

Yaroslavsky, L. P. (2007). Stochastic nonlinear dynamics pattern formation and growth models. *Nonlinear Biomedical Physics*, *1*(4).

This work was previously published in the International Journal of Artificial Life Research, Volume 2, Issue 1, edited by E. Stanley Lee and Ping-Teng Chang, pp. 17-27, copyright 2011 by IGI Publishing (an imprint of IGI Global).

Chapter 2
Propagation of Front Waves in Myelinated Nerve Fibres:
New Electrical Transmission Lines Constituted of Linear and Nonlinear Portions

Aïssatou Mboussi Nkomidio
University of Yaoundé I, Cameroon

Paul Woafo
University of Yaoundé I, Cameroon

ABSTRACT

In this paper, the authors examine the propagation of wave fronts in myelinated nerve fibres and applications as electrical transmission lines constituted of linear and nonlinear portions. Numerical simulations show that the front introduced in the nonlinear portion deforms itself in the linear portion, but recovers its initial profile and velocity in the next nonlinear portion. The phenomenon of deformation and recovery can be used for the development of new and low cost electrical transmission lines that can be used to transport localized excitations.

1. INTRODUCTION

An electrical line is an organ constituted of driver materials serving to the transport of energy or electricity. It can be linear or nonlinear. A nonlinear electrical line has particular importance in communication engineering. Indeed, some recent studies on electrical waves show that they can be interesting as optical wave in communication lines or even dominate these optical sibling since nonlinear electrical lines can be manufactured more easily (as compare to photonic devices) using standard integrated-circuit technology (Malomed, 1992; Hirota & Suzuki, 1970; Ricketts et al., 2006;

DOI: 10.4018/978-1-4666-3890-7.ch002

Lee, 2006; Nguimdo et al., 2008). A line constituted of alternated portions of linear and nonlinear portions could also serve for the propagation of nonlinear signals. Such an electrical line has as major advantage: the ease in the manufacturing of linear components and the reduction of the cost of the electrical line relatively to those constituted only of nonlinear components.

The aim of this study is first to understand the impulse propagation in myelinated nerve fibres considering the Ranvier nodes as nonlinear portions and the myelin sheath as linear portions. The second aim is to extend the results of the analysis to the propagation of front waves in electrical transmission lines with alternated linear and nonlinear portions. Particular attention is paid on the effects of the components of the linear portion on the profile and velocity of the wave.

Reutsky et al. (2003) presented a model of action potential propagation in myelinated nerve fibres. This model combines the single-cable formulation of Goldman and Albus (1968) with a basic representation of the ephaptic interaction among the fibres. The loss of the myelin sheath along tracts of central axons is observed in multi sclerosis, the most common demyelinating disease of central nervous system in humans (Smith et al., 1999). It has been experimentally demonstrated that demyelinated axons may become hyperexcitable and acquire the property of spontaneously generating trains of spurious impulses (Baker et al., 1992; Russell, 1982). The effects of demyelination and remyelination will be analyzed in this paper ant its equivalent for electrical lines is the variation of electrical components of the linear portions.

The next section presents the propagation of impulses in myelinated nerve fibres. The propagation equations are derived and the numerical investigations are carried out in the case of constant electrical components and that of variable electric components. The effects of the disruption of the myelin sheath on the conduction velocity are analyzed. The suggestion of this model as electrical transmission lines with alternated linear and nonlinear portions is discussed. The conclusion appears in the last section.

2. EQUATION OF PROPAGATION

2.1 Description of the Model

Consider a myelinated nerve fibre as it appears in Figure 1a. In this figure, two consecutive Ranvier or active nodes are separated by a myelinated portion. The active node behaves as a nonlinear portion while the myelinated portion is electrically equivalent to a linear portion. The nonlinear and linear portions are represented by their electrical equivalents in Figure 1b for the Ranvier nodes and Figure 1c for the myelinated portion. R is the membrane resistance per unit length; C_1 and G_1 are the capacitance and conductance of the nonlinear portion while C_2 and G_2 represent the capacitance and conductance of the linear portion. The nonlinearity in a fibre resides in the conductance $G1$ of the nonlinear portion as described below. To obtain equations (Equations 1-9) governing the propagation of the wave in any portion of the fibre, we apply Kirchhoff laws in Figure 1.

For the Ranvier nodes, the total transversal membrane current I is given by the following relation

$$I = C_1 \frac{dV}{dt} + I_i \tag{1}$$

where V is the potential across the membrane and I_i is the ionic current through the ionic channels characterizing the movement of charged particles. Following Scott (1999), I_i is a nonlinear function of the potential V and is expressed as

$$I = \frac{G_1}{V_b(V_b - V_a)} V(V - V_a)(V - V_b) \tag{2}$$

Figure 1. a) Axon of a myelinated fibre; b) Corresponding electrical circuit of Ranvier nodes; c) Corresponding electrical circuit of myelinated portion

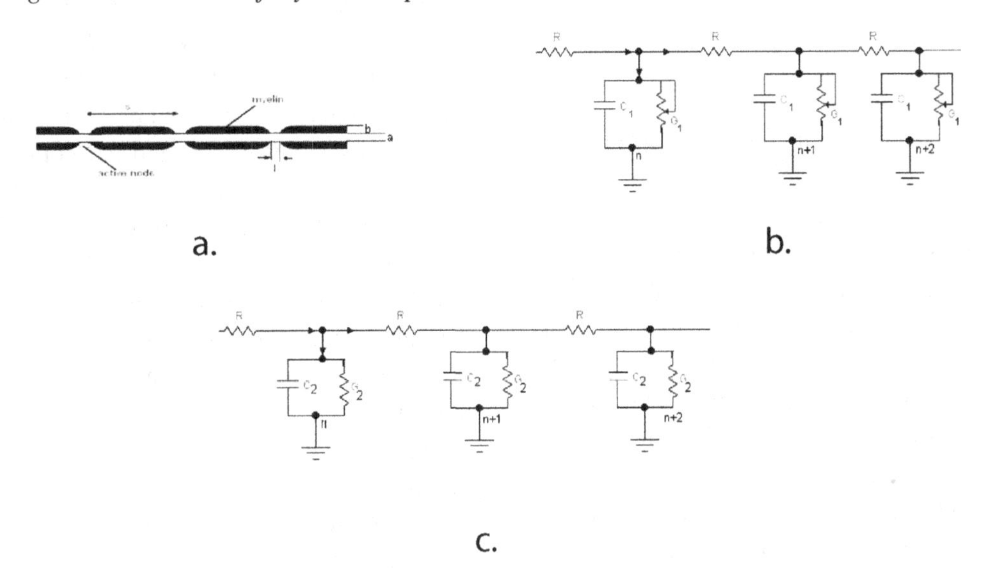

where V_a and V_b are respectively a threshold potential and the diffusion potential. In each Ranvier node, the following equations can be written from Kirchhoff's laws considering the locations *n, n+1*, and *n*

$$I_{n-1} - I_n = I \quad (3)$$

$$V_n - V_{n+1} = RI_n \quad (4)$$

Using Equations (1)-(4) and appropriate dimensionless coefficients (see below), one obtains from the continuum approximation the following equation:

$$\frac{\partial v}{\partial t} = \alpha \frac{\partial^2 v}{\partial x^2} - \beta v(v-a)(v-1) \quad (5)$$

This is a nonlinear partial differential equation describing the propagation of wave front in each Ranvier node of the nerve fibre.

By using the same analysis in the myelinated portion, one can show that the propagation in each myelinated portion sheath is governed by the following linear partial differential equation (assuming that an ionic current is not depend on the line voltage)

$$\frac{\partial v}{\partial t} = \gamma \frac{\partial^2 v}{\partial x^2} - \lambda v \quad (6)$$

The dimensionless coefficients in Equations (5) and (6) are given as

$$\alpha = \frac{l^2}{RC_1\omega_0},\ \beta = \frac{G_1}{C_1\omega_0(1-a)},\ \gamma = \frac{s^2}{RC_2\omega_0},\ \lambda = \frac{G_2}{RC_2\omega_0},\ a = \frac{v_a}{v_b},\ v = \frac{V}{v_b}$$

where *s* and *l* are respectively the length of the myelin sheath and that of Ranvier node.

2.2 Numerical Simulations

To analyze the propagation of localized front waves in the fibre described by Equations (5) and (6), let us consider as initial condition a front wave given as

$$v(x,t) = \frac{1}{1+\exp\left(\frac{(x-ut)}{\sqrt{2}}\right)} \quad (7)$$

where u is the front velocity. This is the front wave solution of Equation (5) (Mboussi & Woafo, 2010). For the numerical simulation, the finite difference scheme is applied for spatial derivatives, and then the fourth order Runge-Kutta algorithm is used to solve the set of discrete differential equations then obtained. The time step is 10^{-3} while the spatial step is 10^{-1}. The simulation is carried out for a line having 5 portions of myelin sheath with length 1 and 5 Ranvier nodes with length 0.5. The initial front is introduced inside the second Ranvier node and periodic boundary conditions are used.

The values of the electrical components are taken from Reutsky et al. (2003) as

$R = 28M\Omega$, $G_1 = 0.57\mu\Omega^{-1}$, $G_2 = 5.6nScm^{-1}$, $C_1 = 1\mu Fcm^{-1}$, $C_2 = 18.7pFcm^{-1}$, l=2.5μm, $s = 2000\mu m$

Consequently, the dimensionless coefficients in Equations (5) and (6) are

$\alpha = 1$, $\beta = 1.7$, γ=0.0002, $\lambda = 0.15$

Figure 2a which presents the variation of the wave velocity as it propagates in the line shows a decrease in the linear portion and then an increase to attain its maximal value in the nonlinear part. Figure 2b stresses this fact in a fibre having a nonlinear part of length 5 instead of 0.5 (even if, for most of myelinated fibres, the length of the Ranvier nodes is very small as compare to that of the myelinated portion). Thus, this propagation indicates the regenerating character of the Ranvier nodes and the slowing down process in the myelin portion of the fibre.

An interesting question dealing with the propagation in myelinated nerve fibres is the effects of demyelination. Several studies have shown that the loss of the myelin can lead to propagation failure which is an important problem both in the field of neurophysiology and cardiophysiology (Migliore, 1996; Manor et al., 1991; Moore & Westerfield, 1983; deCastro et al., 1999; Carpio, 2005). Demyelination is the loss of myelin sheath which surrounds the axon of the nerve fibre. This can take place at a given site or at various sites of the fibre (Figure 3). The myelin sheath consists of a single cell, known as Schwann cell, which is wrapped many times (roughly 100 times) around the axonal membrane. Its presence increases the cross section of the nerve fibre. Consequently, the absence (reduction of the thickness) of the myelin can be modelled by a decrease of the diameter of the cross section of the nerve fibre. Alternatively, the thickening (deposit of more Shawn cells) of the myelin sheath will increase the diameter. One key element that varies because of demyelination or remyelination is the capacitance of the myelin sheath. It is given in Russell (1982) by the following relation

$$C_2 = \frac{2\pi k\varepsilon_0 s}{\log\left(1+\frac{b}{a_1}\right)} \quad (8)$$

where a_1 and b are respectively the radius of the axon and the thickness of the myelin sheath, k is the dielectric constant and ε_0 the permittivity. From the Expression (8), one can approximate the variation of the capacitance by the following equation

$$C_2 = \frac{2\pi k\varepsilon_0 s}{\log\left(1+rq\right)} \quad (9)$$

Figure 2. a) Conduction velocity of the wave as a function of space for the Ranvier node length equal to 0.5; b) Conduction velocity of the wave as a function of space for the Ranvier node length equal to 5.

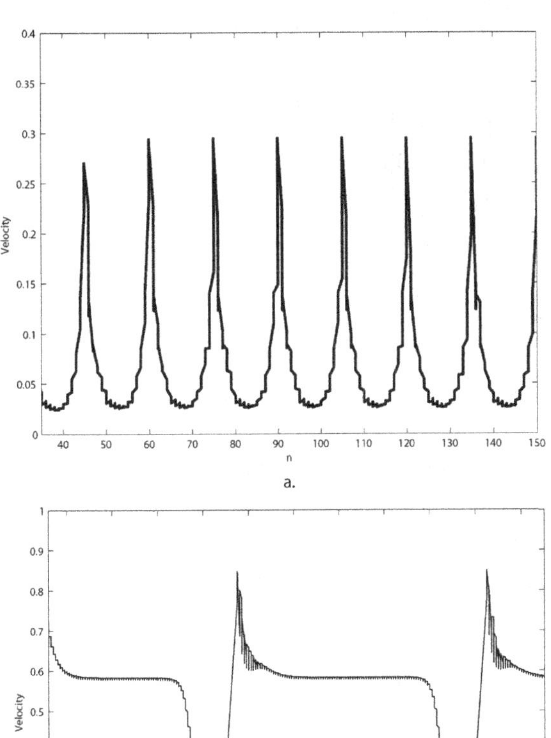

Figure 3. Different representations of the perturbation of myelin sheath on a myelinated fibre. a) Irregular destruction of myelin along the fibre between nodes xi and xj, xk and xp and so on; b) Excess of myelin along the fibre between nodes xi and xj, xk and xp, xq and xn.

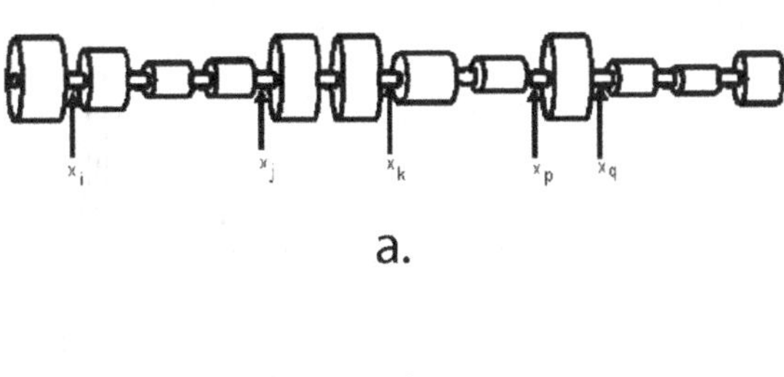

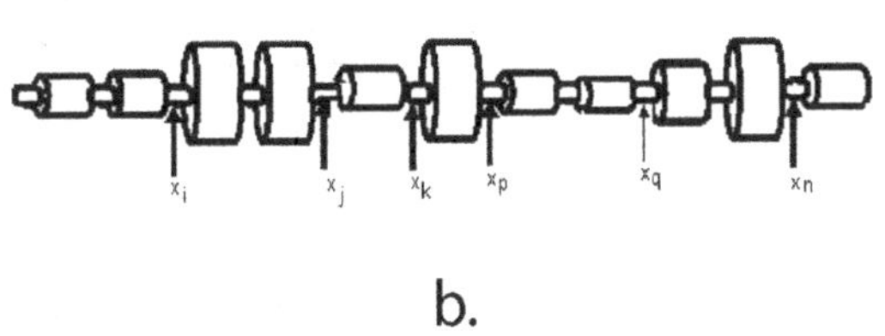

with $r = \frac{b}{a_1}$, q is a random real number with values in the range [0;1]. It characterizes the random variation of the capacitance. When b increases (remyelination), C_2 decreases and the decrease of b (demyelination) corresponds to the increase of $C2$. For the values of a_1 and b used in this work ($a_1 = 10\mu m,\ b = 15\mu m$), one obtains $r = 1.5$. Then, $r > 1.5$ corresponds to remyelination while $0 < r < 1.5$ is the case of demyelination.

An increase of the myelin capacitance leads to Figure 4. It shows the decrease of the nerve conduction velocity when the perturbation parameter r is less than 1.5. We have not found propagation failure in this study meaning that in some worse situations where the fibre is highly damaged, there will be no propagation failure. Figure 5 on the contrary show the increase of the propagation velocity when the myelin sheath capacitance decreases (thickening of the myelin sheath). Results similar to the one presented here were also found from experimental studies (Reutsky et al., 2003) considering a pulse-like profile and using a different mathematical model to analyze the combined effect of demyelination and temperature variation. The absence of propagation failure can be linked to an earlier study of Koles et al. (1972) that shows that the actual progression of the multi sclerosis could remain masked with the absence of clinical symptoms.

We now apply the precedent study to the propagation of front wave to an electrical transmission line with linear and nonlinear portion.

Figure 4. Decrease of the conduction velocity of the wave in the nerve fibre due to the loss of myelin sheath (solid line: r = 1; dashed line: r = 0.1; dashed dot line: r = 0.01)

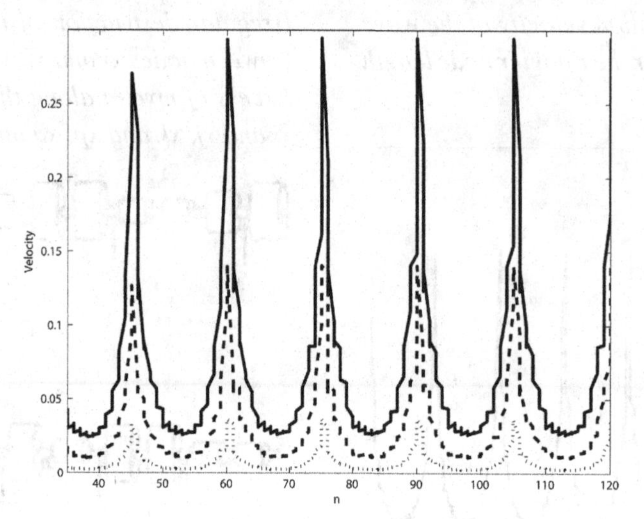

Figure 5. Increase of the conduction velocity of the wave in the nerve fibre due to the thickening of myelin sheath (solid line: r = 5; dashed line: r = 10; dashed dot line: r = 15)

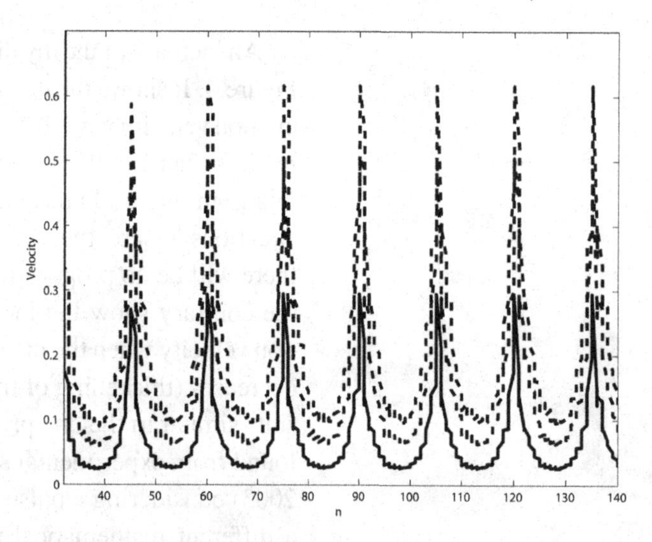

3. APPLICATION TO ELECTRICAL TRANSMISSION LINES WITH LINEAR AND NONLINEAR PORTIONS

Let us consider an electrical transmission line constituted with linear and nonlinear portions equally spaced. Figure 6a depicts such line. As Figure 6b shows, the propagation of the front wave inside the initial nonlinear portion (NL2) is steady. As the front enters the linear portion, a deformation of its shape occurs showing a sort of break-up into two kinks. When the wave goes over the linear portion, it recovers its initial shape

Figure 6. Electrical transmission line

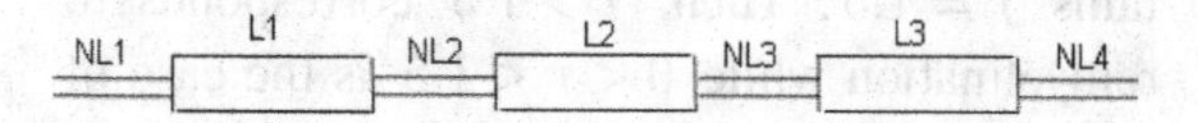

when propagating inside the nonlinear portion. If the length of the linear portion is long enough, the break-up process will continue so that the propagation failure occurs. The front wave will be captured inside the linear portion and will not reach the next nonlinear portion. The portion length leading to such behaviour depends on the values of α, β, γ, λ. The analysis of the effects of remyelination and demyelination carried out above is equivalent to the analysis of the effects of the perturbative variations of the electrical components of the line.

Such electrical transmission line constituted of alternated portions of linear and nonlinear portions could serve for the propagation of nonlinear signals. Its advantages are the ease in the manufacturing of linear components and the reduction of the cost of the electric line relatively to those constituted only of nonlinear components.

4. CONCLUSION

This study has dealt with the propagation of wave fronts in myelinated nerve fibre and its application to the propagation of front wave in electrical transmission lines with alternated linear and nonlinear portions.

The application of the model to an electrical transmission lines showing the regenerative process in the nonlinear portion and slowing down process in the linear portion. The overall propagation suggests that electrical transmission lines for nonlinear signals could be built using portions of linear components alternating with portions of nonlinear components. This will certainly reduce the cost for nonlinear transmission lines. This study contributes not only for the manufacturing of new electrical lines for transmission of high amplitude signals (nonlinear signals), but also for the search of neuronal prosthesis (Mohawald & Douglas, 1991).

REFERENCES

Barker, M., & Bostock, H. (1992). Ectopic activity in demyelinated spiral root axons of the rat. *The Journal of Physiology*, *451*, 539–552.

Carpio, A. (2005). Asymptotic construction of pulses in the discrete Hodgkin-Huxley model for myelinated nerves. *Physical Review Letters E*, *72*.

deCastro, M., Hofer, E., Munuzurri, A. P., Gomez-Gesteira, M., Plamck, G., & Schafferhofer, I. (1999). Comparison between the role of discontinuities in cardiac conduction and in a one-dimensional hardware model. *Physical. Review Letters E*, *59*, 5962–5969.

Goldman, L., & Albus, J. S. (1968). Computation of impulse conduction in myelinated fibres; theoretical basis of the velocity-diameter relation. *Biophysical Journal*, *8*, 596–607. doi:10.1016/S0006-3495(68)86510-5

Hirota, R., & Suzuky, K. (1970). Studies on lattice solitons by using electrical networks. *Journal of the Physical Society of Japan*, *28*(5), 1366. doi:10.1143/JPSJ.28.1366

Hobbie, R. K. (1982). *Intermediate physics for medicine and biology* (3rd ed.). New York, NY: Springer.

Koles, Z. J., & Rasminsky, M. (1972). A computer simulation of conduction in demyelinated nerve fibres. *The Journal of Physiology*, *227*, 351–364.

Lee, T. H. (2006). Device physics: Electrical solitons come of age. *Nature*, *440*, 36–37. doi:10.1038/440036a

Malomed, B. A. (1992). Propagation of solitons in damped ac-driven chains. *Physical Review Letters A*, *45*, 4097–4101.

Manor, Y., Koch, C., & Segev, I. (1991). Effect of geometrical irregularities on propagation delay in axonal trees. *Biophysical Journal*, *60*, 1424–1437. doi:10.1016/S0006-3495(91)82179-8

Mboussi, N. A., & Woafo, P. (2010). Effects of imperfection of ionic channels and exposure to electromagnetic fields on the generation and propagation of front waves in nervous fibre. *Communications in Nonlinear Science and Numerical Simulation*, *15*, 2350–2360. doi:10.1016/j.cnsns.2009.09.040

Migliore, M. (1996). Modeling the attenuation and failure of action potentials in the dendrites of hippocampal neurons protoplasmic. *Biophysical Journal*, *71*, 2394–2403. doi:10.1016/S0006-3495(96)79433-X

Mohawald, M., & Douglas, R. (1991). A silicon neuron. *Nature*, *354*, 515–518. doi:10.1038/354515a0

Moore, J. W., & Westerfield, M. (1983). Action potential propagation and threshold parameters in inhomogeneous regions of squid axons. *The Journal of Physiology*, *336*, 285–300.

Nguimdo, R. M., Nubissie, S., & Woafo, P. (2008). Waves amplification in discrete nonlinear electrical lines: Direct numerical simulation. *Journal of the Physical Society of Japan*, 77.

Reutsky, S., Rossoni, E., & Tirozzi, B. (2003). Conduction in bundles of demyelinated nerve fibres: competer simulation. *Biological Cybernetics*, *89*, 439–448. doi:10.1007/s00422-003-0430-x

Ricketts, D. S., Li, X., & Ham, D. (2006). Electrical soliton oscillator. *IEEE Transactions on Microwave Theory and Techniques*, *54*(1), 373–382. doi:10.1109/TMTT.2005.861652

Scott, A. C. (1999). *Nonlinear science: Emergence and dynamics of coherent structure*. Oxford, UK: Oxford University Press.

Smith, K. J., & McDonald, W. I. (1999). The pathophysiology of multiple sclerosis: The mechanism underlying the production of symptoms and natural history of disease. *Philosophical Transactions of the Royal Society of London*, *354*, 1649–1673. doi:10.1098/rstb.1999.0510

This work was previously published in the International Journal of Artificial Life Research, Volume 2, Issue 1, edited by E. Stanley Lee and Ping-Teng Chang, pp. 34-42, copyright 2011 by IGI Publishing (an imprint of IGI Global).

Chapter 3
What Does Artificial Life Tell Us About Death?

Carlos Gershenson
Universidad Nacional Autónoma de México, México

ABSTRACT

This paper discusses how concepts developed within artificial life (ALife) can help demystify the notion of death. This is relevant because sooner or later we will all die; death affects us all. Studying the properties of living systems independently of their substrate, ALife describes life as a type of organization. Thus, death entails the loss of that organization. Within this perspective, different notions of death are derived from different notions of life. Also, the relationship between life and mind and the implications of death to the mind are discussed. A criterium is proposed in which the value of life depends on its uniqueness, i.e. a living system is more valuable if it is harder to replace. However, this does not imply that death in replaceable living systems is unproblematic. This is decided on whether there is harm to the system produced by death. The paper concludes with speculations about how the notion of death could be shaped in the future.

Every evil leaves a sorrow in the memory,
until the supreme evil, death,
wipes out all memories together with all life.

–Leonardo da Vinci

1. LIFE AS PROCESS/ ORGANIZATION

One of the open problems in artificial life discussed by Bedau et al. (2000) is the establishment of ethical principles for artificial life. In particular:

DOI: 10.4018/978-1-4666-3890-7.ch003

Much of current ethics is based on the sanctity of human life. Research in artificial life will affect our understanding of life and death (...). This, like the theory of evolution, will have major social consequences for human cultural practices such as religion. (Bedau et al., 2000, p. 375).

Focussing on our understanding of death, this will depend necessarily on our understanding of life, and vice versa. Throughout history there have been several explanations to both life and death, and it seems unfeasible that a consensus will be reached. Thus, we are faced with multiple notions of life, which imply different notions of death. However, generally speaking, if we describe life as a process, death can be understood as the irreversible termination of that process.

The general notion of life as a process or organization (Langton, 1989; Sterelny & Griffiths, 1999; Korzeniewski, 2001) has expelled vitalism from scientific worldviews. Moreover, there are advantages in describing living systems from a functional perspective, e.g. it makes the notion of life independent of its implementation. This is an essential aspect of artificial life, where the properties of living systems are studied independently of their substrate. Also, we know that there is a constant flow of matter and energy in living systems, i.e. their physical components can change while the identity of the organism is preserved. On the one hand, not a single atom of an organism is maintained within the organism after a few years (Grand, 2003). The matter changes, but the identity of the organism is maintained. On the other hand, one can make a variation of Kauffman's "blender thought experiment" (Kauffman, 2000): if you put a macroscopic living system in a blender and press "on", after some seconds you will have the same molecules (matter) that the living system had. However, the organization of the living system is destroyed in the blending. Thus, life is an organizational aspect of living systems, not so much a physical aspect. Death occurs when this organization is lost. Given the above arguments, I argue that a physicalist perspective is less suitable than a systemic one for understanding life and death.

2. DIFFERENT NOTIONS OF LIFE AND DEATH

From a systemic perspective, different notions of death can be derived from a non-exhaustive set of different notions of life:

- If we consider life as self-production (Varela et al., 1974; Maturana & Varela, 1980, 1987; Luisi, 1998), then death will the loss of that self-production ability.
- If we consider life as what is common to all living beings (De Duve, 2003, p. 8), then death implies the termination of that commonality, distinguishing it from other living beings.
- If we consider life as computation (Hopfield, 1994), then death will be the end (halting?) of that computing process.
- If we consider life as supple adaptation (Bedau, 1998), death implies the loss of that adaptation.
- If we consider life as a self-reproducing system capable of at least one thermodynamic work cycle (Kauffman, 2000, p. 4), death will occur when the system will be unable to perform thermodynamic work.
- If we consider life as information (i.e. a system) that produces more of its own information than that produced by its environment (Gershenson, 2007), then death will occur when the environment will produce more information than that produced by the system.

3. LIFE, MIND, AND DEATH

One of the main properties of living organization is its self-production (Varela et al., 1974; Maturana & Varela, 1980, 1987; Luisi, 1998; Kauffman, 2000). When death occurs, this self-production cannot be maintained. But is this organization the only thing that is lost with death? What about experience and mind?

The notions of life and death have been much related to those of mind, cognition, awareness, and consciousness[1]. On the one hand, the mind is a property closely related with life. Some even propose that mind and life are essentially the same process (Stewart, 1996; Bedau, 1998). On the other hand, people have speculated since the dawn of civilization on what occurs with the mind after death.

Life is a process described by an observer (Maturana & Varela, 1987), in first or third person perspective. When the process breaks, only description in the third person observer remains. By definition, we can only speak about death from a third person perspective. Since the mind requires a first person perspective, all evidence points to the conclusion that after death the mind is lost together with the organization of the living system.

What can artificial life add to this discussion? Artificial life simulations ("soft" ALife) can be seen as opaque thought experiments (Di Paolo et al., 2000), i.e. one can explore different notions of life and death with them. Robots ("hard" ALife) would also serve this purpose. Artificial life can help us build living systems to be explored from a third person perspective in a synthetic way (Steels, 1993). Can we say that "animats" (Wilson, 1985) have a mind, in the same sense as animals do? Well, there are similarities and differences between the minds of animats and animals. Therefore, we can conclude that animats have *a* mind, but that it is of a different type than that of animals (Gershenson, 2004). The mind of an animal is much more plastic and adaptive, but the mind of an animat can easily be replaced, modified, and restored.

When a digital organism dies (Ray, 1994), what physically changes is the RAM that encoded the organism in bits. When the bits describing the organization of the organism are erased, the only place where the organism prevails is in the observer. The same is for robots. The same is for animals. The same is for humans. If we describe life as an organizational process, and a mind as depending on it (Clark, 1997) when the organization is lost, the life is lost and the mind is lost.

Certainly, the organization of digital organisms is much easier to preserve than that of biological ones. Apart from the ease of copying digital information, digital organisms are generally inhabiting closed environments. Biological organisms face open dynamical environments that constantly threaten their integrity i.e. organisms need to make thermodynamical work (metabolism) (Kauffman, 2000) to maintain themselves. In an open environment such as the biosphere, where different evolving organisms interact, there is no absolutely "best" or "fittest" organism, i.e. independently of the current context, since the fitness depends on the current environment. For example, small animals are fitter when food is scarce. Thus, fitness landscapes change constantly with the environment, since the environment is changed as organisms evolve trying to increase their fitness. In this context, it can be speculated that there is an evolutionary advantage of death. If there was no death, i.e. if an organism somehow managed to maintain its organization indefinitely, evolution would stop. This actually occurs commonly with digital evolution. In fact, death of digital organisms has been used as a measure to introduce novelty (Ray, 1994; Dorin, 2005; Olsen et al., 2008).

The organization represented in a species can survive several lifespans, but is subject to the same pressure as the one just described. The loss of organization gives the opportunity to new forms of organization to develop. Summarizing this argument, we can say that death is beneficial for biological evolution, since it allows the appearance of novelty in an environment.

The development of protocells (Rasmussen et al., 2008) ("wet" ALife) might further contribute to the exploration of the notions of life and death. By chemically producing the organization resulting in living systems, the non-mystical notion of life reviewed here will gain further grounds. Additionally, the non-mystical notion of death explored here will have to be further elaborated. What occurs when a protocell dies? If we can create again a living system with the same organization, did it die in the first place? I think the answer should be affirmative. The fact that an organism—artificial or natural—can easily be replaced or regenerated does not mean that the particular instantiation of its organization is not lost. We will be able to have different instantiations of the same living organization, just like we can have different copies of the same digital organism. Will its death have the same meaning as that of an animal? This brings us to the question of how can we value life and death.

4. THE VALUE OF LIFE

Death can be valued from a first person and from a third person perspective. From a *first person* perspective, one can apply the harm thesis (Luper, 2009a, 2009b): if the welfare of an individual is decreased in the case of death, then the death event was harmful. From a *third person* perspective, I propose that the value of a living system will be given by its uniqueness and replaceability. If a living system is easy to replace, just like a digital organism, then it is less valuable than an irreplaceable living system, e.g. my daughter. Since ALife systems are easy to replicate, they have less value than a biological system which is difficult to replace. Notice, however, that this argument is only for the third person perspective. If any living system—natural or artificial—is harmed by its death, then there is a value to its organization, even if it is easily replaceable. Therefore, it is morally objectable to destroy the organization of a living system even only if the system is harmed by its destruction.

One thing to notice in these questions is that in most biological organisms, the organization lies not only in their genes, but also in their development (epigenesis). Clones can develop different organizations. The same might occur for protocells and other future "wet" artificial living systems. However, on the digital side of artificial life, it is easy again to maintain and reproduce the organization acquired through development (Balkenius et al., 2001). In other words, biological systems are difficult to replace because of their epigenesis, i.e. information acquired in their lifetime through experience, as compared with digital organisms, where epigenetic information can be stored and replicated.

5. THE FUTURE

What will the future bring? Will there be biological systems closer to digital ones, in the sense that living information can be maintained and/or reproduced? Probably. This would decrease the value of life from a third person perspective, but not from a first person perspective. How will death be affected by our scientific and technological developments? We will have more control over it. Will this mark an end to evolution? No, even when some living organization might be more persistent, there will always be new situations where organisms have to adapt. In any case, the cultural attitudes towards death most probably will change. This is not suggesting that we will be less touched by it, or less spiritual towards it. The implication is that we will have a better understanding of the phenomenon, with a broader scientific basis.

ACKNOWLEDGMENT

I should like to thank Javier Rosado and two anonymous referees for their useful comments.

REFERENCES

Balkenius, C., Zlatev, J., Brezeal, C., Dautenhahn, K., & Kozima, H. (Eds.). (2001). *Proceedings of the First International Workshop on Epigenetic Robotics: Modeling Cognitive Development in Robotic Systems*, Lund, Sweden (Vol. 85).

Bedau, M. A. (1998). Four puzzles about life. *Artificial Life*, *4*, 125–140. doi:10.1162/106454698568486

Bedau, M. A., McCaskill, J., Packard, P., Rasmussen, S., Green, D., & Ikegami, T. (2000). Open problems in artificial life. *Artificial Life*, *6*(4), 363–376. doi:10.1162/106454600300103683

Clark, A. (1997). *Being there: Putting brain, body, and world together again*. Cambridge, MA: MIT Press.

De Duve, C. (2003). *Live evolving: Molecules, mind, and meaning*. New York, NY: Oxford University Press.

Di Paolo, E. A., Noble, J., & Bullock, S. (2000). Simulation models as opaque thought experiments. In *Proceedings of the Seventh International Conference on Artificial Life* (pp. 497-506).

Dorin, A. (2005). Artificial life, death and epidemics in evolutionary, generative electronic art. In F. Rothlauf, J. Branke, S. Cagnoni, D. Wolfe Corne, R. Drechsler, Y. Jin et al. (Eds.), *Proceedings of the Evo Workshops on Applications of Evolutionary Computing* (LNCS 3449, pp. 448-457).

Gershenson, C. (2004). Cognitive paradigms: Which one is the best? *Cognitive Systems Research*, *5*(2), 135–156. doi:10.1016/j.cogsys.2003.10.002

Gershenson, C. (2007). The world as evolving information. In *Proceedings of the International Conference on Complex Systems*.

Grand, S. (2003). *Creation: Life and how to make it*. Cambridge, MA: Harvard University Press.

Hopfield, J. J. (1994). Physics, computation, and why biology looks so different. *Journal of Theoretical Biology*, *171*, 53–60. doi:10.1006/jtbi.1994.1211

Kauffman, S. A. (2000). *Investigations*. New York, NY: Oxford University Press.

Korzeniewski, B. (2001). Cybernetic formulation of the definition of life. *Journal of Theoretical Biology*, *209*(3), 275–286. doi:10.1006/jtbi.2001.2262

Langton, C. (1989). Artificial life. In Langton, C. (Ed.), *Artificial life: Santa Fe Institute studies in the sciences of complexity* (pp. 1–47). Reading, MA: Addison-Wesley.

Luisi, P. L. (1998). About various definitions of life. *Origins of Life and Evolution of the Biosphere*, *28*(4-6), 613–622. doi:10.1023/A:1006517315105

Luper, S. (2009a). Death. In Zalta, E. N. (Ed.), *The Stanford encyclopedia of philosophy*. Stanford, CA: Stanford University Press.

Luper, S. (2009b). *The philosophy of death*. Cambridge, UK: Cambridge University Press.

Maturana, H. R., & Varela, F. J. (1980). *Autopoiesis and cognition: The realization of the living* (2nd ed.). Dordrecht, The Netherlands: D. Reidel Publishing.

Maturana, H. R., & Varela, F. J. (1987). *The tree of knowledge: The biological roots of human understanding*. Boston, MA: Shambhala.

Olsen, M., Siegelmann-Danieli, N., & Siegelmann, H. (2008). Robust artificial life via artificial programmed death. *Artificial Intelligence*, *172*(6-7), 884–898. doi:10.1016/j.artint.2007.10.015

Rasmussen, S., Bedau, M. A., Chen, L., Deamer, D., Krakauer, D. C., Packard, N. H., & Stadler, P. F. (Eds.). (2008). *Protocells: Bridging nonliving and living matter bridging nonliving and living matter*. Cambridge, MA: MIT Press.

Ray, T. S. (1994). An evolutionary approach to synthetic biology: Zen and the art of creating life. *Artificial Life*, *1*(1-2), 195–226.

Steels, L. (1993). Building agents out of autonomous behavior systems. In Steels, L., & Brooks, R. A. (Eds.), *The artificial life route to artificial intelligence: Building embodied situated agents*. Mahwah, NJ: Lawrence Erlbaum.

Sterelny, K., & Griffiths, P. E. (1999). *Sex and death*. Chicago, IL: University of Chicago Press.

Stewart, J. (1996). Cognition = life: Implications for higher-level cognition. *Behavioural Processes*, *35*(1-3), 311–326. doi:10.1016/0376-6357(95)00046-1

Varela, F. J., Maturana, H. R., & Uribe, R. (1974). Autopoiesis: The organization of living systems, its characterization and a model. *Bio Systems*, *5*, 187–196. doi:10.1016/0303-2647(74)90031-8

Wilson, S. W. (1985). Knowledge growth in an artificial animal. In *Proceedings of the First International Conference on Genetic Algorithms and Their Applications* (pp. 16-23).

ENDNOTES

[1] It is quite problematic to attempt to define these, but a vague notion will suffice. In the following, "mind" will be used in a broad sense that includes also cognition, awareness, and consciousness.

This work was previously published in the International Journal of Artificial Life Research, Volume 2, Issue 3, edited by E. Stanley Lee and Ping-Teng Chang, pp. 1-5, copyright 2011 by IGI Publishing (an imprint of IGI Global).

Chapter 4
A Constructive Approach to the Evolution of the Planning Ability

Kenichi Minoya
Nagoya University, Japan

Reiji Suzuki
Nagoya University, Japan

Tatsuo Unemi
Soka University, Japan

Takaya Arita
Nagoya University, Japan

ABSTRACT

Human beings have behavioral flexibility based on a general faculty of planning for future events. This paper describes the first stage of a study on the evolution of planning abilities. A blocks world problem is used as a task to be solved by the agents, and encode an inherent planning parameter into the genome. The result of computer simulation shows a general tendency that planning ability emerges when the problem is difficult to solve. When taking social relationships, especially in the collective situation, into account, planning ability is difficult to evolve in the case that the problem is complex because there is a conflict between personal and collective interests. Also, the simulation results indicate that sharing information facilitates evolution of the planning ability although the free rider problem tends to be more serious than the situation where agents do not share information. It implies that there is a strong connection between evolution of the planning ability and symbolic communication.

1. INTRODUCTION

Future-directed behavior can be seen in many animals as well as humans. For example, some hibernators store food for the coming winter just like humans who start building a shelter already in summer preparing for cold winter. So what is the difference in future-directed behavior between animals and humans? It has been said that animal behavior is instinctive but human behavior is flexible. "Mental time travel" is one of the capacities that provide increased behavioral flexibility of humans. Mental time travel is a term to refer to the faculty that allows humans to mentally project themselves backward in time to relive, or forward to prelive, events (Suddendorf & Corballis, 1997). The crucial selective advantage that mental time travel provides is the flexibility in novel situations

DOI: 10.4018/978-1-4666-3890-7.ch004

and the versatility to develop and adopt strategic long-term plans to suit goals (Suddendorf & Corballis, 1997). In this paper, we focus on the mental time travel into the future, especially the evolutionary aspect of the planning ability.

In the sphere of the cognitive science, it has been proposed that episodic memory is part of a more general faculty of planning for future events (Tulving, 1993). Tulving (1993) argued that the owner of an episodic memory system can transport freely into the personal past as well as into the future, a capability not possible for other kinds of memory such as procedural or semantic memory. Baddely (2000) proposed episodic buffer in a working memory system as a function to play an important role in feeding information into and retrieving information from episodic memory. Also, recent studies using functional neuroimaging techniques indicate that the prefrontal cortex plays a crucial role in working memory (Curtis & Esposito, 2003) and episodic memory (Squire et al., 1992). Although distinct elements of information processing of the planning have been gradually elucidated, and several kinds of mental models were proposed, the mechanisms about how these mental models have been shaped through the autonomous developmental of evolutionary process is not fully known.

Natural selection has been considered as one of the most widely held mechanisms to explain the emergence of living creatures' complex characteristics. Evolutionary psychology has attempted to explain psychological traits as adaptations as the functional products of natural selection or sexual selection. It has been proposed that the prefrontal cortex, known to be critically involved in planning abilities, has been especially enlarged through the human evolution than other brain areas (Deacon, 1997). Large brains are extremely costly both to maintain and evolve. Therefore, in a niche where there is little to use planning abilities, it might have a relatively small impact on evolution of it. Recent studies have indicated that ecological pressures drove the evolution of intelligence of human (Byrne, 1997; Darwin, 1871; Hill, 1982; Osvath & Gärdenfors, 2005; Potts, 1998; Tooby & DeVore, 1987). For example, with the global shift to cooler climate after 2.5 million years ago, much of southern and eastern Africa probably became more open and sparsely wooded, and it exposed the hominids to greater risk from predators and drove them into a cognitive niche (Tooby & DeVore, 1987).

Yet, common problems for these ecologically based theories include difficulties with explaining why humans evolved such extraordinary cognitive competencies, considering that many other species hunt, occupy savanna habitats, endured the same climatic fluctuations, and so forth (Flinn et al., 2005). A different approach to the problem of the evolution of intelligence of human involves the consideration of the social aspect (Alexander, 1971, 1990; Brothers, 1990; Dunbar, 1998; Humphrey, 1976; Jolly, 1999). Alexander (1990) argued that it (evolution of the intellect) was rather the necessity of dealing continually with our fellow humans in social circumstances that became ever more complex and unpredictable as the human line evolved (pp. 4-7). Co-operating with other people is considered to be one of the most important factors to deal with our fellows in social circumstances. Furthermore, symbolic communication seems to be indispensable to co-operate smoothly with other individuals. Brinck and Gärdenfors (2003) traced the difference between the ways in which apes and humans co-operate due to differences in communicative abilities, claiming that there is a strong connection between the evolution of planning and symbolic communication. However, there is little known about the specific mechanisms that underlie it.

Considering all of the above factors, this study explores the dynamics inherent in the mechanism of the evolutionary acquisition of the planning abilities, focusing on the benefits of the planning and the costs of it. The first goal is to elucidate the environment which drove the evolution of planning ability. The second goal is to explore the dynamics inherent in the mechanism of evolution of the planning ability in the social circumstances. Our main method consists of a constructive approach which attempts to create not only a symbolic model of a living system, but also a symbolic living object (Moreno, 2002). Accordingly, our models are elaborated without direct and precise reference to empirical biological reality, and allow a new means of computational experimentation to enable us to discover the universal principles of living systems (Moreno, 2002). The next section explains a planner, task, architecture, and fitness of each agent. We then show the basic experiments and describe the evolutionary experiments. Finally, we summarize the paper.

2. THE MODEL

2.1 Planner- Beam Search

Gulz (1991) argued that an organism is planning its actions if it has a representation of a goal and a start situation and it is capable of generating a representation of partially ordered set of actions for itself for getting from start to goal. This criterion presupposes three distinct processes: (1) developing a plan, (2) remembering representations that have been developed, and (3) remembering the set of actions from start to goal. Some kinds of representational space in our mind such as a working memory make possible these processes. In the following model, we define the inherent planning parameter as an attribute value which corresponds to a storage capacity of the working memory system.

A beam search algorithm (Ney et al., 1992) is adopted as the planner of each agent. The beam search utilizes a heuristic value, h, to estimate the approximate steps from the focal state to the goal state, by which partial solutions are evaluated. It also uses a beam width, B, which specifies the number of states that are stored at each level of the breadth-first search. A *BEAM* is used to store the states that are to be expanded in the next loop of the algorithm. Also, a *hash table* is used to store states that have been visited.

At the process of the planning, initially, there is a start state in the *BEAM* and the *hash table*, respectively. Each time through the main loop of the algorithm, the planner expands states connected to the nodes in the *BEAM*, and adds the successor states to the *SET*, which stores all successors of the states in the *BEAM* at the current level, if they are not in the *hash table*, and then adds the best B states ordered by h from the *SET* to the *BEAM*. Note that if the high-priority states in the *SET* have the same heuristic value, some states are randomly chosen, and added to the *BEAM*. If the number of expansion reaches the inherent attribute value of the agent (termed "planning limit"), planner runs through the main loop and then *sub goal* is determined by selecting a state with the best h in the *hash table* other than the start state. If all states in the *hash table* have the same h, *sub goal* is randomly chosen from states in the *hash table*. Finally, solution is obtained tracing the path from a *sub goal* to the start state.

Table 1 shows an example trace of the algorithm on the state space in Figure 1. As presented in Table 1, agents who have long planning limit require large amounts of storage capacity of the hash table. Also, the more the number of times of expanding is, the deeper the search is.

2.2 Task- Blocks World Problem

Planning is important especially when it is necessary to perform actions in the proper sequence to solve problems. We adopted the blocks world problem as a minimal task to deal with such a situation. A blocks world consists of a table with the size T, l rectangular blocks labeled b_l ($l = 1,\ldots, L$), and a grip. The size of the table represents the maximum number of blocks that can be placed on the table. Each space of the table is labeled as t_i ($i = 1, \ldots, T$). An agent is allowed to move a block to the top of another stack of blocks or to the empty space on the table by using a grip. A block can be moved only if there is no block on the top of it. In our model, if the table size is large, agents have many choices to move the block. Given the initial

Table 1. An example trace of the beam search algorithm on the state space in Figure 1 when the B = 3. Each superscript represents the value of corresponding h. Each row shows the trace of the search when the planning limit of the agents is different. For example, agents with a planning limit of 2 choose D^2 as a sub goal. Also, agents with a planning limit of 3 choose D^2 or F^2 as a sub goal.

loop number	planning limit	*SET*	*BEAM*	*hash table*	*sub goal*
		$\{\ \}$	$\{S^0\}$	$\{S^0\}$	$\{\ \}$
1	1	$\{A^0, B^0\}$	$\{A^0, B^0\}$	$\{A^0,B^0,S^0\}$	$\{A^0$ or $B^0\}$
2	2	$\{C^1, D^2\}$	$\{B^0\}$	$\{C^1,D^2,A^0,B^0,S^0\}$	$\{D^2\}$
	3	$\{E^1, F^2, C^1, D^2\}$	$\{D^2\ E^1, F^2\}$	$\{E^1, F^2,C^1,D^2,A^0,B^0,S^0\}$	$\{D^2$ or $F^2\}$
3	4	$\{I^2, J^3\}$	$\{E^1, F^2\}$	$\{I^2,J^3,E^1,F^2,C^1,D^2,A^0,B^0,S^0\}$	$\{J^3\}$
	5	$\{K^1, L^1,I^2, J^3\}$	$\{F^2\}$	$\{K^1,L^1, I^2,J^3,E^1, F^2,C^1,D^2,A^0,B^0,S^0\}$	$\{J^3\}$
	6	$\{M^2, N^2,K^1, L^1,I^2, J^3\}$	$\{J^3, I^2, N^2\}$	$\{M^2,N^2,K^1,L^1, I^2,J^3,E^1, F^2,C^1,D^2,A^0,B^0,S^0\}$	$\{J^3\}$

and target configurations of the blocks, the blocks world problem asks for a sequence of manipulation of the grip to achieve the target configuration with a smaller number of manipulations. In this study, we defined a target state as a configuration in which all blocks are stacked on a predetermined space in descending order as shown in Figure 2. We define the heuristic value h representing the attainment level of the goal state as follows (*on* (b_l, x) indicates that block b_l is on x).

$h = 5$: *on* (b_1, t_3) $\wedge$*on* (b_2, b_1) $\wedge$*on* (b_3, b_2) $\wedge$*on* (b_4, b_3) $\wedge$*on* (b_5, b_4)

Figure 1. An example of the state space. Boxes with alphabets represent distinct states. S represents the start state. The numbers in boxes represent h.

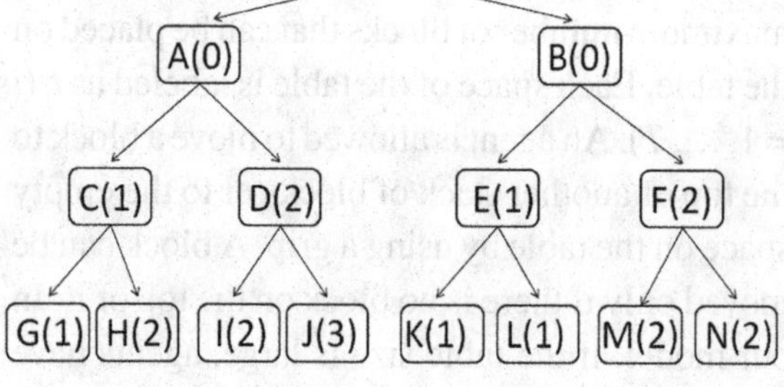

$h = 4$: *on* (b_1, t_3) $\wedge$*on* (b_2, b_1) $\wedge$*on* (b_3, b_2) $\wedge$*on* (b_4, b_3) $\wedge\neg$*on* (b_5, b_4)

$h = 3$: *on* (b_1, t_3) $\wedge$*on* (b_2, b_1) $\wedge$*on* (b_3, b_2) $\wedge\neg$*on* (b_4, b_3)

$h = 2$: *on* (b_1, t_3) $\wedge$*on* (b_2, b_1) $\wedge\neg$*on* (b_3, b_2)

$h = 1$: *on* (b_1, t_3) $\wedge\neg$*on* (b_2, b_1)

$h = 0$: $\neg$*on* (b_1, t_3)

2.3 Architecture

We constructed three models: basic model, nonshared model, and shared model. In the basic model (left in Figure 3), an agent observes the present state at first, and passes it to the planner. Then, the planner makes a plan. Next, the agent moves a block once by the gripper. We define an action step, a, as a movement of a block by a gripper. Action steps are repeated until the target configuration is achieved or the number of performed action steps exceeds the upper limit a_{max}.

As to the nonshared and shared models, we assumed a situation in which two agents (agent A and agent B) participate in a collective task to reach the same goal. In the nonshared model (middle in Figure 3), two agents interact by turn-

Figure 2. A part of the state space of blocks world problem (T-3). "h" represents the heuristic value corresponding to each configuration.

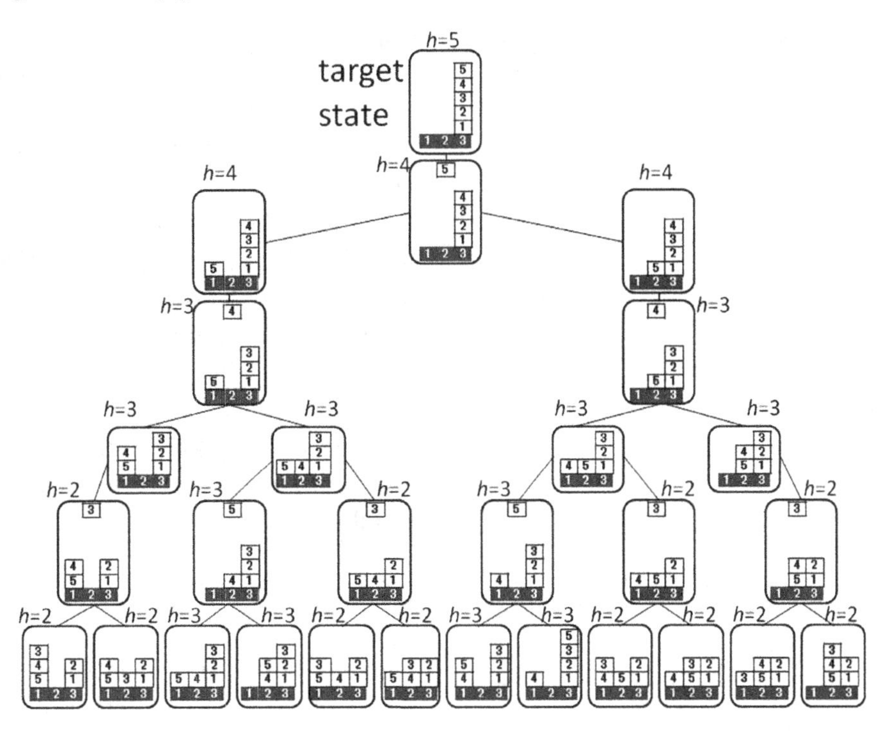

Figure 3. Architecture of an agent (left: basic model; middle: nonshared model; right: shared model)

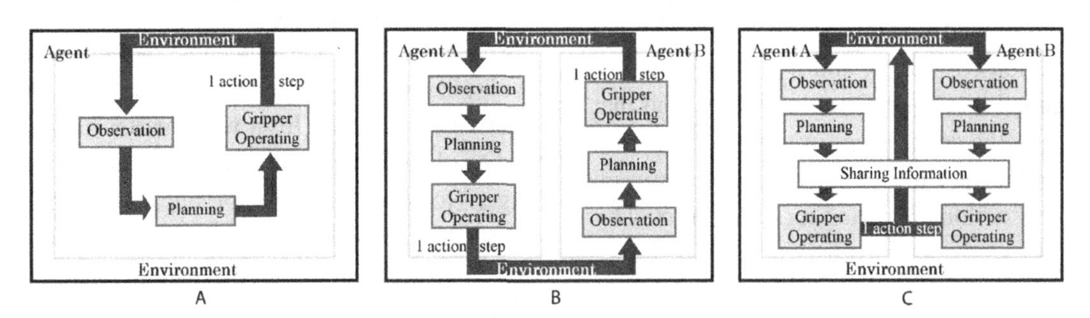

taking: Agent A makes a plan, moves a block once by the gripper, and then agent B changes places with the partner A (makes a plan and moves a block once by the gripper). This cycle is repeated until the agents accomplish a goal, or exceeds an upper limit a_{max}.

In contrast, in the shared model (right in Figure 3), both agents make plan at the same time, and a plan which has the better heuristic value is selected. In the case in which the heuristic values of both agents are the same, the plan with a shorter sequence of actions is selected. When the length of both sequences are the same, either one is randomly selected. After that, both agents move a block once by the gripper. This cycle is repeated until the agents accomplish a goal, or exceeds an upper limit a_{max}. It is plausible to presume that information sharing during the plan selection process is based on symbolic communication.

2.4 Fitness

Each agent solves blocks world problem several times, and was evaluated by the fitness function F:

$$F = \frac{1}{T_{cost}}, \tag{1}$$

$$T_{cost} = w_a \times a_{cost} + w_p \times p_{cost} \left(w_a + w_p = 1\right), \tag{2}$$

$$a_{cost} = \frac{a - a_{min}}{a_{max} - a_{min}}, \tag{3}$$

$$p_{cost} = \frac{p - p_{min}}{p_{max} - p_{min}}, \tag{4}$$

where a is the average number of action steps that each agent performs to reach the target configuration among total trials of each agent, p is the planning limit of each agent, and a_{min}, a_{max}, p_{min}, and p_{max} are fixed numbers. Also, w_a is the weight to the action cost (a_{cost}), and w_p is the weight to the planning cost (p_{cost}). Equation (1) suggests that the greater the action steps or the planning limit is, the lower the fitness is. The balance between the cost of action and planning is determined by w_p and w_a.

3. BASIC EXPERIMENTS

3.1 Experimental Setup

The experiments in the paper focused on the two parameters controlling the complexity of the problem, thereby investigating the conditions of the environment for the planning ability to evolve: The depth of the optimal solution (D) and the size of the table (T). The depth was defined as the shortest path from start to goal. Also, as the size of the table (as we mentioned before) becomes large, the number of optimal paths on the state space increases because agents have many choices when they move a block in the case the table size is large.

We conducted basic experiments (in which the planning limit of agents was not evolved but fixed) to find how obtained solution is influenced by the difference in planning limits, problem complexity or collective manner. The experiments were conducted with 3 different D values varied by changing start configurations (goal configuration was fixed) while fixing T, and the ones with 4 different T values while fixing D (both start and goal configurations were not changed) as shown in Table 2.

3.2 Results

Figure 4 shows the average action steps to solve the problem in various settings of the T, D and planning limit in the basic model. It is shown that the solution became worse (actions steps became larger) when T was smaller and D was higher. This is because the problem became more difficult to solve in those situations. This tendency was stronger when planning limit was smaller. Especially when the planning limit was the minimum, the agent took a random action because the planner explored the states with the same heuristic value next to the present state. On the other hand, when

Table 2. Experimental setup of the basic experiment

L (number of blocks)	5						
a_{max} (upper limit)	500						
B (beam width)	7						
Number of trial run	100						
T (table size)	4			3	4	5	6
D (depth of the problem)	12	14	16	18			

Figure 4. Average action steps to solve a problem in various settings of the table size (T), the depth of the problem (D), and the planning limit in the basic model

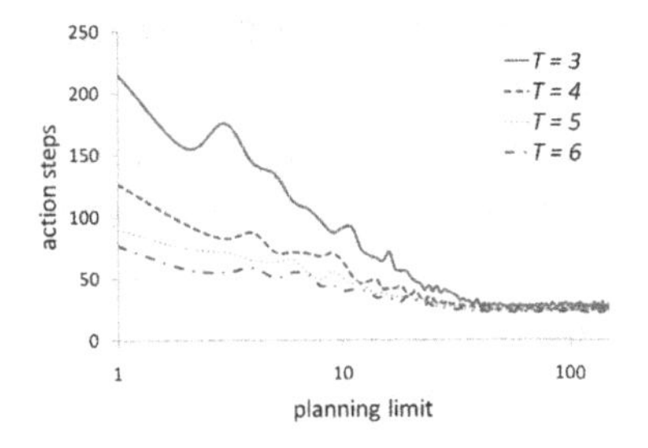

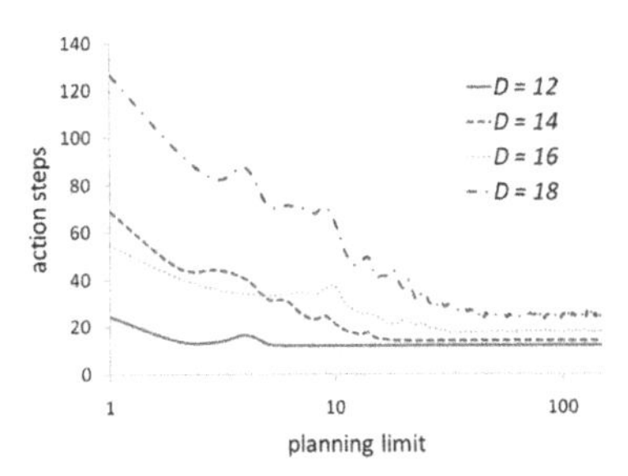

the planning limit was high, the agent took a proper action toward the goal because the planner explored near-goal states far from the present state. Simulation showed the same tendency both in the nonshared and shared models.

Figure 5 shows the action steps averaged over 100 trials of the task evaluation when varying the planning limits of two agents (left: nonshared model; right: shared model). The x and y axes correspond to the planning limit of agent A and that of agent B, and the z axis represents the average action step. We can find that the effects of the planning limits of both agents on the obtained solution were complementary. In other words, it was possible to decrease action steps when the planning limit of either one was long, even if that of the other was very short.

Figure 6 shows the average action steps when two agents took the same planning limit. As shown in Figure 6, the quality of the solution of the shared model was more improved than that of the non shared and basic model for the following reason. At the process of the planning, *sub goal* is randomly chosen from the states in the *hash table* if all states in the *hash table* have the same heuristic value, and it varied in plans. Since a plan with a shorter sequence of actions is selected if the heuristic value of both agents is the same in the shared model, agents in this model could behave more efficiently by comparing both plans.

Figure 7 shows the fitness landscape using the data of the action steps in Figure 5 where w_p was 0.1(left: nonshared model; right: shared model). The x and y axes correspond to the planning limit of agent A and that of agent B, and the z axis represents the fitness of agent B. In the nonshared model (Figure 7 - left), agents who had the middle planning limit (in the range of about 20 to 40) could get the highest fitness when that

Figure 5. Action steps averaged over 100 trials of the task evaluation when varying the planning limits of two agents where F=18 and T=3 (left: nonshared model; right: shared model)

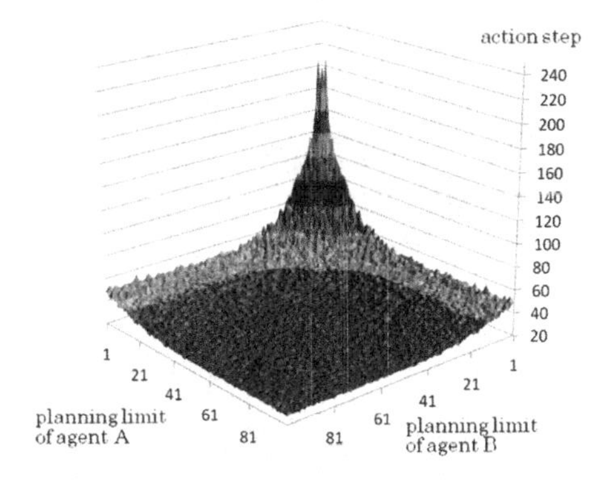

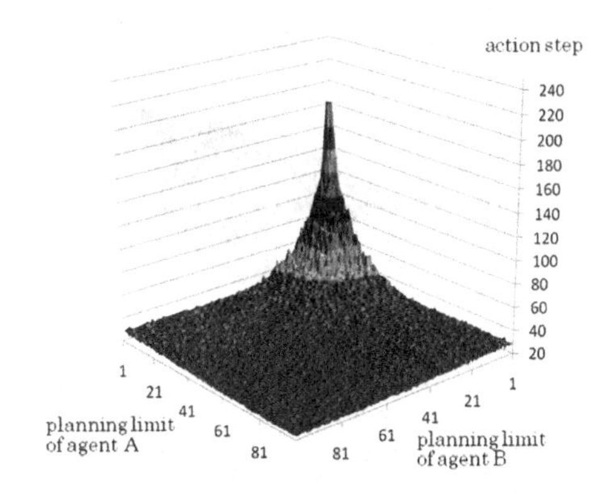

Figure 6. Average action steps of 100 trials of the task evaluation when two agents took the same planning limit where F=18 and T=3

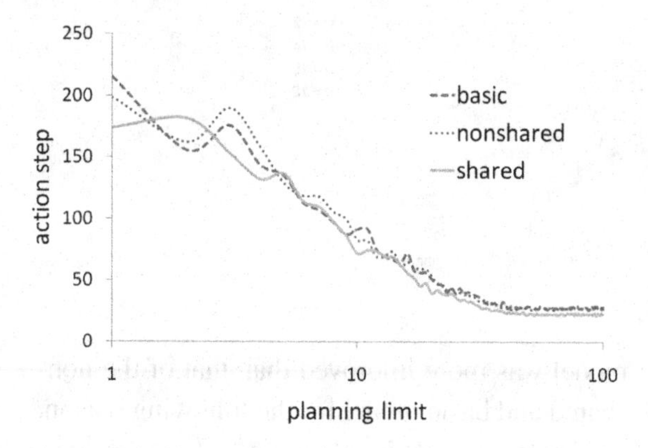

of the other was long (in the range of about 40 to 100). On the other hand, in the shared model (Figure 7 - right), agents who had extremely a short planning limit (in the range of about 1 to 10) could get the highest fitness when that of the other was long (in the range of about 40 to 100) for the following reason. In the shared model, agents who have extremely a short planning limit can decrease action step when that of the other was long because solutions by the agent who has the longer planning limit tend to be adopted in almost all trials. Therefore, they can get the highest fitness because of the low cost for planning.

4. EVOLUTIONARY EXPERIMENTS

4.1 Experimental Setup

We conducted simulations in which the planning limit of agents was evolved by using a genetic algorithm. A chromosome was represented by integer encoding, which determines the planning limit of each agent. We first created N individuals whose planning limits were randomly selected from 1 to 3. As to the basic model each agent solved blocks world problem H times. As to the nonshared and shared model, every pair of agents solved the problem in a round robin manner. Then, action steps were averaged over those games, and agents were evaluated by the fitness function (1).

The offspring in the next generation were selected by the ranking selection as follows. The selection probability p_i is defined using the scaled fitness f_i' as:

$$p_i = \frac{f_i^{'}}{\sum_{j=1}^{N} f_j^{'}}. \quad (5)$$

Here, f_i' is defined as

$$f_i^{'} = \left(N - R_i + 1\right)^2, \quad (6)$$

Figure 7. Fitness landscape of the agent B where action steps were averaged over 100 trials of the task evaluation when varying the planning limits of two agents where F=18, T=3, and w_p=0.1 (left: nonshared model; right: shared model)

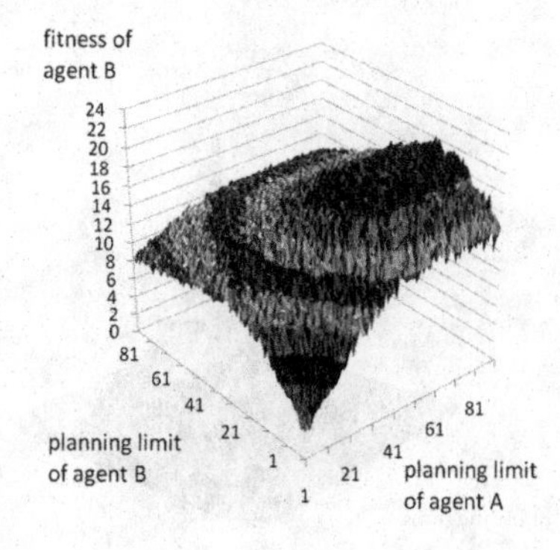

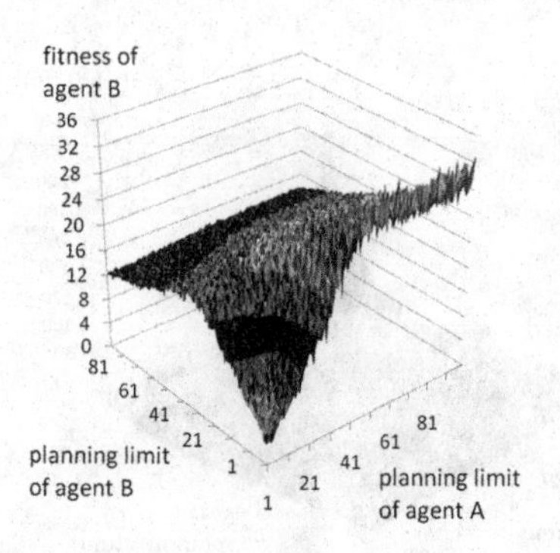

where R_i is the fitness rank of individual i. Then, each gene of all offspring was mutated with a probability P. In the phase of mutation, a random integer digit m was generated from a uniform distribution between $-M$ to $+M$, and added to the original genetic value.

4.2 Results

We conducted evolutionary experiments using parameters of Table 3 where w_p was 0.1, 0.3, and 0.4. Figure 8 shows the difference in transition of the fitness, action steps, and the average planning limit among individuals between basic model, nonshared model, and shared model when w_p=0.1.

Table 3. Experimental setup of the evolutionary experiment

Parameter	Value				
L (number of blocks)	5				
B (beam width)	7				
a_{max}	500				
a_{min}	11				
p_{max}	200				
p_{min}	0				
H(repeat number of times)	5				
N (population size)	20				
P (mutation rate)	0.5				
M (mutation range)	5				
generation	300				
Number of trial run	10				
T (table size)	4	3	4	5	6
D (depth of the problem)	12	14	16	18	

Figure 8. Difference in transition of the fitness, action steps, and planning limits between the basic model, the nonshared model, and the shared model when the complexity of the given problem was changed (w_p=0.1)

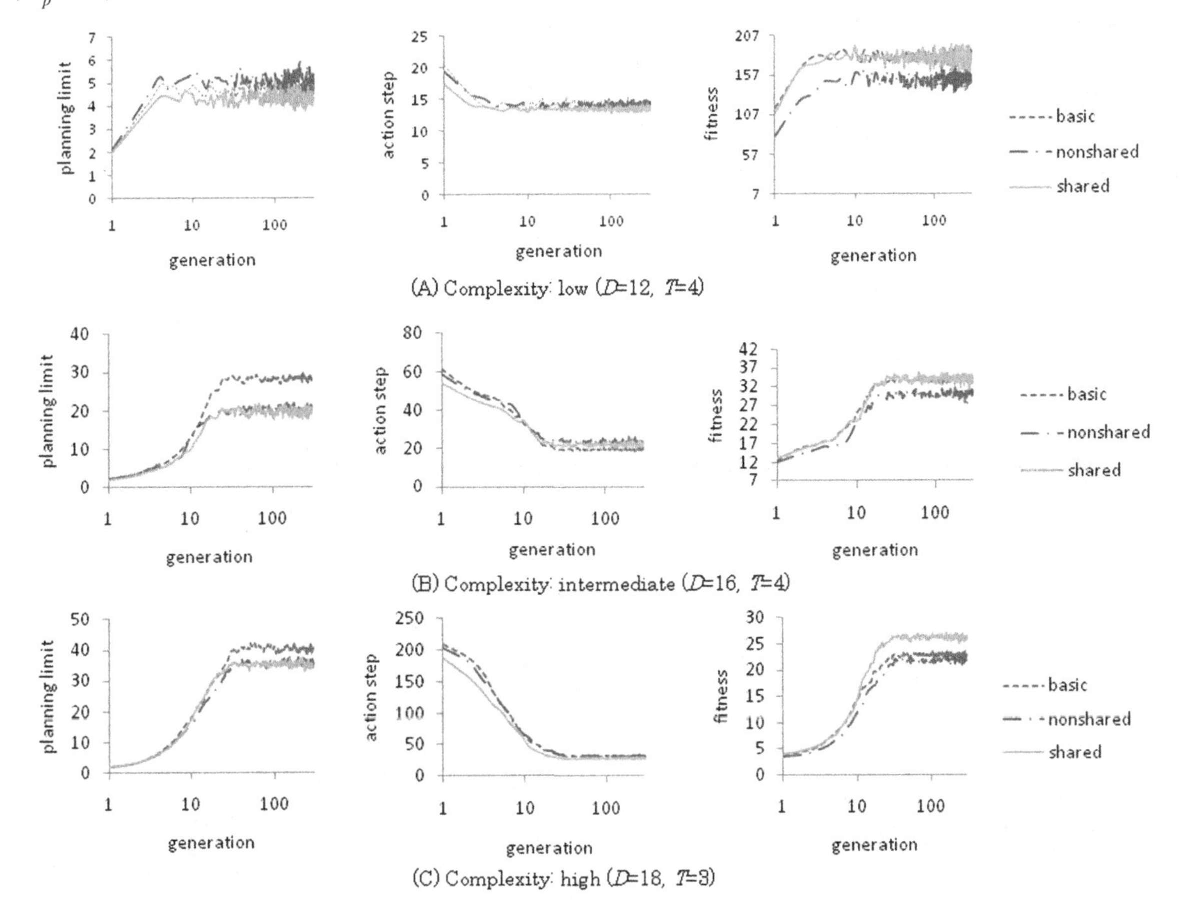

Top row (A), middle row (B) and bottom row (C) show the results of the easiest task (D=12, T=4), an intermediate task (D=16, T=4), and the most difficult task (D=18, T=3), respectively. We can find that the action steps decreased and the planning limit increased through the course of evolution. Also, long planning limits were emerged when the complexity of the problem was intermediate (Figure 8 - (B)) or high (Figure 8 - (C)) because agents could minimize action steps as a reward for increasing the planning limit. On the other hand, long planning limits were not so emerged when the complexity of the problem was low (Figure 8 - (A)) because action steps only slightly changed even if the planning limit increased. Simulation also showed the same tendency when w_p=0.3 and 0.4 as shown in Figure 9 which represents the results of the average planning limits between 200 to 300 generations in the basic model, nonshared model, and shared model when the complexity of the given problem was changed.

Figure 10 shows the transition of the distribution of the planning limits in the population on a certain trial of the basic, nonshared, and shared models when the complexity of the problem was high. As shown in Figure 10, agents who had even the short planning limit (in the range of 10 to 29) could exist in the collective (nonshared and shared) models. Also, the average planning limit in the collective models was less than that in the basic model when the complexity of the problem was intermediate (left in Figure 8 - (B)) or high (left in Figure 8 - (C)). This is because the effects of the planning limit of both agents on the obtained solution were complementary, and it was possible to decrease action steps when the planning limit of either one was long, even if that of the other was short (Figure 5). Therefore, in a case in which agents who had a long planning limit occupied a major part of population, agents who had a short planning limit could enter the population because they could get relatively great fitness because of the low cost for planning. This situation was equivalent to the tragedy of the commons (Hardin, 1968). It explains the reason why the planning ability was difficult to emerge in the collective situation. The reason why difference in planning limits of the most difficult task between the basic and collective (nonshared and shared) model was smaller than that of the intermediate task is that the merit of the planning limit would slightly weaken the force of free rider problem.

The notable point is that planning ability equally evolved both in the shared and nonshared model even though the free rider problem tends to be more serious in the shared model as shown in Figure 7. This may be because the optimization of the planning limit proceeded for the following reasons. First, sharing information improved a quality of the solution. Second, planning of the

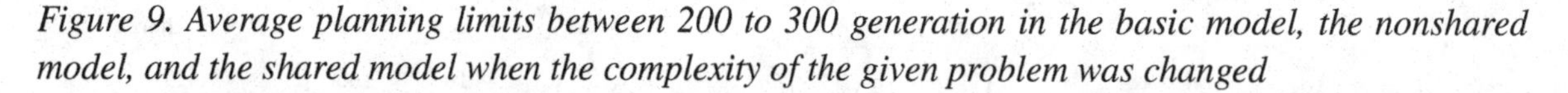
Figure 9. Average planning limits between 200 to 300 generation in the basic model, the nonshared model, and the shared model when the complexity of the given problem was changed

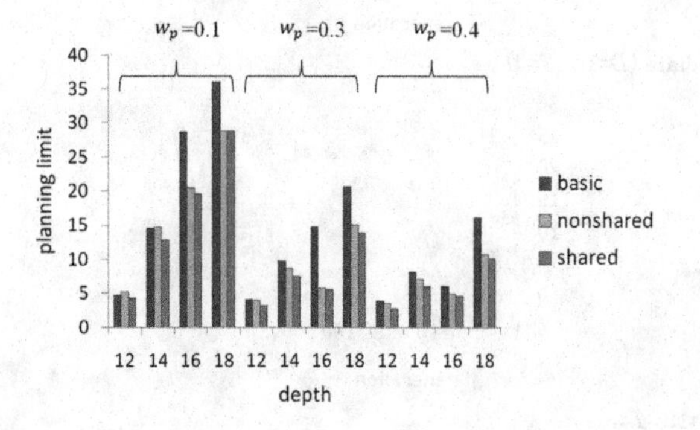

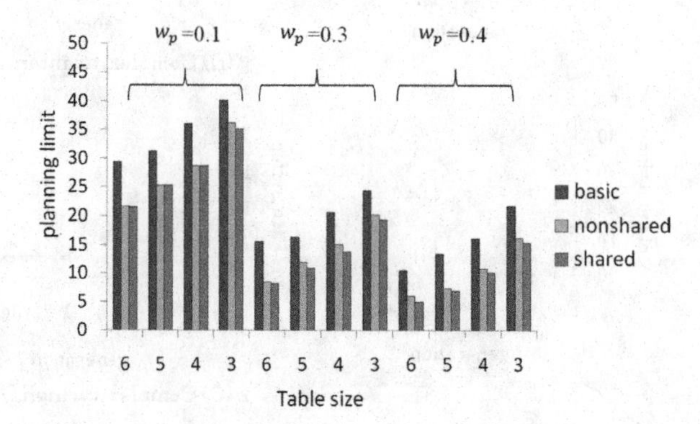

Figure 10. Transition of the distribution of the planning limit in the population on a certain trial

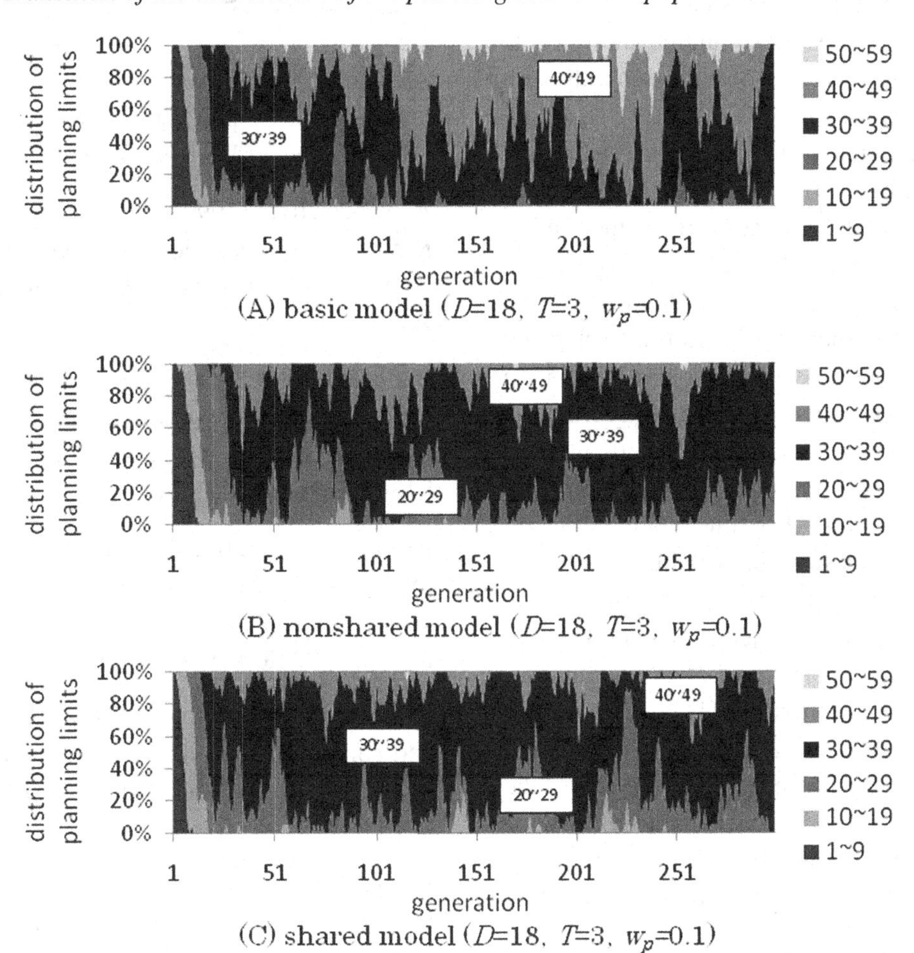

shared model have a greater tendency to have a positive effect on the solution than that of the nonshared model when the own planning limit is longer than other agent. As a result, planning of the shared model equally evolved to that of the nonshared model, and fitness of the shared model was higher than that of nonshared model.

5. CONCLUSION AND DISCUSSION

In this paper, we investigated the mechanism of the evolution of the planning abilities by focusing on the benefits of the planning and the costs of it. Simulation results clarified that planning ability emerged when the complexity of the given problem was high. So what does the complexity of the problem mean in the evolution of the human intelligence? It has been said that ice sheets started to grow in the northern parts of the world, and Africa experienced deforestation and expanding savannas. These conditions in savannah might force the hominid to use a wide variety of food sources which were more transient and scattered than the predominantly vegetarian food sources (Bickerton, 2002). It might work as a selective pressure for more efficient feeding, and thus an increasing need for sophisticated tool use by planning might be selected (Byrne, 1997). Yet, as the problem becomes more complex such as making complicated stone tools, making a tent, farming, or stock raising, working together seems to be important to accomplish a goal efficiently. Furthermore, symbolic communication seems to be indispensable to co-operate smoothly with other individuals.

The simulation results indicated that when the problem was difficult, planning ability was difficult to evolve in the collective situation because there was a conflict between personal and collective interest. So, how could hominids climb the steep cost gradient? The results also clarified that planning ability equally evolved both in the situations where individuals shared information and did not share information even though the free rider problem tended to be more serious in the former situation. Also, fitness of the situation where individuals share information was higher than that of individuals did not share information because of improving a quality of the solution. Considering all of the above factors, we can present the following scenario as: (1) First, a selective pressure for more efficient feeding in savannah made the use of prospective cognition that is the skill to plan for future events and needs, beneficial; (2) As the problem became more complex, increasing need for collectively work would be selected, however; a cost of thinking might be serious at the same time; (3) The select of symbolic communication might be favored because it was an efficient way of solving problems. This result implied that there is a connection between evolution of the planning and symbolic communication. What remains to be done is to clarify the mechanism of the co-evolutionary dynamics of the planning and symbolic communications in the situation where two kinds of groups, sharing and no sharing information, coexist in the same population.

It has been claimed that difference in the ecology of the early hominids and the other apes is important but neglected factor in the discussions of the evolution of language (Bickerton, 2002). Our results showed one of the ecologically based answers to why humans are the only animals who have developed a symbolic communication.

REFERENCES

Alexander, R. D. (1971). The search for an evolutionary philosophy of man. *Proceedings of the Royal Society of Victoria*, *84*, 99–120.

Alexander, R. D. (1990). *How did humans evolve?: Reflections on the uniquely unique species*. Ann Arbor, MI: University of Michigan Museum of Zoology.

Baddeley, A. (2000). The episodic buffer: A new component of working memory? *Trends in Cognitive Sciences*, *4*(11), 417–423. doi:10.1016/S1364-6613(00)01538-2

Bickerton, D. (2002). Foraging versus social intelligence in the evolution of protolanguage. In Wray, A. (Ed.), *The transition to language* (pp. 207–225). Oxford, UK: Oxford University Press.

Brinck, I., & Gärdenfors, P. (2003). Co-operation and communication in apes and humans. *Mind & Language*, *18*, 484–501. doi:10.1111/1468-0017.00239

Brothers, L. (1990). The social brain: A project for integrating primate behavior and neurophysiology in a new domain. *Concepts in Neuroscience*, *1*, 27–51.

Byrne, R. W. (1997). The technical intelligence hypothesis: An additional evolutionary stimulus to intelligence? In A. Whiten & R. W. Byrne (Eds.), *Machiavellian intelligence, vol. II: Extensions and evaluations* (pp. 289-211). Cambridge, UK: Cambridge University Press.

Curtis, C. E., & D'Esposito, M. (2003). Persistent activity in the prefrontal cortex during working memory. *Trends in Cognitive Sciences*, *7*(9), 415–423. doi:10.1016/S1364-6613(03)00197-9

Darwin, C. (1871). *The descent of man, and selection in relation to sex*. London, UK: John Murray.

Deacon, T. (1997). *The symbolic species: The co-evolution of language and the brain*. New York, NY: W. W. Norton.

Dunbar, R. I. M. (1998). The social brain hypothesis. *Evolutionary Anthropology*, *6*(5), 178–190. doi:10.1002/(SICI)1520-6505(1998)6:5<178::AID-EVAN5>3.0.CO;2-8

Flinn, M. V., Geary, D. C., & Ward, C. V. (2005). Ecological dominance, social competition, and coalitionary arms races: Why humans evolved extraordinary intelligence. *Evolution and Human Behavior*, *26*, 10–46. doi:10.1016/j.evolhumbehav.2004.08.005

Gulz, A. (1991). *The planning of action as a cognitive and biological phenomenon*. Lund, Sweden: Lund University Cognitive Studies.

Hardin, G. (1968). The tragedy of the commons. *Science*, *162*(3859), 1243–1248. doi:10.1126/science.162.3859.1243

Hill, K. (1982). Hunting and human evolution. *Journal of Human Evolution*, *11*, 521–544. doi:10.1016/S0047-2484(82)80107-3

Humphrey, N. K. (1976). The social function of intellect. In Bateson, P. P. G., & Hinde, R. A. (Eds.), *Growing points in ethology* (pp. 303–317). Cambridge, UK: Cambridge University Press.

Jolly, A. (1999). *Lucy's legacy: Sex and intelligence in human evolution*. Cambridge, MA: Harvard University Press.

Moreno, A. (2002). Artificial life and philosophy. *Leonardo*, *35*(4), 401–405. doi:10.1162/002409402760181204

Ney, H., Mergel, D., Noll, A., & Paeseler, A. (1992). Data driven organization of the dynamic programming beam search for continuous speech recognition. *IEEE Transactions on Signal Processing*, *40*(2), 272–281. doi:10.1109/78.124938

Osvath, M., & Gärdenfors, P. (2005). Oldwan culture and the evolution of anticipatory cognition. *Lund University Cognitive Science, 122*.

Potts, R. (1998). Variability selection in hominid evolution. *Evolutionary Anthropology*, *7*, 81–96. doi:10.1002/(SICI)1520-6505(1998)7:3<81::AID-EVAN3>3.0.CO;2-A

Squire, L. R., Ojemann, J. G., Miezin, F. M., Petersen, S. E., Videen, T. O., & Raichle, M. E. (1992). Activation of the hippocampus in normalhumans: A functional anatomical study of memory. *Proceedings of the National Academy of Sciences of the United States of America*, *89*, 1837–1841. doi:10.1073/pnas.89.5.1837

Suddendorf, T., & Corballis, M. C. (1997). Mental time travel and the evolution of human mind. *Genetic, Social, and General Psychology Monographs*, *123*, 133–167.

Tooby, J., & DeVore, I. (1987). The reconstruction of hominid behavioral evolution through strategic modelling. In Kinzey, W. (Ed.), *The evolution of human behavior: Primate models* (pp. 183–238). Albany, NY: State University of New York Press.

Tulving, E. (1993). What is episodic memory? *Current Directions in Psychological Science*, *2*, 67–70. doi:10.1111/1467-8721.ep10770899

This work was previously published in the International Journal of Artificial Life Research, Volume 2, Issue 3, edited by E. Stanley Lee and Ping-Teng Chang, pp. 22-35, copyright 2011 by IGI Publishing (an imprint of IGI Global).

Chapter 5
Noise Power Spectrum for Firecrackers

K. B. Patange
Deogiri College, India

A. R. Khan
Maulana Azad College, India

S. H. Behere
Dr. Babasaheb Ambedkar Marathwada University, India

Y. H. Shaikh
Shivaji Arts, Commerce and Science College, India

ABSTRACT

Frequency of noise can affect human beings in different ways. The sound of firecrackers is a type of intensive impulsive noise, which is hazardous. In this paper, the noise produced by firecrackers during celebration festivals in Aurangabad (M.S.), India is measured. The noise is analyzed from the study of power spectra for different types of firecrackers. Noise measurements of firecrackers show that they produce high sound pressure peak levels at their characteristics frequencies. Plots of noise power versus frequency for different crackers are presented and the inferences are discussed. Typical firecracker peak noise levels are given.

1. INTRODUCTION

Sound level, its frequency spectrum and its variation over time characterize noise (Alam, 2006). Noise can also be characterized by its frequency content. This can be assessed by various types of frequency analysis to determine the relative contributions of the frequency components to the total noise (Berglund, 1999). Conventionally the range of frequency of audible sound is considered to be from 20 Hz to about 20,000 Hz (Carl, 2006). There are individual variations in the frequency of sound that can be heard by different individuals. The study of sound related noise includes the study of noise levels at different frequencies. A standard well known method to find out noise power associated with different frequencies is to use the Fourier transform technique (Berglund, 1995).

Fourier transform of sound reveals the amplitude of noise at the constituent frequencies from which one can estimate the corresponding power levels. The main principle of frequency analysis is

DOI: 10.4018/978-1-4666-3890-7.ch005

that any selected frequency range is divided into a number of consecutive and discrete analysis bandwidths, such that the amount of energy present in each analysis bandwidth can be determined.

Firecrackers are used all over the world to celebrate different social as well as religious occasions. The firecrackers traditionally used for celebration are another major source of excessive noise. They are used indiscriminately in residential areas, next to hospitals, schools, with little consideration for the effect on the well being of persons unable for a variety of reasons to bear the high level of noise created. Bombs, chain bombs, etc. are permitted for manufacture provided they do not exceed 125dB – the level of a jet engine taking off at 25 meters (Noise Free Mumbai). Firecrackers generate instantaneous impulsive noise, which when measured in free field condition gives high sound pressure level. An impulse is much more harmful than a continuous noise (Khopkar, 1993). A sudden noise generated with high pitch or intensity but with a life-time of less than one second is called as impulse. Impulses caused by an exploding bomb (190 dB), naval gun-shooting, firing crackers and metal beating, are capable of producing noise to the extent of 140 dB. An unexpected thud of sound with a short life has high impulse and is dangerous. It is quite obvious than a number of crackers when bursting serially can easily form a band of continuous noise in the presence of reflecting surfaces (West Bengal Pollution Control Board, 2005). Noise pollution due to bursting of firecrackers during Diwali was surveyed at 11 cities in the state of Maharashtra by Maharashtra pollution control board in the year 2005 and found that the noise levels in Aurangabad city are higher than the stipulated limits (Deshpande, 2005). To make matters more complicated, the frequency of noise can affect human beings in different ways. Fatigue and nausea often result from low-frequency vibration, while high frequencies are likely to cause pain and hearing loss. A number of adverse effects of noise in general may be greater for low frequency noise than for the same noise energy in higher frequencies (Goldstein, 1994).

Firecrackers may easily produce very high sound levels, and the noise they cause is of an impulsive nature, which means that it has a very short rise time, i.e., a very rapid onset (Miyara, 2010). This impulse type of noise can cause hearing damage. Noise measurements of firecrackers show that they produce high sound pressure peak levels (Tandon, 2003). The emission of a peak sound pressure level at a given location due to an individual firecracker depends mainly on its sound power, the distance from explosion, and the sound pressure distribution in the time and frequency domain. Noise produced by such activities can cause annoyance and concern regarding the health of people as well as impact on wild and domestic animals. Repeated exposure to such explosions can lead to the development of stress-related diseases. Apart from casualties and permanent hearing loss, such explosions cause numerous extra-aural effects, especially to persons not expecting them. Every strong explosion acts as a stress factor, in this way it presents a serious health hazard, especially for more sensitive people. Firecracker explosions belong to a group of intensive impulsive noises, which are particularly hazardous. Their sound pressure level at a distance of few meters can greatly exceed 140 dB, the level adopted for hearing protection (Cudina, 2005).

The present work deals with the analysis of recorded audio data from the point of view of the power levels present at different frequencies.

1.1 Methodology

The study comprises analysis of various types of sound including noise recorded from bursting firecrackers. The sound was recorded in standard Windows wave file format with .WAV extension. This format in addition to all the recorded audible data contains information about the sampling frequency and other related technical details. These sound files in wave format are opened and read in MathCAD program. Most of the data files used for this purpose were recorded at a sampling rate of 44.1 KHz with single channel and 16 bit resolution. This allows for a resolution of 1 part in 65536, a reasonably high resolution. Each sampling point therefore requires two bytes (16 bits of data), this results in a data rate of 88.2 K Bytes per second.

After reading the audio file in wave format, the sampling rate and the time for each sample is estimated and an array of time steps corresponding to the data points is generated.

From the data read, a suitable interval of data is selected so that the number of points chosen complies with the requirement of the FFT that the number of data points should be equal to 2^N where N is an integer. In most of the studies we used 8192 data points which corresponds to a little less than 0.2 seconds of recorded sound. On implementation of the FFT this gives power spectrum in terms of audio powers at different frequencies in the form of amplitudes at different frequencies. The number of frequencies at which the power spectrum is available is half of the frequencies used i.e. 8192 / 2 = 4096, in other words FFT extracts power at 4096 frequencies. The resulting amplitudes at different frequencies are complex quantities. The magnitude of power can be estimated using the modulus of this complex value of amplitude and taking its square.

1.2 Discussions

Typical power spectra obtained from the FFT of various firecrackers (traditionally used) are shown in Figures 1 to 9. Note that power expressed on y-scale is in the order of 10^8 for all the figures.

Figure 1. Noise power spectrum for Phataka chain 1000

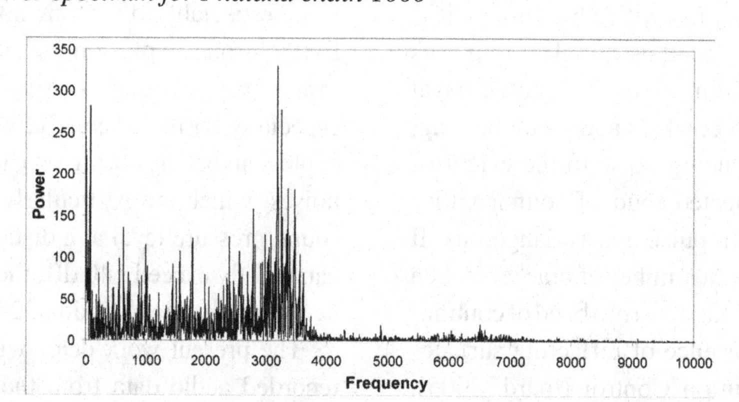

Figure 2. Noise power spectrum for Phataka single

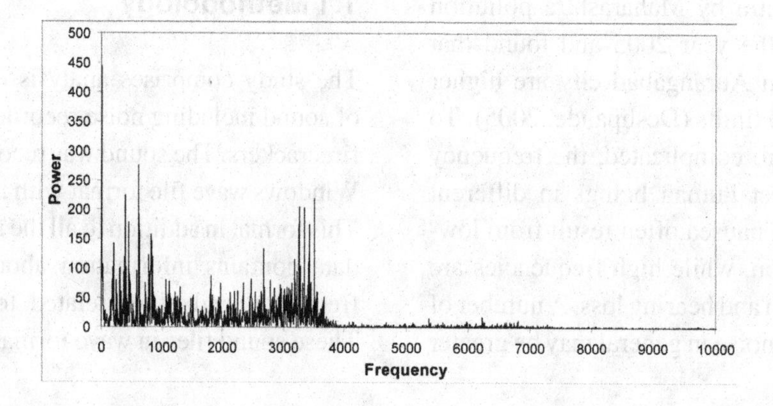

Figures 1 and 2 represents typical noise power spectra of Phataka Chain 1000 recorded during celebration of Diwali Festival. Figure 1 is the recording of crackers (Thousand wala) having one thousand crackers joined in series that explode one after the other. The noise is spread over a broad range of low frequency and shows peaks up to 3.6 KHz. Afterwards noise falls suddenly and shows some small fluctuations up to 7 KHz. It shows a tall spike at 3.2 KHz. Figure 2 represents the power spectrum of single cracker and the gross appearance is very much like those discussed above with average noise level up to about 3.5 KHz and some characteristic spikes with less noise power.

Figure 3 and 4 represents the noise power spectrum for cracker called Rocket. Figure 3 shows clusters of noise around 3.2, 5 and 7 KHz with characteristic peaks and no appreciable noise appears below 3 KHz and above 7.5 KHz. Figure 4 has similar appearance with indication of four clusters in the 3.5 to 6 KHz range, the slight difference in the two patterns is because they are from different make but similar size. The noise

Figure 3. Noise power spectrum for rocket

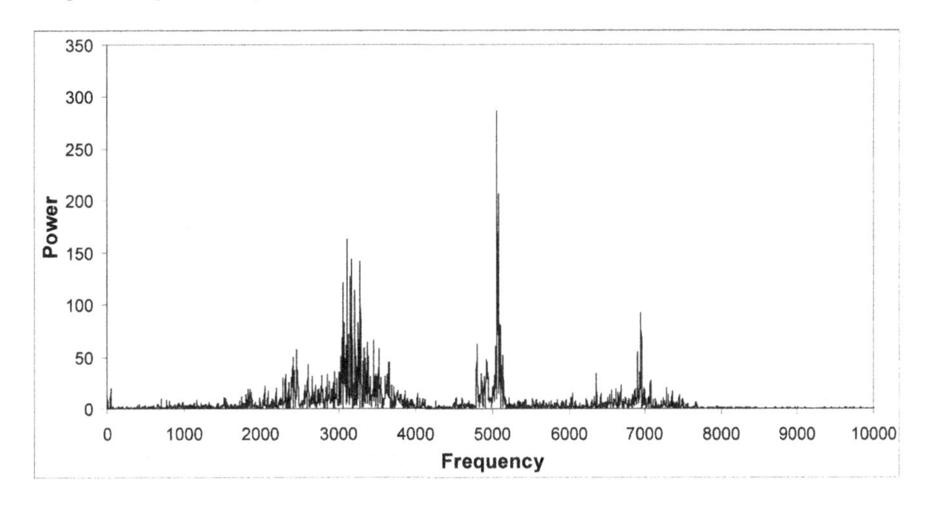

Figure 4. Noise power spectrum for rocket

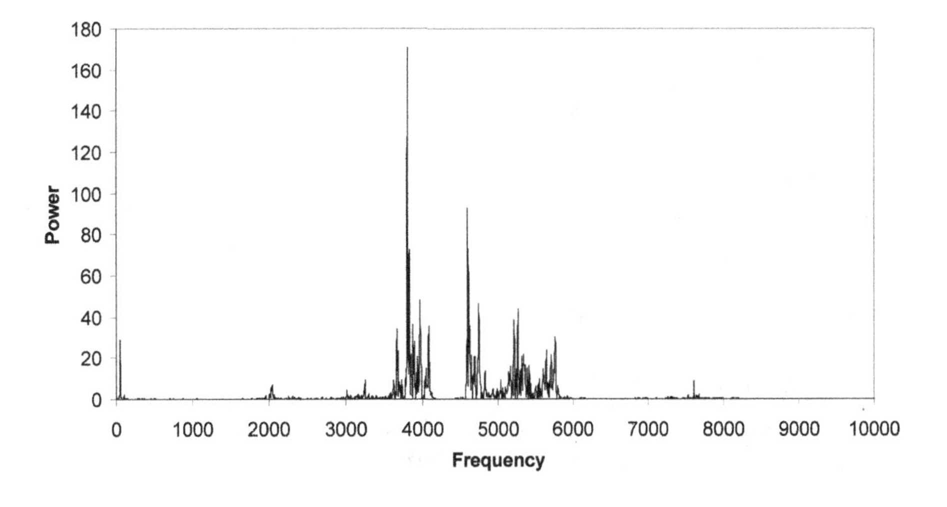

Figure 5. Noise power spectrum for Sutali bomb

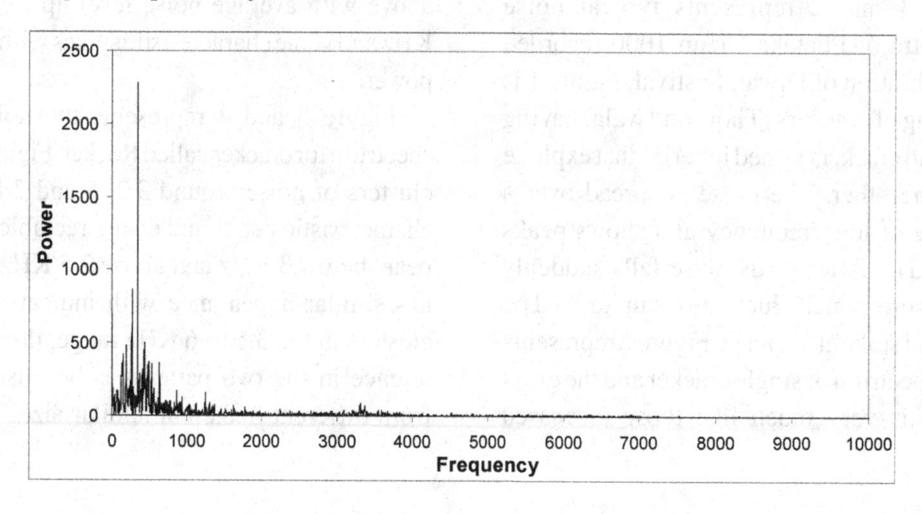

Figure 6. Noise power spectrum for cracker bursting in pieces

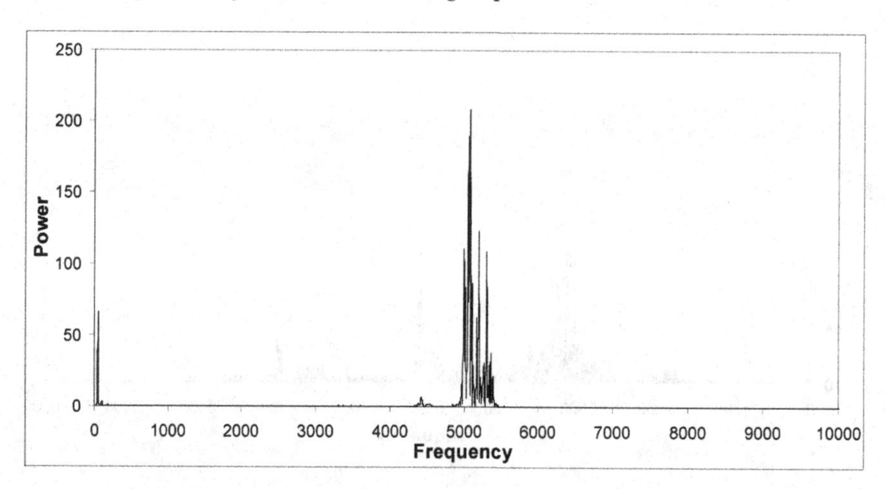

Figure 7. Noise power spectrum for seven shot

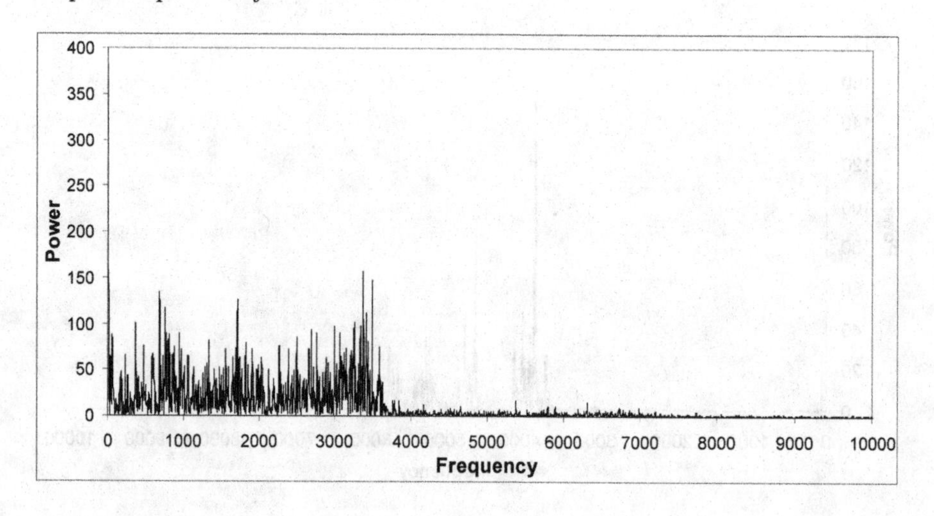

Figure 8. Noise power spectrum for seven shot

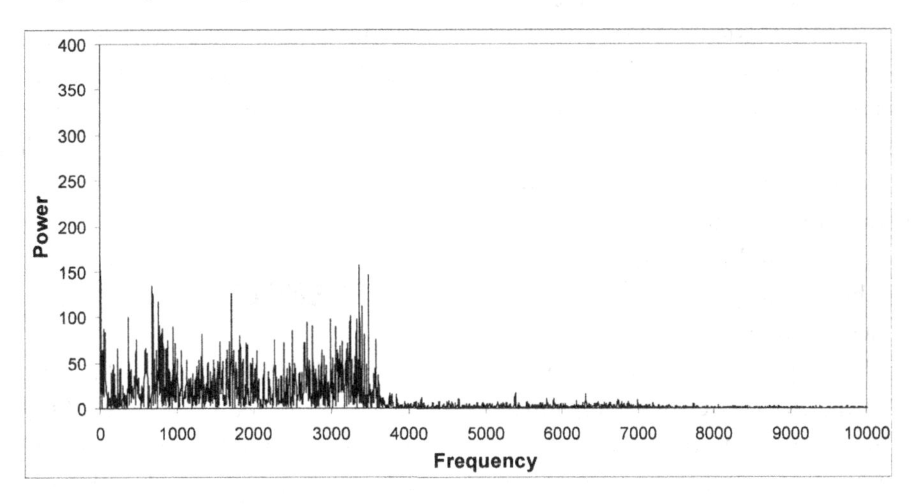

Figure 9. Noise power spectrum for seven shot

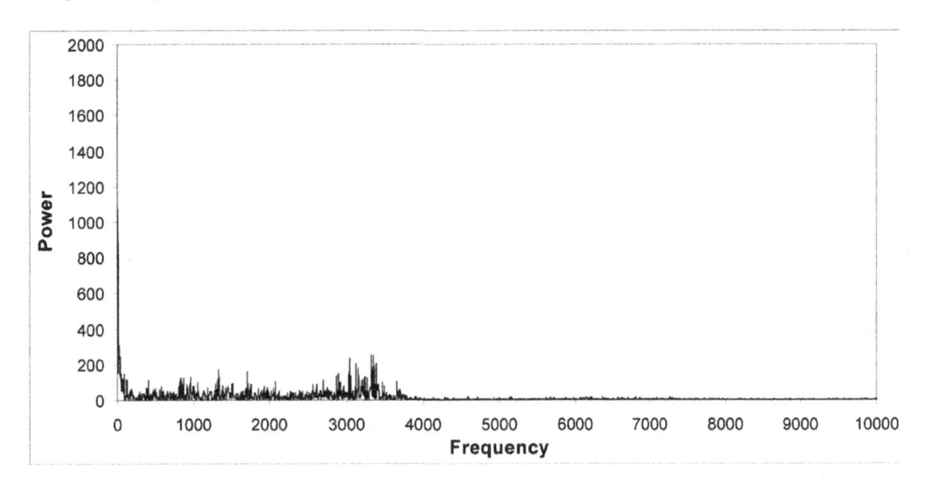

Table 1. Peak noise levels for different types of crackers

Sr. No.	Type of Crackers	Peak Noise Levels dB
1	Rocket	103.7
2	Phataka Lad (1000 wala)	102.1
3	Phataka Lad 1	101
4	Phataka Lad 2 (100 wala Big)	104.4
5	sutali bomb	85.5
6	Cracker bursting in pieces	99.8
7	7 shot	103.7

produced by powerful cracker (bomb) is much different from that in Figures 1 through 4 as seen from Figure 5.

Figure 5 represents the noise power spectrum for sutali bomb (a type of cracker made winding rope around the core of the cracker) and shows a sharp spike in a cluster around 400Hz. Small fluctuations are also seen up to a frequency of 4 KHz and thereafter the noise power drops to insignificant levels.

Figure 6 represents the noise power spectrum for cracker which burst in pieces one after other emerging very fast from a single cracker and produces sharp noise. It is interesting to note that only a single prominent cluster is seen around 5.1 KHz. Excepting for an isolated sharp spike near 50Hz, there is no appreciable noise power at rest of the frequencies.

Figures 7 to 9 represents the noise power spectrum for cracker which burst in 7 parts in the sky one after the other and is called seven shot. All the graphs show broad range of noise spikes up to 4 KHz with a prominent cluster on the higher side of the frequency. It is seen from the plots shown that the cracker devices with higher noise exhibit higher noise power levels at their characteristic frequencies and those with lower noise exhibit lower noise power levels.

The peak noise levels for different types of crackers measured at a distance of 10 m using Data logger sound level meter C322 are given in Table 1. For every type of cracker, level goes beyond 100 dB except for sutali bomb. Long time exposures to such levels are harmful for the human being.

2. CONCLUSION

Noise produced by firecrackers used in various festive and social occasions is studied from the point of view of frequency distribution. It is observed that noise produced by paper wound crackers show a broad frequency distribution and the noise prevails over a wide range (up to about 0 - 3.5 KHz) as is seen from Figures 1, 2, 7, 8, and 9. All these firecrackers i.e. phataka, phataka chain or seven shots are made using paper around the basic cracker core. The sutali bomb that is made of by winding thin rope around core of the cracker is more powerful and produces louder burst of sound and the frequency distribution is also different as seen in Figure 5. The frequency range is very limited at the lower frequency i.e. up to about 500 Hz. Figures 3 and 4 shows that the rockets show characteristic frequencies as three or more clusters where as Figure 6 shows only one cluster at its characteristic frequency of 5.1 KHz. Most of these frequencies are governed by the mechanism the loud noise is produced and the shape and size of materials used dictating their characteristic frequencies.

REFERENCES

Alam, J. B., Alam, M. J. B., Rahman, M. M., Dikshit, A. K., & Khan, S. K. (2006). Study on traffic noise level of sylhet by multiple regression analysis associated with health hazards. *Iranian Journal of Environmental Health Sciences & Engineering*, *3*(2), 71–78.

Awaaz Foundation. (2010). *Noise Free Mumbai*. Retrieved from http://www.awaaz.org/downloads/noise-free-mumbai.pdf

Berglund, B., & Lindvall, T. (1995). Community noise. In Berglund, B., & Lindvall, T. (Eds.), *Archives of the Center for Sensory Research* (*Vol. 2*, pp. 1–195).

Berglund, B., Lindvall, T., & Schwela, D. H. (1999). *Guidelines for community noise*. Paper presented at the World Health Organization Expert Task Force Meeting, London, UK.

Cudina, M., & Prezelj, J. (2005). Noise due to firecracker explosions. *Journal of Mechanical Engineering Science*, *219*(6), 523–537.

Deshpande, A. (2005). *Noise pollution scenario in Maharashtra during Diwali festival*. Mumbai, India: Maharashtra Pollution Control Board. Retrieved from http://mpcb.gov.in/images/pdf/noisereport2005.pdf

Goldstein, M. (1994). *Low-frequency components in complex noise and their perceived loudness and annoyance*. Solna, Sweden: National Institute of Occupational Health.

Hanson, C., Towers, D., & Meister, L. (2006). *Transit noise and vibration impact assessment* (Tech. Rep. No. FTA-VA-90-1003-06). Washington, DC: Department of Transportation Federal Transit Administration Office of Planning and Environment.

Khopkar, S. M. (1993). *Environmental pollution analysis*. New Delhi, India: New Age International.

Miyara, F. (2010). *A note on Firecracker's noise*. Retrieved from http://www.eie.fceia.unr.edu.ar/~acustica/biblio/firecr1.htm

Tandon, N. (2003). Firecrackers noise. *Noise and Vibration Worldwide, 34*, 5.

West Bengal Pollution Control Board. (2005). *Report of assessment of noise pollution survey in Kolkata during Kalipuja and Diwali festivals*. Retrieved from http://www.wbpcb.gov.in/html/downloads/kalipuja_diwali_06.pdf

This work was previously published in the International Journal of Artificial Life Research, Volume 2, Issue 1, edited by E. Stanley Lee and Ping-Teng Chang, pp. 62-70, copyright 2011 by IGI Publishing (an imprint of IGI Global).

Chapter 6
Traffic Noise:
$1/f$ Characteristics

K. B. Patange
Deogiri College Aurangabad, Maharashtra, India

S. H. Behere
Dr. Babasaheb Ambedkar Marathwada University, India

A. R. Khan
Maulana Azad College, India

Y. H. Shaikh
Shivaji Arts, Commerce, Science College, India

ABSTRACT

In this paper, the study of traffic noise is presented from the point of view of 1/f noise. Samples of Traffic Noise are collected from selected locations from busy roads of Aurangabad city in Maharashtra state (India) and data is analyzed. It is observed that in many cases the traffic noise possesses pink noise (1/f noise) prevailing over appreciable range of frequency. The log log plot of noise power versus frequency results in a straight line with a slope approximately equal to unity confirming the presence of pink noise. After certain frequency, the noise power no longer behaves like pink noise (1/f noise) and becomes more or less constant with random fluctuations. Plots of noise power versus frequency on log log basis for different locations studied are presented and the inferences are discussed.

1. INTRODUCTION

Traffic noise is considered as one of the important sources of noise pollution that adversely affects human health in residential urban areas (Onuu, 2000; Martin, 2002; Gambart, Myncke, & Cops, 1976). Low frequency noise is common as background noise in urban environments arising due to many artificial sources like road vehicles, aeroplanes, industrial machinery, artillery and mining explosions and air movement machinery. This includes wind turbines, compressors, and indoor ventilation and air conditioning units etc. (Tempest, 1985; Leventhall, 1988). Low-frequency noise or flicker noise has been found in many systems (Li, 2009). Intense low frequency noise may produce clear symptoms like respiratory impairment and aural pain (Von Gierke & Nixon, 1976). The $1/f$ behavior generally persists over low frequencies (Sinha, 1996). The power spectra

DOI: 10.4018/978-1-4666-3890-7.ch006

of large variety of complex systems exhibit $1/f$ behavior at low frequencies. It is widely accepted that $1/f$ noise and self-similarity are characteristic signatures of complexity (Gilden, Thornton, & Mallon, 1995; Wong, 2003). Self-similarity, scale invariance and fractal nature are found to be characteristics of many natural phenomena (Shaikh, Khan, Pathan, Patil, & Behere, 2009; Shaikh, Khan, Iqbal, Behere, & Bagare, 2008). $1/f$ noise refers to the phenomenon of the spectral density, $S(f)$ of a stochastic process (Ward & Greenwood, 2007) having the form

$$S(f) = \text{constant}/f^{\alpha}$$

where f is frequency, on an interval bounded away from both zero and infinity. Spectral density (power distribution in the frequency spectrum) is such a property, which can be used to distinguish different types of noise (Wikipedia, n. d.). This classification by spectral density is given "color" terminology. The spectral density of white noise is flat ($\alpha = 0$), while pink noise has $\alpha = 1$, and brown noise has $\alpha = 2$. During last 80 years since the first observation by Johnson (1925), long-memory processes with long-term correlations and $1/f^{\alpha}$ (with $0.5 \le \alpha \le 1.5$) behavior of power spectra at low frequencies have been observed in physics, technology, biology, astrophysics, geophysics, economics, psychology, language and even music (Wong, 2003; Press, 1978; Hooge, Kleinpenning, & Vandamme, 1981; Dutta & Horn, 1981; Kogan, 1985; Weissman, 1988; West & Shlesinger, 1990; Van Vliet, 1991; Zhigalskii, 1997; Milotti, 2002) and in traffic flow too (Yale University, n. d.).

The frequency spectrum of pink noise is flat in logarithmic space; it has equal power in bands that arc proportionally widc (Li, 2009; Prcss, 1978; West & Shlesinger, 1990; Milotti, 2002; Gardner, 1978). This means that pink noise would have equal power in the frequency range from 40 to 60 Hz as in the band from 4000 to 6000 Hz. Since humans hear in such a proportional space, where a doubling of frequency is perceived the same regardless of actual frequency (40–60 Hz is heard in the same interval and distance as 4000–6000 Hz), every octave contains the same amount of energy and thus pink noise is often used as a reference signal in audio engineering. That is, the human auditory system perceives approximately equal magnitude on all frequencies. The power density, compared with white noise, decreases by 3 dB per octave (proportional to $1/f$).

1.1 Materials and Methodology

To study the frequency distribution of the noise power, the noise recorded was analyzed using FFT (Fast Fourier Transform) technique. For the implementation of FFT, we used MathCAD 11. The audio files were opened in MathCAD for reading in the data and the characteristics of the recorded file such as sampling rate, number of channels, resolution etc were found. This is necessary for deciding the frequencies present based on sampling rate.

A program was written for this purpose, after reading in the data this displays the plot of amplitude versus time just like the wave representation of sound in any sound editing software like wave editor or wave pad. The amplitude data in time domain was then subjected to Fast Fourier Transform using the built in function FFT of MathCAD. We used 8192 (2^{13}) points for Fourier transform. Total number of frequencies resulting from the Fourier transform of the signal as described above is 4096. This allows a range of frequencies of up to about 22 KHz which is more than sufficient. In fact power at frequencies greater than about 10 KHz is marginal in general as compared to that at lower frequencies.

Fourier transform of the signal gives amplitude of noise at different frequencies. The amplitude resulting from the Fourier transform is a complex quantity with both real and the imaginary parts.

Power can be found from the amplitude by taking its square, resulting data is written to text files for further use.

1.2 Results and Discussion

Having analyzed plenty of data from wave files recorded at selected locations under different traffic conditions it was observed that the noise power is higher at lower frequencies in most of the cases and as one goes to higher frequencies, the noise power rapidly falls down. Later a stage is reached where the noise power is found to be more or less same with random fluctuations. A typical recording of noise level spectrum is shown in Figure 1. The noise was recorded at Highway near Baba Petrol Pump, Aurangabad for normal, general traffic conditions during busy hours. The traffic consisted of light vehicles such as cars, three wheelers, motorcycles and Buses. It is seen from the plot that the power level is on the higher side initially (at lower frequencies), with

Figure 1. Noise amplitude recorded at highway near Baba Petrol Pump Aurangabad for normal, general traffic conditions during busy hours

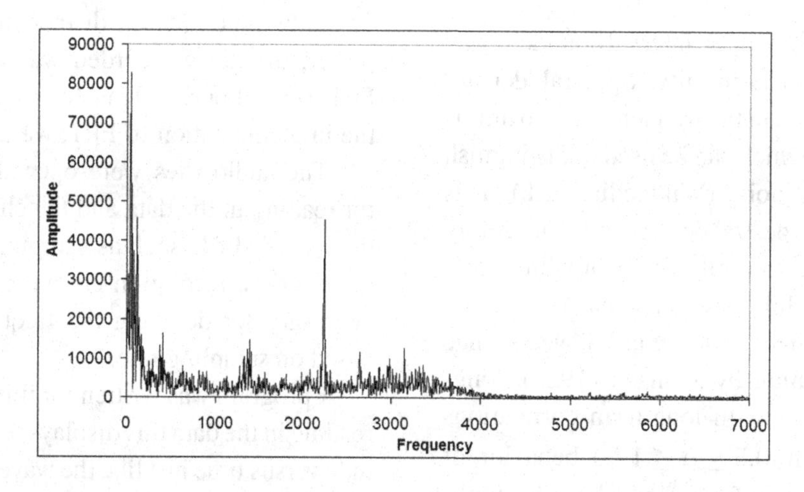

Figure 2. Plot of log of noise power level versus log of frequency for the data presented in Figure 1 (50 – 2000 Hz)

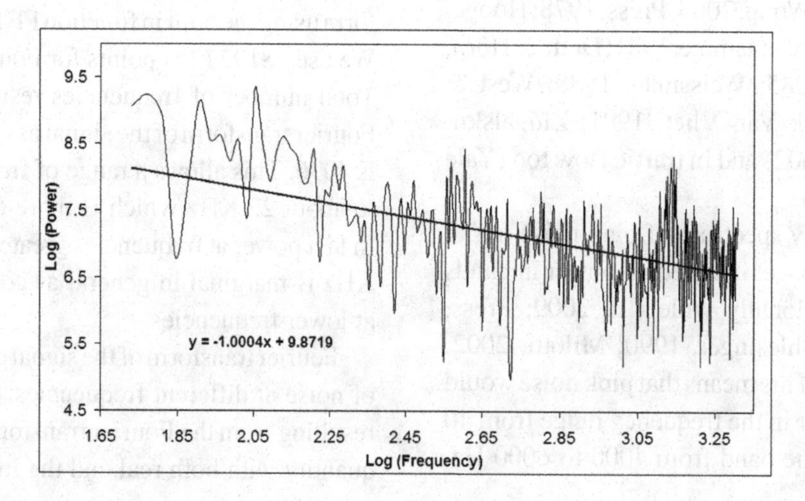

Figure 3. Plot of log of noise power level versus log of frequency for noise recorded at the highway near Baba Petrol Pump

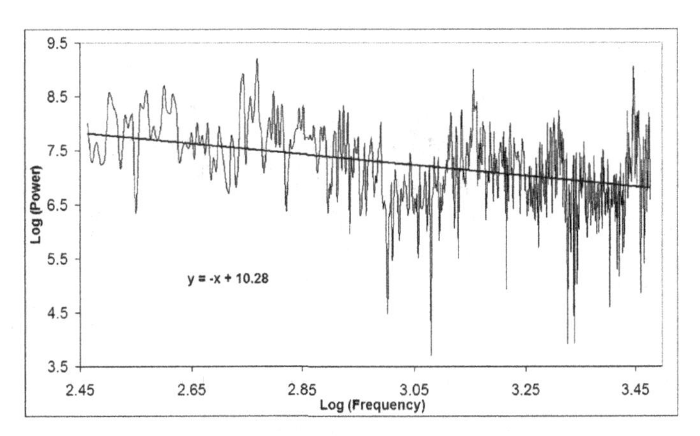

Figure 4. Plot of log of noise power level versus log of frequency for noise recorded at the highway near Aakashwani

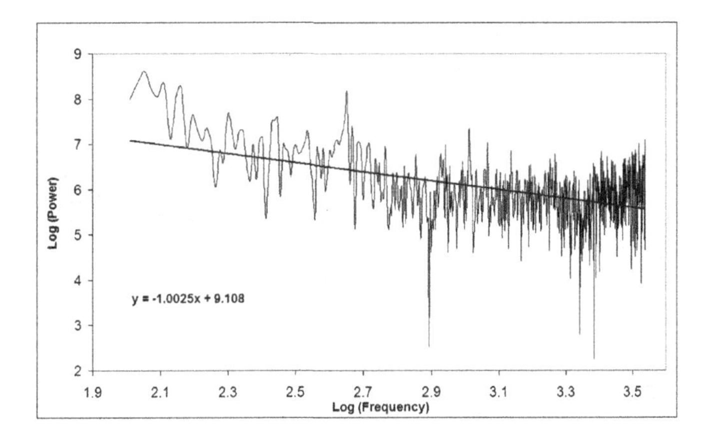

increase in frequency, the power level rapidly falls to a plateau like quasi steady state indicating more or less similar noise level over a wide range of frequencies. In addition to this, there are few sharp peaks and spikes seen at certain frequencies. These spikes or sharp peaks are characteristic of any typical sound like Horn etc being blown there. The sharp peak seen near 2.2 KHz in Figure 1 is due to blowing of horn of a bus and the broad burst near 1.4 KHz is due to blowing of horns by three wheelers.

The noise power spectrum shows fast falling trend with frequency, this exponential decrease in noise power suggests the presence of power law being valid; this also supports existence of 1/f noise. The study of large number of noise power spectra recorded under various conditions show that, not all the fast falling power spectra exhibit 1/f noise but very few of them show real 1/f noise in some frequency range and beyond that the noise ceases to behave like 1/f noise. The traffic conditions are general which in majority includes motorcycles, cars, buses and three wheelers. The percentage of heavy vehicles is around 7%.

It is seen from Figure 1 that the noise amplitude falls rapidly with frequency in the range 50 – 2000 Hz, however when the noise power level

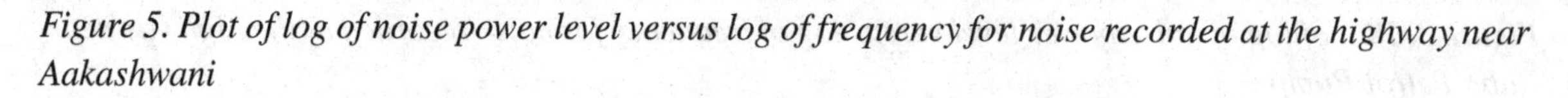

Figure 5. Plot of log of noise power level versus log of frequency for noise recorded at the highway near Aakashwani

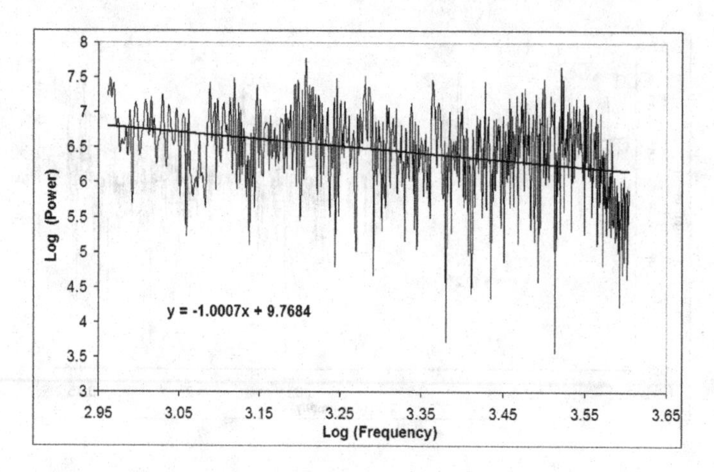

Figure 6. Plot of log of noise power level versus log of frequency for noise recorded at the highway near Amarpreet Hotel

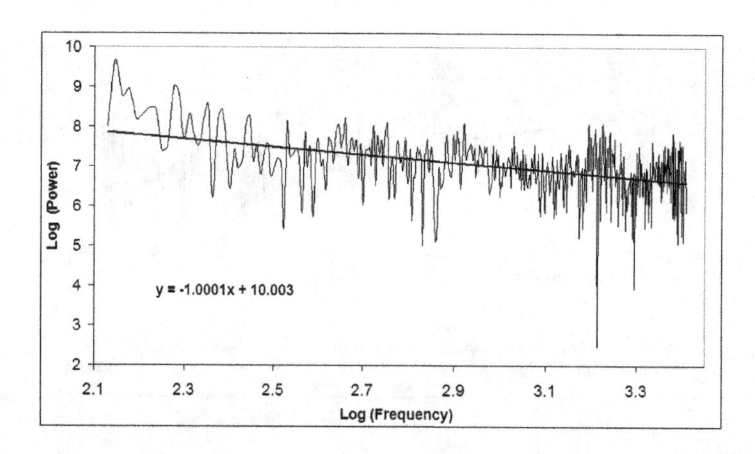

Figure 7. Plot of log of noise power level versus log of frequency for noise recorded at the highway near Amarpreet Hotel

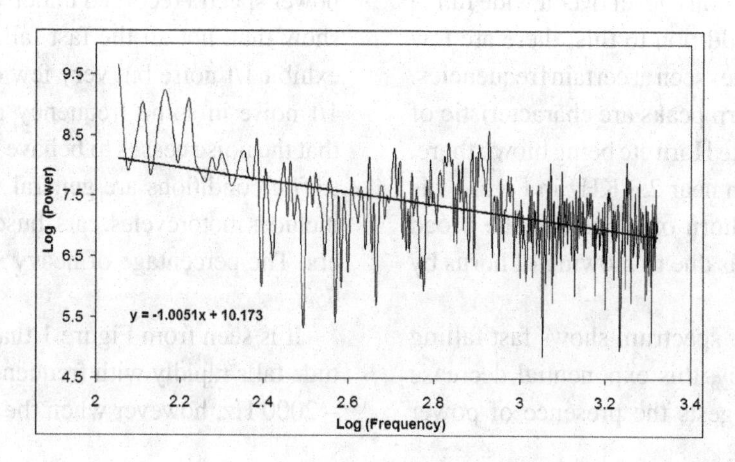

Figure 8. Plot of log of noise power level versus log of frequency for noise recorded at the highway near Kranti Chowk

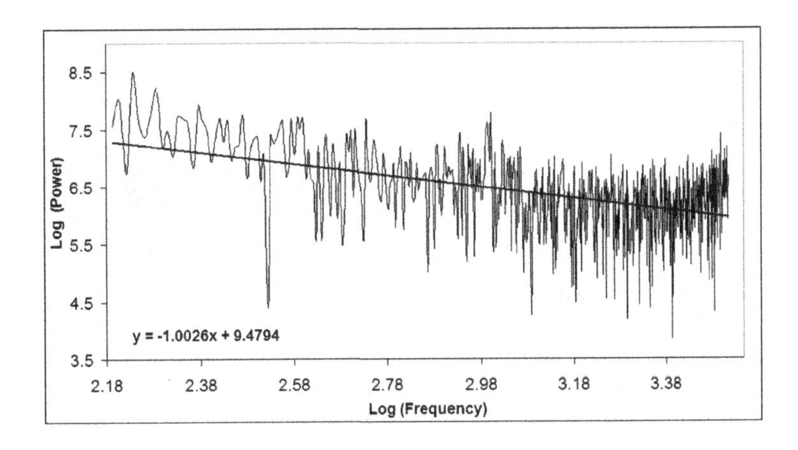

Figure 9. Plot of log of noise power level versus log of frequency for noise recorded at the highway near Mondha Naka

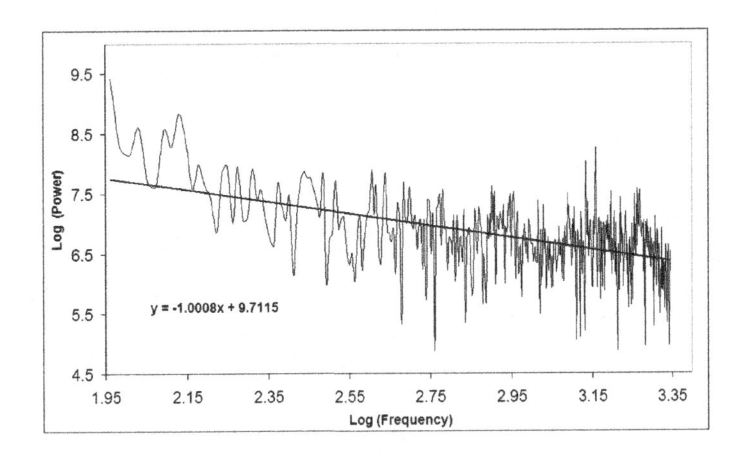

Figure 10. Plot of log of noise power level versus log of frequency for noise recorded at the highway near Mondha Naka

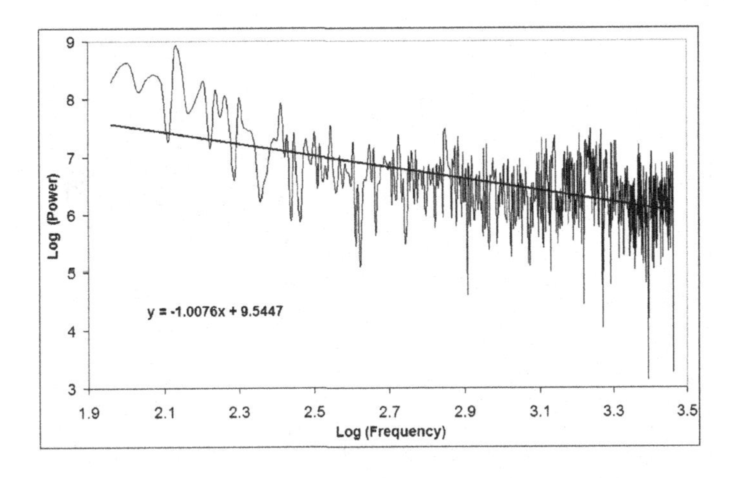

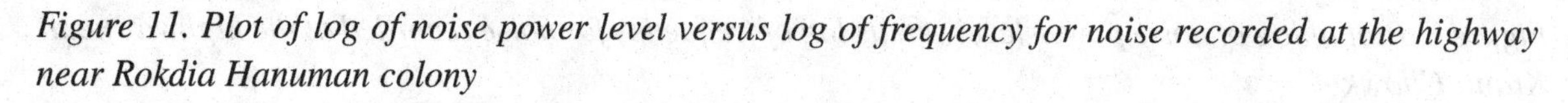

Figure 11. Plot of log of noise power level versus log of frequency for noise recorded at the highway near Rokdia Hanuman colony

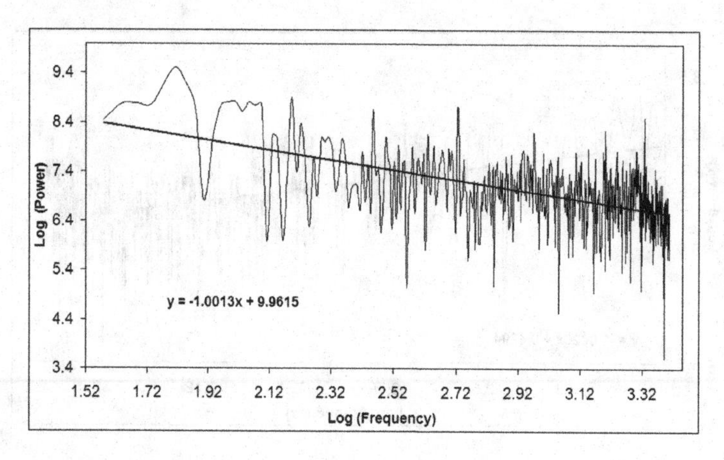

Figure 12. Plot of log of noise power level versus log of frequency for noise recorded at the highway near Rokdia Hanuman colony

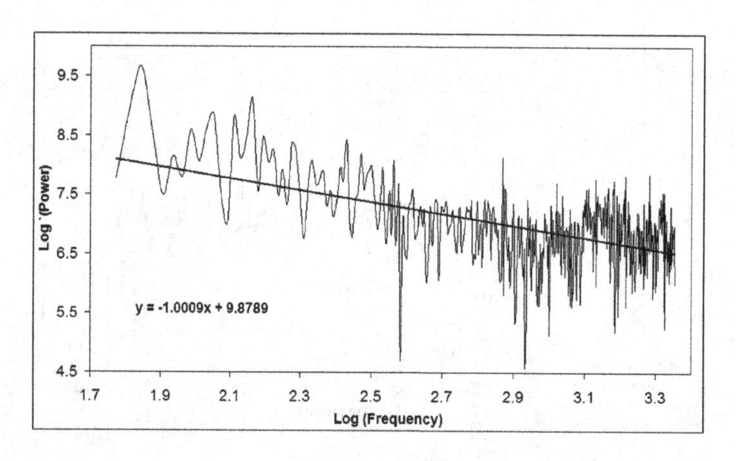

Figure 13. Plot of log of noise power level versus log of frequency for noise recorded at the highway near Rokdia Hanuman colony

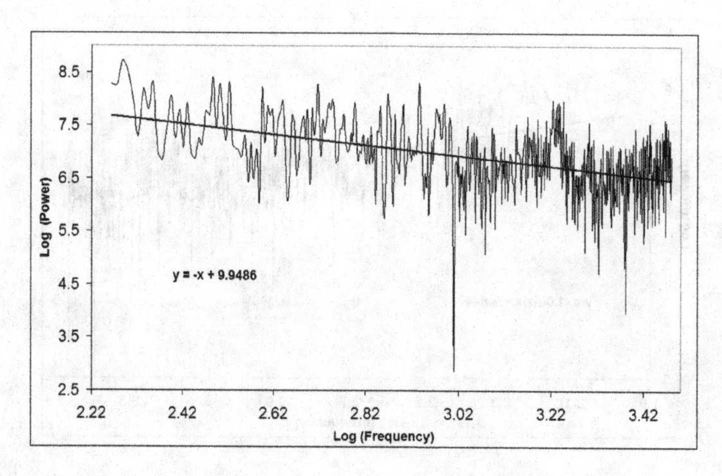

Figure 14. Plot of log of noise power level versus log of frequency for noise recorded at the highway near Rokdia Hanuman colony

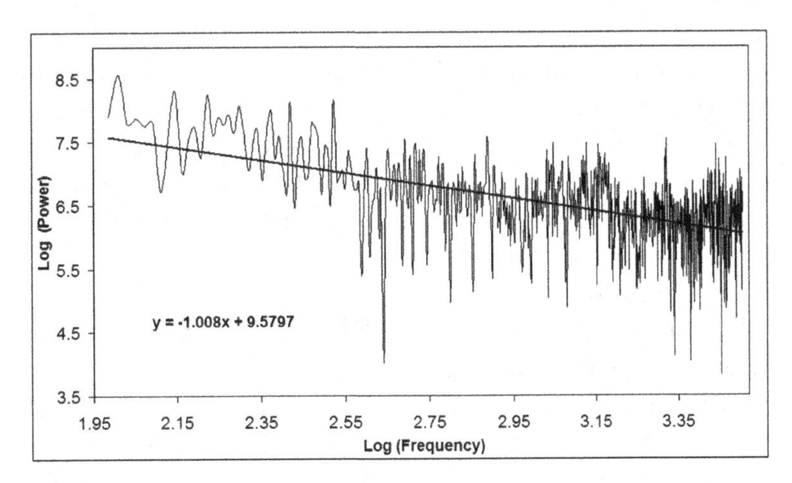

Table 1. Table showing Power law exponent and the range of frequencies for noise presented in Figure 2 through 14

Figure No.	Location	Frequency Range		Slope (α)
		Minimum	Maximum	
2	Baba Petrol Pump	53.83	2035	1.0004
3	Baba Petrol Pump	290.7	3004	1.0000
4	Aakashwani	102.3	3445	1.0025
5	Aakashwani	915.2	4016	1.0007
6	Amarpreet Hotel	134.6	2562	1.0001
7	Amarpreet Hotel	113	2099	1.0051
8	Kranti Chowk	156.1	3284	1.0026
9	Mondha Naka	91.52	2213	1.0008
10	Mondha Naka	91.52	2902	1.0076
11	Rokdia Hanuman colony	37.68	2557	1.0013
12	Rokdia Hanuman colony	59.22	2256	1.0009
13	Rokdia Hanuman colony	183	3015	1.0000
14	Rokdia Hanuman colony	96.9	3192	1.0080

is plotted on log scale the shape is quite different as seen in Figure 2 which is the plot of log of noise power level versus log of frequency for the data presented in Figure 1 for the initial part of frequency. Thin line is the actual data plotted and the thick straight line is the least square fit applied to all these data points. Slope of the straight line fitting the data is found to be -1.0004 confirming that the noise power level follows 1/f trend and thus obeys 1/f law.

As is seen from Figure 2 that the power spectrum confirms presence of 1/f noise, the data from many other locations are found to exhibit the presence of 1/f noise in the initial part of frequencies. Few of such plots are shown in Figures 3 through 14 which indicate the presence of 1/f noise in the initial part of the power spectrum. The noise samples were collected from different locations and under different noise conditions.

Figure 3 shows a plot of log of noise power versus log of frequency for the noise recorded near Baba Petrol pump when there was heavy traffic. 1/f noise is found to prevail over a limited range of frequencies i.e., 290 Hz to 3 KHz.

Figures 4 and 5 show plots for the noise recorded near Aakashwani and Figures 6 and 7 are that for highway near Amarpreet Hotel for two different timings. Slope of the line fitting the data remains close to -1 confirming the presence of 1/f noise. Figure 8 also shows power law exponent close to -1 that confirms presence of 1/f noise over the range of frequencies plotted (Table 1).

Similarly, Figures 9 through 14 shows the noise power spectrum plotted on log log scale for traffic noise recorded at six locations for traffic under free flow (without signals). The Noise power spectrum in this cases is found to have 1/f noise over wide range of frequencies. In some of the plots sharp spikes are present in the power spectrum; however the 1/f nature remains persistent in these cases.

It is confirmed from Figures 2 through 14 and Table 1 that for traffic conditions presented, the noise power spectrum has 1/f noise during the initial part of the frequencies (50 Hz to 4 KHz) and the power law exponent is close to 1.

2. CONCLUSION

It is seen from Figures 3 through 14 that noise originating from highway traffic under certain conditions exhibits the characteristics of pink noise over certain range of frequency. Table 1 summarizes the locations of study, the range of frequency and the power law exponent α. It is seen from the table that the value of α lie close to unity confirming the presence of pink noise (1/f noise) over the frequency range indicated. Most of the power spectra exhibit fast falling power with frequency; however few of them exhibit pink noise (1/f noise) characteristics with power law exponent α equal to unity.

REFERENCES

Dutta, P., & Horn, P. M. (1981). Low-frequency fluctuations in solids: 1/*f* noise. *Reviews of Modern Physics*, *53*, 497. doi:10.1103/RevModPhys.53.497

Gambart, R., Myncke, H., & Cops, A. (1976). Study of annoyance by traffic noise in Leuven (Belgium). *Applied Acoustics*, *9*(3), 193. doi:10.1016/0003-682X(76)90017-7

Gardner, M. (1978). Mathematical games -- white and brown music, fractal curves and one-over-f fluctuations. *Scientific American*, *238*(4), 16. doi:10.1038/scientificamerican0478-16

Gilden, D. L., Thornton, T., & Mallon, M. W. (1995). 1/f noise in human cognition. *Science*, *267*, 1837–1839. doi:10.1126/science.7892611

Hooge, F. N., Kleinpenning, T. G. M., & Vandamme, L. K. J. (1981). Experimental studies on 1/f noise. *Reports on Progress in Physics*, *44*(5), 479. doi:10.1088/0034-4885/44/5/001

Kogan, S. M. (1985). Low-frequency current 1/f-noise in solids. *Uspekhi Fizicheskikh Nauk*, *145*, 285–328. doi:10.3367/UFNr.0145.198502d.0285

Leventhall, H. G. (1988). Low frequency noise in buildings–internal and external sources. *Journal of Low Frequency Noise and Vibration*, *7*, 74.

Li, W. (2009). *A bibliography on 1/f noise*. Retrieved from http://www.nslij-genetics.org/wli/1fnoise/index.html

Martin, S. J. (2002). Numerical modelling of median road traffic noise barrier. *Journal of Sound and Vibration*, *251*(4), 671. doi:10.1006/jsvi.2001.3955

Milotti, E. (2002). *1/f noise: a pedagogical review*. Retrieved from http://arxiv.org/abs/physics/0204033

Onuu, M. U. (2000). Road traffic noise in Nigeria: Measurements, analysis and evaluation of nuisance. *Journal of Sound and Vibration*, *233*(3), 391. doi:10.1006/jsvi.1999.2832

Press, W. H. (1978). Flicker noises in astronomy and elsewhere. *Comments on Astrophysics*, *7*, 103.

Shaikh, Y. H., Khan, A. R., Iqbal, M. I., Behere, S. H., & Bagare, S. P. (2008). Sunspot Data Analysis using time series. *Fractal*, *16*(3), 259. doi:10.1142/S0218348X08004009

Shaikh, Y. H., Khan, A. R., Pathan, J. M., Patil, A., & Behere, S. H. (2009). Fractal pattern growth simulation in electrodeposition and study of the shifting of center of mass. *Chaos, Solitons, and Fractals*, *42*, 2796–2803. doi:10.1016/j.chaos.2009.03.192

Sinha, S. (1996). Transient 1/f noise. *Physics Review Letters E*, *53*, 5.

Tempest, W. (Ed.). (1985). *The noise handbook*. London, UK: Academic Press.

Van Vliet, C. M. (1991). A survey of results and future prospects on quantum 1/f noise and 1/f noise in general. *Solid-State Electronics*, *34*, 1. doi:10.1016/0038-1101(91)90195-5

Von Gierke, H. E., & Nixon, C. W. (1976). Effects of intense infrasound on man. In Tempest, W. (Ed.), *Infrasound and low frequency noise vibration* (p. 115). London, UK: Academic Press.

Ward, L. M., & Greenwood, P. E. (2007). 1/f noise. *Scholarpedia*, *2*(12), 1537. doi:10.4249/scholarpedia.1537

Weissman, M. B. (1988). 1/f noise and other slow, non exponential kinetics in condensed matter. *Reviews of Modern Physics*, *60*(2), 537. doi:10.1103/RevModPhys.60.537

West, B. J., & Shlesinger, M. F. (1990). The noise in natural phenomena. *American Scientist*, *78*, 40.

Wikipedia. (n.d.). *Colors of noise*. Retrieved from http://en.wikipedia.org/wiki/Colors_of_noise

Wong, H. (2003). Low-frequency noise study in electron devices: review and update. *Microelectronics and Reliability*, *43*(4), 585–589. doi:10.1016/S0026-2714(02)00347-5

Yale University. (n. d.). *White noise*. Retrieved from http://classes.yale.edu/fractals/CA/OneOverF/WhiteNoise.html

Zhigalskii, G. P. (1997). 1/f noise and nonlinear effects in thin metal films. *Uspekhi Fizicheskikh Nauk*, *167*, 623–648. doi:10.3367/UFNr.0167.199706c.0623

This work was previously published in the International Journal of Artificial Life Research, Volume 2, Issue 4, edited by E. Stanley Lee and Ping-Teng Chang, pp. 1-11, copyright 2011 by IGI Publishing (an imprint of IGI Global).

Section 2
Mathematical and Analytical Techniques with Applications

Chapter 7
Existence of Positive Solutions for Generalized p-Laplacian BVPs

Wei-Cheng Lian
National Kaohsiung Marine University, Taiwan

Fu-Hsiang Wong
National Taipei University of Education, Taiwan

Jen-Chieh Lo
Tamkang University, Taiwan

Cheh-Chih Yeh
Lunghwa University of Science and Technology, Taiwan

ABSTRACT

Using Kransnoskii's fixed point theorem, the authors obtain the existence of multiple solutions of the following boundary value problem

$$(BVP)\begin{cases} (E)\ \left(\varphi_p\left(u^{(n-1)}(t)\right)\right)' + f\left(t, u(t), \ldots, u^{(n-2)}(t)\right) = 0, \qquad t \in (0,1), \\ (BC)\begin{cases} u^{(i)}(0) = 0, \quad 0 \le i \le n-3, \\ u^{(n-2)}(0) - B_0\left(u^{(n-1)}(\xi)\right) = 0, \\ u^{(n-2)}(1) + B_1\left(u^{(n-1)}(\eta)\right) = 0, \end{cases} \end{cases}$$

where $0 < \xi < \eta < 1$ are given. The authors examine and discuss these solutions.

DOI: 10.4018/978-1-4666-3890-7.ch007

1. INTRODUCTION

In this paper, we concern with the existence of multiple solutions for higher order boundary value problem

$$(BVP)\begin{cases}(E)\ \left(\varphi_p\left(u^{(n-1)}(t)\right)\right)'+f\left(t,u(t),\ldots,u^{(n-2)}(t)\right)=0, & t\in(0,1),\\ (BC)\begin{cases}u^{(i)}(0)=0, \quad 0\le i\le n-3,\\ u^{(n-2)}(0)-B_0\left(u^{(n-1)}(\xi)\right)=0,\\ u^{(n-2)}(1)+B_1\left(u^{(n-1)}(\eta)\right)=0,\end{cases}\end{cases}$$

where $n\geq 3$ is a positive integer, $0<\xi<\eta<1$ are given and $\varphi_p(s)$ is the p-Laplacian operator, that is, $\varphi_p(s)=|s|^{p-2}s$ for $p>1$. Clearly, φ_p is invertible with inverse $\varphi_q(s)=\varphi_p^{-1}(s)$. Here $\frac{1}{p}+\frac{1}{q}=1$.

In recent years, the existence of positive solutions for nonlinear boundary value problems with p-Laplacian operator received wide attention. As we know, two point boundary value problems are used to describe a number of physical, biological and chemical phenomena. Recently, some authors have obtained some existence results of positive solutions of multi-points boundary value problems for second order ordinary differential equations (Wang & Ge, 2007; Yu, Wong, Yeh, & Lin, 2007; Zhao, Wang, & Ge, 2007; Zhou, & Su, 2007). In this paper, we establish the existence of positive solutions of general multi-points boundary value problem (BVP) and related results (Bai, Gui, & Ge, 2004; Guo & Lakshmikantham, 1988; Guo, Lakshmikantham, & Liu, 1996; He & Ge, 2004; Lian & Wong, 2000; Liu, 2002; Ma, 1999; Ma & Cataneda, 2001; Sun, Ge, & Zhao, 2007; Wang, 1997).

In order to abbreviate our discussion, throughout this paper, we assume

(H_1) $f\in C\left([0,1]\times[0,+\infty)^{n-1},[0,+\infty)\right)$;

(H_2) $B_0(s), B_1(s)$ are both nondecreasing continuous and odd functions defined on $(-\infty,+\infty)$ and at least one of them satisfies the condition that there exists $b\geq 0$ such that $0\leq B_i(s)\leq bs$ for all $s\geq 0$, $i=1,2$.

2. PRELIMINARIES AND LEMMAS

Let

$$B=\left\{u\in C^{(n-2)}[0,1]: u^{(i)}=0, 0\leq i\leq n-3\right\}.$$

Then, B is a Banach space with norm $\|u\|=\max_{t\in[0,1]}\left|u^{(n-2)}(t)\right|$. And let

$$K=\left\{u\in B: u^{(n-2)}(t)\geq 0 \text{ is a concave function}, t\in[0,1]\right\}.$$

Obviously, K is a cone in B.

In order to discuss our results, we need the following some lemmas:

Lemma 2.0

Assume that E is a Banach space and $P\subset E$ is a cone in E; Ω_1,Ω_2 are open subsets of E, and $0\in\overline{\Omega}_1\subset\Omega_2$. Furthermore, let $F:P\cap\left(\overline{\Omega}_2\setminus\Omega_1\right)\to P$ be a completely continuous operator satisfying one of the following conditions:

(i) $\|Fx\|\leq\|x\|$, $\forall x\in P\cap\partial\Omega_1$; $\|Fx\|\geq\|x\|$, $\forall x\in P\cap\partial\Omega_2$;

(ii) $\|Fx\|\leq\|x\|$, $\forall x\in P\cap\partial\Omega_2$; $\|Fx\|\geq\|x\|$, $\forall x\in P\cap\partial\Omega_1$.

Then F has a fixed point in $P\cap\left(\overline{\Omega_2}\setminus\Omega_1\right)$.

Lemma 2.1 (Wang, 1997)

Let $u \in K$ and $\theta \in \left(0, \frac{1}{2}\right)$. Then

$u^{(n-2)}(t) \geq \theta \|u\|, t \in [\theta, 1-\theta]$.

Lemma 2.2 Suppose $f \in C([0,1] \times [0,+\infty)^{n-1}, [0,+\infty))$. Then, $u(t) \in B \cap C^{n-1}(0,1)$ is a solution of (BVP) if and only if $u \in B$ is a solution of the integral equation in Box 1.

for some $\sigma_u \in [\xi, \eta]$ with $u^{(n-1)}(\sigma_u) = 0$.

Proof: Let $u(t) \in B \cap C^{n-1}(0,1)$ be a solution of (BVP). Then, it follows from boundary condition (BC) that $u^{(n-1)}(\xi) \geq 0, u^{(n-1)}(\eta) \leq 0$. Thus, there exist $\sigma_u \in [\xi, \eta]$ such that $u^{(n-1)}(\sigma_u) = 0$.

Integrating equation (E) on $(\sigma_u, 1)$,

$$\varphi_p\left(u^{(n-1)}(t)\right) - \varphi_p\left(u^{(n-1)}(\sigma_u)\right) = -\int_{\sigma_u}^{t} f\left(s, u(s), \ldots, u^{(n-2)}(s)\right) ds,$$

which implies

$$u^{(n-1)}(t) = -\varphi_q\left(\int_{\sigma_u}^{t} f\left(s, u(s), \ldots, u^{(n-2)}(s)\right) ds\right),$$

thus,

$$u^{(n-2)}(t) = u^{(n-2)}(\sigma_u) - \int_{\sigma_u}^{t} \varphi_q\left(\int_{\sigma_u}^{s} f\left(\tau, u(\tau), \ldots, u^{(n-2)}(\tau)\right) d\tau\right) ds.$$

Letting $t = 1$,

$$u^{(n-2)}(1) = u^{(n-2)}(\sigma_u) - \int_{\sigma_u}^{1} \varphi_q\left(\int_{\sigma_u}^{s} f\left(\tau, u(\tau), \ldots, u^{(n-2)}(\tau)\right) d\tau\right) ds,$$

this and the boundary condition (BC) imply,

Box 1.

$$u(t) = \int_0^t \int_0^{s_1} \cdots \int_0^{s_{n-3}} w(s_{n-2})\, ds_{n-2} ds_{n-3} \ldots ds_1, \tag{1}$$

where

$$w(t) = \begin{cases} B_0\left(\varphi_q\left(\int_{\xi}^{\sigma_u} f\left(s, u(s), \ldots, u^{(n-2)}(s)\right) ds\right)\right) \\ \qquad + \int_0^t \varphi_q\left(\int_s^{\sigma_u} f\left(\tau, u(\tau), \ldots, u^{(n-2)}(\tau)\right) d\tau\right) ds, & 0 \leq t \leq \sigma_u, \\ B_1\left(\varphi_q\left(\int_{\sigma_u}^{\eta} f\left(s, u(s), \ldots, u^{(n-2)}(s)\right) ds\right)\right) \\ \qquad + \int_t^1 \varphi_q\left(\int_{\sigma_u}^{s} f\left(\tau, u(\tau), \ldots, u^{(n-2)}(\tau)\right) d\tau\right) ds, & \sigma_u \leq t \leq 1, \end{cases}$$

$$u^{(n-2)}(1) = -B_1\left(u^{(n-1)}(\eta)\right) = B_1\left(\varphi_q\left(\int_{\sigma_u}^{\eta} f\left(s,u(s),...,u^{(n-2)}(s)\right)ds\right)\right),$$

hence

$$u^{(n-2)}(t) = B_1\left(\varphi_q\left(\int_{\sigma_u}^{\eta} f\left(s,u(s)\right),...,u^{(n-2)}(s)\right)ds\right)$$

$$+\int_t^1 \varphi_q\left(\int_{\sigma_u}^{s} f\left(\tau,u(\tau),...,u^{(n-2)}(\tau)\right)d\tau\right)ds.$$

Integrating it n-2 times on (0,1)

$$u(t) = \int_0^t\int_0^{s_1}...\int_0^{s_{n-3}} B_1\left(\varphi_q\left(\int_{\sigma_u}^{\eta} f\left(s,u(s),...,u^{(n-2)}(s)\right)ds\right)\right)ds_{n-2}...ds_1$$
$$+\int_0^t\int_0^{s_1}...\int_0^{s_{n-3}}\left(\int_{s_{n-2}}^{1}\varphi_q\left(\int_{\sigma_u}^{s} f\left(\tau,u(\tau),...,u^{(n-2)}(\tau)\right)d\tau\right)ds\right)ds_{n-2}...ds_1.$$

Similarly, integrating equation (E) with respect to $t \in (0,\sigma_u)$,

$$\varphi_p\left(u^{(n-1)}(\sigma_u)\right) - \varphi_p\left(u^{(n-1)}(t)\right) = -\int_t^{\sigma_u} f\left(s,u(s),...,u^{(n-2)}(s)\right)ds,$$

then

$$u^{(n-1)}(t) = \varphi_q\left(\int_t^{\sigma_u} f\left(s,u(s),...,u^{(n-2)}(s)\right)ds\right),$$

thus,

$$u^{(n-2)}(t) = u^{(n-2)}(\sigma_u) - \int_t^{\sigma_u}\varphi_q\left(\int_s^{\sigma_u} f\left(\tau,u(\tau),...,u^{(n-2)}(\tau)\right)d\tau\right)ds.$$

Letting $t = 0$,

$$u^{(n-2)}(0) = u^{(n-2)}(\sigma_u) - \int_0^{\sigma_u}\varphi_q\left(\int_s^{\sigma_u} f\left(\tau,u(\tau),...,u^{(n-2)}(\tau)\right)d\tau\right)ds.$$

By the boundary condition (BC),

$$u^{(n-2)}(0) = B_0\left(u^{(n-1)}(\xi)\right) = B_0\left(\varphi_q\left(\int_{\xi}^{\sigma_u} f\left(s,u(s),...,u^{(n-2)}(s)\right)ds\right)\right).$$

Hence,

$$u^{(n-2)}(t) = B_0\left(\varphi_q\left(\int_{\xi}^{\sigma_u} f\left(s,u(s),...,u^{(n-2)}(s)\right)ds\right)\right)$$

$$+\int_0^t \varphi_q\left(\int_s^{\sigma_u} f\left(\tau,u(\tau),...,u^{(n-2)}(\tau)\right)d\tau\right).$$

Integrating it n-2 times on $(0,1)$,

$$u(t) = \int_0^t\int_0^{s_1}...\int_0^{s_{n-3}} B_0\left(\varphi_q\left(\int_{\xi}^{\sigma_u} f\left(s,u(s),...,u^{(n-2)}(s)\right)ds\right)\right)ds_{n-2}...ds_1$$
$$+\int_0^t\int_0^{s_1}...\int_0^{s_{n-3}}\left(\int_0^{s_{n-2}}\varphi_q\left(\int_s^{\sigma_u} f\left(\tau,u(\tau),...,u^{(n-2)}(\tau)\right)d\tau\right)ds\right)ds_{n-2}...ds_1.$$

This proves u is a solution of (1).

Conversely, suppose that

$$u(t) = \int_0^t\int_0^{s_1}...\int_0^{s_{n-3}} w(s_{n-2})\,ds_{n-2}...ds_1.$$

Then,

$$u^{(n-1)}(t) = \begin{cases} \varphi_q\left(\int_t^{\sigma_u} f\left(\tau,u(\tau),...,u^{(n-2)}(\tau)\right)d\tau\right) \geq 0, t \in [0,\sigma_u], \\ -\varphi_q\left(\int_{\sigma_u}^{t} f\left(\tau,u(\tau),...,u^{(n-2)}(\tau)\right)d\tau\right) \leq 0, t \in [\sigma_u,1]. \end{cases}$$

So,

$$\left(\varphi_p\left(u^{(n-1)}(t)\right)\right)'+f\left(t,u(t),\ldots,u^{(n-2)}(t)\right)=0,\quad t\in(0,1).$$

This completes the proof.

Lemma 2.3 (Sun, Ge, & Zhao, 2007)

Suppose that $u\in K$, then

$$0\le u(t)\le u'(t)\le\ldots\le u^{(n-3)}(t),\ t\in[0,1],$$

and

$$u^{(n-3)}(t)\le\frac{1}{\theta}u^{(n-2)}(t),\ t\in[\theta,1-\theta],$$

where $\theta\in\left(0,\frac{1}{2}\right)$.

Proof: If $u\in K$ and $u^{i}(t)\ge 0, i=1,2,\cdots,n-2,\ t\in[0,1]$, then

$$u^{(i)}(t)=\int_0^t u^{(i+1)}(s)\,ds\le tu^{(i+1)}(t)\le u^{(i+1)}(t),$$

for $i=0,1,\ldots,n-4$,

that is, $u(t)\le u'(t)\le\ldots\le u^{(n-3)}(t),\ t\in[0,1]$.

Next, by Lemma 2.1, for $t\in[\theta,1-\theta]$ we have $u^{(n-2)}(t)\ge\theta\|u\|$ It follows from $u^{(n-3)}(t)=\int_0^t u^{(n-2)}(s)\,ds\le\|u\|$ that

$$u^{(n-3)}(t)\le\frac{1}{\theta}u^{(n-2)}(t)\text{ for }t\in[\theta,1-\theta].$$

Thus, we complete the proof.

Motivated by the above-mentioned results, we can choose a $\sigma\in[\xi,\eta]$ and define an operator $T:K\to C^{(n-1)}[0,1]$ given by the Equation in Box 2.

Lemma 2.4 $T:K\to K$ is completely continuous.

Proof: It follows from the definition of T that is presented in Box 3.

Box 2.

$$(Tu)(t)=$$
$$\begin{cases}\int_0^t\int_0^{s_1}\cdots\int_0^{s_{n-3}}B_1\left(\varphi_q\left(\int_\sigma^\eta f\left(s,u(s),\ldots,u^{(n-2)}(s)\right)ds\right)\right)ds_{n-2}\ldots ds_1\\ +\int_0^t\int_0^{s_1}\cdots\int_0^{s_{n-3}}\left(\int_{s_{n-2}}^1\varphi_q\left(\int_\sigma^s f\left(\tau,u(\tau),\ldots,u^{(n-2)}(\tau)\right)d\tau\right)ds\right)ds_{n-2}\ldots ds_1,\\ \qquad\text{for }t\in(\sigma,1),\\ \int_0^t\int_0^{s_1}\cdots\int_0^{s_{n-3}}B_0\left(\varphi_q\left(\int_\xi^\sigma f\left(s,u(s),\ldots,u^{(n-2)}(s)\right)ds\right)\right)ds_{n-2}\ldots ds_1\\ +\int_0^t\int_0^{s_1}\cdots\int_0^{s_{n-3}}\left(\int_0^{s_{n-2}}\varphi_q\left(\int_s^\sigma f\left(\tau,u(\tau),\ldots,u^{(n-2)}(\tau)\right)d\tau\right)ds\right)ds_{n-2}\ldots ds_1,\\ \qquad\text{for }t\in(0,\sigma].\end{cases}$$

Box 3.

$$(Tu)^{(n-1)}(t) = \begin{cases} \varphi_q\left(\int_t^\sigma f\left(\tau, u(\tau), ..., u^{(n-2)}(\tau)\right) d\tau\right) \geq 0, t \in [0, \sigma], \\ -\varphi_q\left(\int_\sigma^t f\left(\tau, u(\tau), ..., u^{(n-2)}(\tau)\right) d\tau\right) \leq 0, t \in [\sigma, 1], \end{cases}$$

is continuous and decreasing on [0,1], thus, $(Tu)^{(n)}(t) \leq 0$. Moreover, $(Tu)^{(n-2)}(t) \geq 0$, this shows that $TK \subset K$. Furthermore, it is easy to check by Arzela-Ascoli lemma that $T : K \to K$ is completely continuous. Thus, our proof is complete.

Now, we set

$$\max f_0 := \lim_{u_{n-1} \to 0^+} \max_{S(0)} \frac{f(t, u_1, u_2, ..., u_{n-1})}{(u_{n-1})^{p-1}},$$

$$\min f_0 := \lim_{u_{n-1} \to 0^+} \min_{S(\theta)} \frac{f(t, u_1, u_2, ..., u_{n-1})}{(u_{n-1})^{p-1}},$$

$$\max f_\infty := \lim_{u_{n-1} \to \infty} \max_{S(0)} \frac{f(t, u_1, u_2, ..., u_{n-1})}{(u_{n-1})^{p-1}},$$

$$\min f_\infty := \lim_{u_{n-1} \to \infty} \min_{S(\theta)} \frac{f(t, u_1, u_2, ..., u_{n-1})}{(u_{n-1})^{p-1}},$$

where

$$S(\theta) = \left\{(t, u_1, u_2, ..., u_{n-1}) : \theta \leq t \leq 1 - \theta, 0 \leq u_1 \leq ... \leq u_{n-2} \leq \frac{1}{\theta} u_{n-1}\right\}$$

and

$$S(0) = \left\{(t, u_1, u_2, ..., u_{n-1}) : 0 \leq t \leq 1, 0 \leq u_1 \leq ... \leq u_{n-1}\right\}.$$

3. EXISTENCE OF AT LEAST ONE POSITIVE SOLUTION OF (BVP)

Theorem 3.1

Suppose that $(H_1),(H_2)$ hold. Assume that there exist two positive distinct constants r and R satisfing the following assumptions:

(A_1) $f(t, u_1, u_2, ..., u_{n-1}) \geq (k_1 r)^{p-1}$ for $(t, u_1, u_2, ..., u_{n-1}) \in S(\theta)$ and $\theta\, r \leq u_{n-1} \leq r$;

(A_2) $f(t, u_1, u_2, ..., u_{n-1}) \leq (k_2 R)^{p-1}$ for $(t, u_1, u_2, ..., u_{n-1}) \in [0,1] \times \{(u_1, u_2, ..., u_{n-1}) : 0 \leq u_1 \leq ... \leq u_{n-2}$ and $0 \leq u_{n-1} \leq R\}$, where

$k_1 = \frac{2}{\min_{\tau \in [\theta, 1-\theta]} A(\tau)}, k_2 = \frac{1}{1+b}$. Here

$A(\tau) = \int_\theta^\tau \varphi_q(\sigma - s)\, ds + \int_\tau^{1-\theta} \varphi_q(s - \sigma)\, ds$ for $\tau \in [\theta, 1-\theta]$.

Then, (BVP) has a solution u such that $\|u\|$ lying between r and R.

Proof : Without loss of generality, we suppose that $r < R$.

We define two open subsets Ω_1 and Ω_2 of B as follows:

$$\Omega_1 = \{u \in K : \|u\| < r\},$$
$$\Omega_2 = \{u \in K : \|u\| < R\}.$$

For any $u \in \partial\Omega_1$, by Lemma 2.1,

$$r = \|u\| \geq u^{(n-2)}(t) \geq \theta\, \|u\| = \theta\, r, t \in [\theta, 1-\theta].$$

Now, for $t \in [\theta, 1-\theta]$ and $u \in \partial\Omega_1$, we discuss our problem from three perspectives:

(i) If $\sigma \in [\theta, 1-\theta]$, then, for $u \in \partial\Omega_1$,

$$2\|Tu\| = 2(Tu)^{(n-2)}(\sigma)$$

$$= B_0\left(\varphi_q\left(\int_\xi^\sigma f\left(s, u(s), \ldots, u^{(n-2)}(s)\right) ds\right)\right)$$

$$+\int_0^\sigma \varphi_q\left(\int_s^\sigma f\left(\tau, u(\tau), \ldots, u^{(n-2)}(\tau)\right) d\tau\right) ds$$

$$+ B_1\left(\varphi_q\left(\int_\sigma^\eta f\left(s, u(s), \ldots, u^{(n-2)}(s)\right) ds\right)\right)$$

$$+\int_\sigma^1 \varphi_q\left(\int_\sigma^s f\left(\tau, u(\tau), \ldots, u^{(n-2)}(\tau)\right) d\tau\right) ds$$

$$\geq \int_0^\sigma \varphi_q\left(\int_s^\sigma f\left(\tau, u(\tau), \ldots, u^{(n-2)}(\tau)\right) d\tau\right) ds$$

$$+\int_\sigma^1 \varphi_q\left(\int_\sigma^s f\left(\tau, u(\tau), \ldots, u^{(n-2)}(\tau)\right) d\tau\right) ds$$

$$\geq \int_\theta^\sigma \varphi_q\left(\int_s^\sigma f\left(\tau, u(\tau), \ldots, u^{(n-2)}(\tau)\right) d\tau\right) ds$$

$$+\int_\sigma^{1-\theta} \varphi_q\left(\int_\sigma^s f\left(\tau, u(\tau), \ldots, u^{(n-2)}(\tau)\right) d\tau\right) ds$$

$$\geq \int_\theta^\sigma \varphi_q\left(\int_s^\sigma (k_1 r)^{p-1}\, d\tau\right) ds$$

$$+\int_\sigma^{1-\theta} \varphi_q\left(\int_\sigma^s (k_1 r)^{p-1}\, d\tau\right) ds$$

$$= (k_1 r)\left\{\int_\theta^\sigma \varphi_q(\sigma - s)\, ds + \int_\sigma^{1-\theta} \varphi_q(s-\sigma)\, ds\right\}$$

$$= (k_1 r) A(\sigma) \geq (k_1 r) \min_{\tau \in [\theta, 1-\theta]} A(\tau)$$

$$\geq 2r = 2\|u\|$$

(ii) If $\sigma \in (1-\theta, 1]$, then, for $u \in \partial\Omega_1$,

$$\|Tu\| = (Tu)^{(n-2)}(\sigma)$$

$$= B_0\left(\varphi_q\left(\int_\xi^\sigma f\left(s, u(s), \ldots, u^{(n-2)}(s)\right) ds\right)\right)$$

$$+\int_0^\sigma \varphi_q\left(\int_s^\sigma f\left(\tau, u(\tau), \ldots, u^{(n-2)}(\tau)\right) d\tau\right) ds$$

$$\geq \int_0^\sigma \varphi_q\left(\int_s^\sigma f\left(\tau, u(\tau), \ldots, u^{(n-2)}(\tau)\right) d\tau\right) ds$$

$$\geq \int_\theta^{1-\theta} \varphi_q\left(\int_s^\sigma f\left(\tau, u(\tau), \ldots, u^{(n-2)}(\tau)\right) d\tau\right) ds$$

$$\geq \int_\theta^{1-\theta} \varphi_q\left(\int_s^\sigma (k_1 r)^{p-1}\, d\tau\right) ds$$

$$= (k_1 r) \int_\theta^{1-\theta} \varphi_q(\sigma - s)\, ds$$

$$= (k_1 r) A(1-\theta) \geq (k_1 r) \min_{\tau \in [\theta, 1-\theta]} A(\tau)$$

$$\geq 2r \geq r = \|u\|.$$

(iii) If $\sigma \in (0, \theta)$, then, for $u \in \partial \Omega_1$,

$$\|Tu\| = (Tu)^{(n-2)}(\sigma)$$

$$= B_1 \left(\varphi_q \left(\int_\sigma^\eta f\left(s, u(s), \ldots, u^{(n-2)}(s)\right) ds \right) \right)$$

$$+ \int_\sigma^1 \varphi_q \left(\int_\sigma^s f\left(\tau, u(\tau), \ldots, u^{(n-2)}(\tau)\right) d\tau \right) ds$$

$$\geq \int_\sigma^1 \varphi_q \left(\int_\sigma^s f\left(\tau, u(\tau), \ldots, u^{(n-2)}(\tau)\right) d\tau \right) ds$$

$$\geq \int_\theta^{1-\theta} \varphi_q \left(\int_\sigma^s f\left(\tau, u(\tau), \ldots, u^{(n-2)}(\tau)\right) d\tau \right) ds$$

$$\geq \int_\theta^{1-\theta} \varphi_q \left(\int_\sigma^s (k_1 r)^{p-1} d\tau \right) ds$$

$$= (k_1 r) \int_\theta^{1-\theta} \varphi_q (s - \sigma)\, ds$$

$$= (k_1 r) A(\theta) \geq (k_1 r) \min_{\tau \in [\theta, 1-\theta]} A(\tau)$$

$$\geq 2r \geq r = \|u\|.$$

Therefore, $\|Tu\| \geq \|u\|$ for $u \in \partial \Omega_1$. Thus, it follows from Lemma 2.3 that

$$0 \leq u(t) \leq u'(t) \leq \ldots \leq u^{(n-3)}(t) \text{ on } [0,1].$$

If $u \in \partial\Omega_2$, then

$$2\|Tu\| = 2(Tu)^{(n-2)}(\sigma)$$

$$= B_0 \left(\varphi_q \left(\int_\xi^\sigma f\left(s, u(s), \ldots, u^{(n-2)}(s)\right) ds \right) \right)$$

$$+ \int_0^\sigma \varphi_q \left(\int_s^\sigma f\left(\tau, u(\tau), \ldots, u^{(n-2)}(\tau)\right) d\tau \right) ds$$

$$+ B_1 \left(\varphi_q \left(\int_\sigma^\eta f\left(s, u(s), \ldots, u^{(n-2)}(s)\right) ds \right) \right)$$

$$+ \int_\sigma^1 \varphi_q \left(\int_\sigma^s f\left(\tau, u(\tau), \ldots, u^{(n-2)}(\tau)\right) d\tau \right) ds$$

$$\leq B_0 \left(\varphi_q \left(\int_0^1 f\left(s, u(s), \ldots, u^{(n-2)}(s)\right) ds \right) \right)$$

$$+ B_1 \left(\varphi_q \left(\int_0^1 f\left(s, u(s), \ldots, u^{(n-2)}(s)\right) \cdots ds \right) \right)$$

$$+ \int_0^1 \varphi_q \left(\int_0^1 f\left(\tau, u(\tau), \ldots, u^{(n-2)}(\tau)\right) d\tau \right) ds$$

$$\leq 2b \varphi_q \left(\int_0^1 f\left(s, u(s), \ldots, u^{(n-2)}(s)\right) ds \right)$$

$$+ 2 \int_0^1 \varphi_q \left(\int_0^1 f\left(\tau, u(\tau), \ldots, u^{(n-2)}(\tau)\right) d\tau \right) ds$$

$$= 2b\varphi_q\left(\int_0^1 f\left(s, u(s), ..., u^{(n-2)}(s)\right) ds\right)$$

$$+2\varphi_q\left(\int_0^1 f\left(s, u(s), ..., u^{(n-2)}(s)\right) ds\right)$$

$$\le 2b\varphi_q\left(\int_0^1 (k_2 R)^{p-1} ds\right) + 2\varphi_q\left(\int_0^1 (k_2 R)^{p-1} ds\right)$$

$$\le 2k_2 R\{1+b\}$$

$$\le 2R = 2\| u\|$$

Therefore, $\|Tu\| \le \|u\|$ for $u \in \partial\Omega_2$.

Using Kransnoskii's fixed point theorem, we see that T has a fixed point $u \in \left(\overline{\Omega_2} \setminus \Omega_1\right)$ satisfying $r \le \|u\| \le R$. This completes our proof.

Theorem 3.2 Suppose that $(H_1),(H_2)$ hold. Assume that f satisfies

$$(A_3)\ \max f_0 = C_1 \in \left[0, k_2^{p-1}\right)$$

and

$$(A_4)\ \min f_\infty = C_2 \in \left(\left(\frac{k_1}{\theta}\right)^{p-1}, \infty\right].$$

Then, for any two distinct positive numbers r and R, (BVP) has a solution u such that $\|u\|$ lying between r and R.

Proof:First,by (A_3), $\max f_0 = C_1 \in \left[0, k_2^{p-1}\right)$. Taking $\varepsilon - k_2^{p-1} - C_1 > 0$, there exists a $\rho_1 > 0$ (ρ_1 can be chosen arbitrarily small) such that

$$\max_{S(0)} \frac{f(t, u_1, u_2, ..., u_{n-1})}{(u_{n-1})^{p-1}} \le \varepsilon + C_1 = k_2^{p-1}$$

for $u_{n-1} \in [0, \rho_1]$.

Thus,

$$f(t, u_1, u_2, ..., u_{n-1}) \le k_2^{p-1} u_{n-1}^{p-1} \le k_2^{p-1}\rho_1^{p-1} = (k_2\rho_1)^{p-1}$$

for

$(t, u_1, u_2, ..., u_{n-1}) \in [0,1] \times \{(u_1, u_2, ..., u_{n-1}) : 0 \le u_1 \le ... \le u_{n-2} \text{ and } 0 \le u_{n-1} \le \rho_1\}$.

Next, by (A_4), $\min f_\infty = C_2 \in \left(\left(\frac{k_1}{\theta}\right)^{p-1}, \infty\right]$.

Taking $\varepsilon = C_2 - \left(\frac{k_1}{\theta}\right)^{p-1} > 0$, there exists a $\lambda_1 > 0$ (λ_1 can be chosen arbitrarily large) such that

$$\min_{\Omega(\theta)} \frac{f(t, u_1, u_2, ..., u_{n-1})}{(u_{n-1})^{p-1}} \ge -\varepsilon + C_2 = \left(\frac{k_1}{\theta}\right)^{p-1}$$

for $u_{n-1} \in [\theta\lambda_1, \lambda_1]$.

Thus,

$$f(t, u_1, u_2, ..., u_{n-1}) \ge \left(\frac{k_1}{\theta}\right)^{p-1} u_{n-1}^{p-1} \ge \left(\frac{k_1}{\theta}\theta\lambda_1\right)^{p-1} = (k_1\lambda_1)^{p-1}$$

for $(t, u_1, u_2, ..., u_{n-1}) \in S(\theta)$ and $u_{n-1} \in [\theta\lambda_1, \lambda_1]$.

This completes the proof.

Theorem 3.3

Suppose that $(H_1),(H_2)$ hold. Assume that f satisfies

(A_5) $\min f_0 = C_3 \in \left(\left(\frac{k_1}{\theta}\right)^{p-1}, \infty\right]$;

(A_6) $\max f_\infty = C_4 \in \left[0, k_2^{p-1}\right)$.

Then, for any two distinct positive numbers r and R, (BVP) has a solution u such that $\|u\|$ lying between r and R.

Proof: First, by (A_5), $\min f_0 = C_3 \in \left(\left(\frac{k_1}{\theta}\right)^{p-1}, \infty\right]$. Taking $\varepsilon = C_3 - \left(\frac{k_1}{\theta}\right)^{p-1} > 0$, there exists a $\lambda_2 > 0$ (λ_2 can be chosen arbitrarily small) such that

$$\min_{S(\theta)} \frac{f(t, u_1, u_2, \dots, u_{n-1})}{(u_{n-1})^{p-1}} \geq -\varepsilon + C_3 = \left(\frac{k_1}{\theta}\right)^{p-1}$$

for $u_{n-1} \in [\theta\lambda_2, \lambda_2]$.

Thus,

$$f(t, u_1, u_2, \dots, u_{n-1}) \geq \left(\frac{k_1}{\theta}\right)^{p-1} u_{n-1}^{p-1} \geq \left(\frac{k_1}{\theta}\theta\lambda_2\right)^{p-1} = (k_1\lambda_2)^{p-1}$$

for $(t, u_1, u_2, \dots, u_{n-1}) \in S(\theta)$ and $u_{n-1} \in [\theta\lambda_2, \lambda_2]$.

Next, by (A_6), $\max f_\infty = C_4 \in \left[0, k_2^{p-1}\right)$. Taking $\varepsilon = k_2^{p-1} - C_4 > 0$, there exists a $\Theta > 0$ (Θ can be chosen arbitrarily large) such that

$$\max_{S(0)} \frac{f(t, u_1, u_2, \dots, u_{n-1})}{(u_{n-1})^{p-1}} \leq \varepsilon + C_4 = k_2^{p-1}$$

for $u_{n-1} \in [\Theta, \infty]$.

Hence, we have the following two cases.

Case 1. Assume that $\max_{S(0)} f(t, u_1, u_2, \dots, u_{n-1})$ is bounded, that is, there exists $M > 0$ such that

$$f(t, u_1, u_2, \dots, u_{n-1}) \leq M^{p-1}.$$

Taking $\rho_2 = \frac{M}{k_2}$ (since M^{p-1} can be chosen arbitrarily large, ρ_2 can also be chosen arbitrarily large, too),

$$f(t, u_1, u_2, \dots, u_{n-1}) \leq M^{p-1} = (k_2\rho_2)^{p-1}$$

for

$(t, u_1, u_2, \dots, u_{n-1}) \in [0,1] \times \{(u_1, u_2, \dots, u_{n-1}) : 0 \leq u_1 \leq \dots \leq u_{n-2} \text{ and } 0 \leq u_{n-1} \leq \rho_2\}$.

Case 2. Assume that $\max_{S(0)} f(t, u_1, u_2, \dots, u_{n-1})$ is unbounded. Hence, there exists a $\rho_2 \geq \Theta$ (ρ_2 can be chosen arbitrarily large) such that

$$f(t, u_1, u_2, \dots, u_{n-1}) \leq f(t, \rho_2, \dots, \rho_2)$$

for $[0,1] \times [0, \rho_2]^{n-1}$.

It follows from $\rho_2 \geq \Theta$ that

$$f(t, u_1, u_2, \dots, u_{n-1}) \leq f(t, \rho_2, \dots, \rho_2) \leq k_2^{p-1}\rho_2^{p-1} = (k_2\rho_2)^{p-1}$$

for

$(t, u_1, u_2, \dots, u_{n-1}) \in [0,1] \times \{(u_1, u_2, \dots, u_{n-1}) : 0 \leq u_1 \leq \dots \leq u_{n-2} \text{ and } 0 \leq u_{n-1} \leq \rho_2\}$.

This completes the proof.

Corollary 3.4

Suppose that $(H_1), (H_2)$ hold. Assume that f satisfies

(A_7) $\min f_\infty = C_2 \in \left(\left(\frac{k_1}{\theta}\right)^{p-1}, \infty\right]$,

(A_8) $\min f_0 = C_3 \in \left(\left(\frac{k_1}{\theta}\right)^{p-1}, \infty\right]$,

(A_9) there exist a $\lambda^* > 0$ such that $f(t, u_1, u_2, \dots, u_{n-1}) \le (k_2 \lambda^*)^{p-1}$

for

$(t, u_1, u_2, \dots, u_{n-1}) \in [0,1] \times \{(u_1, u_2, \dots, u_{n-1}) : 0 \le u_1 \le \dots \le u_{n-2} \text{ and } 0 \le u_{n-1} \le \lambda^*\}$.

Then (BVP) has at least two positive solutions u_1, u_2 such that $0 < \|u_1\| < \lambda^* < \|u_2\|$.

Corollary 3.5

Suppose that $(H_1), (H_2)$ hold. Assume that f satisfies

(A_{10}) $\max f_0 = C_1 \in \left[0, k_2^{p-1}\right)$,

(A_{11}) $\max f_\infty = C_4 \in \left[0, k_2^{p-1}\right)$,

(A_{12}) there exist a $\eta^* > 0$ such that $f(t, u_1, u_2, \dots, u_{n-1}) \ge (k_1 \eta^*)^{p-1}$

for $(t, u_1, u_2, \dots, u_{n-1}) \in S(\theta)$, $u_{n-1} \in [\theta\eta^*, \eta^*]$.

Then (BVP) has at least two positive solutions u_1, u_2 such that $0 < \|u_1\| < \eta^* < \|u_2\|$.

REFERENCES

Bai, Z., Gui, Z., & Ge, W. (2004). Multiple positive solutions for some p-Laplacian boundary value problems. *Journal of Mathematical Analysis and Applications*, *300*, 477–490. doi:10.1016/j.jmaa.2004.06.053

Guo, D., & Lakshmikantham, V. (1988). *Nonlinear problems in abstract cone*. Orlando, FL: Academic Press.

Guo, D., Lakshmikantham, V., & Liu, X. (1996). *Nonlinear intergal equations in abstract spaces*. Dordrecht, The Netherlands: Kluwer Academic Publishers.

He, X., & Ge, W. (2004). Twin positive solutions for the one-dimensional p-Laplacian boundary value problems. *Nonlinear Analysis, Theory . Method and Application*, *56*, 975–984.

Lian, W. C., & Wong, F. H. (2000). Existence of positive solutions for higher-order generalized p-Laplacian BVPs. *Applied Mathematics Letters*, *13*, 35–43. doi:10.1016/S0893-9659(00)00051-3

Liu, B. (2002). Positive solutions for a nonlinear boundary value problems. *Journal of Computer and Mathematics with Applications*, *44*, 201–217. doi:10.1016/S0898-1221(02)00141-4

Liu, B. (2002). Positive solutions for a nonlinear three point boundary value problems. *Applied Mathematics and Computation*, *132*, 11–28. doi:10.1016/S0096-3003(02)00341-7

Ma, R. (1999). Posiive solutions for a nonlinear three point boundary value problems. *Electronic Journal of Differential Equations*, *34*, 1–8.

Ma, R., & Cataneda, N. (2001). Existence of solutions for nonlinear m-point boundary value problems. *Journal of Mathematical Analysis and Applications*, *256*, 556–567. doi:10.1006/jmaa.2000.7320

Sun, B., Ge, W., & Zhao, D. (2007). Three positive solutions for multipoint one-dimensional p-Laplacian boundary value problems with dependence on the first order derivative. *Mathematical and Computer Modelling*, *45*, 1170–1178. doi:10.1016/j.mcm.2006.10.002

Wang, J. Y. (1997). The existence of positive solutions for the one p-Laplacian. *Proceedings of the American Mathematical Society*, *125*, 2275–2283. doi:10.1090/S0002-9939-97-04148-8

Wang, Y., & Ge, W. (2007). Multiple positive solutions for multipoint boundary value problems with one-dimensional p-Laplacian. *Journal of Mathematical Analysis and Applications*, *327*, 1381–1395. doi:10.1016/j.jmaa.2006.05.023

Yu, S. L., Wong, F. H., Yeh, C. C., & Lin, S. W. (2007). Existence of positive solutions for n+2 order p-Laplacian BVP. *Computers & Mathematics with Applications (Oxford, England)*, *53*(9), 1367–1379. doi:10.1016/j.camwa.2006.05.023

Zhao, D., Wang, H., & Ge, W. (2007). Existence of triple positive solutions to class of p-Laplacian boundary value problems. *Journal of Mathematical Analysis and Applications*, *328*, 972–983. doi:10.1016/j.jmaa.2006.05.073

Zhou, Y., & Su, H. (2007). Positive solutions of four-point boundar value problems for higher-order with p-Laplacian operator. *Electronic Journal of Differential Equations*, 1-14.

This work was previously published in the International Journal of Artificial Life Research, Volume 2, Issue 1, edited by E. Stanley Lee and Ping-Teng Chang, pp. 43-53, copyright 2011 by IGI Publishing (an imprint of IGI Global).

Chapter 8
Mathematical Model to Assess the Relative Effectiveness of Rift Valley Fever Countermeasures

Holly Gaff
Old Dominion University, USA

Colleen Burgess
MathEcology, LLC, USA

Jacqueline Jackson
Old Dominion University, USA

Tianchan Niu
Georgetown University Medical Center, USA

Yiannis Papelis
Old Dominion University, USA

David Hartley
Georgetown University Medical Center and National Institutes of Health, USA

ABSTRACT

Mathematical modeling of infectious diseases is increasingly used to explicate the mechanics of disease propagation, impact of controls, and sensitivity of countermeasures. The authors demonstrate use of a Rift Valley Fever (RVF) model to study efficacy of countermeasures to disease transmission parameters. RVF is a viral infectious disease that propagates through infected mosquitoes and primarily affects animals but also humans. Vaccines exist to protect against the disease but there is lack of data comparing efficacy of vaccination with alternative countermeasures such as managing mosquito population or destroying infected livestock. This paper presents a compartmentalized multispecies deterministic ordinary differential equation model of RVF propagation among livestock through infected Aedes and Culex mosquitoes and exercises the model to study the efficacy of vector adulticide, vector larvicide, livestock vaccination, and livestock culling on livestock population. Results suggest that livestock vaccination and culling offer the greatest benefit in terms of reducing livestock morbidity and mortality.

DOI: 10.4018/978-1-4666-3890-7.ch008

1. INTRODUCTION

Rift Valley fever virus is a mosquito-borne pathogen that causes widespread febrile illness and mortality in domestic animals such as sheep, cattle and goats as well as humans (Gaff, 2007). Rift Valley fever virus was first described in peer-reviewed research in 1930 (Daubney, 1931) and was generally considered a disease primarily of sub-Saharan and southern Africa (Gaff, 2007). Then 1977, the disease moved outside of sub-Saharan and southern Africa with an outbreak occurring in Egypt; since then, outbreaks have occurred in Saudi Arabia and Yemen proving it to be a virus able to invade ecologically diverse regions (Gaff, 2007).

Over the past few decades, significant changes in the distribution and intensity of Rift Valley fever (RVF) have been recorded (WHO, 2007). Since the isolation of the virus in 1931 in the Rift Valley in Kenya, it has been held responsible for several epizootics in small ruminants, causing abortions and stillborns in the ovine species in Eastern and Southern Africa (Gerdes, 2004). Epizootics first occurred in regions of high altitude such as South Africa in 1951 (which resulted in the death of an estimated 100,000 sheep), Zimbabwe in 1958, Nigeria in 1958, and Chad and Cameroon in 1967 (WHO, 2007). Until the 1970s human infection remained low, and the agent mostly affected breeders in contact with affected or dead animals. In 1973, after the first source of infection appeared in the White Nile in Sudan, a human epidemic soon began in South Africa with the first recognized human deaths (Peters, 1994). Human outbreaks then occurred in Egypt in 1977 causing 598 human deaths (Gerdes, 2004), in 1987 causing 200 human deaths, and in Kenya and Somalia in 1997 causing 478 human deaths (CDC, 1998). In 2000, cases of RVF were discovered in Saudi Arabia and Yemen marking the spread of the disease outside of Africa and the Rift Valley (Jupp, 2002). By November 2000, over 500 cases of serious RVF were discovered in Saudi Arabia, with 87 deaths. In Yemen, between August and November 2000, there were over 1,000 suspected occurrences of the disease among humans. The result of the outbreak in Yemen was 121 deaths. Since 2000, outbreaks have occurred in Kenya and Somalia (2006), Tanzania and Sudan (2007) and Madagascar and South Africa (2008).

The potential for an exotic arbovirus to be introduced and widely established across North America can be inferred by the introduction and rapid spread of West Nile viral activity across North America in 1999 (Turell, 2008). Currently, Rift Valley fever is listed as a Category A agent on the Center for Disease Control bioterrorism list and is therefore considered a major threat to the United States. Despite the existence of several vaccines, which can protect against RVF, livestock vaccination is not currently a standardized activity in the United States. This leaves livestock culling of both exposed and infected animals as the only viable after-the-fact countermeasures. The disease is spread by infected mosquitoes whose population can vary widely, and such variance is hard to control as it primarily depends on environment factors. Should RVF be intentionally or unintentionally introduced in the Continental United States, exposed livestock is likely to encompass a significant percentage of all livestock near the vicinity of first appearance. Under such a scenario, the potential economic disruption due to loss of livestock and any subsequent trade restrictions is significant.

Disease control and public health intervention measures for RVF entail three primary countermeasures: the application of insecticide to vector populations targeting either adult mosquitoes or mosquito larvae, livestock vaccination, or finally culling of exposed and/or infected animals. Each of these countermeasures has an associated a-priori cost and corresponding posterior effectiveness; however, there is little research that allows a quantitative cost-benefit analysis among these countermeasures. In this paper, we extend a mathematical model of RVF to include transmission of

RVF between two vector populations and a single livestock population. The resultant model allows experimentation with four specific intervention strategies: vector adulticide, vector larvicide, livestock vaccination, and livestock culling. We utilize the enhanced model to explore each of these systems for various degrees of intensity and efficacy of intervention, and to evaluate the sensitivity of the models to the various disease transmission parameters.

2. METHODS

2.1 The RVF Model

We extend the mathematical model for Rift Valley fever originally presented in (Gaff, 2007). The model itself is a compartmentalized multi-species deterministic ordinary differential equation system, which predicts the spread of Rift Valley Fever in *Aedes* and *Culex* mosquitoes and a livestock population (in this case, cattle). The model is based on a standard framework for epidemiology models using disease states as the population compartments: SEIR. S are the susceptible individuals of a given population with no assumed immunity or previous exposure to the disease. E is the exposed individuals who have been exposed and infected but are not yet infectious. I is the infectious individuals who can now expose others and have symptoms. R are the recovered individuals who are recovered and assumed immune from the disease. In addition to this framework, the mosquito populations have larval compartments, P. The infected *Aedes* mosquitoes are capable of laying infected eggs so there is an additional larval compartment, Q. Movement between compartments for the SEIR model is determined by disease dynamics while age controls flow between the larval compartments (PQ) and the appropriate adult compartment (S or I). Subscripts are used to identify the three populations: (1) *Aedes*, (2) *Culex*, (3) livestock.

The original model has been expanded to accommodate vector control, livestock vaccination and culling. The model has been modified to include logistic growth for the mosquito populations as replacement birth and death does not make sense when modeling vector control measures. The carrying capacity for the populations will eventually be tied to climate factors such as rainfall. The four modified models, one with each type of intervention, are represented schematically in Figures 1(a) through (d); the full set of equations for the general intervention model is provided in Table 1, and represents all interventions simultaneously: adulticide ($v_A(t)$), larvicide ($v_L(t)$), livestock vaccination ($V(t)$) and livestock culling ($c(t)$).

2.2 Model Parameters and Initial Values

Vector and livestock parameters were set based upon species biology and are shown in Table 2. Also shown are disease parameters, which were based on the behavior of RVF in mosquitoes and cattle, and high and low transmission parameters, which were derived previously for the development of the original model. The simulation was run over a 10-year period using Runge-Kutta 4th order methods to solve the differential equations, which have a one-day time step.

For the analysis performed, each intervention was explored singly. Simulations in which vector control and culling are employed, as well as reactive livestock vaccination (defined below), are initiated with all livestock individuals and adult *Culex* mosquitoes in the susceptible compartment, and with the majority of *Aedes* mosquitoes in the susceptible compartment and a single infected *Aedes* adult (Table 3). Simulations employing preventive livestock vaccination are initiated with a given proportion of the livestock population already in the removed state, indicating pre-existing protection against RVF prior to the outbreak of disease in the system (Table 4).

Figure 1. Flow diagram for the model of Rift Valley Fever transmission. Note that red denotes the key differences for each of the models. (a) Vector adulticide intervention. (b) Vector larvicide intervention. (c) Livestock vaccination intervention. (d) Livestock culling intervention.

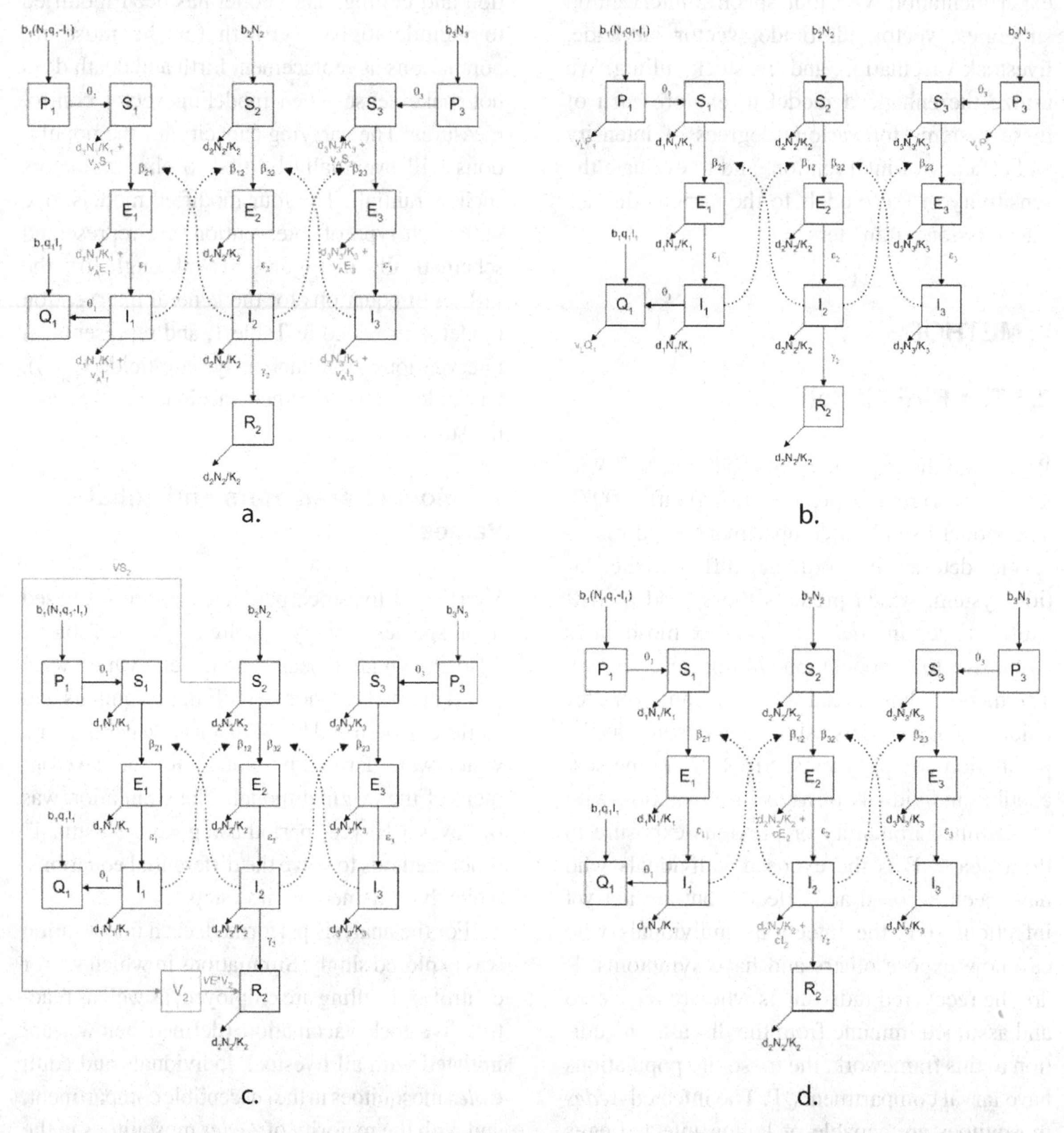

The initial conditions for livestock population numbers under preventive vaccination scenarios were obtained by implementing given combinations of target population and vaccine efficacy values in the absence of disease transmission and allowing the system to reach equilibrium. The resulting proportion of immune livestock (*PROP*) is then employed as the initial proportion of immune for simulations in which a single *Aedes* mosquito is infected, and the initial proportion of susceptible livestock in these runs is defined as 1-*PROP*.

Table 1. Equations for the model of Rift Valley Fever transmission. Subscript 1 denotes Aedes mosquitoes, subscript 2 denotes Culex mosquitoes, and subscript 3 denotes livestock

$$\frac{dP_1}{dt} = b_1(N_1 - q_1 I_1) - \theta_1 P_1 - v_L(t) P_1$$
$$\frac{dQ_1}{dt} = b_1 q_1 I_1 - \theta_1 Q_1 - v_L(t) Q_1$$
$$\frac{dS_1}{dt} = \theta_1 P_1 - \frac{d_1 S_1 N_1}{K_1} - \frac{\beta_{21} S_1 I_2}{N_2} - v_A(t) S_1$$
$$\frac{dE_1}{dt} = -\frac{d_1 E_1 N_1}{K_1} - \frac{\beta_{21} S_1 I_2}{N_2} - \varepsilon_1 E_1 - v_L(t) E_1$$
$$\frac{dI_1}{dt} = \theta_1 Q_1 - \frac{d_1 I_1 N_1}{K_1} + \varepsilon_1 E_1 - v_A(t) I_1$$
$$\frac{dN_1}{dt} = (\theta_1 - v_L(t))(P_1 + Q_1) - \frac{d_1 N_1 N_1}{K_1} - v_A(t) N_1$$

$$\frac{dP_3}{dt} = b_3 N_3 - \theta_3 P_3 - v_L(t) P_3$$
$$\frac{dS_3}{dt} = \theta_3 P_3 - \frac{d_3 S_3 N_3}{K_3} - \frac{\beta_{23} S_3 I_2}{N_2} - v_A(t) S_3$$
$$\frac{dE_3}{dt} = -\frac{d_3 E_3 N_3}{K_3} - \frac{\beta_{23} S_3 I_2}{N_2} - \varepsilon_3 E_3 - v_L(t) E_3$$
$$\frac{dI_3}{dt} = -\frac{d_3 I_3 N_3}{K_3} + \varepsilon_3 E_3 - v_A(t) I_3$$
$$\frac{dN_3}{dt} = (\theta_3 - v_L(t)) P_3 - \frac{d_3 N_3 N_3}{K_3} - v_A(t) N_3$$

$$\frac{dS_2}{dt} = b_2 N_2 - \frac{d_2 S_2 N_2}{K_2} - \frac{\beta_{12} S_2 I_1}{N_1} - \frac{\beta_{32} S_2 I_3}{N_3} - V(t) S_2$$
$$\frac{dE_2}{dt} = -\frac{d_2 E_2 N_2}{K_2} + \frac{\beta_{12} S_2 I_1}{N_1} + \frac{\beta_{32} S_2 I_3}{N_3} - \varepsilon_2 E_2 - c(t) E_2$$
$$\frac{dI_2}{dt} = -\frac{d_2 I_2 N_2}{K_2} + \varepsilon_2 E_2 - \gamma_2 I_2 - \mu_2 I_2 - c(t) I_2$$
$$\frac{dV_2}{dt} = V(t) S_2 - VE * V_2$$
$$\frac{dR_2}{dt} = -\frac{d_2 R_2 N_2}{K_2} + \gamma_2 I_2 + VE * V_2$$
$$\frac{dN_2}{dt} = b_2 N_2 - \frac{d_2 N_2 N_2}{K_2} - \mu_2 I_2 - c(t)(E_2 + I_2)$$

Table 2. Parameter values for the model of Rift Valley Fever transmission

Parameter	Description	Unit	Value	High Transmission	Low Transmission
β_{12}	Adequate contact rate from *Aedes* to livestock	1/day		0.480	0.150
β_{21}	Adequate contact rate from livestock to *Aedes*	1/day		0.395	0.150
β_{23}	Adequate contact rate from livestock to *Culex*	1/day		0.560	0.150
β_{32}	Adequate contact rate from *Culex* to livestock	1/day		0.130	0.050
μ_2	RVF mortality rate in livestock (cattle)	1/day	0.0176		
$1/d_1$	Lifespan of *Aedes* mosquitoes	days	10		
$1/d_2$	Lifespan of livestock	days	3650		

continued on following page

Table 2. Continued

Parameter	Description	Unit	Value	High Transmission	Low Transmission
$1/d_3$	Lifespan of *Culex* mosquitoes	days	10		
b_1	Number of *Aedes* eggs laid per day	1/day	d_1		
b_2	Daily birthrate of livestock	1/day	d_2		
b_3	Number of *Culex* eggs laid per day	1/day	d_3		
$1/e_1$	Disease incubation period in *Aedes* mosquitoes	days	6		
$1/e_1$	Disease incubation period in livestock	days	4		
$1/e_1$	Disease incubation period in *Culex* mosquitoes	days	6		
$1/\gamma_2$	Infectiousness period in livestock	days	4		
q_1	Transovarial transmission rate in *Aedes* mosquitoes	-	0.05		
$1/\theta_1$	Development time of *Aedes* mosquitoes	days	10		
$1/\theta_3$	Development time of *Culex* mosquitoes	days	10		
K_1	Carrying capacity of *Aedes* mosquitoes	-	20,000	(Newton 1992)	
K_2	Carrying capacity of livestock	-	1,000		
K_3	Carrying capacity of *Culex* mosquitoes	-	20,000	(Newton 1992)	
VE	Vaccine efficacy	1/day	Varies (See Tables 4 and 6)		
$v_l(t)$	Larvicide induced mortality	1/day	Varies (See Table 5)		
$v_l(t)$	Adulticide induced mortality	1/day	Varies (See Table 5)		
$V(t)$	Vaccination rate	1/day	Varies (See Tables 4 and 6)		
$c(t)$	Culling rate	1/day	Varies (See Table 7)		

Table 3. Initial conditions for the model of Rift Valley Fever transmission for all but preventive vaccination scenarios

Name	Symbol	Initial Value (General)
Uninfected *Aedes* eggs	P_1	5,000
Infected *Aedes* eggs	Q_1	0
Susceptible *Aedes*	S_1	4,999
Exposed *Aedes*	E_1	0
Infected *Aedes*	I_1	1
Susceptible livestock	S_2	900
Exposed livestock	E_2	0
Infected livestock	I_2	0
Removed livestock	R_2	0
Uninfected *Culex* eggs	P_3	5,000
Susceptible *Culex*	S_3	5,000
Exposed *Culex*	E_3	0
Infected *Culex*	I_3	0

*Table 4. Initial condition multiplicative factors for susceptible and immune livestock for the model of Rift Valley Fever transmission for preventive vaccination scenarios, where Immune = PROP * 900 and Susceptibles = (1-PROP) * 900*

Transmission Level	Target Population	Vaccine Efficacy	*PROP*
High	50%	65%	0.6648
High	75%	65%	0.7496
High	100%	65%	0.7999
Low	50%	85%	0.7227
Low	75%	85%	0.7967
Low	100%	85%	0.8395

2.3 Vector Control Interventions

Vector control is incorporated in the model by proportionally reducing the population numbers for adult and larva vector populations by 30%, 60% and

90% for both adulticide ($v_A(t)$) and larvicide ($v_L(t)$). In the case of adulticide, all adult mosquitoes in the treated populations are affected equally, regardless of current disease state; similarly, in the case of larvicide, all mosquito larvae are impacted uniformly. The application of insecticide is repeated four times on a weekly basis for periodic scenarios, occurs a single time for one-time scenarios, or continues for 28 consecutive days for continuous scenarios. Reactive vector control is initiated once thresholds for a detectable epidemic have been reached in the simulation, defined as at least 1% disease prevalence in the livestock population or two livestock deaths attributable to RVF in a given time period. Preventive vector control is initiated at the beginning of simulations, and prior to any detectable disease in livestock, though with initial conditions including a single infected adult *Aedes* mosquito. Table 5 summarizes the vector control scenarios that have been simulated.

2.4 Livestock Vaccination Interventions

Livestock vaccination is incorporated in the RVF model by allowing all of the adult livestock population to receive a single immunization against the disease. Once vaccinated, a susceptible animal will be protected according the specific vaccine efficacy (modeled here to be 65% or 85%) after a period of 10 days elapses; though all livestock are vaccinated, only susceptible animals receive benefit from it and all others remain in their current disease state.

As with vector control, livestock vaccination scenarios occur on a periodic, one-time or continuous basis, with timing definitions carrying over. Reactive vaccination occurs either in response to a detectable livestock epidemic as defined above, or in response to the detection of disease in a single mosquito. Preventive vaccination occurs prior to the introduction of RVF into the model system, and has the effect of altering the initial conditions for livestock such that a proportion of the animal population shifts from the susceptible disease state into the removed state. Table 6 summarizes the livestock vaccination scenarios that have been simulated.

2.5 Livestock Culling Interventions

Livestock culling is incorporated into the model by removing exposed and infected animals from the cattle population according to a given compliance level (50%, 75% or 100%) when a RVF outbreak is detected in livestock, and thus is employed on

Table 5. Simulated vector control scenarios

Intervention Type	Effectiveness	Timing	Trigger
Adulticide	30%	Periodic (4 weekly apps.)	Preventive
Larvicide	60%	One-Time (single app.)	Reactive
	90%	Constant (28 daily apps.)	

Table 6. Simulated livestock vaccination scenarios

Intervention Type	Effectiveness	Target Population	Campaign	Trigger
Vaccination	65%	50%	Periodic (4 weekly apps.)	Preventive
	85%	75%	One-Time (single app.)	Reactive - vector
		100%	Constant (28 daily apps.)	Reactive - livestock

a reactive level only. Similar to vector control and livestock vaccination, culling occurs on a periodic, one-time or continuous basis (as defined previously). Table 7 summarizes the livestock culling scenarios that have been simulated.

All four types of interventions are triggered only once – either preventively at the beginning of the simulation or reactively in response to a detectable epidemic – and do not repeat with subsequent outbreaks that may occur. The one exception to this is livestock culling, in which we also simulated the impact of employing the removal of exposed and infected animals on an ongoing basis for the duration of the simulation.

3. RESULTS AND INTERPRETATION

In order to measure the effectiveness of the various intervention scenarios, outcomes for each in terms of cumulative livestock cases and deaths over the 10-year duration of the simulation were compared against the base-case of no intervention for both high and low transmission parameter sets. Under the base-case of no intervention, the 10-year cumulative numbers of livestock cases and deaths under high disease transmission are 1436 and 200, respectively; under low disease transmission, the cumulative number of livestock cases and deaths are 1.4 and 0.2, respectively. Percent prevalence and proportion of susceptible and removed livestock for high and low disease transmission are shown in Figures 2(a) and (b), and 3(a) and (b), respectively.

3.1 Impact of Vector Control

The percent change in livestock deaths due to RVF for the various vector control scenarios versus the base-case of no intervention are shown in Figures 4(a) and (b) and 5(a) and (b). Of the vector control scenarios simulated, when transmission of RVF

Table 7. Simulated livestock culling scenarios

Intervention Type	Compliance	Target Population	Timing	Trigger
Culling	50%	Exposed + Infected	Periodic	Reactive
	75%		One-Time	
	100%		Constant	

Figure 2. Percent livestock RVF prevalence for base-case of no intervention in an immunologically naive livestock population. The high transmission parameter set shows multiple minor outbreaks after the initial large outbreak of over 10% prevalence. The low transmission parameter set shows no subsequent outbreaks after the initial outbreak, which reaches 0.005% prevalence. (a) High transmission parameter set. (b) Low transmission parameter set.

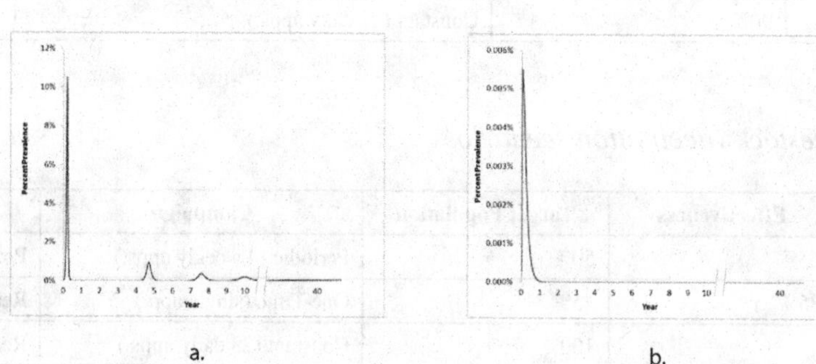

Figure 3. Proportion of susceptible and removed livestock for base-case of no intervention in an immunologically naïve livestock population. The high transmission set shows the cumulative effect of repeated outbreaks while the low transmission set reflects that nearly the entire livestock population remains susceptible.(a) High transmission parameter set. (b) Low transmission parameter set.

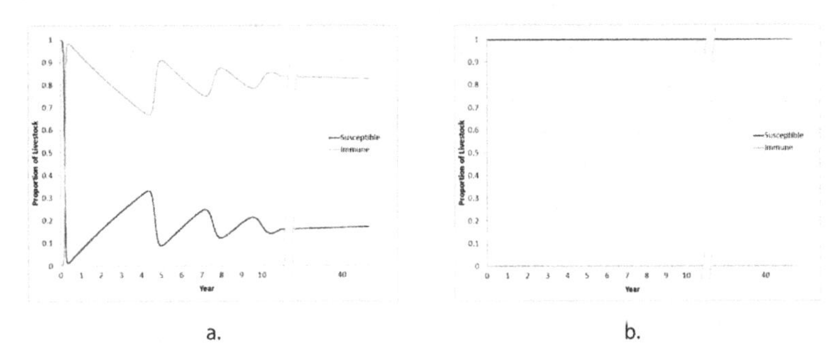

Figure 4. Percent change in livestock deaths over base-case of no intervention for adulticide vector control scenarios with 30%, 60% and 90% effectiveness. It is important to note the scales for these two plots. The greatest reduction for the high transmission set is 2.5% while the low transmission set shows much larger reductions. This is a reflection of the low transmission set having a very small number of deaths, 0.2, for the base case. For both sets, the greatest reductions in deaths for all levels of effectiveness can be seen for the continuous application of adulticide. The low transmission sets never reached infection levels high enough to trigger reactive scenarios. Periodic and one-time applications showed dramatically less reductions.(a) High transmission parameter set. (b) Low transmission parameter set.

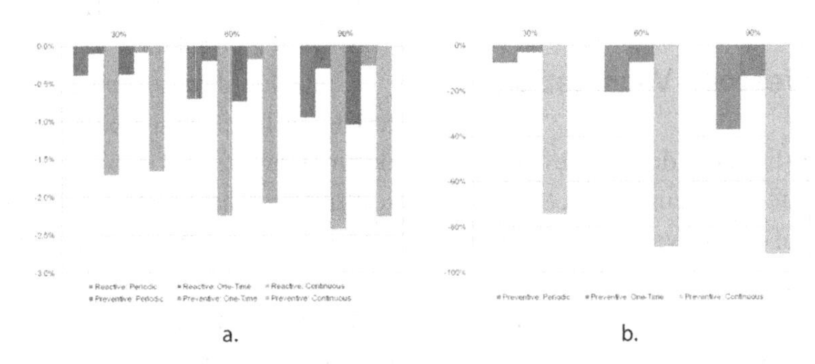

is high, the greatest reduction in livestock deaths (approximately 2.5%) and cases is provided by the reactive application of adulticide for 28 consecutive days ("continuous" application) with 90% effectiveness. Note, however, that preventive vector control for this study implies that no cases or deaths have been detected in the livestock population prior to application of interventions, though mosquito infections and animal exposures have already begun to occur (See Figure 5).

Under low disease transmission, the conditions to trigger reactive interventions fail to occur. Of the preventive vector control scenarios, however, the greatest impact on livestock deaths (greater than 90% reduction) and cases is again provided by preventive continuous adulticide application with 90% effectiveness, though when disease transmission is at these lower levels even under no intervention the predicted number of cases and deaths is nearly low enough to be negligible (1.4 and 0.2, respectively). As shown in the figures, the simulated application of larvicide at low transmission has the effect of increasing the number of livestock deaths. However, these changes are not as significant considering the

Figure 5. Percent change in livestock deaths over base-case of no intervention for larvicide vector control scenarios with 30%, 60% and 90% effectiveness. The application of larvicide shows some more complex results. For the high transmission parameters, reactive continuous and periodic applications showed the greatest reduction in deaths. Proactive applications showed far lower benefits for the high transmission and actually amplified the low transmission set. This is reflective of the fact that fewer uninfected larvae are emerging to dilute the infected adults, and thus more infections are likely. Again, the low transmission base case predicted only 0.2 deaths so even small amplifications can appear dramatic. (a) High transmission parameter set. (b) Low transmission parameter set.

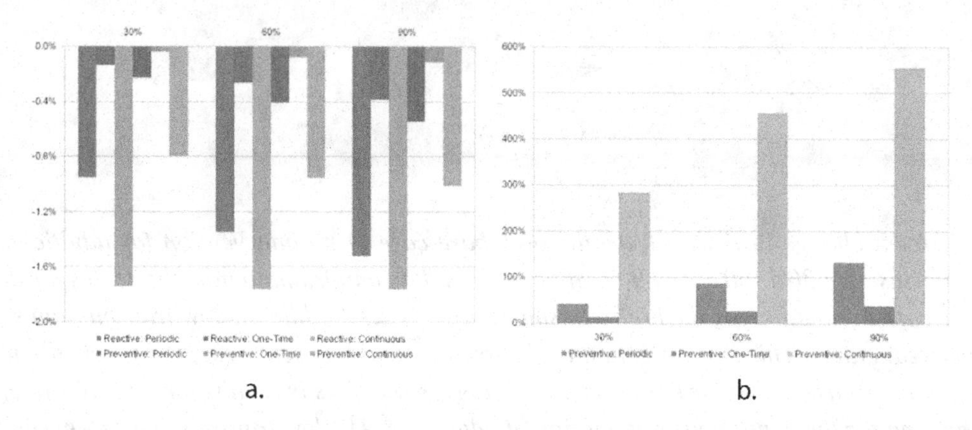

nearly negligible number of cases and deaths in this case.

3.2 Impact of Livestock Vaccination

The percent change in livestock deaths due to RVF for the various livestock vaccination scenarios versus the base-case of no intervention are shown in Figures 6(a) and (b) and 7(a) and (b). Of the vaccination scenarios simulated, when transmission of RVF is high, the greatest reduction in livestock deaths (greater than 50% reduction) and cases is provided by the preventive immunization of 100% of the livestock population extended over 28 consecutive days ("continuous" vaccination) with 85% efficacy. For the case of preventive livestock vaccination, it has been assumed that a given number of animals have already received protection against RVF via immunization prior to the detection of disease in either vector or livestock populations. The vector- and livestock-reactive scenarios that impart the greatest benefit in terms of reduction in cases and deaths (approximately 8% and 5% reduction, respectively) are both also provided by continuous (28-day) vaccination of 100% of the population with 85% efficacy.

Under low disease transmission, the conditions to trigger livestock-reactive interventions fail to occur. However, of the preventive and vector-reactive vaccination scenarios, the greatest impact on livestock deaths (greater than 90% reduction) and cases is again provided by preventive vaccination of 100% of the population with 85% efficacy on a continuous basis for 28 days. The vector-reactive scenario which imparts the greatest benefit in terms of reduction in cases and deaths (approximately 20% reduction) is also provided by continuous (28-day) vaccination of 100% of the population with 85% efficacy, keeping in mind that the total number of livestock cases and deaths even under the base-case of no intervention is still quite low (See Figure 7).

Figure 6. Percent change in livestock deaths over base-case of no intervention for livestock vaccination scenarios with 50%, 75% and 100% coverage and 65% vaccine efficacy. For both high and low transmission sets, any preventive application of vaccination has dramatically better results than the any reactive scenario. Also, it is important to note that in contrast to the previous figures, these reductions are 40% for the high transmission and over 80% for the low transmission. The reactive cases demonstrate that if one waits until a threshold percent of infected livestock or mosquitoes is met, it is too late to have such a dramatic reduction. The reductions of the reactive scenarios are on the order of the adulticide and larvicide scenarios. (a) High transmission parameter set. (b) Low transmission parameter set.

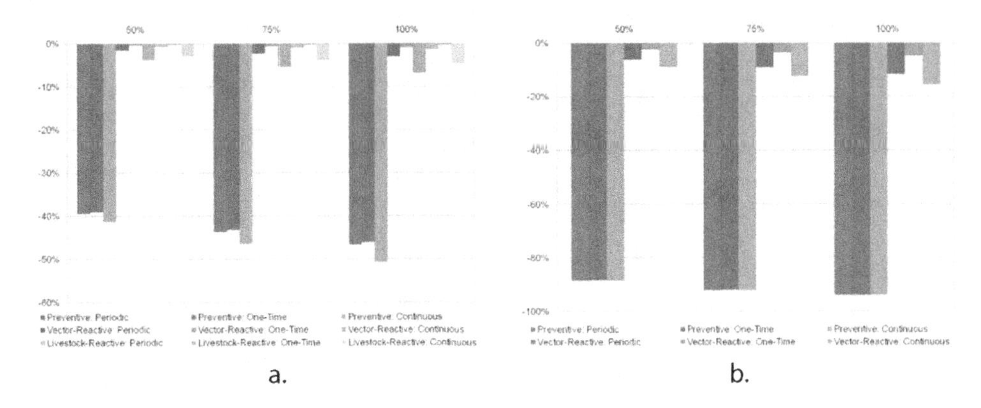

Figure 7. Percent change in livestock deaths over base-case of no intervention for livestock vaccination scenarios with 50%, 75% and 100% coverage and 85% vaccine efficacy. Increasing the efficacy of the vaccination is directly reflected in additional decreases in the livestock deaths. As with the lower efficacy, the best option is still preventive vaccination. (a) High transmission parameter set. (b) Low transmission parameter set.

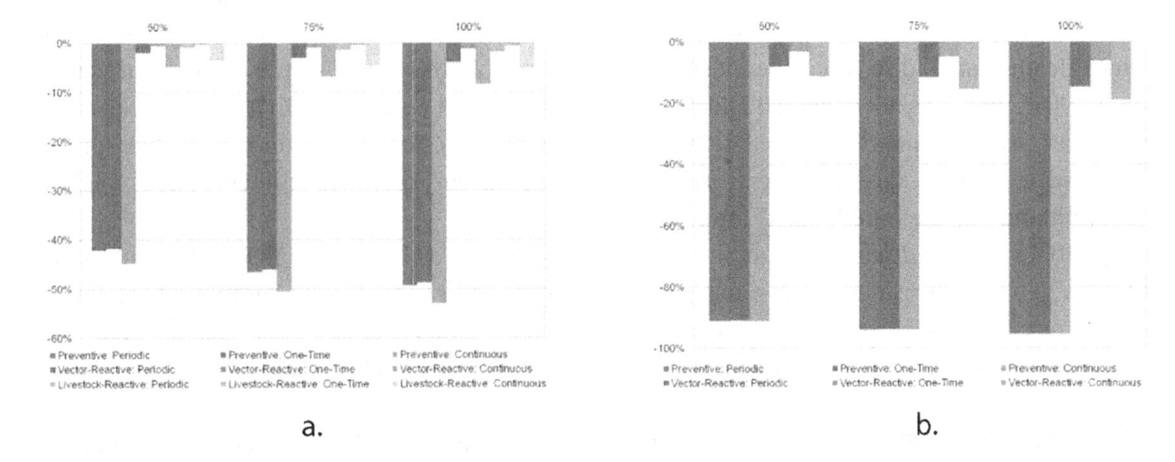

3.3 Impact of Livestock Culling

The percent change in livestock deaths due to RVF for the various livestock culling scenarios versus the base-case of no intervention are shown in Figure 8. Of the culling scenarios simulated, when transmission of RVF is high, the greatest reduction in livestock deaths (greater than 15% reduction) and cases is provided by reactive culling of exposed and infected animals extended

Figure 8. Percent change in livestock deaths over base-case of no intervention for livestock culling scenarios with 50%, 75% and 100% compliance. Results are shown only for the high transmission parameter set. All exposed and infected livestock are culled. Reactive, ongoing culling clearly has a significant impact on reducing the percent of livestock deaths from the disease, but dramatic levels of culling are required to achieve such results.

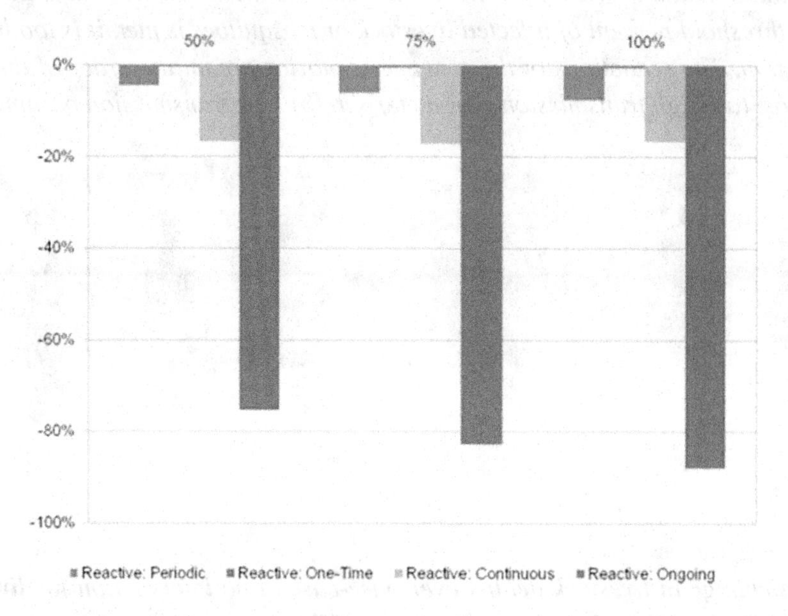

over 28 consecutive days ("continuous" culling) with 100% compliance. If such culling practices are continued for the duration of the simulated period (10 years) ("ongoing" culling), the drop in livestock deaths due to disease can be increased to almost 90%, with the bulk of animal infections occurring at the beginning of the simulation and the disease essentially disappearing from the system (without outside introduction) after year 5 (Figure 9). Under low disease transmission, the conditions to trigger livestock-reactive interventions fail to occur and thus no culling scenarios are implemented under these conditions.

3.4 Parameter Sensitivity

To evaluate the sensitivity of the disease model to transmission parameters, we performed an exploratory analysis in which each of the four inter-species beta terms were varied individually over the range of possible values from a minimum of 0.05 to a maximum of 0.9 or 1.0, while the other three betas were held constant at their low transmission values. This was done in order to explore the degree of transmission at which a "detectable epidemic" (livestock prevalence of 1% or 2 livestock deaths in a given time-period) might occur and reactive interventions might be triggered.

In Figures 10(a) through (d), the total number of livestock cases for the duration of the simulation is plotted against beta values for each of the four livestock-reactive intervention types, plus the base-case of no intervention. The range of beta-conditions under which "detectable epidemic" thresholds are triggered (that is, in a given time-step either 1% of the livestock population is infected or 2 livestock deaths occurred) occurs where variations between interventions are apparent on the graphs.

Figure 9. Disease state time-series plots for Aedes (top left), livestock (top right), and Culex (bottom left), and total livestock deaths and infections for ongoing livestock culling scenario with 100% compliance. All plots have days as the units for the x-axis and numbers of mosquitoes or livestock for the y-axis. The build-up of susceptible livestock can be seen in the upper right hand plot as the disease is clearly eliminated early in the scenario.

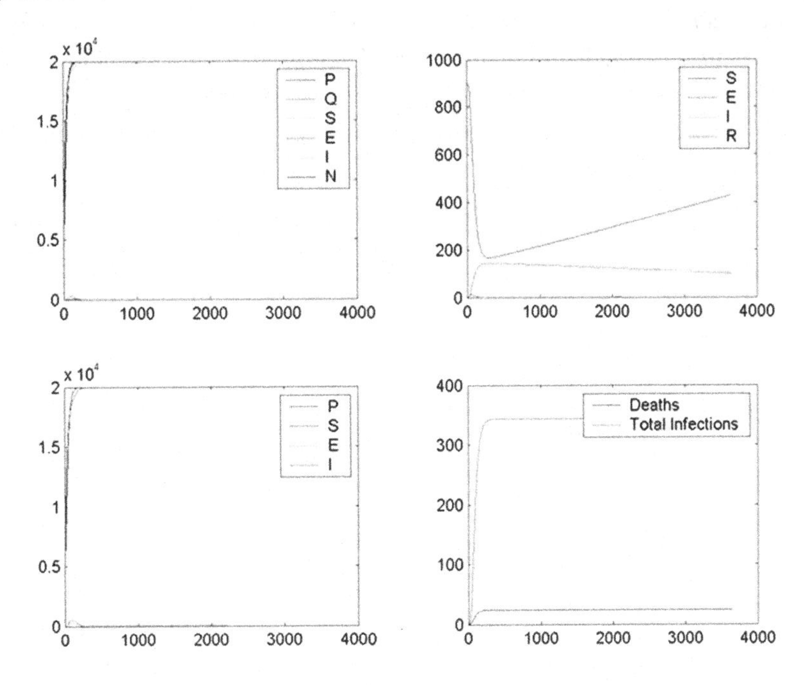

For each of the four betas explored there are definite tipping-points where cases and deaths increase abruptly and then level off, though for β_{23} (transmission from livestock to *Culex* mosquitoes) this happens at the upper end of the range. Within the range of beta values for which "detectable epidemics" occur there is a great deal of switching between which of the four intervention types minimizes total livestock cases, though continuous (28-day) vaccination of 100% of the livestock population at 85% efficacy and continuous livestock culling appear to perform best in general.

4. CONCLUSION

4.1 Main Simulation Results

At high disease transmission levels, all vector control strategies have a positive impact on reducing the 10-year cumulative number of livestock cases and deaths, with reactive application of adulticide offering the greatest reduction in livestock morbidity and mortality overall. However, even at the most effective of the scenarios simulated (reactive application of 90% effective adulticide for 28 consecutive days), adulticide prevents only 36 livestock cases and 5 deaths. Similar results are seen with the application of larvicide, with the most effective larvicide high transmission scenario preventing 26 livestock cases and 4 deaths. Thus, while the application of vector control is beneficial to livestock populations when transmission of

Figure 10. Total number of cases of RVF in livestock when individual transmission parameters are varied while the other three are held constant, under the base-case of no intervention as well as continuous application of adulticide (90% effectiveness), continuous application of larvicide (90% effectiveness), continuous livestock vaccination (100% coverage at 85% vaccine efficacy), and continuous livestock culling (100% compliance). As can be seen, the general trend for all parameters is an increase in the number of cases for an increase in the parameter, which is to be expected. The divergence for the various scenarios demonstrates the nonlinearities in the system. This also demonstrates that increased virulence could swamp even the best intervention strategy. (a) Aedes to livestock transmission. (b) Livestock to Aedes transmission. (c) Livestock to Culex transmission. (b) Culex to livestock transmission.

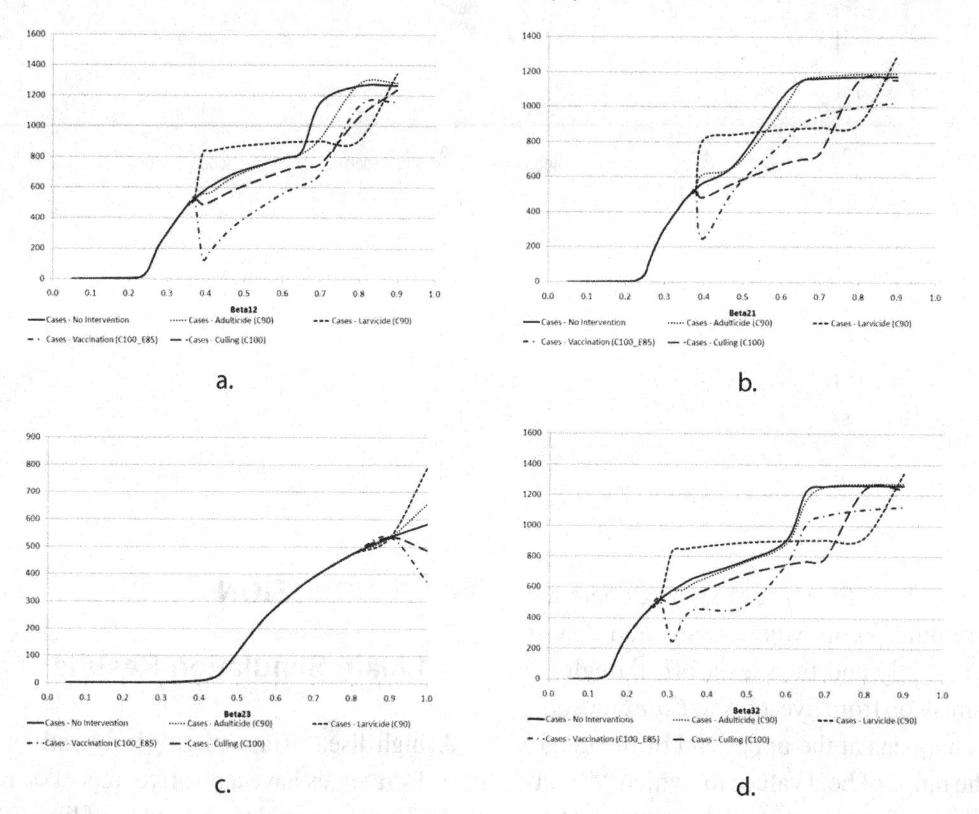

Rift Valley Fever is high, the magnitude of this benefit must be compared to the cost associated with these interventions in order to evaluate their overall effectiveness. However, it is important to keep in mind that vector control interventions occur only once at the beginning of the simulation period, regardless of future outbreaks. Greater benefit to livestock may be seen if subsequent intervention pulses are simulated in response to later outbreaks in addition to the first.

Under conditions of low disease transmission, in which livestock morbidity and mortality is already low, adulticide prevents at most one case and has negligible effect on livestock deaths. The application of larvicide with low disease transmission, however, has the effect of *increasing* the number of cases and deaths in livestock as effectiveness and frequency of application increases. This may be associated with the fact that, since mosquito larvae are being proportionally reduced with each application, adult vector populations are no longer being diluted with un-infected newly emerged mosquitoes and as a result the density of infected adults increases with each larvicide application.

The overall implication is that, under low disease transmission conditions, the application of vector control strategies may not be sufficiently effective to justify the effort, and in the case of larvicide, may actually be contraindicated.

Regardless of disease transmission levels and vaccine efficacy, all livestock vaccination strategies have a positive effect on the reduction of morbidity and mortality in livestock populations. The greatest impact is provided by the preventive vaccination of 100% of livestock with 85% vaccine efficacy via a 28-day continuous campaign, which prevents 1,432 cases and all 200 deaths in the livestock population under high transmission, and prevents all 1.4 cases and 0.2 deaths under low transmission. This is, however, under the assumption that preventive vaccination occurs immediately prior to disease introduction in the system; subsequent delay between vaccination and disease introduction may decrease the degree of benefit as new un-vaccinated animals enter the population over time and dilute the level of herd immunity, since for the current study interventions are only implemented once regardless of future outbreaks.

Due to the nature of livestock culling interventions (killing all exposed and infected animals when detected), these scenarios were simulated only on a reactive level under high disease transmission, since detectable outbreak thresholds were not attained under low transmission conditions. Reactive livestock culling with 100% compliance for 28 consecutive days after the detection of an outbreak in the livestock population has the greatest impact on morbidity and mortality, preventing 189 livestock cases and 33 deaths over the duration of the 10-year simulation period, while 75% and 50% compliance offer similar benefits. Continuing a reactive culling strategy with 100% compliance beyond the initial 28 days through the end of the simulation, however, offers significantly greater benefits and overall has the effect of preventing almost 1,100 cases and 76 deaths. Under this strategy, Rift Valley Fever is eliminated from the three-species system after the final minor outbreaks occur in approximately year 5 of the simulation, even under high disease transmission.

4.2 Sensitivity Results

Model results point to a great deal of variability and switching in terms of which interventions offer the greatest benefit with respect to disease transmission levels. As can be seen in Figures 10(a), (b) and (d), over certain ranges of transmission parameter values, the application of larvicide performs the worst of all four intervention types, yet over a higher parameter range it becomes the best performer for a time, superseding even livestock vaccination which in general prevents a larger number of livestock cases than other interventions. However, in actual epidemic settings it is often difficult to pinpoint precise parameter values associated with the transmission of disease between species, and thus to discern which disease interventions might be optimal within that range. Looking at the range of transmission values in general, continuous livestock vaccination and continuous livestock culling appear to be the best choices in terms of intervention options as a whole.

There are a number of issues, which may have a significant impact on the choice of disease intervention, which have not been included in the current analysis. Associated with each intervention strategy is a series of costs, in terms of executing the interventions as well as the loss of livestock animals either to disease or culling, and in many cases these costs may outweigh the benefits derived to the point that the strategy should not be pursued. At significantly high costs, producers may be unwilling to comply with intervention strategies, regardless of potential benefit in terms of livestock morbidity and mortality reduction.

Additionally, as a vector-borne disease, Rift Valley Fever is driven in large part by environmental and geographic factors, which have not been incorporated into the current model. Factors such as temperature and precipitation may have significant impact on population dynamics in the *Aedes* and *Culex* vector populations, as well as disease dynamics, and may vary dramatically in a spatial context. Future analysis is planned in which environmental parameters will be integrated into the model system and recommendations will be evaluated in a spatially heterogeneous setting. With these enhancements to the model optimal control techniques may be employed to offer policy recommendations on how best to respond to – and prevent – outbreaks of Rift Valley Fever in domestic livestock populations.

ACKNOWLEDGMENT

The authors thank the thoughtful comments of the anonymous reviewers. DMH acknowledges support by the Research and Policy for Infectious Disease Dynamics (RAPIDD) program of the Science & Technology Directorate, Department of Homeland Security and Fogarty International Center, National Institutes of Health. Funding from the DHS Center of Excellence for Foreign Animal and Zoonotic Disease Defense (FAZD Center) is also acknowledged.

REFERENCES

Centers for Disease Control and Prevention (CDC). (1998). Rift Valley Fever-East Africa, 1997-1998. *Morbidity and Mortality Weekly Report*, *47*(13), 261–264.

Daubney, R., Hudson, J. R., & Garnham, P. C. (1931). Enzootic hepatitis of Rift Valley Fever, an undescriptible virus disease of sheep, cattle and man from East Africa. *The Journal of Pathology and Bacteriology*, *34*, 543–579. doi:10.1002/path.1700340418

Gaff, H. D., Hartley, D. M., & Leahy, N. P. (2007). An epidemiological model of Rift Valley Fever. *Electronic Journal of Differential Equations*, *115*, 1–12.

Gerdes, G. H. (2004). Rift Valley Fever. *Revue Scientifique et Technique (International Office of Epizootics)*, *23*(2), 613–623.

Jupp, P. G., Kemp, A., Grobbelaar, A., Leman, P., Burt, F. J., & Alahmed, A. M. (2002). The 2000 epidemic of Rift Valley Fever in Saudi Arabia: Mosquito vector studies. *Medical and Veterinary Entomology*, *15*(3), 245–252. doi:10.1046/j.1365-2915.2002.00371.x

Newton, E. A. C., & Reiter, P. (1992). A model of the transmission of dengue fever with an evaluation of the impact of ultra-low volume (ULV) insecticide applications on dengue epidemics. *The American Journal of Tropical Medicine and Hygiene*, *47*, 709–720.

Peters, C. J., & Linthicum, K. J. (1994). Rift Valley Fever. In G. W. Beran (Ed.), *Handbook of zoonoses, section B: Viral zoonoses* (2nd ed.) (pp. 125-138). Boca Raton, FL: CRC Press.

Turell, M. J., Dohm, D. J., Mores, C. N., Terracina, L., Wallette, D. L., & Hribar, L. J. (2008). Potential for North American mosquitoes to transmit Rift Valley Fever virus. *Journal of the American Mosquito Control Association*, *24*(4), 502–507. doi:10.2987/08-5791.1

World Health Organization (WHO). (2007). *Rift Valley Fever.* Retrieved from http://www.who.int/mediacentre/factsheets/fs207/en/

This work was previously published in the International Journal of Artificial Life Research, Volume 2, Issue 2, edited by E. Stanley Lee and Ping-Teng Chang, pp. 1-18, copyright 2011 by IGI Publishing (an imprint of IGI Global).

Chapter 9
Resource Distribution Strategies for Mitigation of Cross-Regional Influenza Pandemics

Andres Uribe-Sanchez
University of South Florida, USA

Alex Savachkin
University of South Florida, USA

ABSTRACT

As recently acknowledged by the Institute of Medicine, the existing pandemic mitigation models lack dynamic decision support capabilities. This paper develops a simulation optimization model for generating dynamic resource distribution strategies over a network of regions exposed to a pandemic. While the underlying simulation mimics the disease and population dynamics of the affected regions, the optimization model generates progressive allocations of mitigation resources, including vaccines, antivirals, healthcare capacities, and social distancing enforcement measures. The model strives to minimize the impact of ongoing outbreaks and the expected impact of the potential outbreaks, considering measures of morbidity, mortality, and social distancing, translated into the cost of lost productivity and medical expenses. The model was implemented on a simulated outbreak involving four million inhabitants. The strategy was compared to pro-rata and myopic strategies. The model is intended to assist public health policy makers in developing effective distribution policies during influenza pandemics.

1. INTRODUCTION AND MOTIVATION

The history of influenza pandemics is a history of enormous societal calamities aggravated by staggering economic forfeitures. In the U.S. alone, the Spanish flu (1918, virus serotype H1N1), the Asian flu (1957, serotype H2N2), and the Hong Kong flu (1968, serotype H3N2) resulted in the death toll of more than 500,000, 70,000 and 34,000 cases, respectively (Longini, Halloran, Nizam, & Yang, 2004). In recent years, a series of scattered outbreaks of the avian-to-human transmittable

DOI: 10.4018/978-1-4666-3890-7.ch009

H5N1 virus has been mapping its way through Asia, the Pacific region, Africa, the Near East, and Europe (Centers for Disease Control and Prevention - CDC, 2008). As of March 2010, WHO has reported 287 deaths in 486 worldwide cases (World Health Organization, 2010a). In Spring 2009, a mutation of the human-to-human transmissible H1N1 serotype resurfaced and propagated to an ongoing global outbreak; as of March 2010, 213 countries have been affected with a total reported number of infections and mortalities of 419,289 and 16,455, respectively (World Health Organization, 2010b). Nowadays, most experts have an ominous expectation that the next pandemic will be triggered by an emerging pathogenic virus, to which there is little or no pre-existing immunity (Schoenstadt, 2010).

The nation's ability to mitigate influenza pandemics depends on available emergency response infrastructure and resources, and at present, challenges abound. Prediction of the exact virus subtype remains a difficult task, and even when identified, reaching an adequate vaccine supply can take between six and nine months (Aunins, Lee, & Volkin, 1995; Fedson, 2003). Even if the emerged virus has a known epidemiology, the existing stockpiles will be limited due to high manufacturing and inventory costs (WHO Global Influenza Programme, 2009; World Health Organization, 2009). The supply of antiviral drugs, healthcare providers, hospital beds, medical supplies, and logistics will also be significantly constrained. Hence, pandemic mitigation will have to be done amidst a limited knowledge of the virus nature, constrained infrastructure, and limited resource availability. This ongoing challenge has been acknowledged by WHO (2009) and echoed by the HHS and CDC (Centers for Disease Control and Prevention, 2009; U.S. Department of Health & Human Services, 2007).

The existing literature on pandemic influenza (PI) modeling aims to address various complex aspects of the pandemic evolution process including: (1) the underlying spatio-temporal structure, (2) contact dynamics and disease transmission, (3) disease natural history, and (4) analysis and development of mitigation strategies. A comprehensive decision support model for PI containment and mitigation has to invariably consider all of the above aspects: it must incorporate the mechanism of disease progression, from the initial infection, to the asymptomatic phase, manifestation of symptoms, and the final health outcome (Atkinson & Wein, 2008; Handel, Longini, & Antia, 2010; Pourbohloul et al., 2009); it must also consider the population dynamics, including individual susceptibility (Pitzer, Leung, & Lipsitch, 2007; Pitzer, Olsen, Bergstrom, Dowell, & Lipsitch, 2007) and transmissibility (Cauchemez, Carrat, Viboud, Valleron, & Boelle, 2004; Handel et al., 2010; Yang, Halloran, Sugimoto, & Longini, 2007; Yang et al., 2009), and behavioral factors affecting infection generation and disease progression (Colizza, Barrat, Barthélemy, & Vespignani, 2006; Epstein, Parker, Cummings, & Hammond, 2008; Halloran, 2006); finally, it must incorporate the impact of pharmaceutical and non-pharmaceutical measures, including vaccination (Ball & Lyne, 2002; Becker & Starczak, 1997; Carrat, Lavenu, Cauchemez, & Deleger, 2006), antiviral therapy (Ferguson, Mallett, Jackson, Roberts, & Ward, 2003; Lee et al., 2002; Lipsitch, Cohen, Murray, & Levin, 2007), social distancing (Halder, Kelso, & Milne, 2010; Kelso, Milne, & Kelly, 2009; Miller, Randolph, & Patterson, 2008; Milne, Kelso, Kelly, Huband, & McVernon, 2008; Yasuda & Suzuki, 2009) and travel restrictions, and the use of low-cost measures, such as face masks and hand washing (Glass, Glass, Beyeler, & Min, 2006; Lipsitch et al., 2007; Nigmatulina & Larson, 2009; Scharfstein, Halloran, Chu, & Daniels, 2006).

In recent years, the models for PI containment and mitigation have focused on integration of pharmaceutical and non-pharmaceutical measures in search for synergistic strategies, aimed at better resource utilization. Most of these approaches aim to implement a scheme of social distancing to reduce infection exposure, followed by application of pharmaceutical means. Significant contributions in this challenging area include (Chao, Halloran, Obenchain, & Longini, 2010; Colizza, Barrat, Barthelemy, Valleron, & Vespignani, 2007; Cooley et al., 2008; Glass et al., 2006; Longini et al., 2004; Mills, Robins, & Lipsitch, 2004; Patel & Longini, 2005; Wu, Riley, Fraser, & Leung, 2006). One of the most notable among the recent efforts is a 2006-07 initiative by MIDAS (National Institute of General Medical Sciences, 2008), which cross-examined three independent simulation models of large-scale PI spread for rural areas of Asia (Ferguson et al., 2005; Longini et al., 2005), U.S. and U.K. (Ferguson et al., 2006; Germann, Kadau, Longini, & Macken, 2006), and the city of Chicago (Eubank et al., 2004), respectively. MIDAS cross-validated the models by simulating the city of Chicago, with 8.6M inhabitants, and implementing a targeted layered containment (Halloran et al., 2008; Models of Infectious Disease Agent Study, 2004). The research findings of MIDAS and other institutions (Atkinson & Wein, 2008; Glass et al., 2006) were used in a recent "Modeling Community Containment for Pandemic Influenza" report by IOM, to formulate a set of recommendations for mitigating PI at a local level (Committee on Modeling Community Containment for Pandemic Influenza, 2006). These recommendations were also used in a pandemic preparedness guidance developed by CDC (Centers for Disease Control and Prevention, 2007).

At the same time, the IOM report points out several limitations of the MIDAS models, observing that "because of the significant constraints placed on the models" being considered by policy makers, "the scope of models should be expanded." The IOM recommends "to adapt or develop *decision-aid models* that can ... provide *real-time feedback* during an epidemic". The report also emphasizes that "future modeling efforts should incorporate broader outcome measures ... to include the *costs and benefits* of intervention strategies" (Committee on Modeling Community Containment for Pandemic Influenza, 2006). It can indeed be observed that practically all existing models focus on assessment of *apriori* defined strategies; virtually none of the models are capable of "*learning*", i.e., adapting to changes in the pandemic course, yet predicting them, to generate *dynamic* strategies. Such a strategy will be advantageous in its ability to be state-dependent, i.e., being developed dynamically as the pandemic spreads, by selecting a mix of available mitigation options at each decision epoch, based on both the present state of the pandemic and its predicted evolution.

In an effort to address the IOM recommendations, we present a simulation optimization model for developing dynamic predictive distribution strategies of limited resource quantities over a network of regions exposed to the pandemic. The underlying simulation model mimics the disease and population dynamics of the affected regions (Sections Cross-Regional Simulation Model and Single-Region Simulation Model). As the pandemic spreads from region to region, the optimization model generates allocations of mitigation resources, including stockpiles of vaccines and antiviral, healthcare capacities for vaccination and antiviral administration, and social distancing enforcement resources (Section Optimization Model). The model seeks to minimize the impact of ongoing outbreaks and the expected impact of potential outbreaks. The optimality criterion of the model incorporates measures of morbidity, mortality, and social distancing, translated into the societal and economic costs of lost productivity and medical expenses. The methodology was calibrated using historic pandemic data and implemented on a sample cross-regional outbreak in Florida, with over 4 million inhabitants (Section Tesbed). We compared our strategy to two common distribution policies: pro-rata and myopic (U.S. Department of Health & Human Services, 2007). The *pro-rata policy* prescribes an *apriori* allocation of available resources to all network regions, in proportion to their population. The *myopic policy* works in a more dynamic but reactive way:

as the pandemic spreads, the policy implements a resource allocation from one outbreak region to the next, each time trying to cover the *entire* population at risk of the region. The comparison of the three strategies was implemented at different levels of the total resource availability (Section Comparison of the DPO, Pro-rata, and Myopic Strategies).

Compared to our earlier work (Das & Savachkin, 2008), this paper presents the following main advances: (1) the original single-region simulation (Das & Savachkin, 2008) has been expanded to serve as the basis for the cross-regional simulation model presented in this paper; (2) we now propose a novel dynamic optimization model, embedded in the cross-regional simulation, to generate resource distribution strategies; as such, the decision support power of our model has substantially increased; (3) the calibration methodology now incorporates both the basic reproduction number and the infection attack rate; inclusion of the latter adds to a more accurate assessment of the pandemic severity over the entirety of the pandemic period, and (4) our model now incorporates certain socio-behavioral features, such as the target population compliance.

2. SIMULATION OPTIMIZATION METHODOLOGY

Our methodology generates progressive resource distributions over a network of regions exposed to the pandemic. Resources include stocks and capacities for vaccine and antiviral administration, and social distancing enforcement resources, among others. The methodology subsumes a cross-regional simulation model, a set of single-region simulation models, and an embedded optimization model.

The network regions are classified as unaffected, ongoing outbreaks, or contained outbreaks (Figure 1). The cross-regional simulation model connects the regions by air and land travel. The single-region simulation models mimic the population and disease dynamics of each ongoing region, impacted by mitigation measures. The pandemic spreads from ongoing to unaffected regions by infectious travelers passing through regional border control. At every new regional outbreak epoch, the optimization model allocates available resources to the new outbreak region (*actual allocation*) and unaffected regions (*virtual allocation*). Daily statistics is collected for each ongoing region, including the number of infected, deceased, and quarantined cases, for different age groups. As an outbreak is contained, its societal and economic costs are estimated.

Figure 1. Schematic of cross-regional pandemic spread and resource distribution

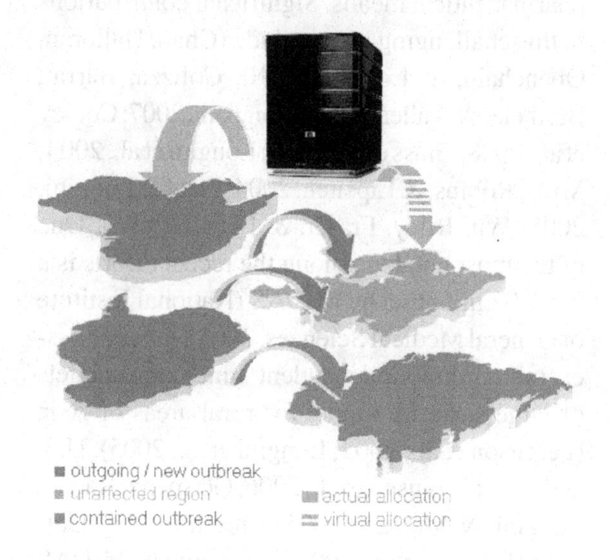

2.1 Cross-Regional Simulation Model

A schematic of the cross-regional simulation model is shown in Figure 2. The model initializes by creating population entities, mixing groups, and entity schedules, for each region. A pandemic is triggered by injecting an infectious case into a randomly chosen region. The details of the resulting contact dynamics and infection transmission inside the region are presented in Section Single-Region

Figure 2. Schematic of cross-regional simulation model

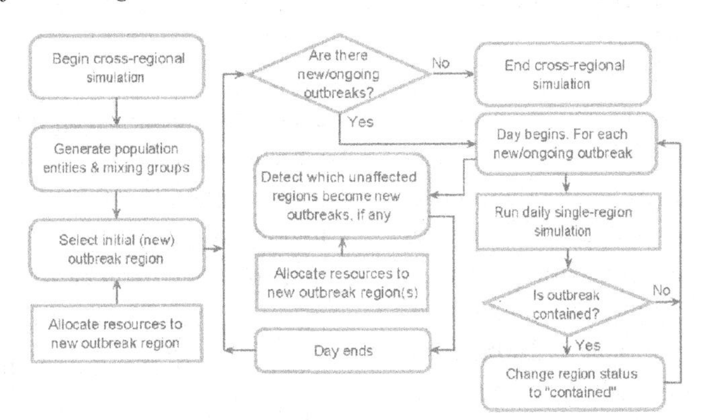

Simulation Model. As the symptomatic cases start seeking medical help, the new regional outbreak is detected. At this point, a resource distribution is generated and passed over to the single-region model. The outbreak can spread to unaffected regions as infectious travelers pass undetected with some probability through the border control. By tracing these travelers, the model determines which of the unaffected regions, if any, become new outbreaks. The model also determines if an ongoing outbreak has been contained. The cross-regional simulation stops when all outbreaks have been contained.

2.2 Single-Region Simulation Model

A schematic of the single-region simulation model is shown in Figure 3 (please also refer to our earlier work (Das & Savachkin, 2008). The model subsumes the following components: (1) population dynamics (mixing groups and schedules), (2) contact and infection process, (3) disease natural history, and (4) mitigation strategies, including social distancing, vaccination, and antiviral application. The model collects detailed regional statistics, including number of infected, recovered, deceased, and quarantined cases, for different age groups. For a contained outbreak, its societal and economic costs are calculated. The societal cost includes the cost of lost lifetime productivity of the deceased; whereas, the economic cost includes the cost of medical expenses of the recovered and deceased and the cost of lost productivity of the quarantined (Meltzer, Cox, & Fukuda, 1999). In what follows, we give the details of each model component.

Figure 3. Schematic of single-region simulation model

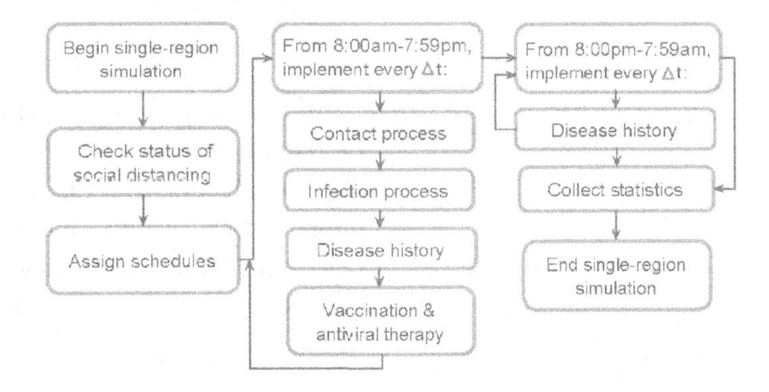

2.3 Population Dynamics (Mixing Groups and Schedules)

We model a region as a set of population centers formed by *mixing groups* or places where inhabitants come into contact with each other during the course of their social interaction. Examples of mixing groups include households, offices, universities, schools, shopping centers, entertainment centers, etc. (Das & Savachkin, 2008). Each inhabitant of the region is assigned a set of attributes such as age, gender, parenthood, workplace, infection susceptibility, and probability of travel, among others. Each person is also assigned Δt time-discrete (e.g., $\Delta t = 1$ hour) weekday and weekend schedules, which depend on: (1) person's age, parenthood, and employment status (2) her disease status, (3) her travel status, (4) social distancing decrees in place and person's compliance to them (Colardo Department of Human Services Division of Mental Health, 2009). As their schedules advance, the inhabitants circulate throughout the mixing groups and come into contact with each other.

It is assumed that at any point of time during the pandemic evolution, any individual of the cross-regional network belongs to one of the following compartments (Figure 4): susceptible, contacted (by an infectious individual), infected (asymptomatic or symptomatic), and recovered or deceased. In the following sections, we present the details of the infection transmission and disease natural history models which delineate the transitions between the above compartments.

Figure 4. Schematic of disease natural history

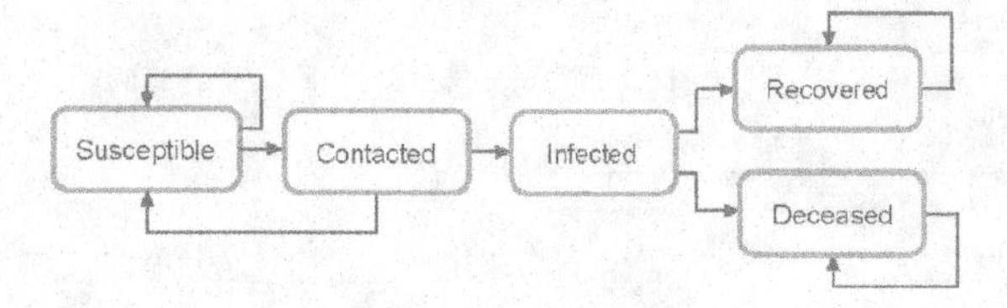

2.4 Contact and Infection Process

Infection transmission occurs during contact events between susceptible and infectious cases which take place in the mixing groups. At the beginning of every Δt period (e.g., one hour), for each mixing group *g*, the simulation tracks the total number of infectious cases, n_g, present in the group. It is assumed that each infectious case makes r_g per Δt unit of time *new contacts* (Germann et al., 2006), chosen randomly (uniformly) from the susceptible. Furthermore, we make the following assumptions about the contact process: (1) during Δt period, a susceptible may come into contact with at most one infectious case and (2) each contact exposure lasts Δt units of time. Once a susceptible has started her contact exposure at time *t*, she will develop infection at time $t + \Delta t$ with a certain probability that is calculated as shown below.

Let $L_i(t)$ be a nonnegative continuous random variable that represents the duration of contact exposure, starting at time *t*, required for contact *i* to become infected. We assume that $L_i(t)$ is distributed exponentially with mean $1 / \lambda_i(t)$, where $\lambda_i(t)$ represents the instantaneous force of infection applied to contact *i* at time *t* (Diekmann & Heesterbeek, 2000; Lawless & Lawless, 1982; Wu, Riley, & Leung, 2007). The probability that susceptible *i*, whose contact exposure has started at time *t*, will develop infection at time $t + \Delta t$ is then given as

$$P\{L_i(t) \leq \Delta t\} = 1 - e^{-\lambda_i(t)\Delta t}. \quad (1)$$

From the previous equation, the infection probability increases with both the instantaneous force of infection (i.e., the amount of virus shedding) and the length of the contact exposure. Once infected, an individual goes through various disease stages the details of which are presented in the next section.

2.5 Disease Natural History

A schematic of the disease natural history is shown in Figure 5. During the incubation phase, the infected individual stays asymptomatic. At the end of the latency phase, she becomes infectious and enters the infectious phase (Germann et al., 2006; Halloran et al., 2008; Longini et al., 2005). She becomes symptomatic at the end of the incubation period. At the end of the infectious phase, she enters the period leading to a health outcome, which culminates in her recovery or death.

Mortality for influenza like diseases is a complex process affected by many factors and variables, most of which have limited accurate data support available from past pandemics. Furthermore, the time of death can sometimes be weeks following the disease episode (which is often attributable to pneumonia related complications (Brundage & Shanks, 2008). Because of the uncertainty underlying the mortality process, we therefore adopted a simplified, age-based form of the mortality probability of infected i, as follows

$$m_i = \mu_i - \tau\rho_i, \quad (2)$$

where μ_i is the age-dependent base mortality probability of infected i, ρ_i is her status of antiviral therapy (0 or 1), and τ is the antiviral efficacy measured as the decrease in the base probability (Longini et al., 2005). We assume that a recovered case develops immunity but continues circulating in the mixing groups.

Figure 5. Schematic of disease natural history model

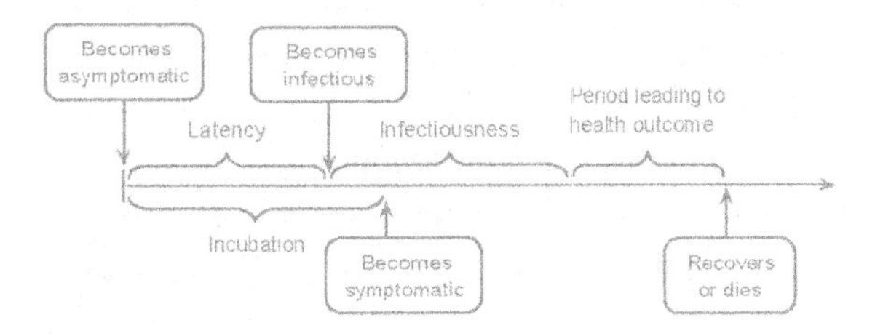

2.6 Mitigation Strategies

Mitigation strategies include pharmaceutical and non-pharmaceutical options. Mitigation strategies are initiated upon detection of a critical number of confirmed infected cases(Centers for Disease Control and Prevention (CDC), 2006), which triggers resource allocation and deployment. The model incorporates a certain delay for deployment of field responders.

Pharmaceutical mitigation includes vaccination and antiviral application. Vaccination is targeted at individuals *at risk* (World Health Organization, 2004) to reduce their infection susceptibility. We assume that a certain fraction of the risk group will not comply with vaccination. The vaccine takes a certain period to become effective (typically, between 10 and 14 days). Vaccination is constrained by the allocated stockpile and administration capacity, measured in terms of the number of immunizer-hours.

We assume that as some symptomatic cases seek medical help (Blendon et al., 2008; Sadique et al., 2007), those *at risk* of them will receive an antiviral, to reduce their infectiousness. The process is constrained by the allocated stockpile and administration capacity, measured in terms of the number of certified providers.

Both vaccination and antiviral application are affected by a number of socio-behavioral factors, including conformance of the target population, degree of risk perception, and compliance of healthcare personnel (Maunder et al., 2003; Pearson, Bridges, & Harper, 2006; Robertson, Hershenfield, Grace, & Stewart, 2004). The conformance level of the population *at risk* can be affected, among other factors, by the demographics and income level (Keane et al., 2005; Niederhauser, Baruffi, & Heck, 2001; Rhodes & Hergenrather, 2002; Rosenthal, Kottenhahn, Biro, & Succop, 1995; Smailbegovic, Laing, & Bedford, 2003) as well as by the quality of public information (Colardo Department of Human

Services Division of Mental Health, 2009). The degree of risk perception can be influenced by the negative experience developed during previous pharmaceutical campaigns (Cummings, Jette, Brock, & Haefner, 1979; Safranek et al., 1991), as well as by public fear and rumors (The New York Times, 2009; The New Yorker, 2009).

Non-pharmaceutical mitigation includes social distancing and travel restrictions. We adopted a CDC guidance (Centers for Disease Control and Prevention, 2007), which establishes five categories of pandemic severity and recommends quarantine and closure options according to the category. The categories are determined based on the value of the case fatality ratio (CFR), the proportion of fatalities in the total infected population. For CFR values lower than 0.1% (Category 1), voluntary at-home isolation of infected cases is implemented. For CFR values between 0.1% and 1.0% (Categories 2 and 3), in addition to at-home isolation, the following measures are *recommended*: (1) voluntary quarantine of household members of infected cases and (2) child and adult social distancing. For CFR values exceeding 1.0% (Categories 4 and 5), all the above measures are *implemented*. As the effectiveness of social distancing is affected by some of the behavioral factors listed above (Colardo Department of Human Services Division of Mental Health, 2009), we assume a certain social distancing conformance level. Travel restrictions considered in the model included regional air and land border control for infected travelers.

2.7 Optimization Model

The optimization model is invoked at the beginning of every n^{th} new regional outbreak epoch $(n = 1, 2, \ldots)$, starting from the initial outbreak region $(\mathrm{n} = 1)$. The objective of the model is to allocate some of the available mitigation resources to the new outbreak region (*actual allocation*) while reserving the rest of the quantities for potential outbreak regions (*virtual allocation*). By doing so, the model seeks to progressively minimize the impact of ongoing outbreaks and the expected impact of potential outbreaks, spreading from the ongoing locations. Mitigation resources can include stockpiles of vaccine and antiviral, administration capacities for their administration, hospital beds, medical supplies, and social distancing enforcement resources, among others. The predictive mechanism of the optimization model is based on a set of regression equations obtained using single-region simulation models. In what follows, we present the construction of the optimization model and explain the solution algorithm for the overall simulation-based optimization methodology.

We introduce the following *general terminology and notation*.

$S =$ set of all network regions,

$A^n =$ *set of regions in which pandemic is contained at the nth* outbreak epoch $(n = 1, 2, \ldots)$,

$B^n =$ *set of ongoing regions at the nth* outbreak epoch,

$C^n =$ *set of unaffected regions at the nth* outbreak epoch,

$R =$ set of available types of mitigation resources $(R = \{1, 2, \ldots, r\})$,

$q_{ik} =$ amount of resource i allocated to region k,

$Q_i^n =$ *available amount of resource $i \in R$* at the nth outbreak epoch,

$H =$ *set of age groups.*

The optimization criterion (objective function) of the model incorporates measures of expected societal and economic costs of the pandemic: the societal cost includes the cost of lost lifetime productivity of the deceased; the economic cost includes the cost of medical expenses of the recovered and deceased and the cost of lost productivity of the quarantined. To compute these costs, the following *impact measures* of morbidity, mortality, and quarantine are used, for each region k:

X_{hk} = total number of infected cases in age group h who seek medical assistance,
Y_{hk} = total number of infected cases in age group h who do not seek medical assistance,
D_{hk} = total number of deceased cases in age group h,
V_{hk} = total number of person-days of cases in age group h who comply with quarantine.

To estimate these measures, we use the following regression models obtained using a single-region simulation of each region h:

$$X_{hk} = \delta^0_{hk} + \sum_{i \in R} \delta^i_{hk} \cdot q_{ik} + \sum_{i,m \in R, i \neq m} \delta^{im}_{hk} \cdot q_{ik} \cdot q_{mk}, \quad (3)$$

where $\delta^i_{..}$ denotes the regression coefficient associated with resource i, and $\delta^{im}_{..}$ is the regression coefficient for the interaction between resources i and m. Similar models are used for Y_{hk}, D_{hk}, and V_{hk}.

The above relationships between the impact measures and the resource allocations ought to be determined *apriori* of implementing a cross-regional scenario. Here, we consider each region k as the initial outbreak region. We assume, however, that as the pandemic evolves, the disease infectivity will naturally subside. Hence, we incorporate a decay factor α^n (Gosavi, Das, & Sarkar, 2004) to adjust the estimates of the regional impact measures at every n^{th} outbreak epoch, in the following way:

$$X^n_{hj} = \alpha^n X_{hj},\ Y^n_{hj} = \alpha^n Y_{hj},\ D^n_{hj} = \alpha^n D_{hj},\ V^n_{hj} = \alpha^n V_{hj}. \quad (4)$$

Alternatively, the regression equations can be re-estimated at every new outbreak epoch, for each region $k \in C^n$, using the single-region simulation models, where each simulation must be initialized to the current outbreak status in region k in the cross-regional simulation.

In addition, we use the following regression model to estimate the *probability of pandemic spread* from affected region l to unaffected region k, as a function of resources allocated to region l, which, in turn, impact the number of outgoing infectious travelers from the region:

$$p_{lk} = \gamma^0_{lk} + \sum_{i \in R} \gamma^i_{lk} \cdot q_{il} + \sum_{\substack{i,m \in R \\ i \neq m}} \gamma^{im}_{lk} \cdot q_{il} \cdot q_{ml}, \quad (5)$$

where $\gamma^i_{..}$ denotes the regression coefficient associated with resource i, $\gamma^{im}_{..}$ is the regression coefficient associated with interaction between resources i and m, and $\gamma^0_{..}$ represents the intercept. Consequently, the total outbreak probability for unaffected region k can be found as $p_k = \sum_{l \in B^n} p_{lk}$. We use a scheme similar to Equation 4 to progressively adjust the estimates of the regional outbreak probabilities as follows:

$$p^n_k = \alpha^n p_k. \quad (6)$$

Finally, we calculate the total cost of an outbreak in region k at the nth decision epoch as follows:

$$TC^n_k = \sum_{h \in H} (m_h + \bar{w}_h) X^n_{hk} + \sum_{h \in H} \bar{w}_h \cdot Y^n_{hk} + \sum_{h \in H} \hat{w}_h \cdot D^n_{hk} + \sum_{h \in H} w_h \cdot V^n_{hk}, \quad (7)$$

where

m_h = total medical cost of an infected case in age group h over his/her disease period,
$\bar{w}_h$ = total cost of lost wages of an infected case in age group h over his/her disease period,

$\hat{w}_h$ = cost of lost lifetime wages of a deceased case in age group h,

w_h = daily cost of lost wages of a non-infected case in age group h who complies with quarantine.

The optimization model has the following form.

$$\textit{Minimize}\ \ TC_j^n(q_{1j}, q_{2j}, \ldots, q_{rj}) + \sum_{s \in C^n} TC_s^n(q_{1s}, q_{2s}, \ldots, q_{rs}) \cdot p_s^n$$

subject to

$$q_{ij} + \sum_{s \in C^n} q_{is} \cdot p_s^n \leq Q_i^n \quad \forall i \in R,$$

$$q_{ij}, q_{is} \geq 0 \ \ \forall i \in R.$$

The first term of the objective function represents the total cost of the new outbreak j, estimated at the nth outbreak epoch, based on the actual resource allocation $\{q_{1j}, q_{2j}, \ldots, q_{rj}\}$ (see Equation 7). The second term represents the total expected cost of outbreaks in currently unaffected regions, based on the virtual allocations $\{q_{1s}, q_{2s}, \ldots, q_{rs}\}$ (Equation 7) and the regional outbreak probabilities p_s^n (Equation 6). The set of constraints assures that for each resource i, the total quantity allocated (current and virtual, both nonnegative) does not exceed the total resource availability at the nth decision epoch. Note that both the objective function and the availability constraints are nonlinear in the decision variables.

2.8 Solution Algorithm

The solution algorithm for our dynamic predictive simulation optimization (DPO) model is given below.

1. Estimate regression equations for each region using the single-region simulation model.
2. Begin the cross-regional simulation model.
3. Initialize the sets of regions: $A^n = \varnothing$, $B^n = \varnothing$, $C^n = S$.
4. Select randomly the initial outbreak region j. Set $n = 1$.
5. Update sets of regions: $B^n \leftarrow B^n \cup \{j\}$ and $C^n \leftarrow C^n \setminus \{j\}$.
6. Solve the resource allocation model for region j. Update the total resource availabilities.
7. If $B^n \neq \varnothing$, do step 8. Else, do step 10.
8.
    a. For each ongoing region, implement a next day run of its single-region simulation.
    b. heck the containment status of each ongoing region. Update sets A^n and B^n, if needed.
    c. For each unaffected region, calculate its outbreak probability.
    d. Based on the outbreak probability values, determine if there is a new outbreak region(s) j.

If there is no new outbreak(s), go to step 7. Otherwise, go to step 9.

9. For each new outbreak region j,
    a. Increment $n \leftarrow n+1$.
    b. Update sets $B^n \leftarrow B^n \cup \{j\}$ and $C^n \leftarrow C^n \setminus \{j\}$.
    c. Re-estimate regression equations for each region $k \in B^n \cup C^n$ using the single-region simulations, where each simulation is initialized to the current outbreak status in the region (optional step; alternatively, use Equation 4 and Equation 6).
    d. Solve the resource allocation model for region j.
    e. Update the total resource availabilities.

10. Calculate the total cost for each contained region and update the overall pandemic cost.

3. TESBED

To illustrate the use of our DPO methodology, we implemented a sample H5N1 outbreak scenario including four counties in Florida: Hillsborough, Miami Dade, Duval, and Leon, containing four major cities and transportation hubs of the state: Tampa, Miami, Jacksonville, and Tallahassee, respectively (Figure 6). The respective county populations are 1.0, 2.2, 0.8, and 0.25 million people. A basic unit of time for population and disease dynamics models was taken as $\Delta t = 1$ hour. Regional simulations were run for a period (up to 180 days) until the daily infection rate approached near zero (see Section Parameters of Mitigation). Below we present the details on selecting model parameter values. Most of the testbed data can be found in the supplement (Savachkin et al., 2010).

3.1 Parameter Values for Population and Disease Dynamics Models

Demographic and social dynamics data for each region (Savachkin et al., 2010) were extracted from the U.S. Census (U.S Census Bureau, 2000) and the National Household Travel Survey (Bureau

Figure 6. Testbed counties in FL

of transportation statistics, 2002). Daily (hourly) schedules (Savachkin et al., 2010) were adopted from (Das & Savachkin, 2008).

Each infected person was assigned a daily travel probability of 0.24% (Table 1) (Bureau of transportation statistics, 2002), of which 7% was by air and 93% by land. The probabilities of travel among the four regions were calculated using traffic volume data (Jacksonville Aviation Authority, 2010; Miami International Airport, 2010; Tallahassee Regional Airport, 2010; Tampa International Airport, 2010). Infection detection probabilities for border control for symptomatic cases were assumed to be 95% and 90%, for air and land, respectively.

The instantaneous force of infection applied to contact i at time t, Equation 1 (Diekmann & Heesterbeek, 2000) was modeled as

$$\lambda_i(t) = -ln(1 - p_i(t)), \text{where}, p_i(t) = \alpha_i - \delta\theta_i(t), \quad (8)$$

where α_i is the age-dependent base instantaneous infection probability of contact i, $\theta_i(t)$ is her status of vaccination at time t (0 or 1), and δ is the vaccine efficacy, measured as the reduction in the base instantaneous infection probability, achieved after 10 days (Pasteur, 2009).

The values of age-dependent base instantaneous infection probabilities were adopted from (Germann et al., 2006) (Table 2). The disease natural history included a latent period of 29 hours (1.21 days), an incubation period of 46 hours (1.92 days), an infectiousness period from 29 to 127 hours (1.21 to 5.29 days), and a period leading to health outcome from 127 to 240 hours (5.29 to 10 days) (Beigel et al., 2005).

Base mortality probabilities (μ_i in Equation 2) were determined using the statistics recommended by the Working Group on Pandemic Preparedness and Influenza Response (Meltzer et al., 1999). This data shows the percentage of mortality for age-based high-risk cases (HRC) (Table 3, columns 1-3). Mortality probabilities (column 4) were estimated under the assumption that high-risk cases are expected to account for 85% of the total number of fatalities, for each age group (Meltzer et al., 1999).

3.2 Calibration of the Single-Region Models

Single-region simulation models were calibrated using two commonly used measures of pandemic severity (Ferguson et al., 2006; Germann et al., 2006; Longini et al., 2004): the basic reproduction number (R_0) and the infection attack rate (*IAR*). R_0 is defined as the average number of secondary infections produced by a typical infected case in a totally susceptible population. *IAR* is defined as the ratio of the total number of infections over the pandemic period to the size of the initial susceptible population. To determine R_0, all infected cases inside the simulation were classified by generation of infection, as in (Ferguson et al., 2006; Glass et al., 2006). The value

Table 1. Inter-regional travel probabilities

Origin\Destination	Inter-Regional Travel Probability			
	Hillsborough	Miami Dade	Duval	Leon
Hillsborough	0.00	0.60	0.27	0.13
Miami Dade	0.74	0.00	0.16	0.10
Duval	0.61	0.29	0.00	0.10
Leon	0.52	0.31	0.17	0.00

Table 2. Instantaneous infection probabilities

Age Group	0-5	6-19	20-29	31-65	66-99
α_i	0.156	0.106	0.205	0.195	0.344

Table 3. Mortality probabilities for different age groups

Age Group	% HRC	% Mortality in HRC	μ_i
0-19	6.4	9.0	0.007
20-64	14.4	40.9	0.069
65+	40.0	34.4	0.162

of R_0 was calculated as the average reproduction number of a typical generation in the early stage of the pandemic, with no interventions implemented (*the baseline* scenario) (Glass et al., 2006). Historically, R_0 values for PI ranged between 1.4 and 3.9 (Ferguson et al., 2005; Mills et al., 2004). To attain similar values, we calibrated the hourly contact rates of mixing groups (Savachkin et al., 2010) (original rates were adopted from (Germann et al., 2006)). For the four regions, the average baseline value of R_0 was 2.54, which represented a high transmissibility scenario. The values of regional baseline *IAR* averaged 0.538.

3.4 Parameters of Mitigation

Mitigation resources included stockpiles of vaccines and antiviral and their administration capacities, and quarantine enforcement resources (required to achieve a target conformance level). We assumed a 24-hour delay for deployment of field responders and resource allocation (Centers for Disease Control and Prevention, 2006).

The vaccination risk group included healthcare providers (Pearson et al., 2006), and individuals younger than 5 years (excluding younger than 12 months old) and older than 65 years (World Health Organization, 2004). The risk group for antiviral included symptomatic individuals below 15 years and above 55 years (Institute of Medicine, 2008; World Health Organization, 2004). The efficacy levels for the vaccine (δ in Equation 8) and antiviral (τ in Equation 2) were assumed to be 40% (Longini et al., 2005; Treanor, Campbell, Zangwill, Rowe, & Wolff, 2006) and 30%, respectively. We did not consider the use of antiviral for a mass prophylactic reduction of infection susceptibility due to the limited antiviral availability (U.S. Department of Health & Human Services, 2007) and the risk of emergence of antiviral resistant transmissible virus strains (Lipsitch et al., 2007). We assumed a 90% target population conformance for both vaccination and antiviral treatment (Maunder et al., 2003). The immunity development period for the vaccine was taken as 10 days (Pasteur, 2009).

A version of the CDC guidance for quarantine and isolation for Category 5 was implemented (Centers for Disease Control and Prevention, 2007). Once the reported CFR value had reached 1.0%, the following policy was declared and remained in effect for 14 days (Centers for Disease Control and Prevention, 2007): (1) individuals below a prespecified age ξ (22 years) stayed at home during the entire policy duration; (2) of the remaining population, a certain proportion ϕ (Blendon et al., 2006) stayed at home and was allowed a one-hour leave, every three days, to buy essential supplies; (3) the remaining $(1-\phi)$ noncompliant proportion followed a regular schedule. All testbed scenarios considered the quarantine conformance level ϕ (a decision variable) bounded between 50% and 80% (Colardo Department of Human Services Division of Mental Health, 2009).

An outbreak was considered contained, if the daily infection rate did not exceed five cases, for seven consecutive days. Once contained, a region was simulated for an additional 10 days for accurate estimation of the pandemic statistics. A 2^5

statistical design of experiment (Montgomery, 2008) was used to estimate the regression coefficient values of the significant decision factors and their interactions; the values of adjusted R^2 ranged from 96.36% to 99.97%). The simulation code was developed in C++. A total of 150 replicates were run for each scenario. The running time for a cross-regional simulation replicate involving over four million inhabitants was between 17 and 26 minutes on a Pentium 3.40 GHz with 4.0 GB of RAM.

3.5 Comparison of the DPO, Pro-Rata, and Myopic Strategies

This section compares the performance of our DPO strategy to that of the pro-rata and myopic policy. The pro-rata policy (U.S. Department of Health & Human Services, 2007) allocates the total available resources to *all* network regions, *apriori*, in proportion to the regional population. The myopic policy allocates the available resources from one actual outbreak region to the next, each time trying to cover the *entire* population at risk of the region.

Table 4 summarizes the total resource requirement for each region, based on the composition of the regional risk groups. Table 5 shows the costs of lost productivity and medical expenses, for different age groups, which were adopted from (Meltzer et al., 1999) and adjusted for inflation for the year of 2010 (Halfhill, 2009).

Comparison of the strategies is done at the levels of 20%, 50%, and 80% of the total resource requirement shown in Table 4. Figures 7 and 8 show the summary comparison of the three strategies in the form of the 95% confidence intervals (CI) for the average number of infected and deceased, respectively. Figure 9 shows the policy

Table 4. Total and regional resource requirements

Region	Resource Requirements by Region				
	Hillsborough	Miami Dade	Duval	Leon	Total
(population)	(1,007,916)	(2,209,702)	(852,168)	(248,761)	(4,318,547)
Resource					
Vaccine stockpile	305,036	679,181	241,522	76,007	1,301,745
Antiviral stockpile	415,294	749,058	460,393	105,307	1,730,052
No. antiv. Nurses	650	1,104	786	166	2,706
No. vacc. nurses	1,059	2,358	839	264	4,520

Table 5. Values of pandemic impact measures (societal and economic costs)

Pandemic Impact Measure (age group, years)	Value US$
Average cost of lost lifetime productivity of a deceased case (0 - 19)	$1,336,347.86
Average cost of lost lifetime productivity of a deceased case (20 - 64)	$1,370,987.28
Average cost of lost lifetime productivity of a deceased case (65 - 99)	$98,959.24
Average cost of lost productivity and medical expenses of a recovered/deceased case (0 -19)	$5,078.48
Average cost of lost productivity and medical expenses of a recovered/deceased case (20 -64)	$10,466.68
Average cost of lost productivity and medical expenses of a recovered/deceased case (65 -99)	$11,566.09
Average daily cost of lost productivity of a non-infected quarantined case (20-99)	$432.54

Figure 7. Average number of infected: Comparison of DPO, pro-rata, and myopic policies

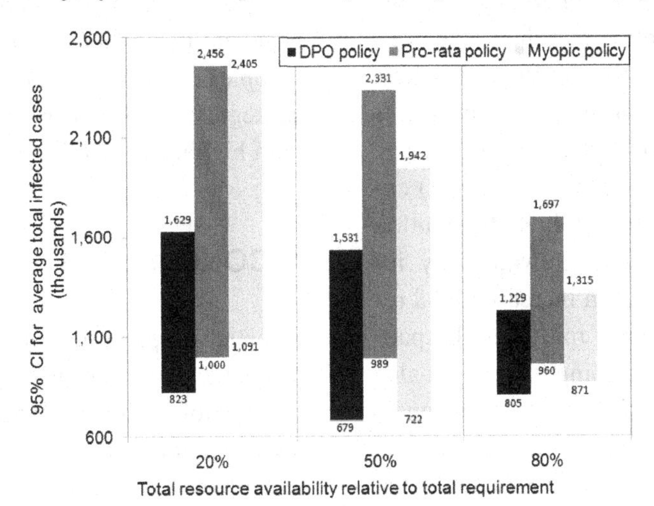

Figure 8. Average number of deaths: Comparison of DPO, pro-rata, and myopic policies

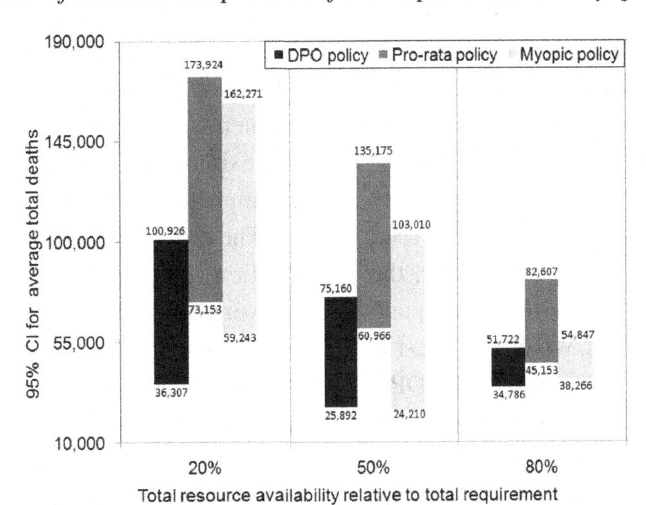

Figure 9. Average total pandemic cost: Comparison of DPO, pro-rata, and myopic policies

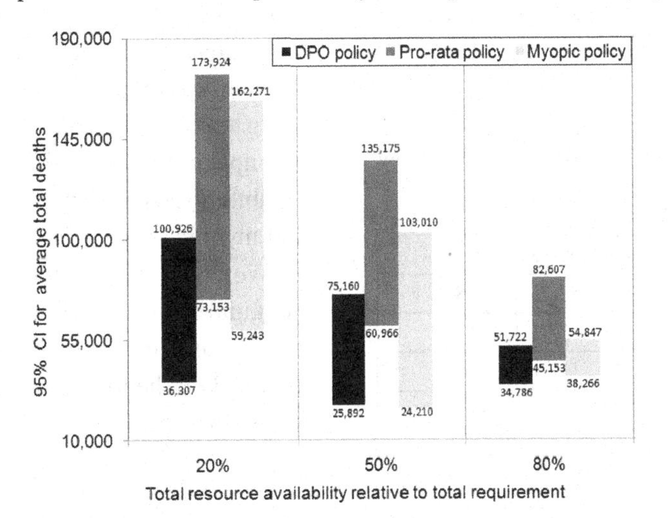

comparison using the 95% CI for the average total pandemic cost. Table 6 also shows the average number of regional outbreaks for each policy.

It can be observed that all impact measures (the average number of infected, the average number of dead, and the average total cost) exhibit a downward trend, for all three distribution policies, as the total resource availability increases from 20% to 80%. An increased total resource availability not only mitigates the pandemic impact inside the ongoing regions but also reduces the probability of spread to the unaffected regions. For all three policies, as the total resource availability approaches the total resource requirement (starting from approximately 60%), the impact curves show a converging behavior, whereby the marginal utility of additional resource availability diminishes. This behavior can be explained by noting that the total resource requirement was determined assuming the worst case scenario when *all* network regions would be affected and provided with adequate resources to cover their respective regional populations at risk.

It can also be observed that on average, the DPO strategy outperforms both the pro-rata and the myopic policy at all levels, which can attest to a more efficient resource utilization by the DPO policy (Table 6). The difference in the policy performance is particularly noticeable at the lower levels of resource availability and it gradually diminishes, as the resource availability increases and as the available quantities become closer to be sufficient to cover the entire populations at risk in all regions. It can also be noted that the variability in the performance of the DPO strategy is generally smaller than that of the pro-rata and myopic policy. In general, for all three distribution strategies, the performance variability decreases with higher availability of resources.

Table 6. Average number of regional outbreaks for each distribution policy

Policy	Total Resource Availability		
	20%	50%	80%
DPO	1.75	1.66	1.44
Myopic	2.40	1.77	1.50
Pro-rata	2.23	2.05	1.83

4. CONCLUSION

Effective preparedness and response to influenza pandemics have to invariably rely on understanding the evolution of disease and population dynamics and intelligent resource distribution strategies. As recently pointed by the IOM, the existing models for PI containment and mitigation fall short of providing *dynamic* decision support, which would incorporate broader outcome measures, including "costs and benefits of intervention" (Committee on Modeling Community Containment for Pandemic Influenza, 2006). In this paper, we present a large-scale simulation-based optimization model which attempts to address the above recommendations.

The model supports development of dynamic predictive resource distribution strategies over a network of regions exposed to the pandemic. The underlying discrete-event agent-based simulation mimics the disease and population dynamics inside and across the affected regions. The simulation embeds an optimization model which generates progressive allocations of mitigation resources. The optimization aims to balance the impact of both ongoing and potential outbreaks, measured in terms of morbidity, mortality, and social distancing, translated into the cost of lost productivity and medical expenses. The model was calibrated using historic pandemic data and implemented on a sample outbreak in Florida, with over 4 million inhabitants, where it was compared to the pro-rata and myopic policy.

We observed that for all three strategies, the marginal utility of additional resource availability was diminishing, as the total resource availability approached the total requirement. We also noted that in the testbed scenario, the DPO strategy on average outperformed both the pro-rata and the myopic policies. Comparing the DPO to the pro-rata, it can be noted that while the latter strategy treats all regions uniformly, regardless of their demographics and composition of the risk groups, the DPO-based approach exploits region-specific effectiveness of mitigation resources and dynamic reassessment of spread probabilities. Hence, we believe that in scenarios involving more regions with heterogeneous demographics, the DPO policy will likely perform even better and with less variability than the pro-rata strategy. At the same time, while the myopic policy on average performed better than the pro-rata, the former was still inferior to the DPO strategy. This can be attributed to the fact that the DPO strategy is advantageous in its ability to consider not only the resource needs of the newly emerged outbreak region, but also the needs of the potential outbreaks. The difference in the model performance was particularly noticeable at lower levels of resources, which is in accordance with a higher marginal utility of additional availability at those levels. Hence, we believe that the DPO model can be particularly useful in scenarios with very limited supply of resources.

The methodology presented in this paper is one of the first attempts to offer *dynamic predictive* decision support for pandemic mitigation, which incorporates measures of societal and economic impact. Our comparison study of the DPO vs. pro-rata vs. myopic cross-regional resource allocations is also novel. In addition, our simulation model represents one of the first of its kind in considering a broader range of social behavioral aspects, including vaccination and antiviral treatment conformance. The simulation features a flexible design which can be particularized to a broader range of pharmaceutical and non-pharmaceutical interventions and even more granular mixing groups.

Based on our methodology, we have also developed a decision-aid simulator with a GUI which is made freely available to general public through our web site at \url{http://imse.eng.usf.edu/pandemics.aspx}. The simulator allows the input of data for regional demographic and social dynamics, and disease related parameters. It is intended to assist public health decision makers in conducting customized what-if analysis for assessment of mitigation options and development of policy guidelines. Examples of such guidelines include targeted risk groups for vaccination and antiviral treatment, social distancing policies (e.g., thresholds for declaration and lifting, closure options (i.e., household-based, schools, etc.), and compliance targets), and guidelines for travel restrictions.

Limitations of the model. Lack of reliable data prevented us from considering geo-spatial aspects of mixing group formation in the testbed implementation. We also did not consider the impact of public education and use of personal protective measures (e.g., face masks) on transmission, again due to a lack of effectiveness data (Bell & World Health Organization Writing Group, 2006). We did not study the marginal effectiveness of individual resources due to a considerable uncertainty about the transmissibility of a future pandemic virus and efficacy of vaccine and antiviral. For the same reason, the vaccine and antiviral risk groups considered in the testbed can be adjusted, as different prioritization schemes have been suggested in the literature, based on predicted characteristics of a future virus. The form of social distancing implemented in the testbed can also be modified as a variety of schemes can be found in the literature, including those based on geographical and social targeting. Effectiveness of these approaches is substantially influenced by the policy compliance, for which limited accurate data support exists. It will thus be vital to gather the most detailed data on the epidemiology of a new virus and the population dynamics early in the evolution of a pandemic, and to expeditiously analyze the data to adjust the interventions accordingly.

REFERENCES

Atkinson, M. P., & Wein, L. M. (2008). Quantifying the routes of transmission for pandemic influenza. *Bulletin of Mathematical Biology*, *70*(3), 820–867. doi:10.1007/s11538-007-9281-2

Aunins, J., Lee, A., & Volkin, D. (1995). Vaccine production. In Bronzino, J. D. (Ed.), *The biomedical engineering handbook* (pp. 1502–1517). Boca Raton, FL: CRC Press.

Ball, F. G., & Lyne, O. D. (2002). Optimal vaccination policies for stochastic epidemics among a population of households. *Mathematical Biosciences*, *177*, 333–354. doi:10.1016/S0025-5564(01)00095-5

Becker, N. G., & Starczak, D. N. (1997). Optimal vaccination strategies for a community of households. *Mathematical Biosciences*, *139*(2), 117–132. doi:10.1016/S0025-5564(96)00139-3

Beigel, J., Farrar, J., Han, A., Hayden, F., Hyer, R., & De Jong, M. (2005). Avian influenza A (H5N1) infection in humans. *The New England Journal of Medicine*, *353*(13), 1374. doi:10.1056/NEJMra052211

Bell, D. M., & World Health Organization Writing Group. (2006). Non-pharmaceutical interventions for pandemic influenza, national and community measures. *Emerging Infectious Diseases*, *12*(1), 88–94.

Blendon, R. J., DesRoches, C. M., Cetron, M. S., Benson, J. M., Meinhardt, T., & Pollard, W. (2006). Attitudes toward the use of quarantine in a public health emergency in four countries. *Health Affairs*, *25*(2), 15. doi:10.1377/hlthaff.25.w15

Blendon, R. J., Koonin, L. M., Benson, J. M., Cetron, M. S., Pollard, W. E., & Mitchell, E. W. (2008). Public response to community mitigation measures for pandemic influenza. *Emerging Infectious Diseases*, *14*(5), 778. doi:10.3201/eid1405.071437

Brundage, J. F., & Shanks, G. D. (2008). Deaths from bacterial pneumonia during 1918–19 influenza pandemic. *Emerging Infectious Diseases*, *14*(8), 1193. doi:10.3201/eid1408.071313

Bureau of Transportation Statistics. (2002). *2001 national household travel survey (NTHS).* Retrieved from http://www.bts.gov/programs/national_household_travel_survey/

Carrat, F., Lavenu, A., Cauchemez, S., & Deleger, S. (2006). Repeated influenza vaccination of healthy children and adults: Borrow now, pay later? *Epidemiology and Infection*, *134*(1), 63–70. doi:10.1017/S0950268805005479

Cauchemez, S., Carrat, F., Viboud, C., Valleron, A., & Boelle, P. (2004). A Bayesian MCMC approach to study transmission of influenza: Application to household longitudinal data. *Statistics in Medicine*, *23*(22), 3469–3487. doi:10.1002/sim.1912

Centers for Disease Control and Prevention. (2006). *CDC influenza operational plan.* Retrieved from http://www.cdc.gov/flu/pandemic/cdcplan.htm

Centers for Disease Control and Prevention. (2007). *Interim pre-pandemic planning guidance: Community strategy for pandemic influenza mitigation in the united states.* Retrieved from http://www.pandemicflu.gov/plan/community/community_mitigation.pdf

Centers for Disease Control and Prevention. (2008). *Avian influenza: Current H5N1 situation.* Retrievedfrom http://www.cdc.gov/flu/avian/outbreaks/current.htm

Centers for Disease Control and Prevention. (2009). *Preparing for pandemic influenza.* Retrieved from http://www.cdc.gov/flu/pandemic/preparedness

Chao, D. L., Halloran, M. E., Obenchain, V. J., & Longini, I. M. (2010). FluTE, a publicly available stochastic influenza epidemic simulation model. *PLoS Computational Biology*, *6*(1). doi:10.1371/journal.pcbi.1000656

Colardo Department of Human Services Division of Mental Health. (2009). *Pandemic influenza: Quarantine, isolation and social distancing.* Retrieved from http://www.flu.gov/news/colorado_toolbox.pdf

Colizza, V., Barrat, A., Barthelemy, M., Valleron, A., & Vespignani, A. (2007). Modeling the world-wide spread of pandemic influenza: Baseline case and containment interventions. *PLoS Medicine*, *4*(1), 95. doi:10.1371/journal.pmed.0040013

Colizza, V., Barrat, A., Barthélemy, M., & Vespignani, A. (2006). The role of the airline transportation network in the prediction and predictability of global epidemics. *Proceedings of the National Academy of Sciences of the United States of America*, *103*(7), 2015. doi:10.1073/pnas.0510525103

Committee on Modeling Community Containment for Pandemic Influenza. (2006). *Modeling community containment for pandemic influenza: A letter report.* Retrieved from http://www.nap.edu/catalog/11800.html

Cooley, P., Ganapathi, L., Ghneim, G., Holmberg, S., Wheaton, W., & Hollingsworth, C. R. (2008). Using influenza-like illness data to reconstruct an influenza outbreak. *Mathematical and Computer Modelling*, *48*(5-6), 929–939. doi:10.1016/j.mcm.2007.11.016

Cummings, K. M., Jette, A. M., Brock, B. M., & Haefner, D. P. (1979). Psychosocial determinants of immunization behavior in a swine influenza campaign. *Medical Care*, *17*(6), 639. doi:10.1097/00005650-197906000-00008

Das, T., & Savachkin, A. (2008). A large scale simulation model for assessment of societal risk and development of dynamic mitigation strategies. *IIE Transactions*, *40*(9), 893–905. doi:10.1080/07408170802165856

Diekmann, O., & Heesterbeek, J. A. P. (2000). *Mathematical epidemiology of infectious diseases: Model building, analysis, and interpretation.* New York, NY: John Wiley & Sons.

Epstein, J. M., Parker, J., Cummings, D., & Hammond, R. A. (2008). Coupled contagion dynamics of fear and disease: Mathematical and computational explorations. *PLoS ONE*, *3*(12), 3955. doi:10.1371/journal.pone.0003955

Eubank, S., Guclu, H., Kumar, V. S. A., Marathe, M. V., Srinivasan, A., & Toroczkai, Z. (2004). Modelling disease outbreaks in realistic urban social networks. *Nature*, *429*(6988), 180–184. doi:10.1038/nature02541

Fedson, D. S. (2003). Pandemic influenza and the global vaccine supply. *Clinical Infectious Diseases*, *36*(12), 1552–1561. doi:10.1086/375056

Ferguson, N. M., Cummings, D. A. T., Cauchemez, S., Fraser, C., Riley, S., & Meeyai, A. (2005). Strategies for containing an emerging influenza pandemic in southeast Asia. *Nature*, *437*(7056), 209–214. doi:10.1038/nature04017

Ferguson, N. M., Cummings, D. A. T., Fraser, C., Cajka, J. C., Cooley, P. C., & Burke, D. S. (2006). Strategies for mitigating an influenza pandemic. *Nature*, *442*(7101), 448–452. doi:10.1038/nature04795

Ferguson, N. M., Mallett, S., Jackson, H., Roberts, N., & Ward, P. (2003). A population-dynamic model for evaluating the potential spread of drug-resistant influenza virus infections during community-based use of antivirals. *The Journal of Antimicrobial Chemotherapy*, *51*(4), 977. doi:10.1093/jac/dkg136

Germann, T. C., Kadau, K., Longini, I. M., & Macken, C. A. (2006). *Mitigation strategies for pandemic influenza in the United States*. Washington, DC: The National Academy of Sciences.

Glass, R. J., Glass, L. M., Beyeler, W. E., & Min, H. J. (2006). Targeted social distancing design for pandemic influenza. *Emerging Infectious Diseases*, *12*(11), 1671–1681.

Gosavi, A., Das, T. K., & Sarkar, S. (2004). A simulation-based learning automata framework for solving semi-markov decision problems under long-run average reward. *IIE Transactions*, *36*(6), 557–567. doi:10.1080/07408170490438672

Halder, N., Kelso, J. K., & Milne, G. J. (2010). Analysis of the effectiveness of interventions used during the 2009 A/H1N1 influenza pandemic. *BMC Public Health*, *10*, 168. doi:10.1186/1471-2458-10-168

Halfhill, T. (2009). *Inflation calculator.* Retrieved from http://www.halfhill.com/inflatation.html

Halloran, M. E. (2006). Invited commentary: Challenges of using contact data to understand acute respiratory disease transmission. *American Journal of Epidemiology*, *164*(10), 945. doi:10.1093/aje/kwj318

Halloran, M. E., Ferguson, N. M., Eubank, S., Longini, I. M., Cummings, D. A. T., & Lewis, B. (2008). Modeling targeted layered containment of an influenza pandemic in the United States. *Proceedings of the National Academy of Sciences of the United States of America*, *105*(12), 4639. doi:10.1073/pnas.0706849105

Handel, A., Longini, I. M., & Antia, R. (2010). Towards a quantitative understanding of the within-host dynamics of influenza A infections. *Journal of the Royal Society, Interface*, *7*(42), 35. doi:10.1098/rsif.2009.0067

Institute of Medicine (Ed.). (2008). *Antivirals for pandemic influenza: Guidance on developing a distribution and dispensing program*. Washington, DC: The National Academies Press.

Jacksonville Aviation Authority. (2010). *Daily traffic volume data*. Retrieved from http://www.jaa.aero/General/Default.aspx

Keane, M. T., Walter, M. V., Patel, B. I., Moorthy, S., Stevens, R. B., & Bradley, K. M. (2005). Confidence in vaccination: A parent model. *Vaccine*, *23*(19), 2486–2493. doi:10.1016/j.vaccine.2004.10.026

Kelso, J. K., Milne, G. J., & Kelly, H. (2009). Simulation suggests that rapid activation of social distancing can arrest epidemic development due to a novel strain of influenza. *BMC Public Health*, *9*, 117. doi:10.1186/1471-2458-9-117

Lawless, J. F., & Lawless, J. (Eds.). (1982). *Statistical models and methods for lifetime data*. New York, NY: John Wiley & Sons.

Lee, P. Y., Matchar, D. B., Clements, D. A., Huber, J., Hamilton, J. D., & Peterson, E. D. (2002). Economic analysis of influenza vaccination and antiviral treatment for healthy working adults. *Annals of Internal Medicine*, *137*(4), 225.

Lipsitch, M., Cohen, T., Murray, M., & Levin, B. R. (2007). Antiviral resistance and the control of pandemic influenza. *PLoS Medicine*, *4*(1), 111. doi:10.1371/journal.pmed.0040015

Longini, I. M. Jr, Halloran, M. E., Nizam, A., & Yang, Y. (2004). Containing pandemic influenza with antiviral agents. *American Journal of Epidemiology*, *159*(7), 623. doi:10.1093/aje/kwh092

Longini, I. M. Jr, Nizam, A., Xu, S., Ungchusak, K., Hanshaoworakul, W., & Cummings, D. A. T. (2005). Containing pandemic influenza at the source. *Science*, *309*(5737), 1083. doi:10.1126/science.1115717

Maunder, R., Hunter, J., Vincent, L., Bennett, J., Peladeau, N., & Leszcz, M. (2003). The immediate psychological and occupational impact of the 2003 SARS outbreak in a teaching hospital. *Canadian Medical Association Journal, 168*(10), 1245.

Meltzer, M. I., Cox, N. J., & Fukuda, K. (1999). The economic impact of pandemic influenza in the united states: Priorities for intervention. *Emerging Infectious Diseases, 5*(5), 659. doi:10.3201/eid0505.990507

Miami International Airport. (2010). *Daily traffic volume data.* Retrieved from http://www.miami-airport.com

Miller, G., Randolph, S., & Patterson, J. E. (2008). Responding to simulated pandemic influenza in San Antonio, Texas. *Infection Control and Hospital Epidemiology, 29*(4), 320–326. doi:10.1086/529212

Mills, C. E., Robins, J. M., & Lipsitch, M. (2004). Transmissibility of 1918 pandemic influenza. *Nature, 432*(7019), 904–906. doi:10.1038/nature03063

Milne, G. J., Kelso, J. K., Kelly, H. A., Huband, S. T., & McVernon, J. (2008). A small community model for the transmission of infectious diseases: Comparison of school closure as an intervention in individual-based models of an influenza pandemic. *PLoS ONE, 3*(12), 4005. doi:10.1371/journal.pone.0004005

Models of Infectious Disease Agent Study (MIDAS). (2004). *Report from the models of infectious disease agent study (MIDAS) steering committee.* Retrieved from http://www.nigms.nih.gov/News/Reports/midas_steering_050404.htm

Montgomery, D. C. (2008). *Design and analysis of experiments.* New York, NY: John Wiley & Sons.

National Institute of General Medical Sciences. (2008). *Models of infectious disease agent study.* Retrieved from http://www.nigms.nih.gov/Initiatives/MIDAS/

Niederhauser, V. P., Baruffi, G., & Heck, R. (2001). Parental decision-making for the varicella vaccine. *Journal of Pediatric Health Care, 15*(5), 236–243.

Nigmatulina, K. R., & Larson, R. C. (2009). Living with influenza: Impacts of government imposed and voluntarily selected interventions. *European Journal of Operational Research, 195*(2), 613–627. doi:10.1016/j.ejor.2008.02.016

Pasteur, S. (2009). *Influenza A(H1N1) 2009 monovalent vaccine.* Retrieved from http://www.fda.gov/downloads/biologicsbloodvaccines/vaccines/approvedproducts/ucm182404.pfd

Patel, R., & Longini, I. M. (2005). Finding optimal vaccination strategies for pandemic influenza using genetic algorithms. *Journal of Theoretical Biology, 234*(2), 201–212. doi:10.1016/j.jtbi.2004.11.032

Pearson, M. L., Bridges, C. B., & Harper, S. A. (2006). Influenza vaccination of health-care personnel. Recommendations of the Healthcare Infection Control Practices Advisory Committee (HICPAC) and the Advisory Committee on Immunization Practices (ACIP). *Morbid and Mortality Weekly Report, 55*(2).

Pitzer, V. E., Leung, G. M., & Lipsitch, M. (2007). Estimating variability in the transmission of severe acute respiratory syndrome to household contacts in Hong Kong, China. *American Journal of Epidemiology, 166*(3), 355. doi:10.1093/aje/kwm082

Pitzer, V. E., Olsen, S. J., Bergstrom, C. T., Dowell, S. F., & Lipsitch, M. (2007). Little evidence for genetic susceptibility to influenza A (H5N1) from family clustering data. *Emerging Infectious Diseases, 13*(7), 1074.

Pourbohloul, B., Ahued, A., Davoudi, B., Meza, R., Meyers, L. A., & Skowronski, D. M. (2009). Initial human transmission dynamics of the pandemic (H1N1) 2009 virus in north America. *Influenza and Other Respiratory Viruses*, *3*(5), 215–222. doi:10.1111/j.1750-2659.2009.00100.x

Rhodes, S. D., & Hergenrather, K. C. (2002). Exploring hepatitis B vaccination acceptance among young men who have sex with men: Facilitators and barriers* 1. *Preventive Medicine*, *35*(2), 128–134. doi:10.1006/pmed.2002.1047

Robertson, E., Hershenfield, K., Grace, S. L., & Stewart, D. E. (2004). The psychosocial effects of being quarantined following exposure to SARS: A qualitative study of Toronto health care workers. *Canadian Journal of Psychiatry*, *49*, 403–407.

Rosenthal, S. L., Kottenhahn, R. K., Biro, F. M., & Succop, P. A. (1995). Hepatitis B vaccine acceptance among adolescents and their parents. *The Journal of Adolescent Health*, *17*(4), 248–254. doi:10.1016/1054-139X(95)00164-N

Sadique, M. Z., Edmunds, W. J., Smith, R. D., Meerding, W. J., De Zwart, O., & Brug, J. (2007). Precautionary behavior in response to perceived threat of pandemic influenza. *Emerging Infectious Diseases*, *13*(9), 1307.

Safranek, T. J., Lawrence, D. N., Kuriand, L. T., Culver, D. H., Wiederholt, W. C., & Hayner, N. S. (1991). Reassessment of the association between guillain-barré syndrome and receipt of swine influenza vaccine in 1976-1977: Results of a two-state study. *American Journal of Epidemiology*, *133*(9), 940.

Savachkin, A., Uribe-Sanchez, A., Das, T., Prieto, D., Santana, A., & Martinez, D. (2010). *Supplemental data and model parameter values for cross-regional simulation-based optimization testbed*. Retrieved from http://imse.eng.usf.edu/pandemic/supplement.pdf

Scharfstein, D. O., Halloran, M. E., Chu, H., & Daniels, M. J. (2006). On estimation of vaccine efficacy using validation samples with selection bias. *Biostatistics (Oxford, England)*, *7*(4), 615. doi:10.1093/biostatistics/kxj031

Schoenstadt, A. (2010). *Spanish flu*. Retrieved from http://flu.emedtv.com/spanish-flu/spanish-flu.html

Smailbegovic, M., Laing, G., & Bedford, H. (2003). Why do parents decide against immunization? The effect of health beliefs and health professionals. *Child: Care, Health and Development*, *29*(4), 303–311. doi:10.1046/j.1365-2214.2003.00347.x

Tallahassee Regional Airport. (2010). *Daily traffic volume data*. Retrieved from http://www.talgov.com/airport/index.cfm

Tampa International Airport. (2010). *Daily traffic volume data*. Retrieved from http://www.tampaairport.com

The New York Times. (2009). *Doctors swamped by swine flu vaccine fears*. Retrieved from http://www.msnbc.msn.com/id/33179695/ns/health-swine_flu/

The New Yorker. (2009). *The fear factor*. Retrieved from http://www.newyorker.com/talk/comment/2009/10/12/091012taco_talk_specter

Treanor, J. J., Campbell, J. D., Zangwill, K. M., Rowe, T., & Wolff, M. (2006). Safety and immunogenicity of an inactivated subvirion influenza A (H5N1) vaccine. *The New England Journal of Medicine*, *354*(13), 1343. doi:10.1056/NEJMoa055778

U.S Census Bureau. (2000). *2001 American community survey*. Retrieved from http://www.census.gov/prod/2001pubs/statab/sec01.pdf

U.S. Department of Health & Human Services. (2007). *HHS pandemic influenza plan*. Retrieved from http://www.hhs.gov/pandemicflu/plan/

World Health Organization. (2004). *WHO guidelines on the use of vaccine and antivirals during influenza pandemics*. Retrieved from http://www.who.int/csr/resources/publications/influenza/11_29_01_A.pdf

World Health Organization. (2009). *Pandemic influenza preparedness and response*. Retrieved from http://www.who.int/csr/disease/influenza/pipguidance2009/en/index.html

World Health Organization. (2009). *Pandemic (h1n1) 2009 vaccine deployment update - 17 December 2009*. Retrieved from http://www.who.int/csr/disease/swineflu/vaccines/h1n1_vaccination_deployment_update_20091217.pdf

World Health Organization. (2010a). *Cumulative number of confirmed human cases of avian InfluenzaA(H5N1) reported to WHO*. Retrieved from http://www.who.int/csr/disease/avian_influenza/country/cases_ table_2010_08_31/en/index.html

World Health Organization. (2010b). *Pandemic (H1N1) 2009 - update 112*. Retrieved from http://www.who.int/csr/don/2010_08_06/en/index.html

Wu, J. T., Riley, S., Fraser, C., & Leung, G. M. (2006). Reducing the impact of the next influenza pandemic using household-based public health interventions. *PLoS Medicine*, *3*(9), 361. doi:10.1371/journal.pmed.0030361

Wu, J. T., Riley, S., & Leung, G. M. (2007). Spatial considerations for the allocation of pre-pandemic influenza vaccination in the united states. *Proceedings. Biological Sciences*, *274*(1627), 2811. doi:10.1098/rspb.2007.0893

Yang, Y., Halloran, M. E., Sugimoto, J. D., & Longini, I. M. Jr. (2007). Detecting human-to-human transmission of avian influenza A (H5N1). *Emerging Infectious Diseases*, *13*(9), 1348.

Yang, Y., Sugimoto, J. D., Halloran, M. E., Basta, N. E., Chao, D. L., & Matrajt, L. (2009). The transmissibility and control of pandemic influenza A (H1N1) virus. *Science*, *326*(5953), 729. doi:10.1126/science.1177373

Yasuda, H., & Suzuki, K. (2009). Measures against transmission of pandemic H1N1 influenza in Japan in 2009: Simulation model. *Euro Surveillance: European Communicable Disease Bulletin*, *14*, 44.

This work was previously published in the International Journal of Artificial Life Research, Volume 2, Issue 2, edited by E. Stanley Lee and Ping-Teng Chang, pp. 19-41, copyright 2011 by IGI Publishing (an imprint of IGI Global).

Chapter 10
A Computational Model of Mitigating Disease Spread in Spatial Networks

Taehyong Kim
State University of New York at Buffalo, USA

Aidong Zhang
State University of New York at Buffalo, USA

Kang Li
State University of New York at Buffalo, USA

Surajit Sen
State University of New York at Buffalo, USA

Murali Ramanathan
State University of New York at Buffalo, USA

ABSTRACT

This study examines the problem of disease spreading and containment in spatial networks, where the computational model is capable of detecting disease progression to initiate processes mitigating infection spreads. This paper focuses on disease spread from a central point in a 1 x 1 unit square spatial network, and makes the model respond by trying to selectively decimate the network and thereby contain disease spread. Attention is directed on the kinematics of disease spreading with respect to how damage is controlled by the model. In addition, the authors analyze both the sensitivity of disease progression on various parameter settings and the correlation of parameters of the model. As the result, this study suggests that the radius of containment process is the most critical parameter and its best values with the computational model would be a great help to reduce damages from disease spread of a future pandemic. The study can be applied to controlling other virus spread problems in spatial networks such as disease spread in a geographical network and virus spread in a brain cell network.

1. INTRODUCTION

Networks are encountered in a wide variety of contexts. For example, computer networks, social networks, circuit networks, networks of neurons, and terrorist networks (Gupta & Kumar, 1998; Kurland & Pelled, 2000; Wetterling, 2001; Newman et al., 2002; Fortunato, 2005; Malarz et al., 2007). In this work we consider the generic problem of damage spreading of in two-dimensional fixed radius random networks. Fixed radius random networks are spatial networks in which

DOI: 10.4018/978-1-4666-3890-7.ch010

the range of connectivity of individual nodes is limited and this network model has been used for simulation of wireless communication networks and geographical networks.

The control strategies that we design in our model are potentially relevant in diverse contexts that span a range of spatial and temporal scales. These include culling during epidemics in farm animals, fire fighting to wildfires and social bullying in community networks. In biology, mechanisms such as programmed cell death and scar formation are activated to control the spread of pathological processes. Hence, the results presented here could be of relevance in many application areas because we address the problem of how a computational model can detect damage progression and control the spreading of the damage. We show that aggressive defense strategy is not always the best way for the network to control damages, but selectively balanced defense system with best values of parameters is a successful way for controlling spreads with minimum damages.

One of the most critical and intolerable threats for human health could be an unknown influenza pandemic in the near future. Efficient control of potential influenza pandemics would be an important strategy in minimizing their adverse economic and public health impacts. A study shows that stochastic epidemic model can be used to investigate the effectiveness of targeted antiviral prophylaxis, quarantine, and pre-vaccination in containing an emerging influenza strain at the source (Longini et al., 2005). Other studies show that simulations on epidemic models can predict a pattern of reduced and lagged epidemics post vaccination (Ferguson et al., 2005; Kim, 2008, 2009; Pitzer et al., 2009; Kim et al., 2010). Mathematical models also can help determine and quantify critical parameters and thresholds in the relationships of those parameters, even if the relationships are nonlinear and obscure to simple reasoning (Menach, 2006; Smith, 2006; Epstein, 2009). A contact pattern model on smallpox spread could be used to contain outbreaks by a strategy of targeted vaccination combined with early detection without resorting to mass vaccination of population (Eubank et al., 2004). In addition to avian influenza (H1N5), influenza A (H1N1) virus has spread rapidly across the world (Neumann, Noda et al. 2009; Smith, Vijaykrishna et al. 2009). Several papers analyze the virus spreading patterns and effective vaccination strategies for maximizing the H1N1 containment (Fraser et al., 2009; Munster et al., 2009).

In this paper, we focus on numerical analyses on experimentation results with a computational model of mitigating disease spread in spatial networks. This work is organized as follows. First we focus on describing the model system considered here and the details of our calculations. We present the results of this study next and briefly discuss the properties of fixed-radius random networks considered here. This is followed by details of calculations on the spatial and temporal features of our computational model of the disease spread process and containment process. We analyze both the effect of the disease spreading process and the effect of the containment process with model parameters. We show the spatial features of the model using a real world geological network as a possible application and conclude with summary and future works of the results.

2. THE SYSTEM MODEL

Many relational and geographical networks have similar properties of fixed-radius random networks. Particularly, communication network models and transportation network models comprise graphs through a set of vertices V embedded randomly in a two-dimensional plane (a geographic map) such that the edges E connecting these vertices exist if the distance between two vertices is less than some maximum range parameter. The set of edges and the resultant network connectivity are emergent properties of the graph that are determined by the locations and communication capability of the vertices. We will consider the connectivity properties of such random graphs for the purpose of evaluating the ability of the communication network to route messages between pairs of vertices. We will evaluate graphs with fixed vertices, as a model of disease spreading and containment.

The fixed-radius random network is modeled as follows. The networks are generated on a 1 x 1 unit square by seeding the square with n points with x and y coordinates independently drawn from a uniform random distribution, $U(x,y)$, where $0 \leq x \leq 1$ and $0 \leq y \leq 1$. Two nodes v_i and v_j are connected with an edge or bond if the distance $d(v_i, v_j)$ between them is δ or less (Gupta & Kumar, 1998; Liben-Nowell et al., 2005). In other words, the definition of a fixed-radius random network is formalized by an undirected space-sensitive graph in a 1 x 1 unit square; $G = \{(V,E) \mid V$ *is a set of nodes and E is a set of edges,* $E \subseteq V \times V$, *an edge* $e = (i, j)$ *connects two nodes i and j,* $i, j \in V$, $e \in E$, $d(i,j) \leq \delta\}$.

The network geometry of this model is theoretically analyzed with the definition of fixed-radius random network as follows (Dowell & Bruno, 2000; Krishnamachari et al., 2001; Bose & Morin, 2004; Liagkou et al., 2006; Pompilia et al., 2008). We denote nodes as N points which are randomly placed in a 1 x 1 unit area, A, with uniform probability per unit area. Each edge, e, is assigned a weight, $w(e)$, indicating its state of health. The weights, $w(e)$, are 0 or 1 which means dead or healthy status of an edge, respectively. The status of health on a node i at time t, $s(i, t)$, is defined via

$$s(i,t) = \frac{\sum_{k=1}^{n_i} w(e_k)}{n_i}, \quad (1)$$

where $w(e_k)$ is the weight of the edge, e_k, linked to the node i, and n_i is the total number of edges linked to the node i.

Thus, the fractional loss of health on a node i at time t, $u(i, t)$, can be calculated via

$$u(i,t) = 1 - s(i,t). \quad (2)$$

For simplicity, in all the simulations considered here, we assume $w(e) \in \{0, 1\}$. At $t = 0$, the start of the simulation, weights of all edges in the network are set to be 1, healthy status.

In our simulations, we suppose that the disease spread process starts in a localized region of radial distance $r_0 = 0.1$ around the center of the unit square at (0.5, 0.5). In each time-step, the model invokes three sequential steps of: i) infection by disease spread process, ii) alarm generation and propagation, and iii) culling or immunization by containment process.

During each unit time step of t, all the healthy edges of the network in a radius, $r(t)$, are targeted by the disease spread process and damaged with probability p_a. The edges damaged by the disease spread process have their weights set to zero which means "dead" status. The spatial spread of the disease spread process at time t, $r(t)$, is calculated by

$$r(t + \Delta t) = \alpha r_0 (1 + f_a(t)) r(t) \text{ for } t > 0. \quad (3)$$

In Equation (3), $r(t+\Delta t)$ and $r(t)$ are the radius of the circular regions of disease spread process at time-steps $t+\Delta t$ and t, respectively. The rate of change on $r(t)$ at each time step is linearly dependent on the fraction $f_a(t)$, which can be calculated by the number of damaged edges divided by the number of healthy edges in the circular region of radius $r(t)$ during time-step t. The term α is a positive constant controlling the pace of the change rate on $r(t)$. The $r(t)$ is a key variable that determines the spatial-temporal extent of edge damage.

Nodes damaged by the disease spread process generate alarm signals when their fractional loss of health are greater than the alarm generation threshold τ_a, i.e., when $u(i, t) > \tau_a$. The alarm signals, generated by node i, are propagated to all its nearest neighbors. The strength of an alarm signal arriving at a node *i* via an edge is equal to the weight of the edge, *w(e)*: healthy edges send one to the connected node; and damaged (dead) edges send zero to the connected node. The alarm signal *a(i, t)* remains at each node *i* for one time-step. The strength of alarm signals arriving at a node *i*, a*(i, t),* is defined as

$$a(i,t) = \frac{\sum_{k=1}^{m_i} w(e_k)}{n_i}, \tag{4}$$

in which n_i is the total number of edges linked to the node *i*, m_i is the total number of signaling edges linked to the node *i*, and $w(e_k)$ is the weight of the edge e_k.

A containment process is initiated at the nodes where the *a(i, t)* is greater than a containment activation threshold value τ_d, i.e., $a(i, t) > \tau_d$. There are two different containment process options: i) culling edges and ii) immunizing edges. If the culling edge option is selected, the containment process reduces the weight of an edge to zero when the shortest Euclidean distance from the node *i* with $a(i,t) > \tau_d$ to the edge is less than the radius of containment process region, ϕ. If the immunizing edge option is selected, the containment process sets the period of edge immunity to ω_e at the same condition of the culling edge option. The probability of culling and immunizing edges in ϕ by the containment process is set to $p_d = 1$, which means that every edge in the region of ϕ is culled or immunized. All alarm signal values in nodes are reset to zero at the end of each time step. In contrast to the disease spread process, nodes with edges damaged by the containment process do not generate alarm signals. Time is incremented upon completion of the containment process.

Before we start to analyze the simulation results on the following section, understanding the characteristics of networks is needed. The generation process of a fixed radius random network, the computational model algorithm and parameters used in this study, are presented in Figures 1A through 1C, Table 1, and Table 2, respectively.

3. THE RESULTS

3.1 Network Topology and Properties

Here we present numerical experiments of the properties of fixed radius random graphs. In particular, we focus on the distribution of number of edges directly linked to a node (node degree) and the distance between directly connected nodes. The average node degree is the number of edges per node and the average distance between directly

Figure 1. Figure A shows the generation process of a fixed-radius random network with radius of connectivity of δ. The process starts with creating nodes randomly located in a 1 x 1 unit square, then visits a node v creating edges linked to other nodes where the circular region of radius δ from a node v reaches until every node is visited. Figure B is schematic representation of the disease spreads process (grayed circular region) and containment processes (dashed circular regions). Figure C is the simplified flow chart for our disease spread process and containment process with parameters used in the model.

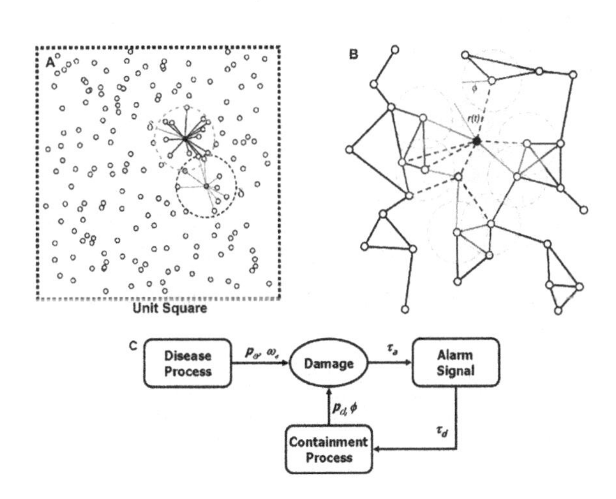

Table 1. Parameter descriptions and values of the model are shown

Parameter	Description	Value
τ_a	Threshold for alarm generation probability	0.2, 0.3, 0.4, 0.5
τ_d	Threshold for containment process activation probability	0.2, 0.3, 0.4, 0.5
ϕ	Radius of containment process region	0.02, 0.12, 0.22 unit
δ	Maximum length for node to node connection	0.05, 0.1 unit
t	Maximum time step for simulation	50 times
r_0	Initial radius for disease spread process	0.1 unit
p_a	Probability for disease spread process	0.33 (0.1~1.0)
p_d	Probability for containment culling process	1.0 (0.1~1.0)
ω_e	Period of time unit for edge immunity	0~10 times
α	Constant for disease spread process rate	0.8

connected nodes is a characteristic length scale among neighbor nodes for this network.

Figure 2A and Figure 2B show two representative fixed radius random graphs *G(1000,δ)*, where each graph contains 1000 points in the unit square and with values of $\delta = 0.05$ and 0.1, respectively. The number of edges in the network increases strongly for the higher value of the radius of connectivity, $\delta = 0.1$.

We examined the distributions of network properties of over 1000 network realizations and summarized the results in Figures 2C and 2D. The node degree (Figure 2C) is approximately Poisson in shape whereas distribution of edge lengths in accumulation is a quadratic both in 0.05 and 0.1 radius of connectivity (Figure 2D). Figure 2E shows that the average edge length in the network increases linearly with the increment of δ whereas the average node degree increases almost quadratically with an exponent of 1.96. By chang-

Table 2. The disease spread and containment process algorithm is described

Disease Spread and Containment Process Algorithm
1: *G*: a spatial network in a 1 x 1 unit square
2: damageEdgeCandidates: edge list in disease progression region *r(t)* at the center of network *G*.
3: While (loop < 1000)
4: While(timeStep ≤ *t*)
5: Reset *a(i, t)* and calculate *r(t)*
6: While(each edge, *e*)
7: If (ω_e <= 0)
8: Add edges into damageEdgeCandidates by selecting an edge, *e*, in disease progression region, *r(t)*
9: Set *w(e) = 0* on selected edge, *e*, from damageEdgeCandidates based on probability, p_a
10: End if
11: $\omega_e = \omega_e - 1$
12: End while
13: While(each node)
14: Calculate fractional loss of health on each node, *u(i, t)*
15: Initiate alarm signals to neighbor nodes from damaged nodes, if $u(i, t) > \tau_a$
16: Calculate strength of alarm signals arriving at a nodes, *a(i, t)*
17: Activate defense process by culling edges linked to nodes, if $a(i, t) > \tau_d$
18: End while
19: timeStep = timeStep + 1
20: End while
21: Calculate the number of damaged edge, culled edge and healthy edge.
22: loop = loop + 1
23: End while
24: Return the average number of damaged edges, culled edges and healthy edges.

Figure 2. Figures A and B show two representative fixed-radius random networks with 1000 nodes and radius of connectivity δ of 0.05 and 0.1, respectively. Figure C shows the distribution of node degree for networks with radius of connectivity of 0.05 and 0.1. Figure D is the normalized histogram of edge lengths. Figure E shows the average edge length (gray line, left axis) and the average node degree (black line, right axis) as a function of the radius of connectivity δ.

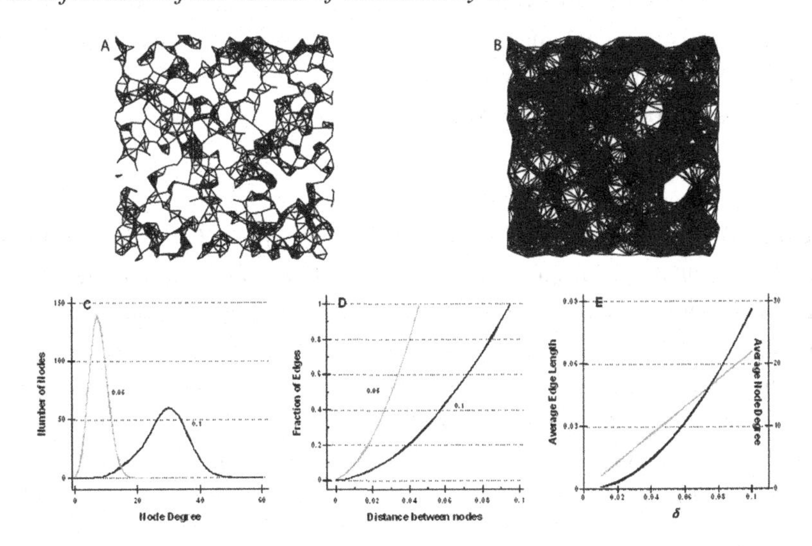

ing δ from 0.05 to 0.1, the number of edges is increased from 3,845 to 14,386 and the average node degree is increased from 7.52 to 28.77.

4. SPATIAL AND TEMPORAL FEATURES OF THE DAMAGE DYNAMICS

The total number of damages in the system is the sum of the edge damages caused by both the disease spread process and containment process mechanisms. Figure 3 shows a representative fixed radius random graph with *G(1000, 0.05)*, consisting of 1,000 nodes and 3,845 edges. The disease spread process is initiated in a localized region of radial distance $r_0 = 0.1$ around the center at (0.5, 0.5) and the value of p_a is set to 0.33. Then the model randomly generates a real number, z, between 0 ~ 1 with pseudo random number generator and it decides whether an edge is damaged or not based on z and p_a values. When $z \leq p_a$, an edge is damaged by the disease spreading process; otherwise, an edge stays in a healthy status. To highlight the spatial-temporal patterns and the underlying mechanisms dominating the dynamics, the damage caused by disease spread process is shown in red and that caused by the containment process is shown in blue in Figures 3A, 3D, and 3G.

Figures 3A through 3C show a case in which the disease spread process is relatively uncontrolled and disease progresses to the boundary of the network at $t = 50$. Figures 3D through 3F show a case where the containment process is activated but the damage from both the disease spread and containment process is comparable. Figures 3G through 3I show a case in which the disease spread process is controlled by the containment process. These figures demonstrate that the outcomes of damage progressions are sensitive to the details of the parameter set used in the analysis and also to the random variation embedded in the calculations.

Another feature, that is most apparent in Figures 3B, 3E, and 3H, is the two-peak and three-regime nature of the dynamics: there is a prominent sharp peak at the beginning of simulations, $t = 4$, in Figures 3E and 3H, representing damages by the containment process activation which shows that the containment process culls selected edges in advance to mitigate damages of disease spread process. The second peak shown in Figure 3E explains that additional edges are still damaged by the disease spread process after the initial containment process activation because the containment process is not strong enough to fully control disease spread. At $t = 40$, the cumulative fraction edge damages converge into 1.0, 0.87 and 0.36 shown in Figures 3C, 3F, and 3I, respectively.

Figure 3. The figures show the spatial and temporal dynamics for a G(1000, 0.05) network for three different values of ϕ. The top panel (Figures A-C) is for $\phi = 0.02$; the middle panel (Figures D-F) is for $\phi = 0.12$ and the lowest panel (Figures G-I) is for $\phi = 0.22$. The network status at time $t = 50$ is summarized in Figures A, D and G with green, red and blue denoting the healthy edges, edges damaged by the disease spread process and the containment process, respectively. The time course of edge damage is shown in Figures B, E and H with the black, gray and dashed lines, representing total damages, damages caused by disease spread process and damages caused by containment process, respectively. Figures C, F and I are cumulative fraction damage plots corresponding to Figures B, E and H, respectively.

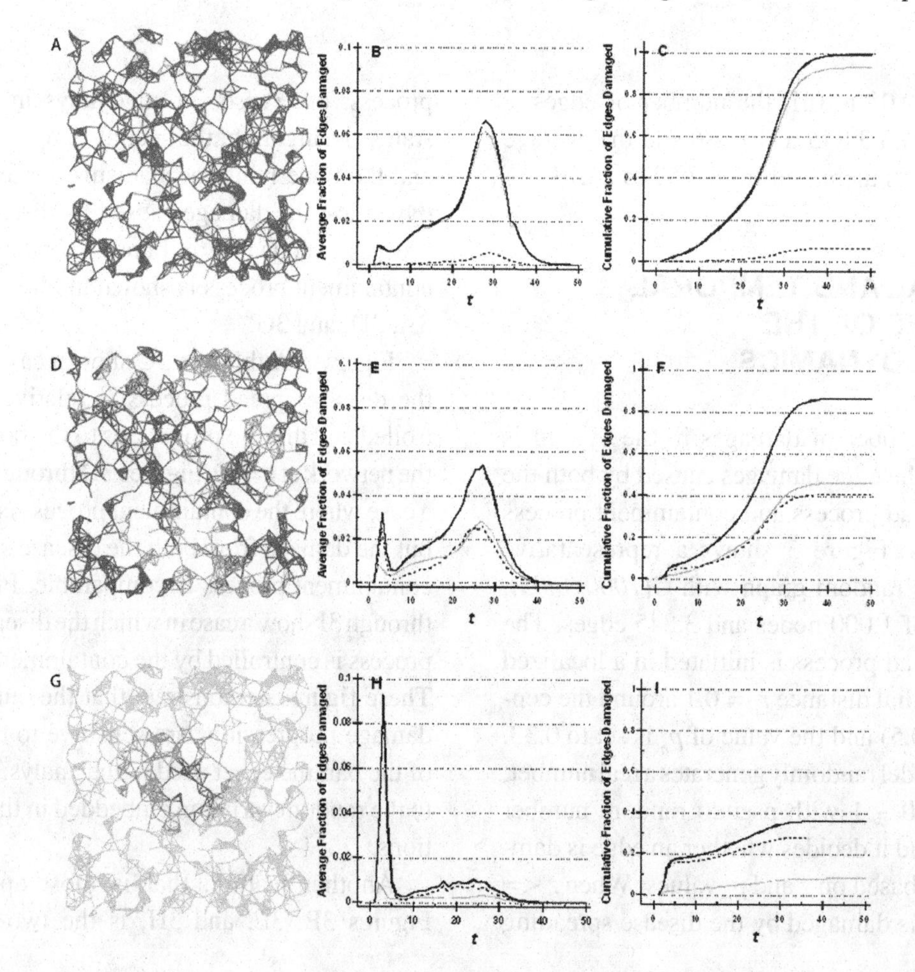

5. RESULT ANALYSIS ON MODEL PARAMETERS

Effects of Radius for Containment Process. The radius of containment process region ϕ is an important parameter that determines the overall damage in the model. Figure 4A shows the time profiles of damage on the same *G(1000, 0.05)* network containing 3,845 edges with low (ϕ = 0.02), intermediate (ϕ = 0.12) and high values of ϕ (ϕ = 0.22). The value of r_0 = 0.1 and alarm generation and containment process activation thresholds, τ_a, and τ_d, are each set to 0.5.

The time profiles have three regimes and the maximal damage fraction is reached in approximately t = 40 time units. In Figure 4B, the overall average fractional damages are minimized at ϕ = 0.24. For these parameter settings, approximately 32% of edges are damaged at the best value of ϕ, ϕ_{best}. At low values of ϕ, the disease spread process causes the majority of the damage whereas at high values of ϕ, the majority of the damage is the result of the containment process shown in Figures 3C and 3I, respectively. The damage vs. ϕ plot for the system in Figure 4B exhibits a characteristic shape with a prominent minimum.

Effects of Alarm Generation and Containment Activation Thresholds. In the next step, we assessed the effects of different alarm generation thresholds, τ_a, and containment process activation thresholds, τ_d, on network damage on the same *G(1000, 0.05)* network with r_0 = 0.1. We also analyzed the average fractional edge damages on 1000 simulations of our model.

Figure 5A is a plot of the total network damage at t = 50 time units vs. ϕ with alarm generation threshold τ_a as a parameter; the value of the containment process activation threshold τ_a is kept constant at 0.5. The overall damage decreases as alarm generation threshold τ_a decreases from 0.5 to 0.2. The best values of ϕ decrease sharply at decreased values of τ_a. At large and small values of ϕ, the overall damage is relatively independent of the values of τ_a. Figure 5B is the corresponding plot of damage at time t = 50 time units vs. ϕ with containment process activation threshold τ_d as a parameter; in this case, the value of the alarm generation threshold τ_a is constant at 0.5.

Figure 6 is a five-graph panel highlighting the relative contributions of the disease spread and containment process as a function of ϕ with values of τ_a and τ_d as parameters. The results show that as ϕ increases the relative contribution of the disease spread process decreases monotonically.

Figure 4. The figures highlight the effects of varying ϕ on the time course. Figure A shows the cumulative fraction of edges damaged as a function of time t. ϕ values of 0.02, 0.12, 0.22 in Figure A correspond to the gray, dashed and black lines, respectively. Figure B shows the average fraction of edges damaged at time t = 50 as a function of ϕ.

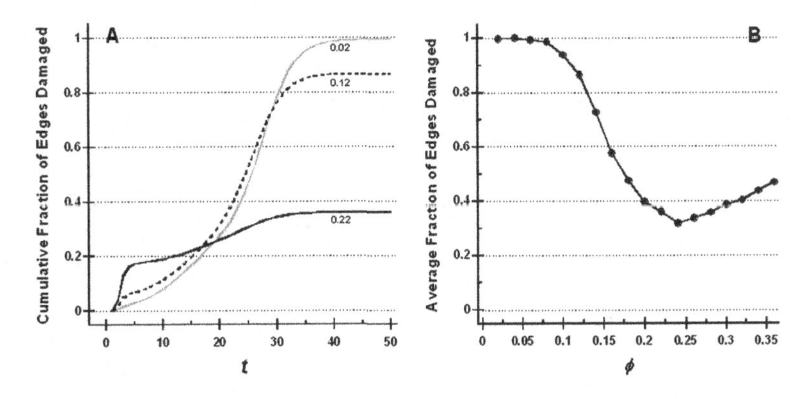

Figure 5. Figure A shows the effects varying ϕ on the time course on the fraction of edges damaged at time $t = 50$ with the threshold for alarm generation τ_a values of 0.2, 0.3, 0.4 and 0.5 (correspond to the mini-dashed, dashed, gray and black lines, respectively) with the threshold for containment process activation fixed at $\tau_d = 0.5$. Likewise, Figure B shows the effects of varying ϕ on the time course on the fraction of edges damaged at time $t = 50$ with the threshold for containment process activation τ_d values of 0.2, 0.3, 0.4 and 0.5 (correspond to the mini-dashed, dashed, gray and black lines, respectively) with the threshold for alarm generation fixed at $\tau_a = 0.5$.

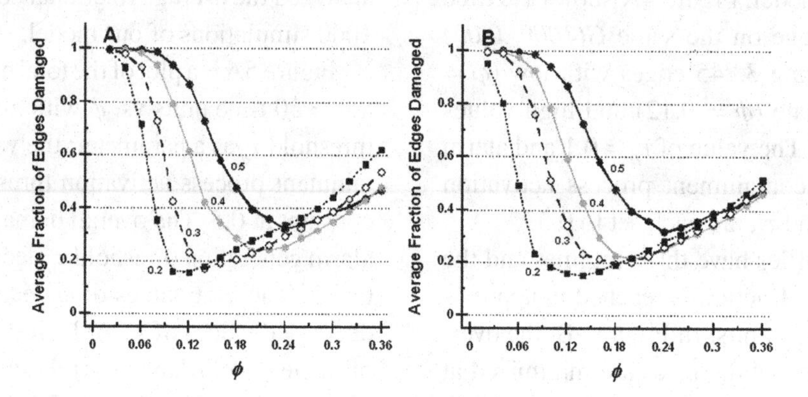

Figure 6. Figures A-E plot the total fraction of edges damaged (black line), damage caused by disease spread process (gray line) and the damage caused by containment process (dashed line) as a function of ϕ. The alarm generation threshold τ_a values for Figures A, B and C are fixed at 0.5 and the containment process activation thresholds are $\tau_d = 0.5$, 0.4 and 0.3 for Figures A, B and C, respectively. Likewise, containment process activation threshold τ_d values for Figures D and E are fixed at 0.5 and the alarm generation thresholds are $\tau_a = 0.4$ and 0.3 for Figures D and E, respectively.

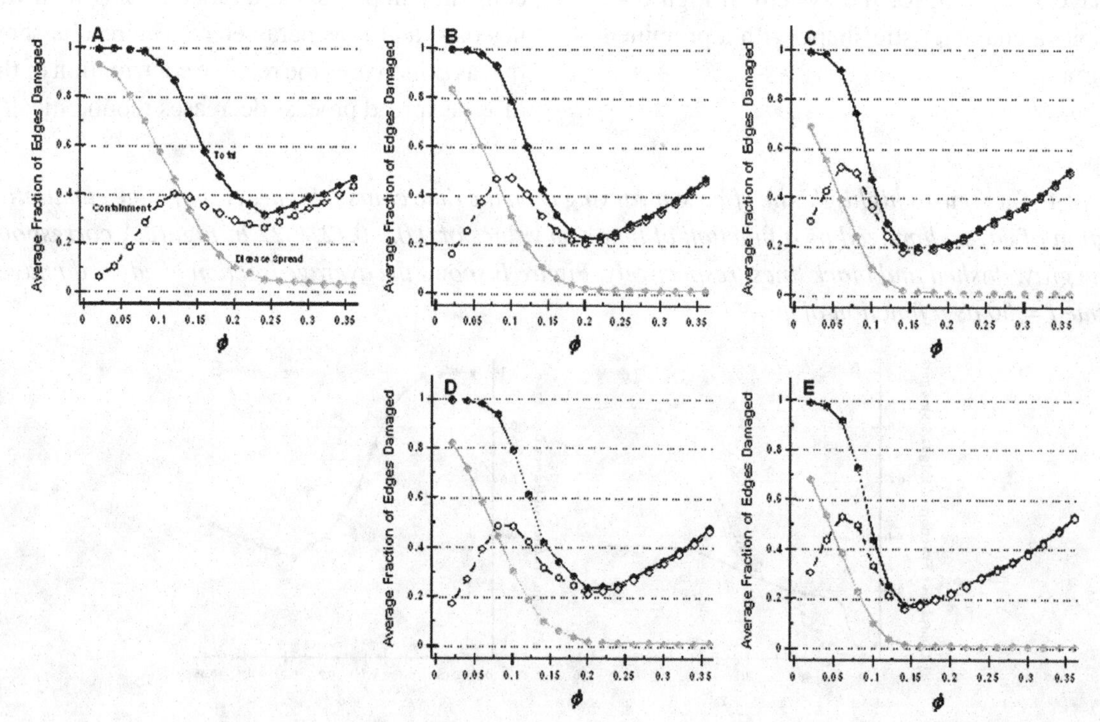

As expected, the disease spread process dominates the overall damage at low ϕ values. However, the dependence of overall fractional edge damages on ϕ has complex relations between disease spread process and containment process containing a minimum. As ϕ is increased from low values, the damage due to the containment process increases and offsets the benefits due to decreased damage from the disease spread process; the radial progression of the disease spread process, $r(t)$, is not limited. The damage contribution of the containment process is maximal at intermediate values of ϕ when the damage contributions of the disease spread process and the containment process are approximately equal. Further increment of ϕ leads to the decrement of overall damage because the radial progression of disease spread process, $r(t)$, is limited. This is responsible for the occurrence of a minimum. Increasing ϕ beyond the minimum fraction edge damage causes unnecessary culling damages by containment process. Thus, it is important to set the best value of ϕ, ϕ_{best}, to effectively mitigate disease spread.

Figure 7 shows the ϕ_{best} values for each choice of alarm generation and containment process activation parameter settings. The ϕ_{best} values (Figure 7A) decrease with decreasing values of τ_a because the network is more sensitive at generating alarms signals. The overall damage depends on the logistic function, $\frac{1}{1+e^{-c\tau}}$, where τ is τ_a or τ_d, and c is the coefficient value for the slope of the logistic function. The slope of the logistic curves is greater at lower values of τ_d because the defensive process is more sensitive to mobilizing. This indicates that sensitive responses to alarm signals reduce damage.

Figures 7A through 7D show the almost same behaviors on the ϕ_{best} and the number of damaged edges on the function of τ_a and τ_d. For the simplification of our model, we can combine two parameters, τ_a and τ_d, into one composite parameter standing for geometric mean of alarm generation threshold and defense activation threshold with $\sqrt{\tau_a \times \tau_d}$, τ_g. Figures 7E and 7F show the ϕ_{best} and the average fractional edges damaged as a function of τ_g respectively, which also depends on the logistic function.

Effects of Edge Immunity. In real world epidemics, it is very important to understand the effectiveness of different immunization strategies, such as, number of immunized population, and length of immunization period. In this respect, we studied effects on fractional edge damages of different length of edge immunity period in the model. We assume that an edge, e, is immunized to the disease spread process in a certain period of time, ω_e, by containment process. Instead of culling the edges, containment process immunizes edges when the distance to the edges from the node i with $a(i,t) > \tau_d$ is less than the radius of ϕ. If an edge, e, is immune, the immunity of an edge can last for the period of ω_e, and the edge cannot be damaged by the disease spread process. Since the immunity of edge decays by one after one time step, an edge becomes vulnerable again to the disease spread process after ω_e time step. While an edge is in the immunity status, it does not get immune again by the defensive process until it loses the immunity.

Figure 8A shows the time profiles of cumulative edge damages with $\omega_e = 1, 3, 5, 7$, and 9. Alarm generation and containment process activation thresholds, τ_a, and τ_d, are set to 0.5 and ϕ is set to 0.22. Other parameters are set as shown in Table 1. When $\omega_e = 1$, the cumulative edge damages is flatten at around $t = 50$. The time of saturation on cumulative edge damages is faster as ω_e increases. In Figure 8B, the average fractional healthy, immune, and damaged edges are shown as a function of ω_e. To study effects on different periods of edge immunity in this system, average fractional damaged edges as a function of ω_e are studied with the various values of ω_e from 1 to 10. As expected, the number of fractional edge damages are decreased as ω_e increased; however,

Figure 7. Figures A and B plot ϕ_{best} and average fraction of edges damaged at ϕ_{best} as a function of the alarm generation threshold τ_a. The containment process activation threshold values are $\tau_d = 0.7$ (dashed line), 0.5 (gray line) or 0.3 (black line). Figures C and D plot ϕ_{best} and average fraction of edges damaged at ϕ_{best} as a function of the containment process activation threshold τ_d. The alarm generation threshold values are $\tau_a = 0.7$ (dashed line), 0.5 (gray line) or 0.3 (black line). Figures E and F plot ϕ_{best} and fraction of edges damaged at ϕ_{best} as a function of geometric mean of τ_a and τ_a which is τ_g.

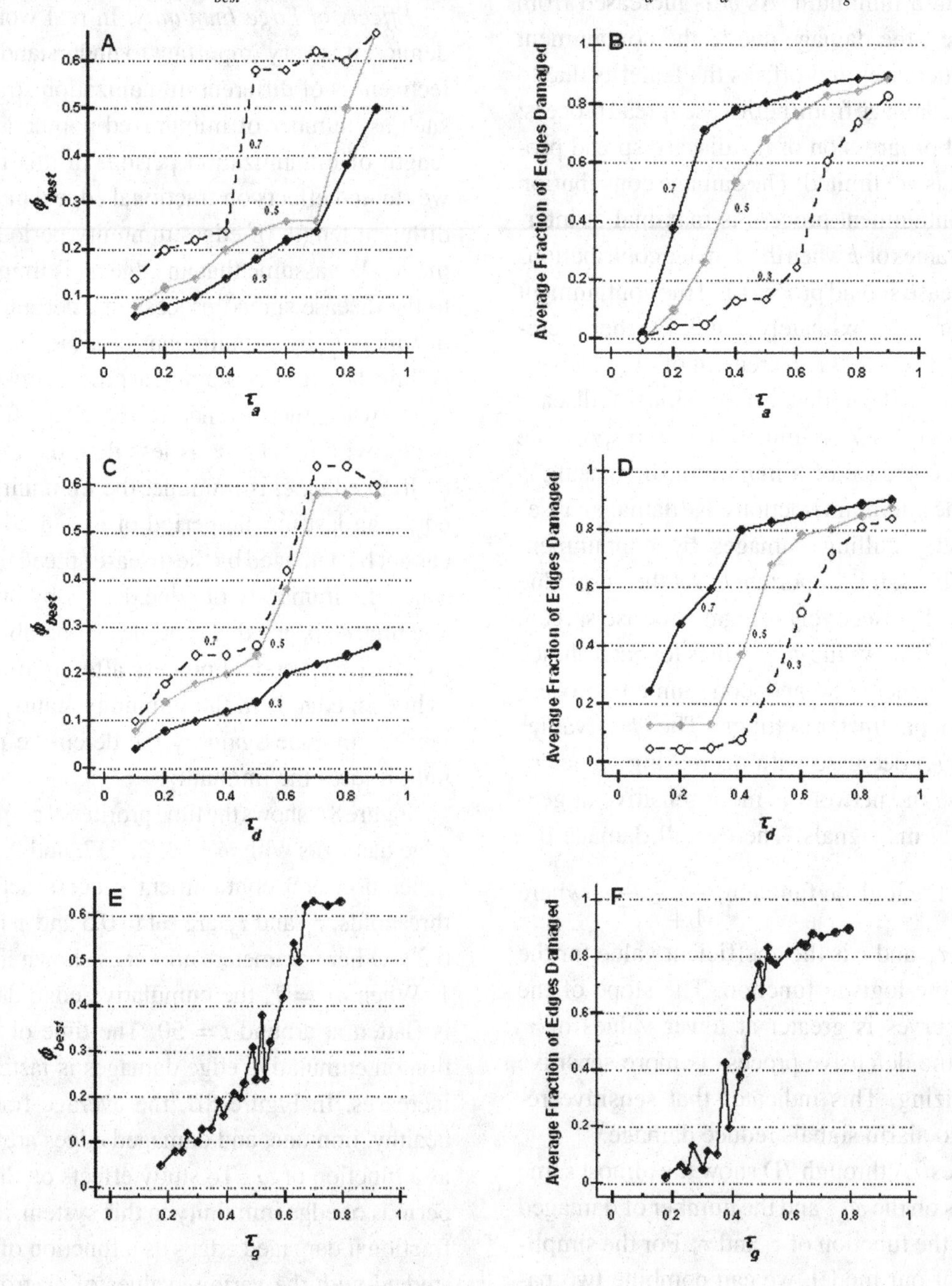

Figure 8. Figure A highlights the effects of different ω_e as a function of time t with ω_e = 1 (black solid line), 3 (gray solid line), 5 (black small dotted line), 7 (black large dotted line), and 9 (gray small dotted line). Figure B shows the average fraction of healthy (gray solid line), immune (black dotted line), and damaged (black solid line) edges at time t = 80 as a function of ω_e.

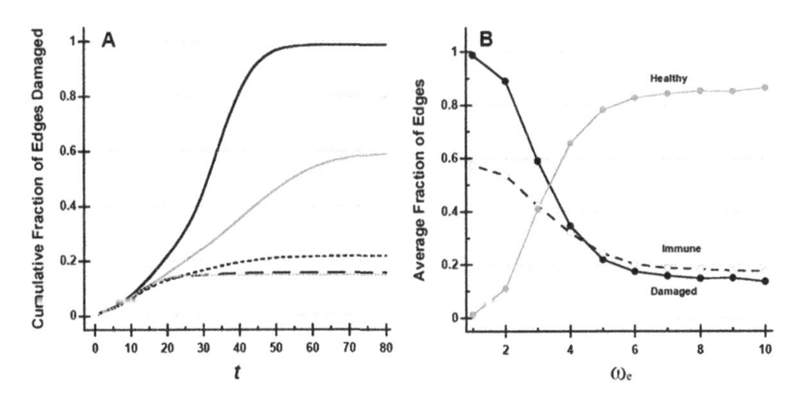

the number of fractional edge damages are no longer decreased after a certain value of ω_e = 6. This can explain that 6 time periods of edge immunity are good enough for retracting the disease spread process by immunizing resources to the disease spread. Thus, the period of edge immunity, ω_e, is needed to be set at least 6 in this system to have the effective protection from the disease spread process.

Sensitivity of Damage on Other Parameters. Figure 9 shows ϕ_{best} and average fractional edge damaged as a function of initial radius for disease spread process region r_0. We assessed the dependence of the extent of damage and the ϕ_{best} values required for minimizing damage on the value of r_0. The ϕ_{best} values require linear increments with increasing r_0 as shown in Figure 9A. The damage by the disease spread process increases only modestly with increased r_0, whereas the overall damage is disproportionately increased with increasing r_0 because the contributions of the containment process increase nonlinearly.

The results for p_a and p_d variation are summarized in Figure 10A and 10B, respectively. Figure 10A shows that the damage by the containment process effectively buffers overall damage until moderately high values p_a. At the highest values of p_a, however, the damage by the disease spread process progresses rapidly. The overall damage fraction decreases monotonically with increase in the value of p_d. However, the contribution on overall damage of the containment process reaches a maximum at intermediate values of p_d because the control of the damage by the disease spread process is delayed.

6. SIMULATION ON A REAL-WORLD NETWORK

To see the effects of the parameter ϕ on a real-world data, the road network at the city of Oldenburg, Germany, is incorporated instead of our synthetic spatial network, the fixed-radius random network (Brinkhoff, 2002). The road network consists of 6105 nodes and 7029 edges with the average node degree, 2.30, and the maximum and minimum node degree, 5 and 1, respectively. The scale of the road network is adjusted to fit into a 1 x 1 unit square.

Similar to the simulation on fixed-radius random networks, we assume that the disease spread process is initiated in a localized region of radial distance $r_0 = 0.1$ around the center at (0.5, 0.5) as an example of a possible biological attack at the center of the city. Other parameter values are set as the same parameter values of the simulation on fixed-radius random networks except $\tau_a = 0.4$,

Figure 9. Figures A and B plot ϕ_{best} and average fraction of edges damaged at ϕ_{best} as a function of the initial radius of the disease spread process region r_0. In Figure B, the gray line, the dashed line, and the black line are for fractional edge damages by disease spread process, containment process, and both processes, respectively.

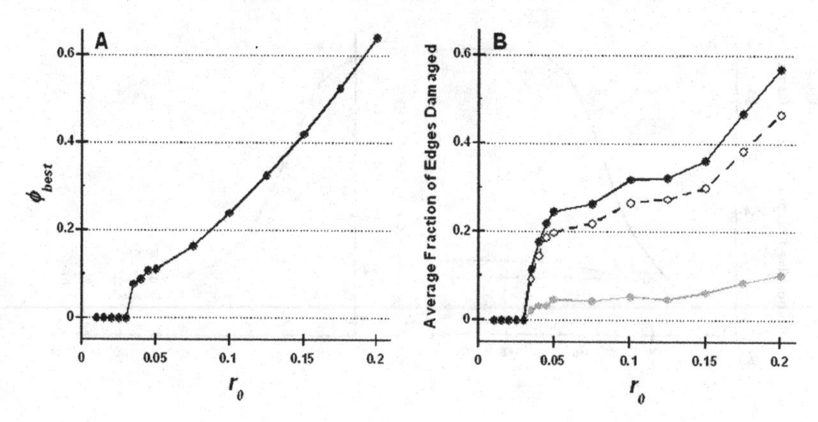

and $\tau_d = 0.4$. Figure 11 shows the spatial patterns of the road network to highlight the underlying mechanisms dominating the dynamics.

Figure 11A shows a case with $\phi = 0.12$ in which the disease spread process is relatively uncontrolled and progresses to the boundary of the network at $t = 50$. Figure 11B shows a case with $\phi = 0.22$ in which the containment process is activated but the damage from both the disease spread and containment process is comparable. Figure 11C shows a case with $\phi = 0.32$ in which the disease spread process is controlled by the containment process. Like Figures 3A, 3D, and 3G, these figures show similar spatial and temporal patterns of the disease spread. With $\phi = 0.32$, the disease spread process is successfully controlled with the average fraction of edges damaged at $t = 50$, 0.47 on 1000 simulations of our model. However, other two cases with $\phi = 0.22$ and $\phi = 0.12$, the disease spread process is

Figure 10. Figures A and B show average fraction of edges damaged as a function of the damage probability p_a on the disease spread process and the damage probability p_d on the containment process at the ϕ value of 0.26, respectively. In Figure A, the value of p_d is fixed at 1 whereas in Figure B, the value of p_a is fixed at 0.33. The gray line, the dashed line, and the black line are for fractional edge damages by disease spread process, containment process, and both processes, respectively.

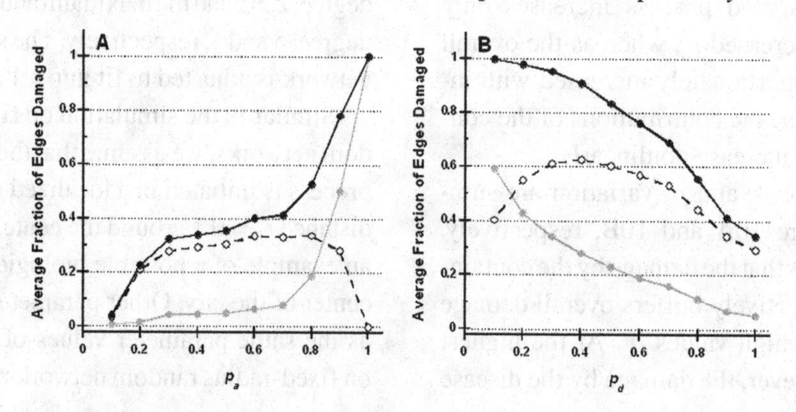

Figure 11. Figures show the spatial and temporal dynamics for the road network at the city of Oldenburg for three different values of ϕ. Figures A, B and C are a representation of results for $\phi = 0.12$, $\phi = 0.22$ and $\phi = 0.32$, respectively. The network status at time $t = 50$ is summarized in green, red and blue denoting the healthy edges, edges damaged by the disease spread process and the containment process, respectively. All parameter values are set as the same values we used in the fixed-radius random network except $\tau_a = 0.4$, and $\tau_d = 0.4$.

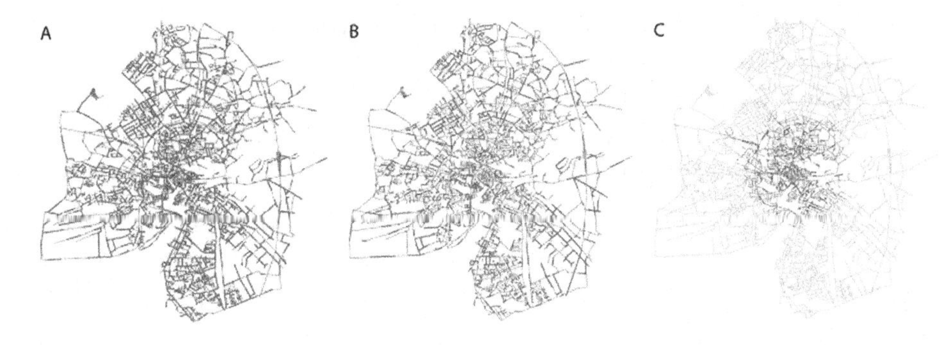

not controlled resulting the average fraction of edges damaged at $t = 50$, 0.80 and 0.99, respectively.

7. SUMMARY AND FUTURE WORK

In this paper, we have evaluated the damage spreads process and containment process in spatial networks with various parameter settings. In the model, we incorporated features of a disease spread process, alarm generation, containment process activation, culling process, and immunizing process to control disease spreads.

We explained the sensitivity of the model to each of the parameters and we found that the kinematics of the spatial-temporal patterns is particularly sensitive to the radius of the containment process region, ϕ. At low values of ϕ, the overall magnitude and extent of damage become global and are dominated by disease spread process. The overall damage is minimized at ϕ_{best} value when the culling of edges by the containment process is capable of preventing progression of the disease spread process without unnecessary culling damages.

We also showed that the overall edge damages depend on the logistic curve as a function of the alarm generation threshold and the containment process activation threshold parameters, τ_a and τ_d, respectively. As expected, this indicates that sensitive responses to alarm signals and the containment process activation threshold reduce damage significantly. Thus immediate action against the disease spread is critical to mitigate disease spread at the early stage of a possible pandemic.

With this model, we can analyze results of containment plans on different environmental disease spread conditions with various parameter settings. For example, strengths of disease progressions can be simulated by parameter controls of r_0 and p_a, and strategies for containment plans can also be simulated by parameter controls of τ_a, τ_d, ϕ, ω_e and p_d. However, just putting a simple aggressive sensitivity on parameters of the containment process is unlikely to be realistic as well as efficient in terms of predicting a disease spread progression. Indeed, finding critical parameters and best values with the computational model would be a useful mean to reduce damages from disease spreads of a future pandemic.

For the next step of this study, we will extend the model by adding more complex properties for in-depth understanding of real world disease spreads. Specifically, the different levels of edge weight will be incorporated to reflect complex health status. In the current model, an edge weight can only be zero or one representing for a dead and a healthy edge, which limits to represent possible different status of health. In addition, parameters, such as the number of population, the density of population, and geographical characteristics, will be considered in the next model.

Parameters and values we have used in this study, such as alarm generation threshold, τ_a, defense activation threshold value, τ_d, and radius of defense process region, ϕ, would not be reliably selected. To reduce these uncertainties, it critically depends on reliable information of high-quality epidemiological data as well as properties of geographical networks. Such increases in our understanding will extend both our model to accurately predict disease spreads by epidemics and our ability to plan an effective containment strategy. Regardless of uncertainties, well-preparedness with this type of studies would be a great help in effective planning and modeling to mitigate disease spreads.

This paper is presently limited to the study of mitigating disease spreads in two-dimensional fixed-radius random networks; however, this is seen as an important first step towards the analyses of various disease propagations and containments as a useful tool for understanding a new disease progression and for modeling a containment plan. As a long term impact, our model has the potential for substantial health care system with the prevention of future problems and multi-billion dollars of health care savings. In conclusion, a feasible strategy on mitigating disease spreads with a computational model could offer potential to reduce disease transfer, minimize economic impact on industry, and prevent possible threats on the human population in the near future.

ACKNOWLEDGMENT

This work was supported by grants from National Science Foundation.

REFERENCES

Bose, P., & Morin, P. (2004). Competitive online routing in geometric graphs. *Online Algorithms in Memoriam . Steve Seiden*, *324*(2-3), 273–288.

Brinkhoff, T. (2002). A framework for generating network-based moving objects. *GeoInformatica*, *6*(2), 153–180. doi:10.1023/A:1015231126594

Dowell, L. J., & Bruno, M. L. (2000). Connectivity of random graphs and mobile networks: Validation of Monte Carlo simulation results. In *Proceedings of the ACM Symposium on Applied Computing* (pp. 77-81).

Epstein, J. M. (2009). Modeling to contain pandemics. *Nature*, 406.

Eubank, S., Guclu, H., Kumar, A. V. S., Marathe, M., Srinivasan, A., & Toroczkai, Z. (2004). Modeling disease outbreaks in realistic urban social networks. *Nature*, *429*, 180–184. doi:10.1038/nature02541

Ferguson, N. M., Cummings, D. A., Cauchemez, S., Fraser, C., Riley, S., & Meeyai, A. (2005). Strategies for containing an emerging influenza pandemic in Southeast Asia. *Nature*, *437*(7056), 209–214. doi:10.1038/nature04017

Fortunato, S. (2005). Damage spreading and opinion dynamics on scale-free networks. *Physica A*, *348*, 683–690. doi:10.1016/j.physa.2004.09.007

Fraser, C., Donnelly, C. A., Cauchemez, C., Hanage, W. P., Van Kerkhove, M. D., & Hollingsworth, T. D. (2009). Pandemic potential of a strain of influenza A (H1N1): Early findings. *Science*, *324*(5934), 1557–1561. doi:10.1126/science.1176062

Gupta, P., & Kumar, P. R. (1998). Critical power for asymptotic connectivity in wireless networks. In McEneany, W. M., Yin, G. G., & Zhang, Q. (Eds.), *Stochastic analysis, control, optimization and applications* (pp. 547–566). Boston, MA: Birkhausen.

Kim, T., Hwang, W., Zhang, A., Ramanathan, M., & Sen, S. (2009). Damage isolation via strategic self-destruction: A case study in 2d random networks. *Europhysics Letters*, *86*(2), 24002. doi:10.1209/0295-5075/86/24002

Kim, T., Hwang, W., Zhang, A., Sen, S., & Ramanathan, M. (2008). Multi-agent model analysis of the containment strategy for avian influenza (AI) in South Korea. In *Proceedings of the IEEE International Conference on Bioinformatics and Biomedicine* (pp. 353-356).

Kim, T., Hwang, W., Zhang, A., Sen, S., & Ramanathan, M. (2010). Multi-agent modeling of the South Korean avian influenza epidemic. *BMC Infectious Diseases*, *10*(236).

Krishnamachari, B., Wicker, S. B., & Bajar, R. (2001). Phase transition phenomena in wireless ad-hoc networks. In *Proceedings of the Global Telecommunications Conference*.

Kurland, N. B., & Pelled, L. H. (2000). Passing the word: Toward a model of gossip and power in the workplace. *Academy of Management Review*, *25*(2), 428–438.

Liagkou, V., Makri, E., Spirakis, P., & Stamatiou, Y. C. (2006). The threshold behaviour of the fixed radius random graph model and applications to the key management problem of sensor networks. In S. E. Nikoletseas & J. D. P. Rolim (Eds.), *Proceedings of the Second International Workshop on Algorithmic Aspects of Wireless Sensor Networks* (LNCS 4240, pp. 130-139).

Liben-Nowell, D., Novak, J., Kumar, R., Raghavan, P., & Tomkins, A. (2005). Geographic routing in social networks. *Proceedings of the National Academy of Sciences of the United States of America*, *102*, 11623–11628. doi:10.1073/pnas.0503018102

Longini, I. M. Jr, Nizam, A., Xu, S., Ungchusak, K., Hanshaoworakul, W., & Cummings, D. A. (2005). Containing pandemic influenza at the source. *Science*, *309*, 1083–1087. doi:10.1126/science.1115717

Malarz, K., Szvetelszky, Z., Szekfu, B., & Kulakowski, K. (2007). *Gossip in random networks.* Retrieved from http://arxiv.org/pdf/physics/0601158.pdf

Menach, A. L. (2006). Key strategies for reducing spread of avian influenza among commercial poultry holdings: lessons for transmission to humans. *Proceedings. Biological Sciences*, *273*(1600), 2467–2475. doi:10.1098/rspb.2006.3609

Munster, V. J., de Wit, E., van den Brand, J. M. A., Herfst, S., Schrauwen, E. J. A., & Bestebroer, T. M. (2009). Pathogenesis and transmission of swine-origin 2009 A(H1N1) influenza virus in ferrets. *Science*, *325*(5939), 481–483.

Neumann, G., Noda, T., & Kawaoka, Y. (2009). Emergence and pandemic potential of swine-origin H1N1 influenza virus. *Nature*, *459*(7249), 931–939. doi:10.1038/nature08157

Newman, M. E. J., Forrest, S., & Balthrop, J. (2002). Email networks and the spread of computer viruses. *Physical Review E: Statistical, Nonlinear, and Soft Matter Physics*, *66*, 035101. doi:10.1103/PhysRevE.66.035101

Pitzer, V. E., Viboud, C., Simonsen, L., Steiner, C., Panozzo, C. A., & Alonso, W. J. (2009). Demographic variability, vaccination, and the spatiotemporal dynamics of rotavirus epidemics. *Science*, *325*(5938), 290–294. doi:10.1126/science.1172330

Pompilia, D., Scoglio, C., & Lopez, L. (2008). Multicast algorithms in service overlay networks. *Computer Communications*, *31*(3).

Smith, D. J. (2006). Predictability and preparedness in influenza control. *Science*, *312*(5772), 392–394. doi:10.1126/science.1122665

Smith, G. J., Vijaykrishna, D., Bahl, J., Lycett, S. J., Worobey, M., & Pybus, O. G. (2009). Origins and evolutionary genomics of the 2009 swine-origin H1N1 influenza A epidemic. *Nature*, *459*(7250), 1122–1125. doi:10.1038/nature08182

Wetterling, F. L. (2001). The Internet and the spy business. *International Journal of Intelligence and CounterIntelligence*, *14*(3), 342–365. doi:10.1080/08850600152386846

This work was previously published in the International Journal of Artificial Life Research, Volume 2, Issue 2, edited by E. Stanley Lee and Ping-Teng Chang, pp. 77-94, copyright 2011 by IGI Publishing (an imprint of IGI Global).

Chapter 11
Simulating the Spread of an Epidemic in a Small Rural Kansas Town

Todd Easton
Kansas State University, USA

Joseph Anderson
U.S. Army, USA

Kyle Carlyle
J. B. Hunt Transportation, USA

Matthew James
Kansas State University, USA

ABSTRACT

Substantial research has been dedicated to simulating the spread of infectious diseases. These simulation models have focused on major urban centers. Rural people have drastically different interaction and travel patterns than urban people. This paper describes a generic simulation package that can simulate the spread of an epidemic on a small rural town. This simulation package is then used to test the effectiveness of various mitigation strategies.

1. INTRODUCTION

Epidemics have played and will play a critical role in human history. Advancing the understanding of how epidemics spread has numerous benefits including: improving healthcare systems, increasing life spans, and reducing the impact of biological warfare. Here, an epidemic is defined as an outbreak of a disease that spreads rapidly and widely.

Various epidemics have hampered society for centuries including such famous cases as the Bubonic Plague, Avian Flu, and SARS. Of particular interest to this paper is the 1918 Spanish Flu, because it began in rural Kansas near the location of this research. This virus spread to all corners of the earth and is estimated to have killed 50 million people (Taubenberger & Morens, 2006), which is more than double the death toll of World War I.

DOI: 10.4018/978-1-4666-3890-7.ch011

Viruses and bacteria continue to evolve and grow stronger, providing new challenges for mankind. In 2007, a strain of evolved drug resistant staph infection spread to 94,000 people throughout the United States and killed 19,000 (Manier, 2007). It is clearly imperative for governmental agencies to understand how diseases spread and, more importantly, how to limit their spread.

After an outbreak occurs, the government is responsible for creating and enforcing a mitigation strategy. These mitigation strategies have immense social, ethical and economic impacts. Some previously used mitigation strategies include closing schools, establishing quarantines, limiting travel, radio and TV announcements, and encouraging healthy habits (washing hands, wearing masks, avoiding unnecessary contact, etc.) (CDC, 2010).

For years, epidemics have been modeled mathematically as a way to safely understand them (Gordis & Saunders, 2000; Rothman, 2002). Mathematically analyzing an epidemic typically requires the modeling of complex biological systems such as viruses, bacteria, parasites, and hosts. These models must mimic how the epidemic affects the individual with respect to age, sex, and recovery rate. In addition, it necessitates an understanding of how epidemics spread, which includes, how the disease is transmitted, what travel patterns that the subjects exhibit, and the geographical layout of the area.

Mathematical models of epidemics can be largely divided into two classes: host and spread. The host class models (Kumar, 2002; Olsson, 1996) focus on the effect of the disease on the individual, while the spread class models predict how diseases proceed through a group of individuals. The focus of this paper is on the spread of infectious diseases and the impact of mitigation strategies. Clearly, a fundamental understanding of how a disease spreads from host to host must be understood to provide an accurate dispersion model.

The most typical mathematical model for the spread of a disease is a state model. This model assumes that time is divided into periods and each individual is classified in a particular state for an entire period. The most basic state model is the SIR model, where each individual is classified in one of three states: susceptible, infectious, and recovered (Fuks et al., 2006; Newman, 2002; Ogren & Martin, 2000; Piqueira, Navarro, & Monteiro, 2005). From this model, researchers have branched out by adding new and more complex states, such as exposed and quarantined states (Cristea, Zaharia, Deutsch, Bunescu, & Blujdescu, 1992; Kolesin, 2007; Neal, 2008; Nuno, Castillo-Chaves, Feng, & Marcheva, 2008; Rosenberg, 1997; Yu, Luo, Gao, & Ai, 2006).

Many of these mathematical models have been used to derive interesting theoretical results, such as how fast a disease dies out. To determine these results, a typical assumption is that every individual is in direct contact with all other individuals. However, this underlying assumption is unrealistic in the real world; yet, with more realistic assumptions on how individuals interact, deriving nice theoretical results becomes challenging. Thus, many researchers have turned to simulation (Britton, Janson, & Martin-Lof, 2007; Huang, Hsieh, Sun, & Cheng, 2006; Guimaraes et al., 1979; Rvachev, 1968; Barrett, Eubank, & Smith, 2005).

This paper describes a simulation tool to analyze the spread of infectious diseases in a small rural town. This simulation models Clay Center, Kansas (Population 4,600) and is used to test various mitigation strategies. Computational results demonstrate that a rapid reaction from the government is vital to the containment of a disease and that the government should focus efforts on eliminating both contacts between individuals and the rate that an individual passes the disease to another individual.

2. SIMULATING THE SPREAD OF A DISEASE USING CONTACT NETWORKS

When modeling an epidemic a contact network is frequently used. A contact network models the chances that an individual can infect another individual. That is, given a set of n people, $N = \{1,\ldots,n\}$ and an n x n probability matrix P where p_{ij} equals the probability that infectious person i spreads the disease to an uninfected person j. The contact network is constructed as follows. Let $G_C = (V_C, A_C)$ be the contact network where $v_i \in V_C$ for $i=1,\ldots,n$ and $(v_i, v_j) \in A_C$ with a weight of p_{ij}. For simplicity, if $p_{ij}=0$, then the arc is not considered in A_C.

Accurate contact networks are difficult to generate, because estimating the probability of an infected person spreading the disease to each other person in a specified region is nearly impossible to quantify. One would need to know the type of disease and the habits and travel patterns of each individual in the area.

Once a contact network is generated, it can be used as input for a simulation and optimization software (Anderson, 2009; Carlyle, 2009). Again, the biggest drawback to this approach is generating the necessary assumptions so that the contact network is accurate. One of the primary research goals was to have a versatile simulation core that could be rapidly modified to fit any disease or scenario.

2.1. Contact Network, Disease States and Tracks

The input to the simulation core is a contact network, a set of infected nodes and how a disease impacts the individuals. This impact upon an individual follows the classical disease state space model. Thus, the simulation uses discrete time intervals. For example, in the simulation below a person will remain exposed for a certain number of periods and then transition into another state. This time period can be adjusted to represent any time unit such as minutes, hours or days.

In a given period, an *infectious* subject has the possibility to spread the disease to individuals in contact with the subject. To simulate this, every node in the graph that is in an *infectious* state generates a random number for every *susceptible* node that has an arc from the *infectious* node. This number is then compared to the number from the contact network associated with the arc between these nodes. If the random number is less than the number on the arc, then the disease spreads to the *susceptible* node. If not, the *susceptible* person remains *susceptible*, but may still become infected during this period from another individual.

This simulation can easily be adapted to any disease due to the idea of a disease track model. A disease track is comprised of multiple different states. For instance, in a given disease an older person may follow a *SID* path, where *D* stands for *dead*, while a teenager could follow a *SIR* path where *R* represents *recovered.*

This simulation allows a specific disease to follow multiple disease tracks based upon different probabilities. For instance with typhoid fever (Typhoid Fever (2007)), a subject could become a *carrier* (e.g. Typhoid Mary) while others could die or recover from this disease. If a subject becomes a carrier 10% of the time for the disease being modeled, then a uniform random number is generated when a node contracts the disease and this random number is compared to .1 to determine whether or not the subject follows that track.

The time spent in each state can also be assigned randomly. As in real life, one person can be infectious with a cold for 3 periods while another is only infectious for 1 period. This randomness is captured in the program and can be adjusted to any discrete probability distribution that models a specific number of periods in a disease state.

2.2. Disease State Transition Example

Figure 1 is an example of a small contact network that is used to describe the simulation core. For this example, assume there are two disease tracks that have an occurrence probability of .4 and .6, respectively, with the following states: *susceptible*, *infectious*, *susceptible*, and *susceptible*, *infectious*, *dead*. In Figure 1 *susceptible* is represented by green, *infectious* is represented by purple, and *dead* is represented by black. Also, subjects can spend anywhere from 1 to 3 time periods in each state that is not either *susceptible* or *dead*. At time period 0, assume subject *A* starts out in the *SID* track and in the *infectious* state. A random number for the time in this state is generated between 1 and 3 and happens to be 2. Therefore, subject *A* transitions into the *dead* state at the end of 2 time periods.

Also during iteration period 0, a uniform random number between 0 and 1 is generated for all contacts from an *infectious* node, *A*, to a *susceptible* node, *B*. This number is compared to the arc weight. If the random number is less than the arc weight, the node contracts the disease.

For example, if the number generated for the arc between node *A* and *B* is 0.5, then node *B* would remain *susceptible* until the next time period. The next time period is simulated in the same manner as the first. Since node *A* has one more time period in the *infectious* state, it can still infect any *susceptible* adjacent nodes. To simulate this, another random number is generated, say 0.1. Since this value is less than the arc weight of 0.3, node *B* has contracted the disease. To determine what disease track *B* follows, a random number is generated, say 0.2. Since this number is below 0.4, node *B* follows the *SIS* track. Thus, node *B* transitions into the *infectious* state for the next period. Additionally, a uniform random number between 1 and 3 is generated to determine the number of periods in this *infectious* state. For this example, assume the number generated is 1. Figure 2 represents the states during the second time period.

In the next time period node *A* transitions into its terminal state, which is *dead*. Also in this time period, node *B* has a chance of infecting node *C*. To simulate this, a uniform random number is generated between 0 and 1 and is 0.7. Since this number is larger than the arc weight between node *B* and *C*, node *B* does not infect node *C*, Figure 3.

Since the duration of node *B*'s infection is only 1 time period, the next time period yields the *susceptible* state for node *B*. Finally no nodes are *infectious* and all are either *dead* or *susceptible*, so there is no way for the disease to spread. This means all nodes are in their terminal state, Figure 4.

Figure 1. Iteration 0

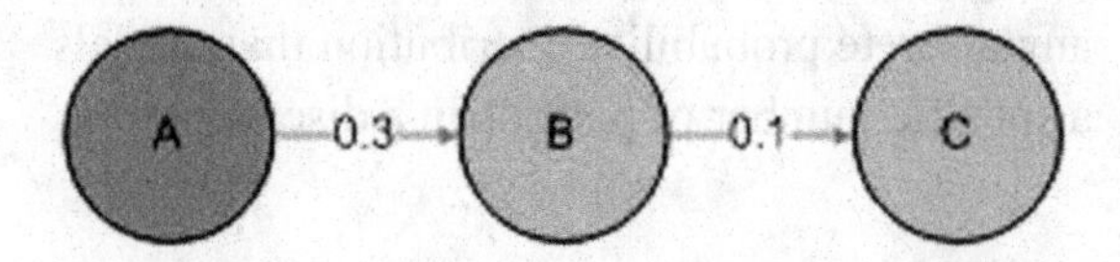

Figure 2. Iteration 1

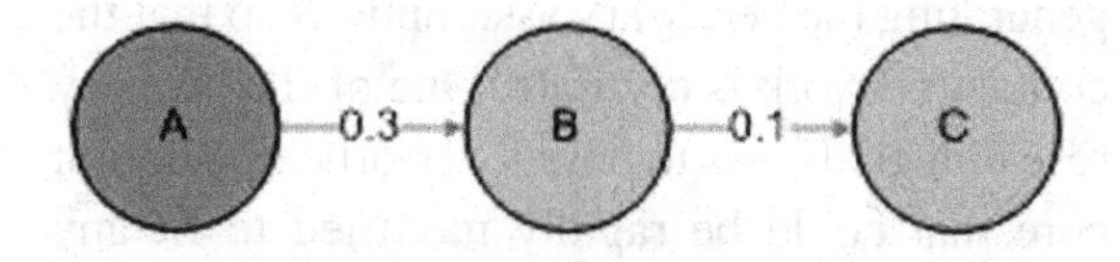

Figure 3. Iteration 2

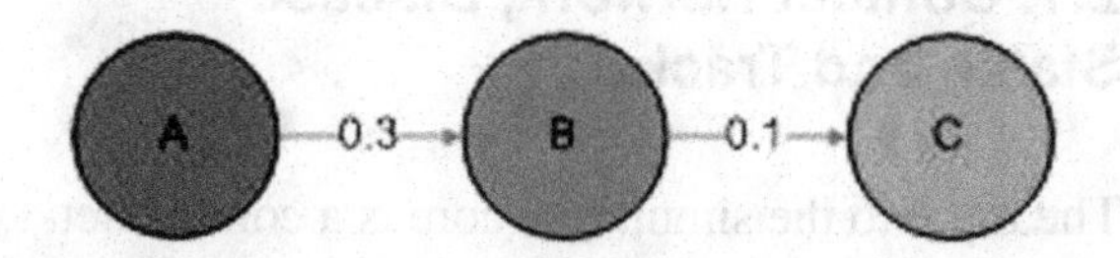

Figure 4. Iteration 3

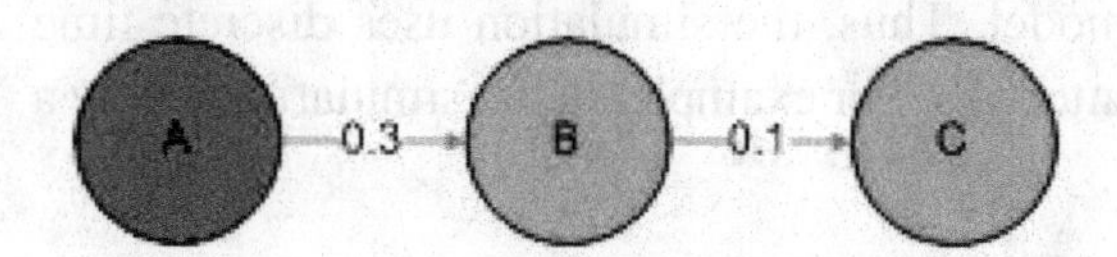

Typically diseases are modeled with a basic reproductive number, R_0, which represents the average number of individuals that a single infectious person would infect with the disease in a completely susceptible society. Furthermore, diseases also have death rates and immunity rates and periods. Applying these parameters uniformly across a population may not accurately reflect how a disease actually travels through a society.

The primary advantages of the simulation core are the concepts of disease tracks and the contact network. The disease tracks enable different groups of individuals to have distinct disease paths. Thus, each disease track could have a different R_0 parameter. Furthermore, the contact network enables the diseases to spread according to the individual's habits.

The concept of state transitions and disease tracks is a vital concept of this disease simulation core. The following section provides a detailed example of how these concepts can be applied to a network of rural people.

Table 1. WAJEC disease tracks

State	Track 1 (probability .3)	Track 2 (probability .5)	Track 3 (probability .2)
1	*Susceptible*	*Susceptible*	*Susceptible*
2	*Exposed*	*Exposed*	*Exposed*
3	*Symptoms not Contagious*	*Dormant*	*Carrier*
4	*Symptoms Contagious*	*Carrier*	*Recovered*
5	*Symptoms not contagious*	*Dead*	
6	*Recovered*		

Table 2. Color legend for the disease states

Node Color	State
Green	*Susceptible*
Yellow	*Exposed*
Red	*Carrier*
Purple	*Symptoms Contagious*
Blue	*Symptoms not Contagious*
Teal	*Immune*
Black	*Dormant*
Olive	*Recovered*

3. SIMULATION EXAMPLES

This section discusses how the simulation core is applied to a contact graph that resembles the population and density of Clay Center, Kansas (population 4,600). For this research, a hypothetical example disease, WAJEC, was created to show how a disease can be simulated. As with most diseases, WAJEC has multiple disease tracks. A person can be in any of the disease tracks in Table 1 according to the associated probability.

A graphical user interface displays how WAJEC spreads. The graphics in the figures represent the town of Clay Center (Figure 5). There are colored nodes that represent each person and their disease state. The disease states and corresponding color for WAJEC are listed in Table 2.

The square blocks of the residential areas of Clay Center are separated by primary streets of the town and are represented by white space in the graph. The large white area in the graph is the town center and is not a residential area. Although not depicted in the figures, the simulation has some nodes that live far away from the town center. Carlyle (2009) provides more information regarding these remote individuals.

Before WAJEC is simulated, a contact network is built. To build the contact network for Clay Center, approximately 4,600 nodes were given locations. These locations were random selected according to the population densities of the town. Also, a random family size was generated between 1 and 5.

Figure 5. Base case: Spread .25; Edges .25, .025, .0025

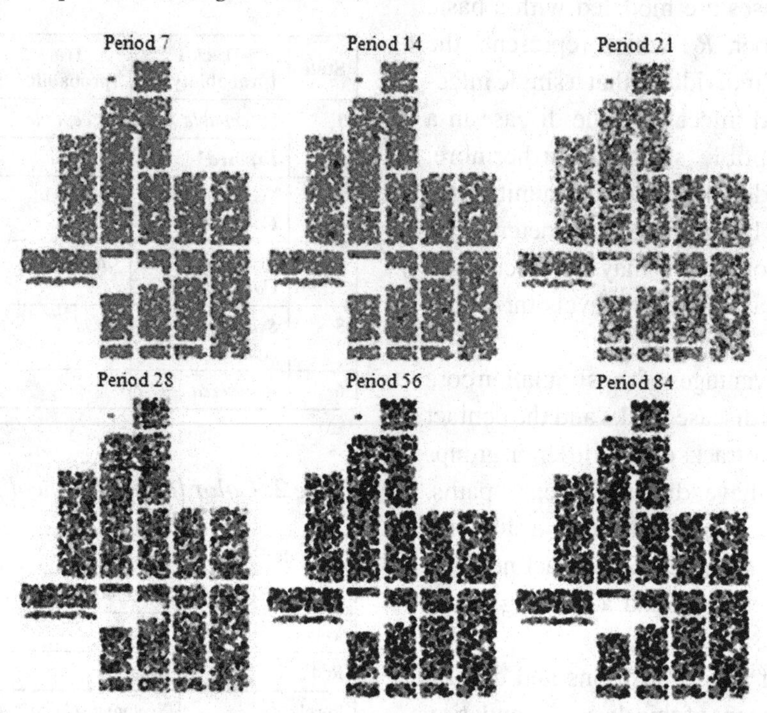

Deriving parameters for a base case required numerous iterations with varying parameter settings. In most instances, the disease spread rapidly through the town and the mitigation strategies described in the next section failed to contain the disease. Occasionally, the disease spread so slowly that it died out and no mitigation strategy was necessary.

The parameters of WAJEC are set right at the middle ground. Without mitigation, the disease infects almost everyone in town, but certain mitigation strategies are useful. This also helps to identify which diseases should have mitigation strategies and where such efforts would have a minimal cumulative impact.

The contact between one node and another is represented by an arc. For this simulation there are three levels of arc existence based upon the distance between the nodes in question. The edge probabilities in the base case are given as .25, .025 and .0025. This indicates that the individual has contact with 25% of the people living in their household or an immediate neighbor during a period. For individuals living within a three block radius, there is a 2.5% chance of contact, and for any individual outside this range there is a .25% contact rate. The contact network has an average degree of 25.6 with 56 maximum number of contacts and 3 for a minimum number of contacts.

Along with the probability that an arc exists, there is a maximum probability of .25 of spreading the disease. This number means that if the arc exists between two people, then there is a probability that an infected person spreads the disease to the uninfected person. Thus, if edge $\{i,j\}$ exists, then the probability that an infectious person i spreads the disease to person j is uniformly distributed between 0 and .25.

The following example is the base case for this simulation. For this study, the time spent in each state is fixed at 3 periods, and a single individual starts the disease. The parameters for the contact network are presented in the next section along with the graphical representation of how these parameters affect the spread of WAJEC after various periods.

4. ASSESSING MITIGATION TECHNIQUES

Accurately assessing a mitigation strategy is challenging and raises various ethical questions. For instance, one mitigation strategy may allow 100 fewer individuals to become infected, but results in the maximum number of individuals infected at one time. In such a situation, medical services would struggle to keep up and a better goal may have been to limit the maximum number of individuals suffering from the disease at one time. Thus, important ethical questions should be answered to provide a responsible goal.

In the absence of the government providing ethical guidelines as a goal for a mitigation strategy, the obvious measurement to determine the efficiency of a mitigation policy is the fewest number of people contracting the disease. In this discussion, define the terminal state to be the first period when all transitions are done. Furthermore define the critical state to be the period when the disease begins to rapidly spread. The exact period where this occurs is open for debate and different people may identify this period earlier or later.

In this section, the pictures indicate a single replication, but the data is gathered over 50 simulation runs. The mean and confidence intervals are reported for each case. The run time for these 50 iterations took less than 30 seconds.

Figure 5 depicts the baseline application of WAJEC for various periods. The probability that an individual passes the disease on to another person is called the *Max Probability* and is .25. The existence of an edge in the contact network is reported as *Edge Probablities* .25, .025, .0025, which corresponds to the probability of an edge existing between individuals that live close, near and far away. The average R_0 parameter for WAJEC for these 50 runs is 17 with a 95% confidence interval between 8.5 and 25.5. An R_0 parameter of 17 is fairly high with a total infectious period between 6 and 12 periods depending upon the disease track. These parameters were selected in order to examine an extremely dangerous situation.

Figure 5, the base case, demonstrates that the entire town eventually caught WAJEC. The critical state occurred around period 17 and the terminal state occurred at period 65. For the 50 iterations, on average only one person did not catch the disease in the entire town and the 95% confidence interval is (-1, 3). Clearly, a mitigation strategy should have been in place to prevent the spread of this disease.

The first two mitigation strategies analyzed compare whether it is better to reduce the probability of one person passing the disease to another person or to reduce the number of contacts. For instance, if each person has regular contact with 20 people and a probability of spreading the disease to any of their contacts of .20. Would a strategy that a strategy that maintains all 20 contacts, but limits the probability of spreading the disease to .05 be superior to a strategy where each person has regular contact with 5 people and a probability of spreading the disease of .20?

The first strategy may occur when the government encourages hand washing, wearing masks and businesses are encouraged to install antibacterial hand sanitizer. The second strategy can be encouraged through advertisements and announcements encouraging individuals to eliminate unnecessary contact with other people or closing schools and businesses. Ideas like these occurred with the recent H1N1 outbreak (CDC, 2010).

Since Clay Center is assumed to be the first occurrence of this disease, eventually some mitigation strategy is going to be implemented. Thus, the contact network does not change through period 14, but in this period, the government takes some action and the contact network changes.

The first scenario cuts each individual's probability of contracting a disease to a quarter (*Max Probability* is .0625) and is shown in Figure 6.

Figure 6. Spread reduction to 25% of base case at period 14

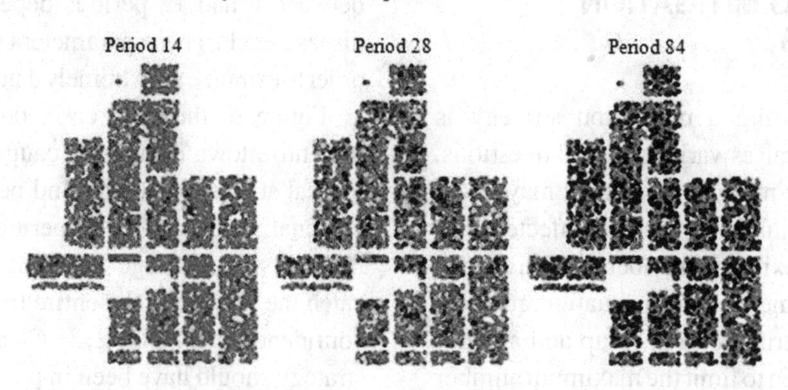

It should be noted that this decrease in the probability significantly slows the progression of the disease. The terminal state occurs at period 86. However, the disease still infects 95% of the town and on average only 228 people are not infected with the disease. The 95% confidence interval for the number of uninfected people is (196, 260).

Figure 7 provides the analysis of a 25% decrease in the number of contacts after period 14. This simulation eliminates the lowest probability contacts first, after 14 time periods have elapsed. This more accurately models how closing schools and businesses would affect the spread of a disease. Here individuals drop their low-level contacts first, such as people passed on the street or other shoppers, while keeping their higher level contacts such as family members and close friends.

A recent survey (Scoglio et al., 2010) performed in Clay Center, KS determined that 40% of the people would ignore governmental advice and visit people outside of their own household even if told the risks of a contracting a serious disease. This strategy performs slightly worse than the mitigation strategy in Figure 6 and about 96.3% of the town becomes infected with WAJEC. The terminal state also occurs earlier at period 80.

Clearly, these mitigation strategies are unacceptable and so the mitigation strategy should improve. As the government applies additional news and guidelines, possibly exaggerating the seriousness of the disease, the probabilities change. Now assume that these mitigation strategies decrease the probabilities to 10% of their base values. Figures 8 and 9 represent each of these cases.

Figure 7. Edge reduction to 25% of base case at period 14

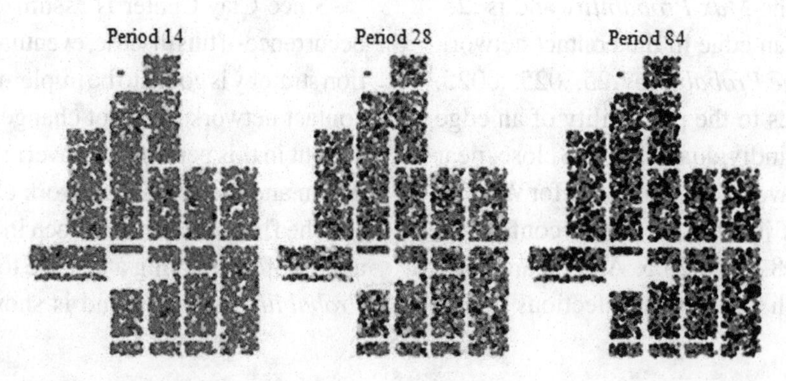

Figure 8. Spread reduction to 10% of base case at period 14

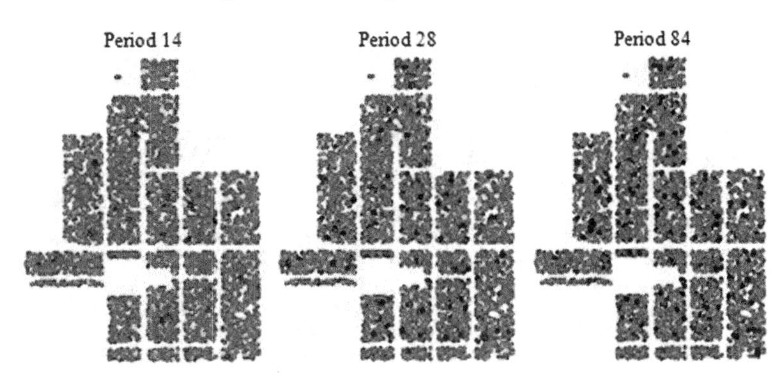

Figure 9. Edge reduction to 10% of base case at period 14

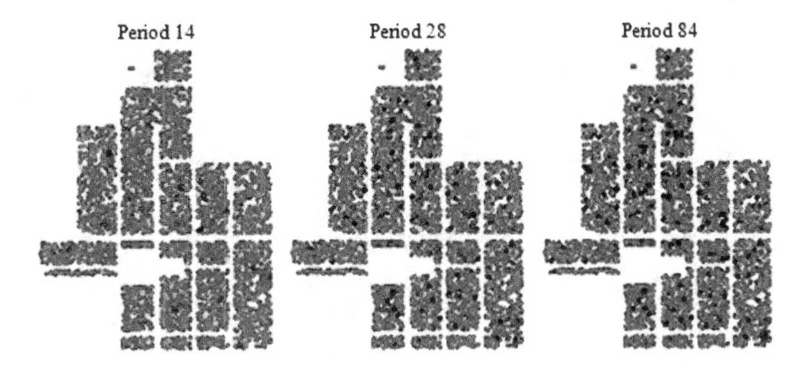

Clearly, this is a much more effective mitigation strategy as about a fifth of the people contract WAJEC in this particular instance. The spread reduction results in 63.2% of the population becoming infected, while the edge reduction only has 64.3% of the population catching this disease.

In both of these instances, the standard deviation was incredibly large leading to 95% confidence intervals that incorporated -1,500 to 4,800 people. The number of healthy people at the end was bimodal with almost everyone gaining the disease or only about 1/5 of the population contracting the disease. Thus the mitigation strategy was either effective or it failed to contain the disease. An interesting observation is that in all 50 replications there was never a middle value where about ½ of the population became infected.

From these previous four cases, an effective mitigation strategy requires a dramatic change to either people's contacts or people's ability to contract the disease. In such a situation, it appears that reducing the rate at which people spread the disease is slightly more successful.

For the remainder of the paper, only the case that reduces the probability of the spread is examined as a mitigation strategy. It is anticipated that the version that eliminates contacts should see fairly similar results.

To demonstrate the sensitivity of the impact on the spread reduction parameter, consider a reduction to 15% of the base cases spread (Figure 10). In this situation 88.2% of the town becomes infected. This extra 5% increases the number of cases by about 25%, thus it has a fairly significant effect on the population as a whole.

The next question to consider is how fast the government should act to keep the disease from becoming an epidemic. Figure 11 demonstrates the same mitigation technique as Figure 8, except the mitigation strategy is enacted in period 7 in-

Figure 10. Spread reduction to 15% of base case at period 14

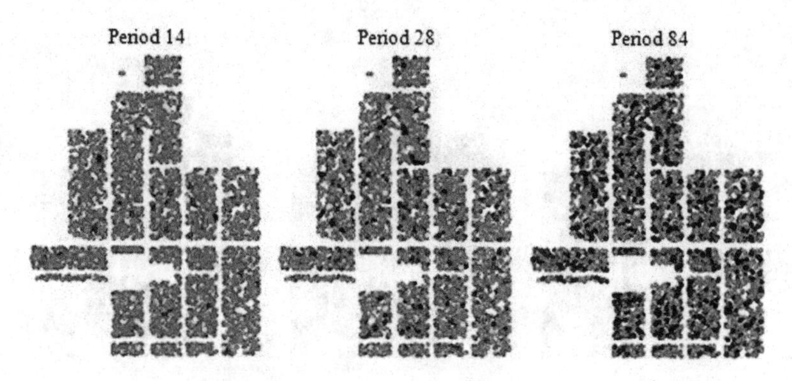

Figure 11. Spread reduction to 10% of base case at period 7

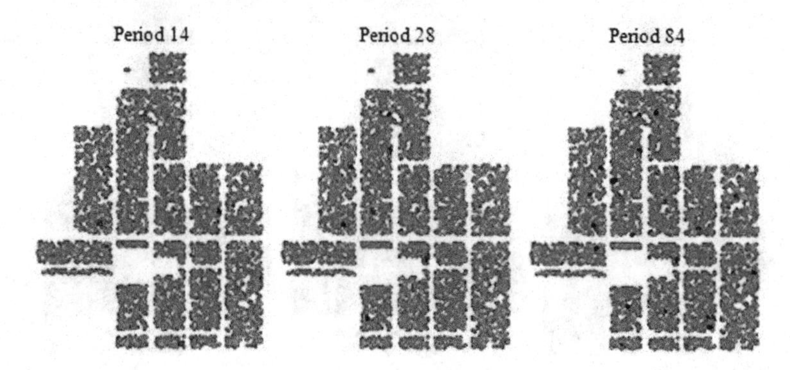

stead of period 14. This rapid response significantly impacts the disease and only 36.5% of the population contracts the disease with a 95% confidence interval having (1352, 5572) individuals that didn't contract the disease. This was again bimodal and several of the iterations had less than 200 people contracting the disease.

Figure 12 shows the impact of an indecisive government. Here the agency fails to enact any policies until there is clearly an epidemic (period 21). At this point is it too late and results in the town having 79.2% of its inhabitants contracting the disease. Again this was a bimodal distribution with few people contracting the disease or about

Figure 12. Spread reduction to 10% of base case at period 21

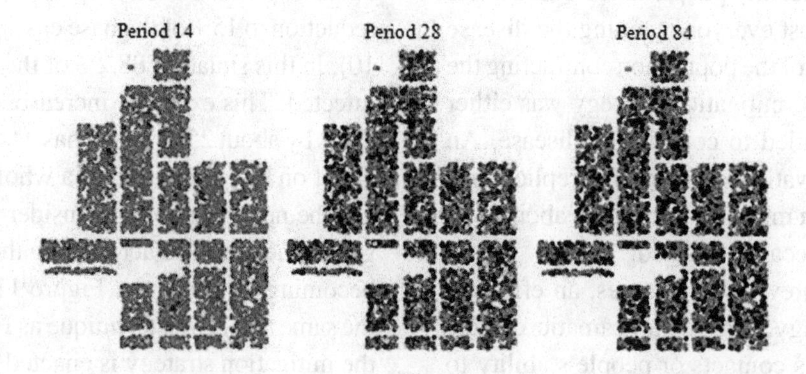

40% of the population. Thus, there is little point in implementing such a mitigation strategy at this point in time.

In summary, whether or not a disease impacts an entire town and becomes a pandemic is extremely sensitive. A slight change in probabilities or mitigation strategy can have an immense impact on the number of infected individuals.

In order to prepare for an epidemic, governmental agencies should focus upon the timeliness of response. The faster a mitigation strategy is in place, the better chance that the disease can be contained. Additionally, the government should also focus on limiting contacts between individuals. Closing schools and governmental agencies is expensive and causes many economic and societal problems; however, such action limits the number contacts and could greatly reduce the chance that a disease turns into an epidemic.

5. CONCLUSION AND FUTURE WORK

This paper describes a generic simulation core that can simulate the spread of an infectious disease in a small rural town. Due to the long time friendships and family connections in a small town, these individuals are likely to visit each other even during a pandemic. Thus, this software provides a new tool to help governmental agencies develop effective policies for rural areas.

This paper also examines mitigation strategies. It determines that the critical factors in controlling an outbreak are the responsiveness of the mitigation strategy and the overall reduction on the spread of the disease due to the mitigation strategy. Additionally, limiting people's ability to spread the disease is slightly more effective than just reducing the number of contacts.

There still remains ample research in this area. What would the impact of a quarantine region within a town be? What is the likelihood that people follow the government's advice during a pandemic? Can this simulation be extended to a rural region that would have multiple towns and how do the distances that people in rural areas travel impact the dispersion of an epidemic?

ACKNOWLEDGMENT

This research was partially funded by the National Science Foundation grant number NSF SES-084112: "SGER: Exploratory research on complex network approach to epidemic spreading in rural regions"

REFERENCES

Anderson, J. (2009). *Simulating epidemics in rural areas and optimizing preplanned quarantine areas using a clustering heuristic*. Unpublished master's thesis, Kansas State University, Manhattan, KS.

Barrett, C. L., Eubank, S. G., & Smith, J. (2005). If smallpox strikes Portland (simulation of the spread of disease in social networks). *Scientific American, 292*(3), 54–61. doi:10.1038/scientificamerican0305-54

Britton, T., Janson, S., & Martin-Lof, A. (2007). Graphs with specified degree distributions, simple epidemics, and local vaccination strategies. *Advances in Applied Probability, 39*, 922–948. doi:10.1239/aap/1198177233

Carlyle, K. (2009). *Optimizing quarantine regions through graph theory and simulation*. Unpublished master's thesis, Kansas State University, Manhattan, KS.

CDC. (2010). *The Center for Disease Control and Prevention's flu website.* Retrieved from http://flu.gov/

Cristea, A., Zaharia, C., Deutsch, I., Bunescu, E., & Blujdescu, M. (1992). Mathematical modeling of measles epidemics and the optimization of corresponding antiepidemic programs. In *Proceedings of the 4th International Symposium on Systems Analysis and Simulation* (pp. 623-628).

Fuks, H., Lawniczak, A., & Duchesne, R. (2006). Effects of population mixing on the spread of SIR epidemics. *The European Physical Journal B, 50*(1-2), 209–214. doi:10.1140/epjb/e2006-00136-7

Gordis, L., & Saunders, W. B. (2008). *Epidemiology* (4th ed.). Amsterdam, The Netherlands: Elsevier.

Guimaraes, P., de Menezes, M., Baird, R., Lusseau, D., Guimaraes, P., & dos Reis, S. (2007). Vulnerability of a killer whale social network to disease outbreaks. *Physical Review E: Statistical, Nonlinear, and Soft Matter Physics, 76*, 1–4. doi:10.1103/PhysRevE.76.042901

Huang, C.-Y., Hsieh, J.-L., Sun, C.-T., & Cheng, C.-Y. (2006). Teaching epidemic and public health policies through simulation. *WSEAS Transactions on Information Science and Applications, 3*, 899–904.

Kolesin, I. D. (2007). Mathematical model of the development of an epidemic process with aerosol transmission. *Biophysics, 52*(1), 92–94. doi:10.1134/S0006350907010150

Kumar, R. (2002). Models of competing host-parasite pairs. *Indian Journal of Pure and Applied Mathematics, 33*(10), 1515–1528.

Manier, J. (2007). *How staph became drug-resistant threat 94,000 infections a year, many occurring outside of hospitals*. Chicago Tribune.

Neal, P. (2008). The SIS great circle epidemic model. *Journal of Applied Probability, 45*(2), 513–530. doi:10.1239/jap/1214950364

Newman, M. E. J. (2002). Spread of epidemic disease on networkds. *Physical Review E: Statistical, Nonlinear, and Soft Matter Physics, 66*(1).

Nuno, M., Castillo-Chaves, C., Feng, Z., & Marcheva, M. (2008). Mathematical models of influenza: The role of cross-immunity, quarantine and age structure. *Mathematical Epidemiology*, 349-364.

Ogren, P., & Martin, C. F. (2000). Optimal vaccination strategies for the control of epidemics in highly mobile populations. In *Proceedings of the 39th IEEE Conference on Decision and Control* (Vol. 2, pp. 1782-1787).

Olsson, B. (1996). Optimization using a host-parasite model with variable-size distributed populations. In *Proceedings of the IEEE Conference on Evolutionary Computation* pp. 295-299).

Piqueira, J., Navarro, B., & Monteiro, L. (2005). Epidemiological models applied to viruses in computer networks. *Journal of Computer Sciences, 1*(1), 31–34. doi:10.3844/jcssp.2005.31.34

Rosenberg, R. (1997). *Typhoid Mary: The sad story of a woman responsible for several typhoid outbreaks.* Retrieved from http://history1900s.about.com/od/1900s/a/typhoidmary.htm

Rothman, K. (2002). *Epidemiology: An introduction*. Naperville, IL: Sourcebooks.

Rvachev, L. A. (1968). Simulation of large-scale epidemics on a digital computer. *Soviet Physics, Doklady, 13*(5), 384–386.

Scoglio, C., Schumm, W., Schumm, P., Sydney, A., Chowdhury, S., & Youssef, M. (2010). Efficient mitigation strategies to control epidemics. *PLoS ONE, 5*(7), 11569. doi:10.1371/journal.pone.0011569

Taubenberger, J., & Morens, D. (2006). *1918 influenza: The mother of all pandemics.* Retrieved from http://www.cdc.gov/ncidod/EID/vol12no01/05-0979.htm

World Health Organization. (2007). *Typhoid fever.* Retrieved from http://www.who.int/vaccine_research/diseases/diarrhoeal/en/index7.html

Yu, Y., Luo, X., Gao, G., & Ai, S. (2006, November 1-4). Research of a potential worm propagation model based on pure P2P principle. In *Proceedings of the International Conference on Communication Technology.*

This work was previously published in the International Journal of Artificial Life Research, Volume 2, Issue 2, edited by E. Stanley Lee and Ping-Teng Chang, pp. 95-104, copyright 2011 by IGI Publishing (an imprint of IGI Global).

Chapter 12
A Structural Model to Investigate Factors Affect Patient Satisfaction and Revisit Intention in Jordanian Hospitals

Abbas Al-Refaie
University of Jordan, Jordan

ABSTRACT

By measuring to what extent hospitals meet or exceed patient's expectations, hospital managers can determine the needed service design and delivery improvements that contribute to patient satisfaction and revisit intention. It is necessary to evaluate quality of health care services from patient perspective. This research investigates the factors, including hospital performance, hospital stay, hospital facilities, interaction with patients, service quality, and patient security culture, that affect significantly patient satisfaction and revisit intention in Jordanian hospitals using structural equation modeling. Data were collected from five main hospitals. The results showed that hospital performance has no significant effect on patient satisfaction and revisit intention. This result indicates that the patients are facing troubles in admission, registration, waiting time, and response time for results of medical tests. Also, the hospital stay, hospital facilities, service quality, and patient security culture are found significantly important in achieving patient satisfaction and revisit intention. Further, the interaction with patients' requirements and needs significantly related to service quality and hospital stay. These results shall provide policy and planning manager a great assistance in determining the factors that improve hospital performance, maintain quality medical services, and plan future improvements in the design and development of medical health care services in Jordan.

DOI: 10.4018/978-1-4666-3890-7.ch012

1. INTRODUCTION

Following the increases in the number of hospitals in Jordan, the quality of health care services provided to patients affects hospital performance under severe competition. Customer defines how a firm determines requirements, expectations and customer performance and investigates the procedures undertaken by the firm to acquire information about current and future customers (Lee *et al.*, 2003). ISO 9000 certification enforces firms to meet or exceed customer expectations and have activities designed to increase customer focus (Al-Refaie *et al.*, 2011). Patient satisfaction is considered an important measure of the quality of health care services and a key determinant of patients' behavioral intention. By including the patient perspective as to how well the health care services meet or exceed patient's expectations, managers can identify the service design and delivery improvements that contribute to patient satisfaction and revisit intention. For this purpose, it is necessary to evaluate quality of health care services from patient perspective.

Recently, several studies have been conducted to investigate the factors affecting patient satisfaction and revisit intention in health care services. For example, Carden and DelliFraine (2004) examined the factors that predict overall hospital satisfaction with blood suppliers. A total of 1325 blood-utilizing hospitals were included in the final study database. The measurement of hospital satisfaction with its blood supplier encompasses the five composites of service quality, including tangibles, reliability, responsiveness, assurance, and empathy. Significant predictors of hospital satisfaction with blood suppliers are satisfaction with medical and clinical support provided by the blood center, satisfaction with the routine delivery schedule, and price (service fee) of red cells. Chang *et al.* (2006) proposed a mathematical model for evaluating the quality of hospital services from the customers' satisfaction perspective. The model was able to determine how much budget must be allocated to each quality element in order to maximize the customer satisfaction under budget constraint. Donini (2008) examined the course of the quality of the institutional catering service over a five year period to verify the effectiveness of the quality improvement process used by objective and subjective quality control. Objective quality control was measured by meal order accuracy, proper distribution of food in trolleys, route time from the kitchen to the ward and time of food distribution, food weight and temperature, waste assessment. While, subjective quality control was measured by giving the patients a questionnaire after meals. A significant amount of qualitative errors, such as lack of respect for patient preferences or at the moment of supplying the food trolley, have been found. Also, results showed that patient satisfaction with menu variability, portion size, temperature and cooking quality were improved over time. Kim *et al.* (2008) explored the factors affecting the value of care and patient satisfaction, and tested the correlations among the value of care, patient satisfaction and intention to revisit in large-sized hospitals located in Seoul, Korea with approximately 1000 hospital beds. Their study revealed that the value of care had a significant influence on patient satisfaction and revisit intention. Kucukarslan (2008) determined the relationship between disconfirmation of expectations with medication-related services and patient satisfaction with medical care. Satisfaction with medical care and the likelihood of positive word of mouth regarding the medical care were measured. Patient satisfaction and the behavioral intentions measures were significantly related. It was concluded that the disconfirmation of expectations had a role in a post service experience response expressed by the patient, but not as a direct antecedent to patient satisfaction. Chang and Chang (2008) implemented the service encounters evaluation model and concluded that: (1) technology-based service encounters had a positive impact on service quality, but not patient satisfaction, (2) after experiencing technology-based service encounters, the cognition of the service quality was positively related to patient satisfaction; and (3) network security contributed a positive moderating effect on service quality and patient satisfaction. McFadden (2009) investigated the existence of a patient safety chain for hospitals and developed a model for improving patient safety using data from a nationwide survey of over 200 hospitals. The proposed patient's safety chain model would enhance operations in hospital settings. Hsiao (2009) assessed the outsourcing situation in Taiwanese hospitals and compared the differences in hospital ownership and in accreditation levels. The results for non-medical items showed medical waste and common trash both have the highest rate of being outsourced, while the lowest rate of outsourcing is in utility maintenance. For medical items, the highest rate of outsourcing corresponds to the ambulance units, whereas the outsourcing rates for departments of nutrition, pharmacy, and nursing were very low.

In Jordan, however, little research has been conducted to evaluate the quality of health care services from patient perspective. To help policy makers and hospital manager identify needed improvements that sharpen hospitals competitive strength, this research investigates the structural relationships between service quality, hospital facilities, hospital performance, interaction with medical staff, hospital stay, and patient safety culture and their effects on patient satisfaction and revisit intention from the perspective of patients in Jordanian hospitals. The remainder of this research is outlined in the following sequence. Section 2 develops conceptual framework. Section 3 provides data collection and analysis. Section 4 presents research results. Section 5 summarizes conclusions.

2. CONCEPTUAL FRAMEWORK

Based on published research and hospital experts, several important factors affect patient satisfaction and revisit intention including: hospital performance, hospital stay, interaction with hospital staff, service quality, hospital facilities, patient safety culture, patient satisfaction and revisit intention. These factors are briefly introduced as follows.

2.1 Hospital Performance (HP)

HP represents the degree to which hospitals are improving/deteriorating over time measured by the level of coordination, waiting time to check in, and registration procedures. It also assesses how hospitals achieve and maintain that improvement. Table 1 displays the measures adopted to measure HP.

It is believed that hospital performance is positively related to patient satisfaction and revisit intention. Thus, the following hypothesis is constructed:

$H_1$1: HP has a positive influence on patient satisfaction and revisit intention.

2.2 Hospital Stay (HS)

HS involves providing sufficient care of patients by physicians, nurses, physical therapists, nutritional support, pharmacists, and others, until a patient is able to take care of himself/herself at home. The

Table 1. Measures of HP

	Item
A1	You could get admitted to this hospital without any trouble.
A2	I am satisfied with the time spent to check in.
A3	I am satisfied with the financial matters explanation.
A4	The person handling registration is skilled.
A5	The registration representative fully answered all your questions.
A6	The speed and efficiency of registration process was convenient.
A7	I am satisfied with the time it took to see the physician.
A8	Waiting time for test results.
A9	I am satisfied with the time it took to get the tests results.

HS will be measured by clarity of information, information about the medicine, explanation, and answering patient's questions as shown in Table 2. Considering these measures during patient's stay will lead to improving patient satisfaction and revisit intention. Consequently, the following hypothesis will be tested:

$H_1$2: Hospital stay positively influences patient satisfaction and revisit intention.

2.3 Interaction with Hospital Staff (INTER)

INTER is a very important aspect of the treatment process, which evaluates to what effectiveness level responses to patients' behaviors are handled to promote healthy growth and wellness during patient's stay. Table 3 displays the measures of INTER.

Conceptually, INTER affects positively both SQ and HS and is measured by the performance level of staff, including nurses, physicians, and other staff. Thus, the following two hypotheses are proposed:

$H_1$3: INTER is positively related to HS.
$H_1$4: INTER has a positive influence on SQ.

Table 2. Measures of HS

	Item
B1	The frequency of visits received from your doctor is convenient.
B2	Your doctor fully explained the procedures of your treatment.
B3	Your doctor provided you with high quality of care.
B4	The area around your room was "always" quiet at night.
B5	The visitor's policy at the hospital is adequate.
B6	The doctor answered all my questions.

Table 3. Measures of INTER

	Item
C1	Nursing staff is skilled.
C2	The nurses communicate well.
C3	Nurses are able to solve your problem.
C4	Nurses respond to call button on time.
C5	Nurses do their job in a professional way.
C6	Physicians are willing to listen and answer your questions.
C7	Physicians keep you informed.
C8	Accessibility to the physicians is easy.
C9	Physicians keep your family informed with your condition.

2.4Service Quality (SQ)

Delivering high SQ has been recognized as the most efficient way of ensuring that a company's offerings are competitive. SQ represents the patient overall impression of the relative inferiority or superiority of the hospital and its services. In hospitals, up-to-date equipments, meals arrival, quality and variety of food are among the main factors to achieve quality service. The main measures for SQ in hospitals are summarized in Table 4.

Commonly, maintaining efficient SQ leads to better patient satisfaction and revisit intention. Thus, the following hypothesis is proposed:

$H_1$5: SQ has a positive effect on patient satisfaction and revisit intention.

2.5 Hospital Facilities (HF)

HF is represented by patient's movement, finding facility, convenient facilities, waiting room for care, and cleanliness of interior and up-to-date equipment. Table 5 displays the items for HF. The location of the hospital, the availability of parking, rooms' quality, and easy use of facilities play a significant role in the total patient satisfaction during the patient's stay and revisit intention. As a result, it is hypothesized that:

Table 4. Measures of SQ

	Item
D1	You can trust staff of this hospital.
D2	This hospital has up to date equipment.
D3	When this hospital promises to do something by a certain time, it does so.
D4	Meals arrive on time.
D5	Quality food.
D6	You received what you ordered.
D7	Food variety.

Table 5. Measures of HF

	Item
E1	The facility is conveniently located.
E2	Valet parking- Easy parking.
E3	Adequate internal & external signs.
E4	Intercom by bedside.
E5	Quietness of the room.
E6	Cleanness of the room.
E7	Temperature and lighting are comfortable.
E8	In general, to what level are satisfied you with hospital facilities?.

$H_1$6: HF is positively related to patient satisfaction and revisit intention.

2.6 Patient Safety Culture (PSC)

PSC of a hospital is a key determinant of its ability to achieve high levels of patient safety, through: (*1*) caring about patient safety concerns, (*2*) commitment and drive to be a safety-centered institution, (*3*) communication, and action taken on patient safety suggestions when communicated, and (*4*) collegiality and openness about errors. Table 6 displays the items used to PSC.

Typically, higher level of safety consideration leads to better the patient's satisfaction and sense of being secure. Therefore, the following hypothesis is proposed:

Table 6. Measures of PSC

	Question
F1	I am satisfied with the safety procedures in this hospital.
F2	I am satisfied with the security measures for maintaining my personal belongings.
F3	I feel secure in this hospital.

$H_1$7: PSC positively influences patient satisfaction and revisit intention.

2.7 Patient Satisfaction and Revisit Intention (PS&IR)

PS&IR is determined by the degree to which a patient exhibits repeat visit behavior for the same hospital and considers dealing with only this hospital, possesses a positive attitudinal disposition toward the hospital and its services when a need for medical care arises. Table 7 displays the items used to measure PS&IR.

Figure 1 shows the hypothesized relations between the factors affecting PS&IR.

2.8 Data Collection and Analysis

The questionnaire used is mainly composed of two parts. The aim of the first part is to obtain personal information about the patient. It contains the demographic parameters about each patient, including age, gender, nationality, and education level. The second part reflects the patient's response of each factor element in the proposed model and consists

Table 7. Measures of PS and IR

	Question
G1	Would you recommend this hospital to other patients?.
G2	I am going to tell my friends about the quality of care I received here.
G3	This hospital is my first choice in emergency.
G4	Would you come to this hospital in the future?.

Figure 1. The proposed model

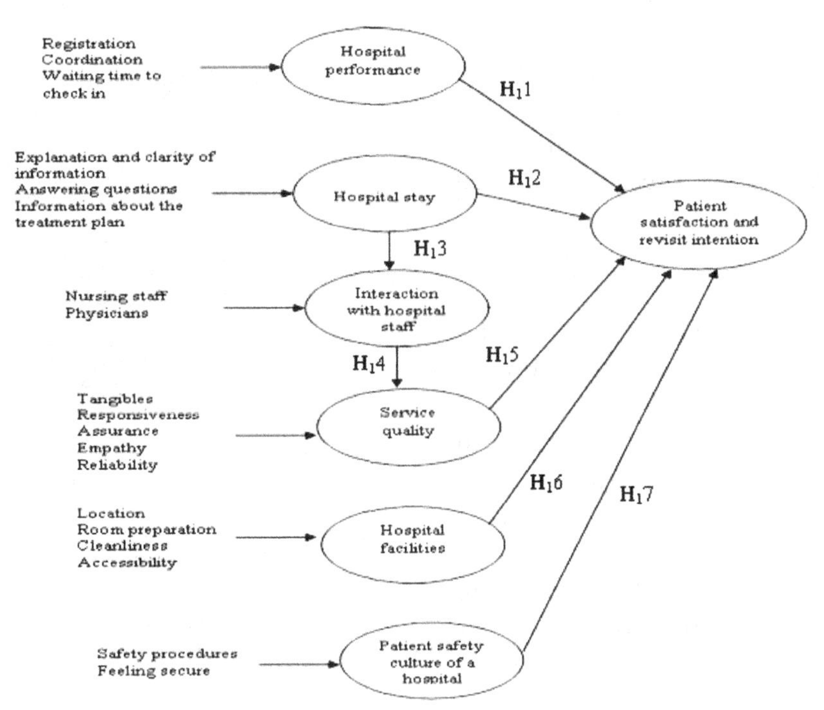

of measure items regarding HP, HS, INTER, SQ, HF, and PSC. Table 8 displays the demographic parameters. The questions were created with the help of expert opinions and previous literature (Krowinski & Steiber, 1996). A five-point Likert scale will be used to evaluate the measures of model elements (Chi & Gursoy, 2009).

The model and the questionnaire items were reviewed by quality managers and experts in the healthcare domain. A pilot study was conducted to check for the appropriateness of the tools and item measures. Then, the required modifications were made accordingly before distributing the final version.

2.9 Sample Survey

Random sampling is performed in a manner that each member of the population has some nonzero chance for being selected in the sample. A random sample is usually a representative sample. The represented sample was from five Jordanian private hospitals, out of 300 questionnaires distributed to patients, 275 responses were obtained. Out of those, 20 were dismissed because of incomplete reply. Thus, 255 questionnaires were ultimately used for the study. Consequently, a response rate of 85% was attained.

Table 8. Demographic parameters

Demographic Parameter	Alternatives
Age	Less than 20 years 20-30 years 31-40 years 41-50 years Above 50 years
Gender	Female Male
Nationality	Filled by the patient
Education Level	High School University degree Higher Education Others

2.10 Demographic Profile Analysis

As mentioned earlier, the general profile of the respondents is looked up in terms of the age, gender, nationality, and level of education. Figures 2 through 5 show the percentages for each demographic parameter. The following remarks are noticed:

- The highest percentage of 24% corresponds to patients who are above 50 years old.
- Female patients' percentage contributed 58% comparing with males patients 42%.
- The majority of patients (= 66%) were Jordanian.
- Patients with university degree represent 37% followed by high-school patients of 34%.

2.11 Validity Analysis

To check the validity of the model test of reliability of the measurement variables and model fit were conducted. The results of analysis for the model are presented as follows.

Figure 2. Age percentages

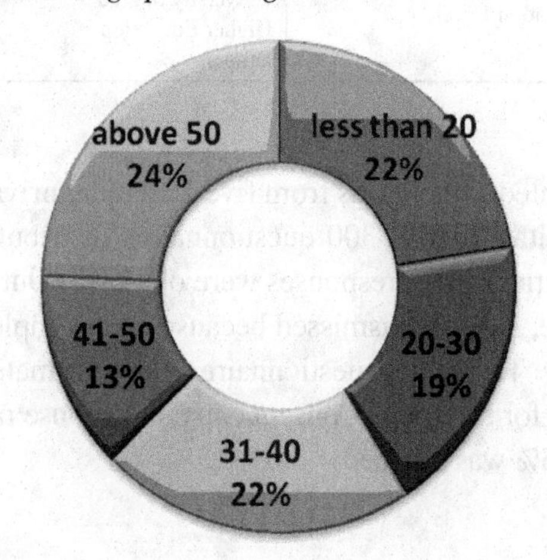

Figure 3. Gender percentages

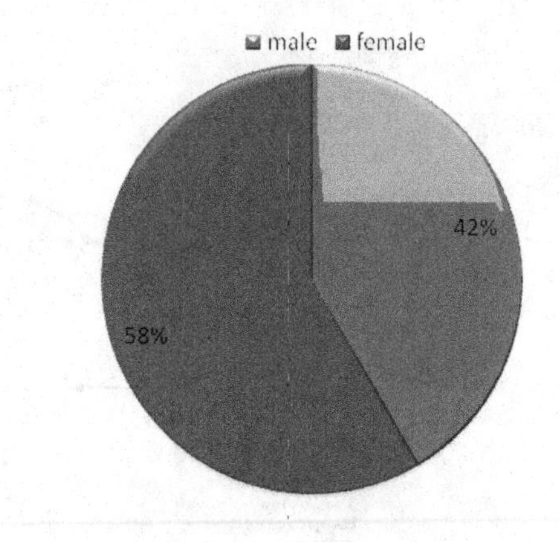

Figure 4. Nationality percentages

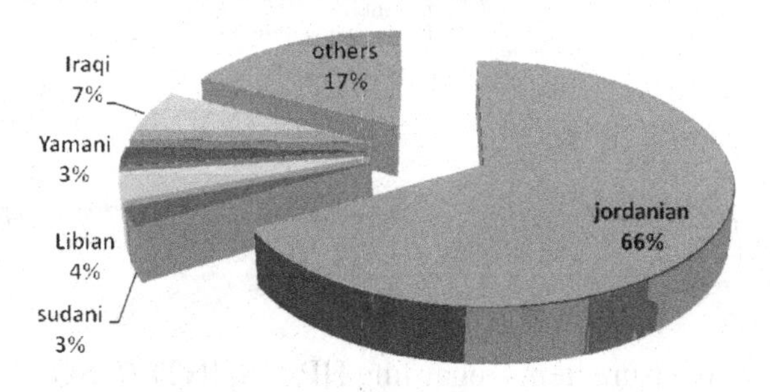

Figure 5. Level of education percentages

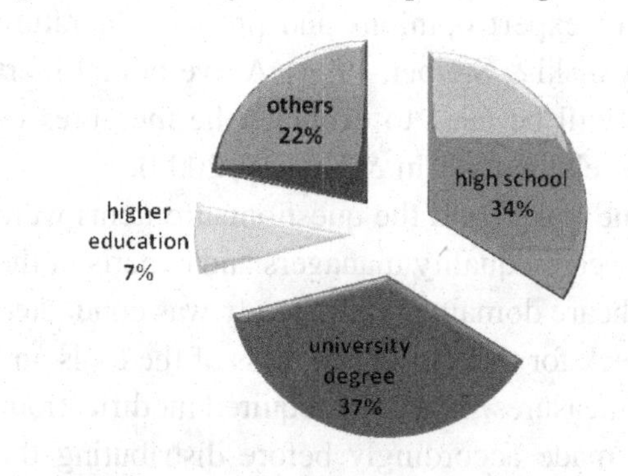

2.11.1 Multicollinearity Test

Multicollinearity measures the degree by which items measure the same entity, and a value of 0.9 or above indicates the possibility that two or more items measure the same entity (Hair *et al.*, 1995). The samples of inter-item correlation matrix and inter-item covariance matrix for the Model items were calculated then provided in Tables 9 and 10, respectively. It is found that all the inter-item correlations and covariance do not exceed the threshold of 0.9. Hence, the multicollinearity type problems do not exist.

2.11.2 Test of Reliability

The Cronbach's alpha, α, is adopted to assess internal consistency between the item measures of each model element as well as all item measures. Table 11 shows the Cronbach's α individual coefficient values for all model element. Then, a value of Cronbach's α greater than 0.6 indicates internal consistency in item measures, and hence reflects high reliability. In Table 11, the value of lowest Cronbach's α (= 0.6538) corresponds to the patient safety culture, which is larger than 0.6. In addition, the overall Cronbach's α is calculated 0.9342. These values indicate the reliability of the model.

2.11.3 Evaluating the Model

Structural equation modeling (SEM) is used to provide empirical support for the effectiveness of the model. The fit of the model can be carried out using chi-square. A value of the Chi-square measure divided by the degrees of freedom is less than two or larger than 5.00 indicates a reasonable fit for the model (Byrne, 2001). The chi-square test is sensitive to sample size that may bias the results. Therefore, other indices were used to test the goodness of fit. A value of 0.08 or lower

Table 9. Sample data of inter-correlation matrix

	A1	A2	A3	A4	A5	A6	A7	A8
A2	0.609							
A3	0.190	0.156						
A4	0.239	0.317	0.457					
A5	0.325	0.289	0.485	0.571				
A6	0.597	0.603	0.263	0.476	0.551			
A7	0.226	0.323	0.252	0.240	0.275	0.268		
A8	0.278	0.294	0.256	0.408	0.418	0.378	0.370	
A9	0.243	0.255	0.306	0.240	0.274	0.188	0.449	0.458

Table 10. Sample data of inter-covariance matrix

	A1	A2	A3	A4	A5	A6	A7	A8	A9
A1	0.6741								
A2	0.4567	0.8347							
A3	0.1439	0.1322	0.6553						
A4	0.1520	0.2243	0.3275	0.5994					
A5	0.2075	0.2051	0.3488	0.3437	0.6054				
A6	0.3936	0.4428	0.1952	0.2963	0.3447	0.6455			
A7	0.1673	0.2657	0.2097	0.1672	0.1924	0.1937	0.8110		
A8	0.1753	0.2045	0.1826	0.2437	0.2509	0.2342	0.2570	0.5948	
A9	0.1754	0.1008	0.2487	0.1631	0.1870	0.1325	0.3555	0.3104	0.7718

Table 11. Cronbach's α values

Model element	Cronbach's α value
HP	0.8238
HS	0.7019
INTER	0.8492
SQ	0.7739
HF	0.7478
PSC	0.6538
PS&RI	0.7579

for the Root Mean Square Error of Approximation (RMSEA) indicates a good model (Chang and Chang, 2008). The Goodness of Fit Index (GFI), Adjusted Goodness of Fit Index (AGFI) can be classified as absolute indexes of fit. Both indexes range from zero to 1.00, with values close to 1.00 being indicative of good fit. Solving the SEM model, the Minimum was achieved with a Chi-square of 867.8 and Degrees of Freedom of (DF=440) and Probability level of (p-value=0.00).

The measurement model is shown in Figure 6. Although the minimum was achieved, still the values of model fit are unacceptable. Table 12 represents the values of fit measures for both the original and revised measurement model.

Figure 6. Measurement model

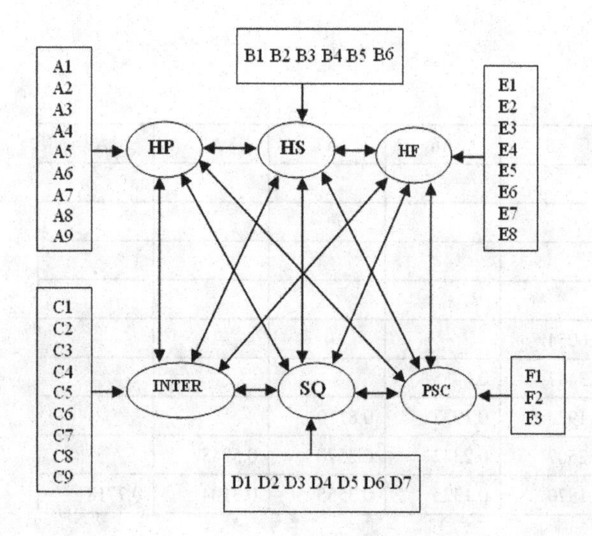

Table 12. Model fit summary for measurement model

Model	GFI	AGFI	RMSEA	χ^2/ DF
Original	0.769	0.729	0.079	2.356
Revised	0.84	0.80	0.071	2.083

In Table 12, the model fit measures do not indicate reliable measurement model. Thus, considering the error covariance having the largest Modification Index (MI) then omitting the item which has the highest MI in the regression weight table, acceptable values are obtained regarding the model fitness as also shown in Table 12. Cronbach's α was recalculated to make sure that the data is still reliable. Table 13 represents the Cronbach's α individual coefficient for the model in the revised phase, where the total Cronbach's α value is 0.8989.

2.11.4 The Structural Model

In order to draw conclusions about model hypotheses, the structural model, shown in Figure 7, is analyzed using AMOS.

The results of fit summary of GFI=0.823, AGFI=0.784, RMSEA=0.07, χ^2/ DF=2.356, for the structural model are found acceptable.

Table 13. Cronbach's α individual variable coefficient for the model (revised phase)

Measure	Cronbach's α coefficient
HP	0.7350
HS	0.6567
INTER	0.7698
SQ	0.7030
HF	0.6471
PSC	0.6538
PS& RI	0.7579
Total Cronbach's α	0.8989

Figure 7. Structural model

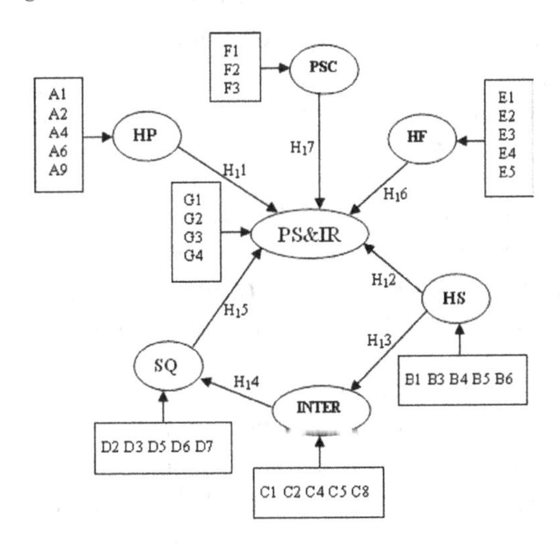

3. RESEARCH RESULTS

The structural model is analyzed and the results are displayed in Table 14.

From Table 14, the following results are obtained:

1. According to responding patients, the HP (p value = 0.841) has no significant effect on patient satisfaction and revisit intention (PS&RI); that is, $H_1 1$ is not supported. This result indicates that the patients are facing troubles in admission, registration, waiting time, response time for results of medical tests. Thus, improvement tools such as six-sigma, value stream mapping, statistical quality control, can be utilized to improve performance.
2. The HS, HF, SQ, and PSC are found significantly (α =0.1) important in achieving PS&RI. Therefore, in order to improve hospital performance measured by patient satisfaction and revisit intention, manager should continually develop improvement plans related hospital stay, hospital facilities and service quality.
3. The INTER significantly related to SQ and HS with estimates of 0.407 and 0.560, respectively. This result highlights the positive relations between interaction within/ across the health care services and service quality, and hospital stay in order to achieve patient satisfaction and revisit intention. The p-values (= 0.00) indicate that these hypotheses are supported at any alpha risk value. Consequently, decision makers should maintain effective interactions within and between all levels of health care services.

4. CONCLUSION

In Jordan, the quality of hospital services provided to patients affects patient satisfaction and revisit intention under severe competition. Hospital performance, hospital stay, hospital facilities, interaction

Table 14. Hypotheses testing for structural model

Hypothesis	Estimate	Standard Error	p value	Result (α =0.1)
$H_1 1$: HP has a positive influence on PS&RI.	0.009	0.046	0.841	Unsupported
$H_1 2$: HS positively influences PS&RI.	0.104	0.060	0.084	Supported
$H_1 3$: INTER is positively related to HS.	0.560	0.093	0.000	Supported
$H_1 4$: INTER has a positive influence on SQ.	0.407	0.099	0.000	Supported
$H_1 5$: SQ has a positive effect on PS&RI.	0.431	0.140	0.002	Supported
$H_1 6$: HF is positively related to PS&RI.	0.474	0.209	0.021	Supported
$H_1 7$: PSC positively influences PS&RI.	0.469	0.278	0.092	Supported

with patients, service quality, and patient security culture are the main factors that influence patient satisfaction and revisit intention. This research investigates the relationship between these factors and determines the factors affect significantly patient satisfaction and revisit intention. Data were collected from five main hospitals. According to responding patients, hospital performance is not significantly affecting patient satisfaction and revisit intention. However, hospital stay, hospital facilities, service quality, and patient security culture are significantly influencing patient satisfaction and revisit intention. Finally, the interaction with patient needs significantly related to service quality and hospital stay. The results of this research are valuable to policy maker and planning managers in hospitals for determining the factors that improves patient satisfaction and revisit intention and designing and developing competitive medical services business.

REFERENCES

Al-Refaie, A., Ghnaimat, O., & Ko, J.-H. (2011). The effects of quality management practices on customer satisfaction and innovation: a perspective from Jordan. *International Journal of Productivity and Quality Management*, *8*(4).

Byrne, B. (2001). *Structural Equation Modeling with Amos: Basic Concepts, Applications, and Programming*. Mahwah, NJ: Lawrence Erlbaum.

Carden, R., & DelliFraine, J. (2004). An examination of hospital satisfaction with blood suppliers. *Transfusion*, *44*, 1648–1655. doi:10.1111/j.0041-1132.2004.04184.x

Chang, H., & Chang, C. (2008). An assessment of technology-based service encounters & network security on the e-health care systems of medical centers in Taiwan. *BMC Health Services Research*, *8*(87).

Chang, W. K., Wei, C., & Huang, N. T. (2006). An approach to maximize hospital service quality under budget constraints. *Total Quality Management*, *17*(6), 757–774. doi:10.1080/14783360600725040

Chi, C. G., & Gursoy, D. (2009). Employee Satisfaction, Customer Satisfaction and Financial Performance: An Empirical Examination. *International Journal of Hospitality Management*, *28*(2), 245–253. doi:10.1016/j.ijhm.2008.08.003

Donini, L. M., Castellaneta, E., & Guglielmi, S. (2008). Improvement in the quality of the catering service of a rehabilitation hospital. *Clinical Nutrition (Edinburgh, Lothian)*, *27*, 105–114. doi:10.1016/j.clnu.2007.10.004

Hair, J. F., Anderson, R. E., Tatham, R. L., & Black, W. C. (1995). *Multivariate data analysis*. Upper Saddle River, NJ: Prentice Hall.

Hsiao, C., Pai, J., & Chiu, H. (2009). The study on the outsourcing of Taiwan's hospitals: a questionnaire survey research. *BMC Health Services Research*, *9*, 78. doi:10.1186/1472-6963-9-78

Kim, Y., Cho, C., & Ahn, S. (2008). A study on medical services quality and its influence upon value of care and patient satisfaction. *Total Quality Management*, *19*(11), 1155–1171. doi:10.1080/14783360802323594

Krowinski, W. J., & Steiber, S. R. (1996). *Measuring and managing patient satisfaction* (2nd ed.). Chicago, IL: Hospital Publishing.

Kucukarslan, S. (2008). Evaluating medication-related services in a hospital. *Research in Social & Administrative Pharmacy*, *4*, 12–22. doi:10.1016/j.sapharm.2007.01.001

Lee, S. M., Rho, B. H., & Lee, S. G. (2003). Impact of Malcolm Baldrige National Quality Award criteria on organizational quality performance. *International Journal of Production Research*, *41*(9), 2003–2020. doi:10.1080/0020754031000077329

McFadden, K., Henagan, S., & Gowen, C. (2009). The patient safety chain: Transformational leadership's effect on patient safety culture, initiatives, and outcomes. *Journal of Operations Management*, *27*, 390–404. doi:10.1016/j.jom.2009.01.001

This work was previously published in the International Journal of Artificial Life Research, Volume 2, Issue 4, edited by E. Stanley Lee and Ping-Teng Chang, pp. 43-56, copyright 2011 by IGI Publishing (an imprint of IGI Global).

Chapter 13 Generating Fully Bounded Chaotic Attractors

Zeraoulia Elhadj
University of Tébessa, Algeria

ABSTRACT

Generating chaotic attractors from nonlinear dynamical systems is quite important because of their applicability in sciences and engineering. This paper considers a class of 2-D mappings displaying fully bounded chaotic attractors for all bifurcation parameters. It describes in detail the dynamical behavior of this map, along with some other dynamical phenomena. Also presented are some phase portraits and some dynamical properties of the given simple family of 2-D discrete mappings.

1. INTRODUCTION

Generating chaotic attractors from nonlinear dynamical system is quite important, because of their applicability in sciences and engineering. The discreet mathematical models are gotten directly via scientific experiences, or by the use of the Poincaré section for the study of continuous-time models. This type of applications is used in secure communications using the notions of chaos (Tsonis, 1992; Andreyev, Belsky, Dmitriev, & Kuminov, 1996; Newcomb & Sathyan, 1983). Many papers have described chaotic systems, one of the most famous being a two-dimensional discrete map which models the original Hénon map $(x,y) \rightarrow (1 - a\,x^2 + by, x)$ studied in (Hénon, 1976; Benedicks & Carleson, 1991; Sprott, 1993; Zeraoulia & Sprott, 2008). This map has been widely studied because it is the simplest example of a dissipative map with chaotic solutions. It has a single quadratic nonlinearity and a constant area contraction over the orbit in the *xy*-plane. However, the Hénon map is unbounded for the almost values of its bifurcation parameters. Thus, constructing a fully bounded chaotic map

DOI: 10.4018/978-1-4666-3890-7.ch013

is a very important result. In the literature, there is some cases where the boundedness of a map was proved rigorously in some regions of the bifurcation parameters space, for example in (Zeraoulia & Sprott, 2008) it was proved that the two-dimensional, C^{∞} discrete mapping given by $(x,y) \to (1 - a\sin x + by, x)$ is bounded for all $|b| < 1$ and unbounded for all and $|b| > 1$. This map is capable to generating "multi- fold" strange attractors via period-doubling bifurcation routes to chaos. This partial boundedness of the above map is due to the presence of the terms by and x. To avoid this problem, we will consider maps of the form $(x,y) \to (f(\mu, X), g(\mu, X))$ where $\mu \in R^k$ is the vector of bifurcation parameters space and $X \in R^n$ is the vector of the state space. The simplest form of this map is obtained when the functions *f* and *g* are linear and the resulting map displays chaotic attractors.

In this paper we present some phase portrait and some dynamical properties of the following simple family of 2-D discrete mappings:

$$\begin{pmatrix} x \\ y \end{pmatrix} \to \begin{pmatrix} f(a_0, a_3, \sin x, \cos y, \cos x, \sin y) \\ g(a_0, a_3, \sin x, \cos y, \cos x, \sin y) \end{pmatrix} \quad (1)$$

where a_0, a_3 makes a part of the bifurcation parameters space and *f* and *g* are linear functions in their corresponding arguments. Equation (1) is an interesting minimal system, similar to the 2-D linear quadratic mapping but with the functions $\sin x, \cos y, \cos x,$ and $\sin y$.

2. SOME DYNAMICAL PROPERTIES

The different cases of map (1) includes the following maps:

$$\begin{pmatrix} x \\ y \end{pmatrix} \to \begin{pmatrix} a_0 + a_1 \sin x + a_2 \cos y \\ a_3 + a_4 \cos x + a_5 \sin y \end{pmatrix},$$

$$\begin{pmatrix} x \\ y \end{pmatrix} \to \begin{pmatrix} a_0 + a_1 \sin y + a_2 \cos x \\ a_3 + a_4 \cos y + a_5 \sin x \end{pmatrix},$$

$$\begin{pmatrix} x \\ y \end{pmatrix} \to \begin{pmatrix} a_0 + a_1 \cos x + a_2 \sin y \\ a_3 + a_4 \sin x + a_5 \cos y \end{pmatrix},$$

$$\begin{pmatrix} x \\ y \end{pmatrix} \to \begin{pmatrix} a_0 + a_1 \cos y + a_2 \sin x \\ a_3 + a_4 \sin y + a_5 \cos x \end{pmatrix}, \ldots \text{etc.}$$

It is clear that the system (1) is not symmetric in general, but it is bounded for all bifurcation parameters values. Generally, this main property avoids some problems related to the existence of unbounded solutions. For each map there are 10 terms (six parameters and four functions $\sin x, \cos y, \cos x,$ and $\sin y$) and the map is simple if it displays chaotic attractor with the minimum number of terms. In this paper we focus our attention to the following map

$$\begin{pmatrix} x \\ y \end{pmatrix} \to \begin{pmatrix} a_0 + a_1 \sin x + a_2 \cos y \\ a_3 + a_4 \cos x + a_5 \sin y \end{pmatrix} \quad (2)$$

in which the same logic apply for the other cases. The computing of fixed points and the determination of their stability type required some algebra and some numerical resolution of the corresponding equations. We omit this method and we will consider an approach based on the notion of Lyapunov exponents. Indeed, it was shown in Li and Chen (2004) that if we consider a system $x_{k+1} = f(x_k)$, $x_k \in \Omega \subset \Re^n$, such that $\|h'(x)\| = \sqrt{\lambda_{\max}\left((h'(x))^T h'(x)\right)} \le N < +\infty$, with a smallest eigenvalue of $(h'(x))^T h'(x)$ that satisfies: $\lambda_{\min}\left((h'(x))^T h'(x)\right) \ge \theta > 0$, where $N^2 \ge \theta$, then, for any $x_0 \in \Omega$, all the Lyapunov exponents at x_0 are located inside $\left[\frac{\ln \theta}{2}, \ln N\right]$, that is, $\frac{\ln\theta}{2} \le l_i\left(x_0\right) \le \ln N, \quad i = 1,2,...,n,$ where $l_i\left(x_0\right)$ are the Lyapunov exponents for the map h and $\| \ \|$ is the Euclidian norm in $\Re^n$. The application of this result to the map (2) show that if $N = \sqrt{\alpha + \frac{1}{2}\beta} < 1$, that is $\alpha + \frac{1}{2}\beta < 1$, then the map (2) is not chaotic, where $\alpha = \frac{a_1^2 + a_2^2 + a_4^2 + a_5^2}{2}$ and $\beta = a_5^2 + a_1^2 + 2\left|a_1 a_5\right| + a_2^2 + a_4^2 + 2\left|a_2 a_4\right|$. The inequality $\alpha + \frac{1}{2}\beta < 1$ helps us to finding numerically the chaotic regions of the map (2) in the bifurcation parameters space by taking the compliment of the resulting set. Notice that the determination of chaotic attractors using the above result is not possible since the expression $\lambda_{\min}\left(\left(h'\left(x\right)\right)^T h'\left(x\right)\right)$ has no lower bound.

3. CLASSIFICATION BASED ON THE NUMBER OF NONLINEARITIES

In this paper, we are only interested to finding simplest cases of the map (2). The nonlinearities here are $\sin x, \cos y, \cos x,$ and $\sin y$. The classification is based on the number of these nonlinearities appeared in the formula of the map (2):

3.1. One Nonlinearity

There four cases with only one nonlinearity in one component and the other component is constant, the resulting 1-D map displays polynomial like dynamics (Douady & Hubbard, 1985; Feigenbaum, 1979). This case was deeply studied in the literature and cannot be considered here.

3.2. Two Nonlinearities

There are six cases:

$$\begin{pmatrix} x \\ y \end{pmatrix} \to \begin{pmatrix} a_0 + a_1 \sin x + a_2 \cos y \\ a_3 \end{pmatrix},$$

$$\begin{pmatrix} x \\ y \end{pmatrix} \to \begin{pmatrix} a_0 \\ a_3 + a_4 \cos x + a_5 \sin y \end{pmatrix},$$

$$\begin{pmatrix} x \\ y \end{pmatrix} \to \begin{pmatrix} a_0 + a_1 \sin x \\ a_3 + a_4 \cos x \end{pmatrix}, \begin{pmatrix} x \\ y \end{pmatrix} \to \begin{pmatrix} a_0 + a_1 \sin x \\ a_3 + a_5 \sin y \end{pmatrix},$$

$$\begin{pmatrix} x \\ y \end{pmatrix} \to \begin{pmatrix} a_0 + a_2 \cos y \\ a_3 + a_4 \cos x \end{pmatrix}, \begin{pmatrix} x \\ y \end{pmatrix} \to \begin{pmatrix} a_0 + a_2 \cos y \\ a_3 + a_5 \sin y \end{pmatrix}.$$

The first three cases and the last one are also displays polynomial-like dynamics (Douady & Hubbard, 1985; Feigenbaum, 1979). Thus, the most important cases are given by:

$$\begin{pmatrix} x \\ y \end{pmatrix} \to \begin{pmatrix} a_0 + a_1 \sin x \\ a_3 + a_5 \sin y \end{pmatrix} \tag{3}$$

$$\begin{pmatrix} x \\ y \end{pmatrix} \to \begin{pmatrix} a_0 + a_2 \cos y \\ a_3 + a_4 \cos x \end{pmatrix} \tag{4}$$

Figure 1. A new chaotic attractor obtained from system (3) for: $a_0 = 1.0,\ a_1 = 4.0,\ a_3 = 0.0,\ a_5 = -3.0$.

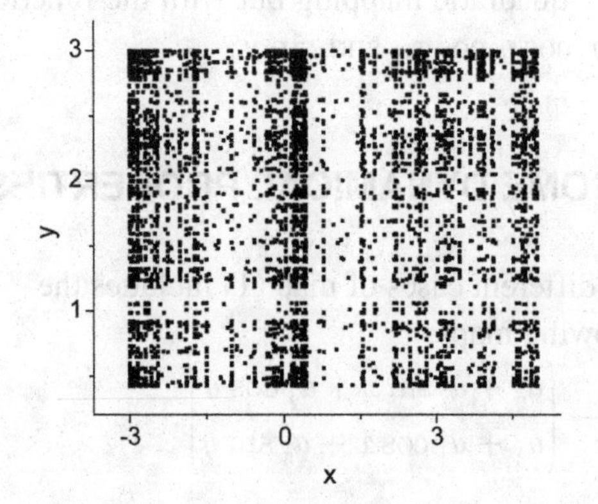

Figure 2. A new chaotic attractor obtained from system(5)for: $a_0 = -2.0,\ a_1 = -1.0,\ a_2 = -2.0,\ a_3 = 1.0,\ a_4 = 0.0,\ a_5 = -3.0.$

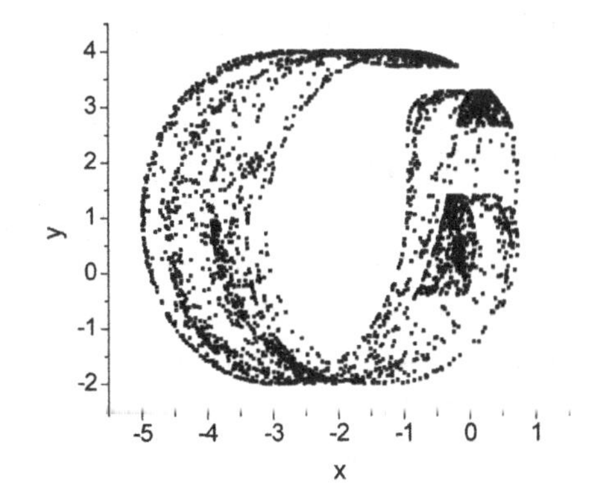

Figure 3. A new chaotic attractor obtained from system(6)for: $a_0 = -2.0,\ a_1 = -1.0,\ a_2 = -2.0,\ a_3 = 1.0,\ a_4 = -3.0,\ a_5 = 0.0.$

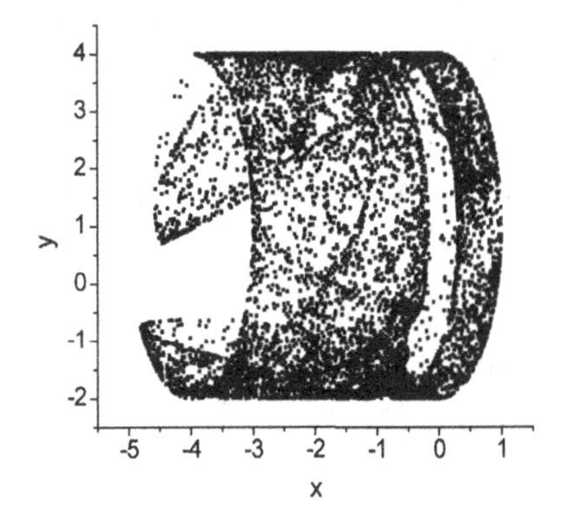

Figure 4. A new chaotic attractor obtained from system(7)for: $a_0 = -2.0,\ a_1 = 0.0,\ a_2 = -2.0,\ a_3 = 0.0,\ a_4 = -3.0,\ a_5 = -1.0.$

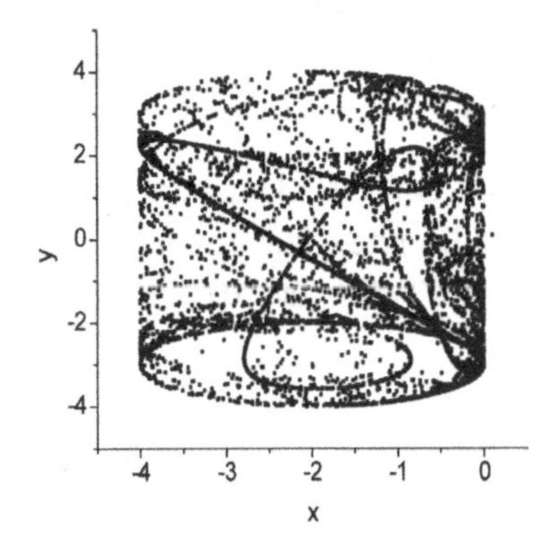

Figure 5. A new chaotic attractor obtained from system(8)for: $a_0 = -2.0,\ a_1 = -4.0,\ a_2 = 0.0,\ a_3 = 0.0,\ a_4 = -3.0,\ a_5 = -1.0.$

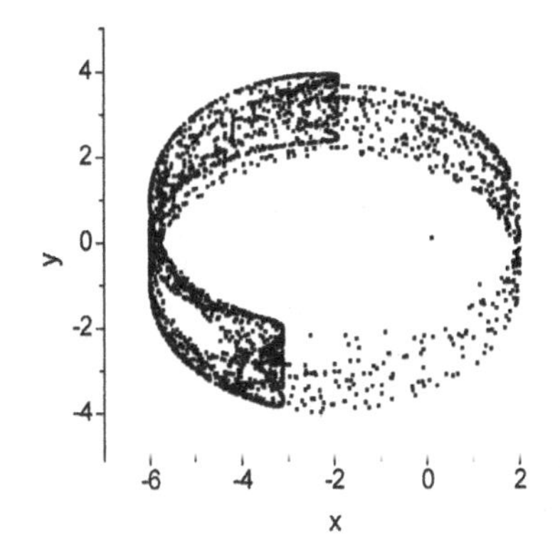

Figure 6. A new chaotic attractor obtained from system(2)for: $a_0 = -2.0,\ a_1 = -4.0,\ a_2 = 2.0,\ a_3 = 0.0,\ a_4 = -3.0,\ a_5 = -1.0.$

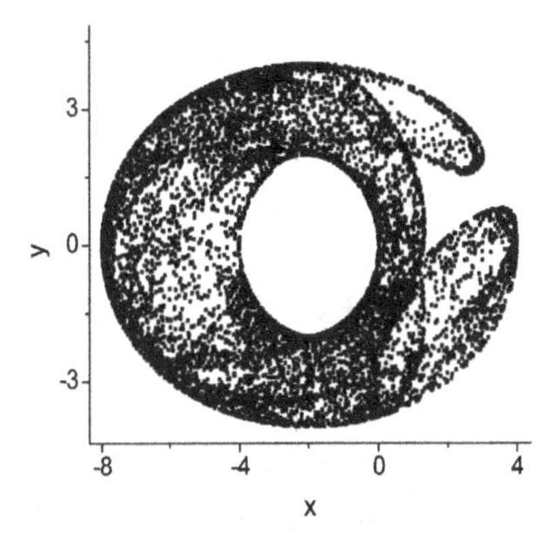

Figure 7. A new chaotic attractor obtained from system (2) for: $a_0 = 0.0,\ a_1 = 1.0,\ a_2 = 2.0,\ a_3 = 0.0,\ a_4 = -3.0,\ a_5 = -1.0.$

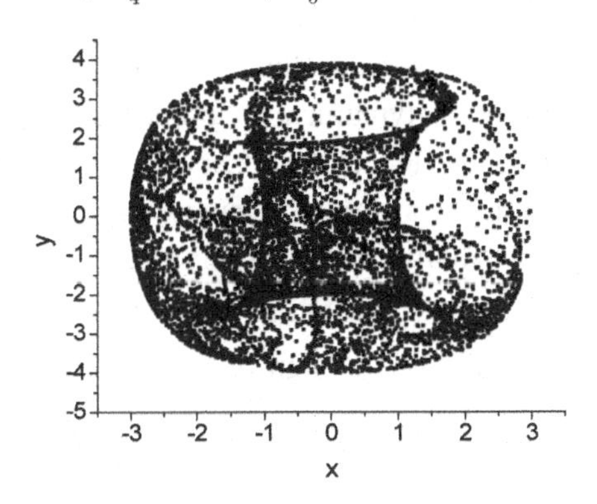

3.3. Three Nonlinearities

There are four cases:

$$\begin{pmatrix} x \\ y \end{pmatrix} \to \begin{pmatrix} a_0 + a_1 \sin x + a_2 \cos y \\ a_3 + a_5 \sin y \end{pmatrix} \quad (5)$$

$$\begin{pmatrix} x \\ y \end{pmatrix} \to \begin{pmatrix} a_0 + a_1 \sin x + a_2 \cos y \\ a_3 + a_4 \cos x \end{pmatrix} \quad (6)$$

$$\begin{pmatrix} x \\ y \end{pmatrix} \to \begin{pmatrix} a_0 + a_2 \cos y \\ a_3 + a_4 \cos x + a_5 \sin y \end{pmatrix} \quad (7)$$

$$\begin{pmatrix} x \\ y \end{pmatrix} \to \begin{pmatrix} a_0 + a_1 \sin x \\ a_3 + a_4 \cos x + a_5 \sin y \end{pmatrix} \quad (8)$$

3.4. Four Nonlinearities

There is only the case (2)

3.5 Observation of New Chaotic Attractors

There are several possible ways for a dynamical system to make a transition from regular behaviour to chaos. According to the bifurcation diagrams with respect to each parameter $(a_i)_{0\le i\le 5}$, , one can observe the chaotic regions for system (1) and this effect permits to finding the possible corresponding chaotic attractors. We will illustrate some newly observed chaotic attractors, along with some other dynamical phenomena. From our numerical experiences we observe that these new chaotic attractors shown in Figures 1 through 7 persist for small values of their bifurcation parameters $(a_i)_{0\le i\le 5}$ and it seem that their orientation depend mainly of the sign of these parameters.

The aperiodicity of these attractors can be seen from the calculation of the power spectrum of the time series (one can chose the x-component which confirm that the spectrum is broadband).

4. CONCLUSION

We have reported the finding of several new chaotic attractors obtained from a fully bounded 2-D mapping. We have then briefly discussed some dynamical properties.

REFERENCES

Andreyev, Y. V., Belsky, Y. L., Dmitriev, A. S., & Kuminov, D. A. (1996). Information processing using dynamical chaos: Neural networks implantation. *IEEE Transactions on Neural Networks*, *7*, 290–299. doi:10.1109/72.485632

Benedicks, M., & Carleson, L. (1991). The dynamics of the Hénon maps. *The Annals of Mathematics*, *133*, 1–25. doi:10.2307/2944326

Douady, A., & Hubbard, J. H. (1985). On the dynamics of polynomial-like mappings. *Annales Scientifiques de l'E.N.S, 18*(2), 287-343.

Feigenbaum, M. (1979). The universal metric properties of nonlinear transformation. *Journal of Statistical Physics*, *21*, 669–706. doi:10.1007/BF01107909

Hénon, M. (1976). A two dimensional mapping with a strange attractor. *Communications in Mathematical Physics*, *50*, 69–77. doi:10.1007/BF01608556

Li, C., & Chen, G. (2004). Estimating the Lyapunov exponents of discrete systems. *Chaos (Woodbury, N.Y.)*, *14*(2), 343–346. doi:10.1063/1.1741751

Newcomb, R. W., & Sathyan, S. (1983). An RC op amp chaos generator. *IEEE Transactions on Circuits and Systems*, *30*, 54–56. doi:10.1109/TCS.1983.1085277

Sprott, J. C. (1993). *Strange attractors: Creating patterns in chaos*. New York, NY: M & T Books.

Tsonis, A. A. (1992). *Chaos: From theory to applications*. New York, NY: Plenum Press.

Zeraoulia, E., & Sprott, J. C. (2008). A minimal 2-D quadratic map with quasi-periodic route to chaos. *International Journal of Bifurcation and Chaos in Applied Sciences and Engineering*, *18*(5), 1567–1577. doi:10.1142/S021812740802118X

Zeraoulia, E., & Sprott, J. C. (2008). A two-dimensional discrete mapping with C^∞-multifold chaotic attractors. *Electronic Journal of Theoretical Physics*, *5*(17), 111–124.

This work was previously published in the International Journal of Artificial Life Research, Volume 2, Issue 3, edited by E. Stanley Lee and Ping-Teng Chang, pp. 36-42, copyright 2011 by IGI Publishing (an imprint of IGI Global).

Chapter 14
Orbit of an Image Under Iterated System II

S. L. Singh
Pt. L. M. S. Govt. Autonomous Postgraduate College Rishikesh, India

S. N. Mishra
Walter Sisulu University, South Africa

Sarika Jain
Amity University, India

ABSTRACT

An orbital picture is a mathematical structure depicting the path of an object under Iterated Function System. Orbital and V-variable orbital pictures initially developed by Barnsley (2006) have utmost importance in computer graphics, image compression, biological modeling and other areas of fractal geometry. These pictures have been generated for linear and contractive transformations using function and superior iterative procedures. In this paper, the authors introduce the role of superior iterative procedure to find the orbital picture under an IFS consisting of non-contractive or non-expansive transformations. A mild comparison of the computed figures indicates the usefulness of study in computational mathematics and fractal image processing. A modified algorithm along with program code is given to compute a 2-variable superior orbital picture.

1. INTRODUCTION

Let (X, d) be a complete metric space. A map $g: X \to X$ is a Banach contraction (also called strictly contractive transformation by Barnsley (2006)) if

$d(gx, gy) \leq qd(x, y)$ for all $x, y \in X$, where $0 \leq q < 1$.

The map g is non-expansive if $q = 1$. A translation map, identity map and isometry are simple examples of non-expansive maps. Some of the properties of contractive maps do not carry over to non-expansive maps. A non-expansive map may not have a unique common fixed point. For example, the identity map on a metric space has every point fixed. Even in a compact space, the sequence of iterates of a non-expansive map

DOI: 10.4018/978-1-4666-3890-7.ch014

sometimes does not converge to a fixed point. If T^n for some positive integer n, has a fixed point, it does not necessarily imply that T has a fixed point. For the theory of contractive and non-expansive operators in non-linear analysis, one may refer to Agarwal, Meehan, and Regan (2001) and Goebel and Kirk (1990).

Orbital pictures are ubiquitous in fractal geometry, as they are always expressed in terms of transformations of an IFS. An orbital picture of an IFS is developed by traversing a defined path, which may be referred to an orbital path. Indeed, a recursive application of an iterative procedure results in an orbital path. However, some issues are of vital concern, while computing the orbit of the picture. For example, there may be some cases when the orbit overlaps. In such cases, to determine the correct orbital picture, one may use the concept of tops union which nicely describes the union of two pictures. Indeed, tops union describes the methodology to take the union of two pictures so as to define a new picture. In other words, tops union simply means, a picture on the top will remain on the top. Moreover, certain real objects are better described by V-variability, as no two clouds are ever the same or two leaves of the same plant differ. V- variability gives a wide range of fractals to the existing deterministic and random fractals, within the framework of Iterated Function Systems. Vast families of homeomorphic objects can be generated with little variation, which may be used to model and study a wide range of phenomena across many areas of science and technology. The concept of V-variability may be used to model almost same looking objects but not exactly the same. For a descriptive knowledge of fractals, V-variable fractals, new generation of fractals and their properties, refer to Barnsley (1993, 2006, 2009), Barnsley, Hutchinson, and Stenflo (2005, 2008), Devaney (1986, 1992), Hutchinson (1981), Mandelbrot (1982), Encarnacao, Peitgen, Sakas, and Englert (1992) and Peitgen, Jürgens, and Saupe (2004).

Orbital pictures for linear and non-linear contractive transformations have initially been studied by Barnsley (2006). Indeed, a semi-group of transformations is needed to generate these beautiful pictures. Orbital pictures are the new mathematical structures, which have been constructed by using one-step feedback process namely, the function iterative procedure. Although, one-step process works very well for contractive transformations, sometimes the problem arises when the transformations are non-contractive, in particular non-expansive transformations but still making a semi-group. In this paper, we generate orbital pictures of different variability for non-contractive and non-expansive transformations using two-step feedback process namely, the superior iterative procedure. It has been observed that the superior iterative procedure generally works very well to construct orbital pictures in case of non-contractive transformations, and converge smoothly wherein one-step process does not converge (Rani, 2010a, 2010b; Rani & Agarwal, 2009a, 2009b, 2010a, 2010b; Rani & Chandra, 2009; Rani & Goel, 2009, 2010; Rani & Kumar, 2002, 2005, 2009, 2003, 2004a, 2004b, 2004c, 2005; Rani & Negi, 2008a, 2008b, 2008c; Rani, Negi, & Mahanti, 2008; Rani & Prasad, 2010; Singh, Jain, & Mishra, 2009, 2011; Singh, Mishra, & Sinkala, 2012; Haq, Sulaiman, & Rani, 2010).

2. PRELIMINARIES

We follow the following notations and definitions from Barnsley (2006), Peitgen et al. (2004), and Singh et al. (2011).

Let X be a linear metric space. An Iterated Function System (IFS) on X can be represented by $F = \left(X; f_1, f_2, \ldots, f_M\right)$ with $f_m : X \rightarrow X$ being transformations for $m = 1, 2, \ldots, M$ and $m \geq 1$.

Now consider the space X together with a collection of iterated function system IFS $\{F_m : m=1,2,\ldots,M\}$ with probabilities

$$F_m = \{X; f_1^m, f_2^m, \ldots, f_l^m;\ p_1^m, p_2^m, \ldots, p_l^m\}$$

Then $\{X; F_1, F_2, \ldots, F_M\}$ (also denoted by $\{X; F_1, F_2, \ldots, F_M; P_1, P_2, \ldots, P_M\}$) is called a superIFS, where the P_m's are probabilities with $\sum P_m = 1,\ P_m \geq 0$ for all $m \in \{1, 2, \ldots, M\}$.

Let $H(X)$ be the collection of all non-empty compact subsets of X, and let the hyper space $(H(X), h)$ is Hausdorff metric space induced by the metric d.

The space $H(X)$ has a remarkable importance in measuring the distance between images. As a consequence, we can also talk about the limit of the sequence of images, i.e., $h(A_\infty, A_k)$ tends to zero as k tends to ∞. For details of measuring the Hausdorff distance, one may refer to Huttenlocher, Klanderman, and Rucklidge (1993), Shonkwiler (1989, 1991), and Singh et al. (2011).

A superIFS can be used to define IFSs acting on higher order spaces such as

$$H(X)^V = \underbrace{H(X) \times H(X) \times \ \ldots \times H(X)}_{V\ times},$$

where $V \in \{1, 2, \ldots\}$. In computer graphics, the parameter V is referred to as the case of having V-buffers in place of 1-buffer. It is a control parameter, which keeps an account on the number of configurations, i.e., at each subsequent iteration, we will have V distinct images. V-variability can be used in fractal theory to find a V-variable fractal set.

Consider a set S together with a binary operation $*$. Then S is known as a semi-group if it satisfies the associative law, i.e., $a*(b*c) = (a*b)*c \quad \forall a,b,c \in S.$ Let $\{f_1, f_2, \ldots, f_N\}$ be a set transformations with $f_i : X \to X$. The semi-group generated by such transformations is the IFS semi-group. It can be represented as $S_{\{X; f_1, f_2, \ldots, f_N\}}$. An orbital picture is constructed from the orbit of pictures under IFS semi-group with the help of tops union. For details one may refer to Barnsley (2006).

The following definition is due to Barnsley (2006, p. 192). Let $\Pi = \Pi_C(X)$ denote the space of pictures with colour values from space C, where the colour space C is a subset of R^3 such as $C = [0,\infty)^3 \subset R^3$ or $C = [0,255]^3 \subset R^3$ or $C = \{0,1,\ldots,255\}^3 \subset R^3$ and the components of a point $c = (c_1, c_2, c_3) \in C$ may be called the colour components with c_1 named red component, c_2 named the green component and c_3 named the blue component.

The tops union $\beta_1 \cup \beta_2$ such that $\beta_1, \beta_2 \in \Pi$ is the picture $\beta_1 \cup \beta_2 \in \Pi$ defined as

$\beta_1 \cup \beta_2 : D_{\beta_1 \cup \beta_2} \subset X \to C$, where $D_{\beta_1 \cup \beta_2}$ is the domain of $\beta_1 \cup \beta_2$ and

$$\beta_1 \cup \beta_2(x) = \begin{cases} \beta_1(x)\ if\ x \in D_{\beta_1} \\ \beta_2(x)\ if\ x \in D_{\beta_2} \setminus D_{\beta_1} \end{cases} \quad \text{for all}$$

$x \in D_{x_1 \cup \beta_2}$ and for all $\beta_1, \beta_2 \in \Pi$.

Indeed, the tops union is used to choose the colour of the subsequent picture by adhering to the colour of its predecessor.

Let $\beta_0 \in \Pi$ be a picture having its domain in X, where X can be taken as R^2. The orbit of the picture β_0 under the semi-group $S(X)$ is the set of pictures $O(\beta_0) = \{f(\beta_0 : f \in S(X))\}$.

Let x_0 be an arbitrary element of real or complex numbers. Construct a sequence $\{x_n\}$ such that $x_n = s f(x_{n-1}) + (1 - s) x_{n-1}$, where $0 < s \leq 1$. The sequence $\{x_n\}$ essentially studied by Mann (1953) in nonlinear analysis, is called superior *sequence of iterates* in computer graphics Rani & Kumar (2004b, 2004c). Notice that the superior iterates $\{x_n\}$ with $s = 1$ are the well-known function iterates (also called Picard iterates). It may be noticed that the function iteration is one-step feedback process while the superior iteration is two-step feedback process, since it requires two inputs. Superior iterations have been found very useful in generation, analysis and applications of superior fractals (Rani et al., 2002, 2003, 2004a, 2004b, 2004c, 2005, 2008, 2008a, 2008b, 2008c, 2009, 2009a, 2009b, 2010, 2010a, 2010b; Singh et al., 2009; Haq et al., 2010).

3. SUPERIOR ORBITAL PICTURES FOR NON-CONTRACTIVE TRANSFORMATIONS

An orbital picture is the picture of the orbit of an image. In general, an orbit is the path, which an image traverses under an Iterated Function System. A feedback process is applied to an IFS in which the outcome of the first transformation works as input for the next transformation. An orbital picture which is generated using function iterations will be called Picard orbital picture, and it will be called superior orbital picture when it is generated using superior iterations. Although, orbital pictures are generated using IFS acting on images, it is important to mention here that the theory behind the orbital pictures is basically mathematics having its roots in fixed-point theory. We may consider them as mathematical entities without regard for pictures. To study the diverse nature of these beautiful pictures, one may refer to Barnsley (2006).

The idea of the Picard orbit of an image may be understood by considering an IFS consisting of a single transformation.

Example 3.1: Suppose $I(x, y)$ is the initial image (inset picture of Figure 1), where x and y are the pixel coordinates and $f = (0.7x + 0.3, 0.7y + 0.3)$ is the transformation, which we have taken to compute the orbit. Here $S_{\{X;f\}}$ is the semi-group generated by f. Then the orbit of f is the sequence of images obtained by the repeated application of f, i.e., the sequence $\{I, f(I), f^2(I), \cdots\}$ (Figure 1).

There are some more important details relating to orbital pictures like picture tiling, over-lapping orbit, non over-lapping orbit and much more. For the details, one may refer to Barnsley (2006).

In a similar manner, superior orbital path of a non-contractive transformation $f = (x + x^2 - 1, y + y^2 - 1)$ can be found. Orbit of f is the sequence of images obtained by the repeated application of $sf(I) + (1-s)f(I)$ which is a two-step feedback process. We remark here that Picard iterates of this map do not follow

Figure 1. Picard orbit of a picture

the required path wherein superior iterates follow the desired path of moving in a diagonal line (Figure 2).

Function iterative procedures are found very useful for describing the path when the transformations are contractive. Since the function iterates of non-contractive, in particular non-expansive, transformations need not converge (Agarwal et al., 2001). Our study suggests that in such cases if we apply the superior iterative procedure, the problem is resolved to a good extent. In this paper, we extend our work (Singh et al., 2011) by applying the superior iterative procedure to non-contractive and non-expansive transformations. We make an attempt to develop a working model so that the objects obtained for non-contractive and non-expansive transformations are manageable. For this, we compare the results obtained by applying the function iterative procedure with those obtained by the superior iterative procedure. We find that the sequence of objects obtained by superior iterations gives an attractive orbital picture, while the sequence of objects obtained by the function iterative procedure oscillates and does not converge towards a regular pattern. The following algorithm describes the computational steps to obtain an orbital picture by the superior iterative procedure:

Algorithm 3.1: To start the experiment we need an input/base image $I(x, y)$, where x and y are the pixel coordinates. We can normalize the pixel coordinates to make the image fit on the computer screen. The following steps illustrate the steps used in the program to construct the superior orbital picture.

1. Consider the base image $I(x, y)$ and apply the superior iteration on I, i.e., compute $sf_1\left(I\left(x,y\right)\right)+\left(1-s\right)\left(x,y\right)$. Store this result in a temporary memory O_1, say.
2. Now apply $sf_2\left(I\left(x,y\right)\right)+\left(1-s\right)\left(x,y\right)$ and store the result in another temporary memory O_2, say.
3. Take the union of I, O_1 and O_2 using tops union and store this image as output image O.
4. Clear the temporary images O_1 and O_2. Make output image O to work as new input image for the next iteration.
5. Repeat steps 1 through 4 for sufficient number of times.

Using the above algorithm, we generate some orbital pictures (see Examples 3.2 and 3.3).

Example 3.2: Consider the IFS semi-group generated by f_1 and f_2, where

$$f_1 = \left(1-x, 1-y\right),$$
$$f_2 = \left(\frac{1+2x^2}{3}, \frac{1+2y^2}{3}\right)$$

are non-contractive or non-expansive transformations.

Figure 2. Superior orbit of a picture for non-contractive map $f = \left(x+x^2-1,\ y+y^2-1\right)$

For the sake of comparison, we generate the orbital picture by using both the procedures, i.e., function iterative procedure and superior iterative procedure. Resultant pictures are shown in Figures 3 and 4.

It is clear from Figures 3 and 4 that the object obtained by function iterations (Figure 3) does not converge but oscillates, while the object obtained by superior iteration (Figure 4) converges to an object.

Example 3.3: Consider a non contractive IFS consisting of the transformations f_1 and f_2, where

$$f_1(x,y) = \left(\frac{41x-19y+19}{19x+19y+41}, \frac{-19x+41y+19}{19x+19y+41}\right),$$
$$f_2(x,y) = \left(\frac{-10x-y+19}{10x+10y+10}, \frac{-10x+21y+1}{10x+10y+10}\right).$$

Example 3.3 is a typical example of a non-contractive IFS. It may be shown that function iterates of this IFS do not converge, wherein superior iterates converge. In a similar manner, the sequence of orbital pictures of the above IFS does not converge on applying the function iterative procedure (*cf.* Figures 5 and 7). On the other

Figure 3. Picard orbital picture: non-converging

Figure 4. Superior orbital picture (converging) with s = 0.1 and n = 15

Figure 5. Picard Orbital picture for n = 2

hand, if we apply the superior iterative procedure, the sequence of orbital pictures converges (*cf.* Figures 6 and 8).

In an analogous manner, we generate orbital pictures for two variability. The following algorithm has been used for the same:

Algorithm 3.2: Consider the above-defined superIFS $\{F_1, F_2\}$. Take any two input images, say I_1 and I_2. At each step, we get two output images. In the programming, we use the following procedure:

1. Choose one of the IFS F_1 or F_2 randomly, say F_1 and one of the images, again chosen randomly, say I_1.
2. Compute $sf_1^1\left(I_1\left(x, y\right)\right) + \left(1 - s\right)\left(x, y\right)$ and $sf_1^2\left(I_1\left(x, y\right)\right) + \left(1 - s\right)\left(x, y\right)$ and store both of these images as temporary output images O_1 and O_2 respectively.
3. Take tops union of I, O_1 and O_2 and store the resulting image as one of the final output image FO_1.
4. Clear the temporary output images O_1 and O_2.

Figure 7. Picard orbital picture for n = 20

Figure 6. Superior orbital picture with s = 0.1 and n = 2

Figure 8. Superior orbital picture with s = 0.1 and n = 20

5. Repeat steps 1 through 4 to calculate the second final output image FO_2.
6. Now FO_1 and FO_2 will work as new input images. For the same, switch over the input and final output images and clear the output screens.
7. Repeat steps 1 through 6 for sufficient number of times.

Example 3.4: To illustrate the above algorithm, we consider the superIFS $\{F_1', F_2'\}$.

$F_1 = \{\square;\ f_1^1, f_2^1\}$ and $F_2 = \{\square;\ f_1^2, f_2^2\}$, where $\square := [0,1] \times [0,1] \subset R^2$ and

$$f_1^1(x,y) = \left(1-x, 1-y\right), \quad p_1^1 = \frac{1}{2},$$

$$f_2^1(x,y) = \left(\frac{1+2x^2}{3}, \frac{1+2y^2}{3}\right), \quad p_2^1 = \frac{1}{2},$$

$$f_1^2(x,y) = \left(0.2x\left(1-x\right), 0.2y\left(1-y\right)\right), \quad p_1^2 = \frac{1}{2},$$

$$f_2^2(x,y) = \left(x+x^2-1, y+y^2-1\right), \quad p_2^2 = \frac{1}{2},$$

where $f_1^1, f_2^1, f_1^2, f_2^2$ are non-contractive or non-expansive transformations. It can be verified that the set of transformations $\{f_1^1, f_2^1, f_1^2, f_2^2\}$ generates a semi-group.

Indeed, using the above preliminaries and algorithms, we have taken some illustrations which require computer programming to visualize the outcome on the screen. For this purpose, we have written a program in Visual C++ using Visual Studio 2005. For the sake of illustration, we provide the program code (given at the end) to calculate a 2-variable orbital picture, which is relatively more complicated than others.

4. CONCLUDING REMARKS

Orbital pictures and *V*-variable orbital pictures have been generated using one-step feedback process namely, the function iterative procedure in Barnsley (2006). These attractive pictures could also be generated using a two-step feedback process namely, superior iterative procedure. Some restrictions are noticed for generating the orbital pictures by the function iterative procedure, specifically when the transformations are non-linear non-contractive or non-expansive. In this regard, we compute orbital and *V*-variable orbital pictures using the superior iterative procedure and compare the results obtained by using the function iterative procedure. Indeed, non-expansive transformations generally do not converge when using the function iterative procedure, as compared to contractive transformations. We develop an algorithm to compute orbital and *V*-variable orbital pictures using the superior iterative procedure and the results thus obtained have a vital role in advanced non-linear analysis.

In our experimental study, we start with a simple example to understand the orbital nature of an object (*cf.* Figures 1 and 2). Afterwards, we generate pictures using the function iterative procedure as well as the superior iterative procedure and make a comparison between the objects obtained by these two procedures. We find that the sequence of objects computed by using the superior iterative procedure converges in a regular and smooth fashion, while the objects corresponding to the function iterative procedure do not converge towards a regular pattern. Figures 3 through 8 show the difference between two images of orbital pictures. Recall that a superior iterative procedure is the function iterative procedure with $s = 1$. Moving towards *V*-variability, we include an algorithm to compute a 2-variable orbital picture. Figures 9 and 10 are computed using the given algorithm. On the basis of the

Figure 9. 2-variable Picard orbital picture with n = 20

Figure 10. 2-variable superior orbital picture with s = 0.5 and n = 20

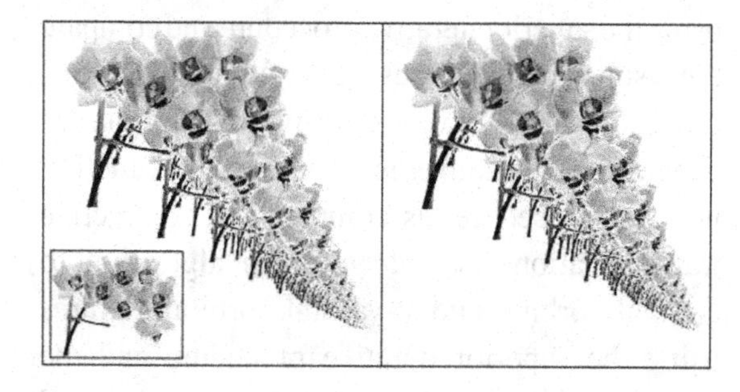

foregoing discussion, we see that the sequence of superior orbital pictures converges to a regular pattern while Picard orbital pictures obtained by using the function iterative procedure show non-converging pattern.

5. PROGRAM CODE

```
\ * CChildView class is a CWnd de-
rivative in Microsoft Visual C++ de-
clared in childView.h. */
void CChildView::ImageConvergFun (CI-
mage * srcImg, CImage * destImg, char
funCase)
\ * In this function, first argument
refers to the input image. Second
argument is the output image after
applying the function given in third
argument. */
{
Double xScaleDown, yScaleDown, tempX,
tempY, s = 0.1, ImageX = 0, ImageY =
0;
int absX, absY;
byte r, g, b;
COLORREF pixel;
int MaxY = srcImg → GetHeight (),
MaxX = srcImg → GetWidth ();
for (int x = 0; x < MaxX; x++)
{
for (int y = 0; y < MaxY; y++)
{
pixel = srcImg → GetPixel(x, y);
r = GetRValue (pixel);
g = GetGValue (pixel);
b = GetBValue (pixel);
if (pixel ≠ m_ transparentColor)
{
xScaleDown = (x - MaxX/2.0)/
(MaxX/2.0);
yScaleDown = (y - MaxY/2.0)/
(MaxY/2.0);
switch (funCase)
{
case 'a':
tempX = 1 - xScaleDown;
tempY = 1 - yScaleDown;
ImageX = s * tempX + (1 - s) * xS-
caleDown;
ImageY = s * tempY + (1 - s) *
yScaleDown;
break;
case 'b':
tempX = (1 + 2 * xScaleDown * xScale-
Down)/ 3;
tempY = (1 + 2 * yScaleDown * yScale-
Down)/ 3;
ImageX = s * tempX + (1 - s) * xS-
caleDown;
ImageY = s * tempY + (1 - s) *
yScaleDown;
break;
case 'c':
tempX = 0.2 * xScaleDown * (1 - xS-
```

```
caleDown);
tempY = 0.2 * yScaleDown * (1 -
yScaleDown);
ImageX = s * tempX + (1 - s) * xS-
caleDown;
ImageY = s * tempY + (1 - s) *
yScaleDown;
break;
case 'd':
tempX = xScaleDown + (xScaleDown *
xScaleDown) - 1;
tempY = yScaleDown + (yScaleDown *
yScaleDown) - 1;
ImageX = s * tempX + (1 - s) * xS-
caleDown;
ImageY = s * tempY + (1 - s) *
yScaleDown;
break;
default:
;
}
absX = (int)(ImageX * (MaxX/2.0) +
(MaxX/2.0));
absY = (int)(ImageY * (MaxY/2.0) +
(MaxY/2.0));
if (absX > 0 & absX < MaxX & absY > 0
& absY < MaxY)
{
destImg → SetPixelRGB (absX, absY, r,
g, b);
}
}
}
}
}
void CChildView::OnToolsOptiimagecon
v ()
\ * In this function, we have taken
number of iterations, a blank image
and two initial images form the user.
Blank Image serves as a temporary
storage. At the end, resultant images
will be stored as img1 and img2. */
{
UserInputDlg InputDlg;
InputDlg.SetDlgMode (4);
InputDlg.m_Iteration = 5;
CImage *pImage1, *pImage2;
pImage1 = new CImage();
pImage2 = new CImage();
InputDlg.m_blankImage = _T ("C:\
temp\ ImageConverg\ Blank.bmp");
InputDlg.m_Image1 = _T ("C:\ temp\
ImageConverg\ InitImg1.bmp");
InputDlg.m_Image2 = _T ("C:\ temp\
ImageConverg\ InitImg2.bmp");
if (InputDlg.DoModal () ≠ IDOK)
return;
m_strImageName = InputDlg.m_blankIm-
age;
CString strImg1 = InputDlg.m_Image1;
CString strImg2 = InputDlg.m_Image2;
int iterations = InputDlg.m_Itera-
tion;
pImage → Destroy ();
pImage → Load (m_strImageName);
pImage1 → Destroy ();
pImage2  → Destroy ();
pImage1 → Load (strImg1);
pImage2  → Load (strImg2);
m_transparentColor = pImage → Get-
Pixel (0, 0);
pImage → Destroy ();
m_nImageSize = SIZE_ORIGINAL;
OnToolsRefresh ();
CWaitCursor wait;
CImage *ptempImage1, *ptempImage2,
*ptempOutImage1, *ptempOutImage2;
ptempImage1 = new CImage ();
ptempImage2 = new CImage ();
int functiongroup1, functiongroup2;
int Imageno1, Imageno2;
for (int k = 0; k < iterations; k++)
{
ptempOutImage1 = new CImage();
ptempOutImage2 = new CImage();
ptempImage1 → Destroy ();
ptempImage1 → Load (m_strImageName);
```

```
ptempImage2 → Destroy ();
ptempImage2 → Load (m_strImageName);
ptempOutImage1 → Destroy ();
ptempOutImage1 → Load (m_strImage-
Name);
ptempOutImage2 → Destroy ();
ptempOutImage2 → Load (m_strImage-
Name);
functiongroup1 = rand() %2;
Imageno1 = rand() %2;
if(functiongroup1 == 0)
{
if(Imageno1 == 0)
{
ImageConvergFun(pImage1, ptempImage1,
'a');
ImageConvergFun(pImage1, ptempImage2,
'b');
MakeUnionOfImage(ptempImage1, ptem-
pImage2);
MakeUnionOfImage(ptempOutImage1, pIm-
age1);
MakeUnionOfImage(ptempOutImage1,
ptempImage1);
OnToolsRefresh();
}
else
{
ImageConvergFun(pImage2, ptempImage1,
'a');
ImageConvergFun(pImage2, ptempImage2,
'b');
MakeUnionOfImage(ptempImage1, ptem-
pImage2);
MakeUnionOfImage(ptempOutImage1, pIm-
age2);
MakeUnionOfImage(ptempOutImage1,
ptempImage1);
OnToolsRefresh();
}
}
else
{
if(Imageno1 == 0)
{
ImageConvergFun(pImage1, ptempImage1,
'c');
ImageConvergFun(pImage1, ptempImage2,
'd');
MakeUnionOfImage(ptempImage1, ptem-
pImage2);
MakeUnionOfImage(ptempOutImage1, pIm-
age1);
MakeUnionOfImage(ptempOutImage1,
ptempImage1);
OnToolsRefresh();
}
else
{
ImageConvergFun(pImage2, ptempImage1,
'c');
ImageConvergFun(pImage2, ptempImage2,
'd');
MakeUnionOfImage(ptempImage1, ptem-
pImage2);
MakeUnionOfImage(ptempOutImage1, pIm-
age2);
MakeUnionOfImage(ptempOutImage1,
ptempImage1);
OnToolsRefresh();
}
}
\* (Second Half) */
ptempImage1 → Destroy ();
ptempImage1 → Load (m_strImageName);
ptempImage2 → Destroy ();
ptempImage2 → Load (m_strImageName);
functiongroup2 = rand() %2;
Imageno2 = rand() %2;
if(functiongroup2 == 0)
{
if(Imageno2 == 0)
{
ImageConvergFun(pImage1, ptempImage1,
'a');
ImageConvergFun(pImage1, ptempImage2,
'b');
MakeUnionOfImage(ptempImage1, ptem-
```

```
pImage2);
MakeUnionOfImage(ptempOutImage2, pIm-
age1);
MakeUnionOfImage(ptempOutImage2,
ptempImage1);
OnToolsRefresh();
}
else
{
ImageConvergFun(pImage2, ptempImage1,
'a');
ImageConvergFun(pImage2, ptempImage2,
'b');
MakeUnionOfImage(ptempImage1, ptem-
pImage2);
MakeUnionOfImage(ptempOutImage2, pIm-
age2);
MakeUnionOfImage(ptempOutImage2,
ptempImage1);
OnToolsRefresh();
}
}
else
{
if(Imageno2 == 0)
{
ImageConvergFun(pImage1, ptempImage1,
'c');
ImageConvergFun(pImage1, ptempImage2,
'd');
MakeUnionOfImage(ptempImage1, ptem-
pImage2);
MakeUnionOfImage(ptempOutImage2, pIm-
age1);
MakeUnionOfImage(ptempOutImage2,
ptempImage1);
OnToolsRefresh();
}
else
{
ImageConvergFun(pImage2, ptempImage1,
'c');
ImageConvergFun(pImage2, ptempImage2,
'd');
MakeUnionOfImage(ptempImage1, ptem-
pImage2);
MakeUnionOfImage(ptempOutImage2, pIm-
age2);
MakeUnionOfImage(ptempOutImage2,
ptempImage1);
OnToolsRefresh();
}
}
pImage1 → Destroy ();
pImage2 → Destroy ();
delete pImage1;
delete pImage2;
pImage1 = ptempOutImage1;
pImage2 = ptempOutImage2;
OnToolsRefresh();
}
pImage1 → Save("C:\ temp\ ImageCon-
verg\ Img1.bmp");
pImage2 → Save("C:\ temp\ ImageCon-
verg\ Img2.bmp");
ptempImage1 → Destroy ();
ptempImage2 → Destroy ();
delete ptempImage1;
delete ptempImage2;
delete pImage1;
delete pImage2;
}
void CChildView::MakeUnionOfImage(CI
mage * unionImage, CImage * imageTo-
Merge)
\ * In this function, we are taking
the tops union of two images passed
in two arguments and the resultant
image is stored in the first argu-
ment. */
{
int MaxY = unionImage → GetHeight (),
MaxX = unionImage → GetWidth ();
byte r, g, b;
COLORREF pixel;
COLORREF unionBaseImage;
for (int x = 0; x < MaxX; x++)
{
```

```
for (int y = 0; y < MaxY; y++)
{
unionBaseImage = unionImage → Get-
Pixel (x, y);
if (unionBaseImage = m_transparent-
Color)
{
pixel = imageToMerge → GetPixel (x,
y);
if (pixel ≠ m_transparentColor)
{
r= GetRValue (pixel);
g = GetGValue (pixel);
b = GetBValue (pixel);
unionImage → SetPixelRGB (x, y, r, g,
b);
}
}
}
}
}
```

REFERENCES

Agarwal, R. P., Meehan, M., & Regan, D. O. (2001). *Fixed point theory and applications*. Cambridge, UK: Cambridge University Press. doi:10.1017/CBO9780511543005

Barnsley, M. F. (1993). *Fractals everywhere* (2nd ed.). Boston, MA: Academic Press Professional.

Barnsley, M. F. (2006). *Superfractals*. Cambridge, UK: Cambridge University Press.

Barnsley, M. F. (2009). Transformations between Self-Referential Sets. *The American Mathematical Monthly*, *116*(4), 291–304. doi:10.4169/193009709X470155

Barnsley, M. F., Hutchinson, J. E., & Stenflo, Ö. (2005). A fractal valued random iteration algorithm and fractal hierarchy. *Fractals*, *13*(2), 111–146. doi:10.1142/S0218348X05002799

Barnsley, M. F., Hutchinson, J. E., & Stenflo, Ö. (2008). Fractals with partial self-similarity. *Advances in Mathematics*, *218*(6), 2051–2088. doi:10.1016/j.aim.2008.04.011

Devaney, R. L. (1986). *An introduction to chaotic dynamical systems*. Menlo Park, CA: Benjamin/Cummings.

Devaney, R. L. (1992). *A first course in chaotic dynamical systems, theory and experiment*. Reading, MA: Addison-Wesley.

Encarnacao, J. L., Peitgen, H. O., Sakas, G., & Englert, G. (1992). *Fractal geometry and computer graphics*. Berlin, German: Springer-Verlag.

Goebel, K., & Kirk, W. A. (1990). *Topics in metric fixed point theory* (pp. 27–35). Cambridge, UK: Cambridge University Press. doi:10.1017/CBO9780511526152.004

Haq, R. U., Sulaiman, N., & Rani, M. (2010, June 28-29). Superior fractal antennas. In *Proceedings of the Malaysian Technical University Conference on Engineering and Technology*, Melaka, Malaysia (pp. 23-26).

Hutchinson, J. E. (1981). Fractals and self-similarity. *Indiana University Mathematics Journal*, *30*(5), 713–747. doi:10.1512/iumj.1981.30.30055

Huttenlocher, D. P., Klanderman, G. A., & Rucklidge, W. J. (1993). Comparing Images using the Hausdorff Distance. *IEEE Transactions on Pattern Analysis and Machine Intelligence*, *15*(9), 850–863. doi:10.1109/34.232073

Mandelbrot, B. B. (1982). *The fractal geometry of nature*. San Francisco, CA: W. H. Freeman.

Mann, W. R. (1953). Mean value methods in iteration. *Proceedings of the American Mathematical Society*, *4*, 506–510. doi:10.1090/S0002-9939-1953-0054846-3

Peitgen, H. O., Jürgens, H., & Saupe, D. (2004). *Chaos and fractals: New frontiers of science* (2nd ed.). New York, NY: Springer.

Rani, M. (2010a, February 26-28). Superior Antifractals. In *Proceedings of the IEEE International Conference on Computer and Automation Engineering*, Singapore (Vol. 1, pp. 798-802).

Rani, M. (2010b, March 23-25). Superior tricorns and multicorns. In *Proceedings of the 9th WSEAS International Conference on Application of Computer Engineering*, Penang, Malaysia (pp. 58-61).

Rani, M., & Agarwal, R. (2009a). A new experimental approach to study the stability of logistic map. *Chaos, Solitons, and Fractals*, *41*(4), 2062–2066. doi:10.1016/j.chaos.2008.08.022

Rani, M., & Agarwal, R. (2009b). Generation of fractals from complex logistic map. *Chaos, Solitons, and Fractals*, *42*(1), 447–452. doi:10.1016/j.chaos.2009.01.011

Rani, M., & Agarwal, R. (2010a, February 26-28). Effect of noise on Julia sets generated by logistic map. In *Proceedings of the 2nd IEEE International Conference on Computer and Automation Engineering*, Singapore (Vol. 2, pp. 55-59).

Rani, M., & Agarwal, R. (2010b). Effect of stochastic noise on superior Julia sets. *Journal of Mathematical Imaging and Vision*, *36*, 63–68. doi:10.1007/s10851-009-0171-0

Rani, M., & Chandra, M. (2009). Categorization of fractal plants. *Chaos, Solitons, and Fractals*, *41*(3), 1442–1447. doi:10.1016/j.chaos.2008.05.024

Rani, M., & Goel, S. (2009). Categorization of new fractal carpets. *Chaos, Solitons, and Fractals*, *41*(2), 1020–1026. doi:10.1016/j.chaos.2008.04.056

Rani, M., & Goel, S. (2010). A new approach to pattern recognition in fractal ferns. *International Journal of Artificial Life Research*, *1*(2). doi:10.4018/jalr.2010040102

Rani, M., & Kumar, M. (2002). A Fractal hedgehog theorem. *Journal of the Korean Society of Mathematical Education Series B: Pure & Applied Mathematics*, *9*(2), 91–105.

Rani, M., & Kumar, M. (2005). A new approach to superior Julia sets. *Journal of Natural and Physical Science*, *19*(2), 148–155.

Rani, M., & Kumar, M. (2009, December 14-16). Circular saw Mandelbrot sets. In *Proceedings of the WSEAS 14th International Conference on Recent Advances in Applied Mathematics* (pp. 131-136).

Rani, M., & Kumar, V. (2003, Oct 18-20). Fractals in Vedic heritage and fractal carpets. In *Proceedings of the National Seminar on History, Heritage and Development of Mathematical Science*, Allahabad, India (pp. 110-121).

Rani, M., & Kumar, V. (2004a). New fractal carpets. *Arabian Journal for Science and Engineering: Section B. Engineering*, *29*(2), 125–134.

Rani, M., & Kumar, V. (2004b). Superior Julia set. *Journal of the Korean Society of Mathematical Education*, *8*(4), 261–277.

Rani, M., & Kumar, V. (2004c). Superior Mandelbrot set. *Journal of the Korean Society of Mathematical Education*, *8*(4), 279–291.

Rani, M., & Kumar, V. (2005). A new experiment with the logistic function. *Journal of the Indian Academy of Mathematics*, *27*(1), 143–156.

Rani, M., & Negi, A. (2008a). Midgets of superior Mandelbrot set. *Chaos, Solitons, and Fractals*, *36*(2), 237–245. doi:10.1016/j.chaos.2006.06.059

Rani, M., & Negi, A. (2008b). New Julia sets for complex CarotidKundalini function. *Chaos, Solitons, and Fractals*, *36*(2), 226–236. doi:10.1016/j.chaos.2006.06.058

Rani, M., & Negi, A. (2008c). A new approach to dynamic noise on superior Mandelbrot set. *Chaos, Solitons, and Fractals*, *36*(4), 1089–1096. doi:10.1016/j.chaos.2006.07.026

Rani, M., Negi, A., & Mahanti, P. K. (2008). Computer simulation of the behavior of Julia sets using switching processes. *Chaos, Solitons, and Fractals*, *37*(4), 1187–1192. doi:10.1016/j.chaos.2006.10.061

Rani, M., & Prasad, S. (2010). Superior Cantor sets and superior Devils Staircases. *International Journal of Artificial Life Research*, *1*(1), 78–84. doi:10.4018/jalr.2010102106

Shonkwiler, R. (1989). An image algorithm for computing the Hausdorff distance efficiently in linear time. *Information Processing Letters*, *30*(2), 87–89. doi:10.1016/0020-0190(89)90114-2

Shonkwiler, R. (1991). Computing the Hausdorff set distance in linear time for any Lp point distance. *Information Processing Letters*, *38*(4), 201–207. doi:10.1016/0020-0190(91)90101-M

Singh, S. L., Jain, S., & Mishra, S. N. (2009). A new approach to Superfractals. *Chaos, Solitons, and Fractals*, *42*, 3110–3120. doi:10.1016/j.chaos.2009.04.052

Singh, S. L., Jain, S., & Mishra, S. N. (2011). Orbit of an Image under iterated system. *Communications in Nonlinear Science and Numerical Simulation*, *16*, 1469–1482. doi:10.1016/j.cnsns.2010.07.012

Singh, S. L., Mishra, S. N., & Sinkala, W. (2012). A new iterative approach to fractal models. *Communications in Nonlinear Science and Numerical Simulation*, *17*, 521–529. doi:10.1016/j.cnsns.2011.06.014

This work was previously published in the International Journal of Artificial Life Research, Volume 2, Issue 4, edited by E. Stanley Lee and Ping-Teng Chang, pp. 57-74, copyright 2011 by IGI Publishing (an imprint of IGI Global).

Chapter 15
Superior Koch Curve

Sanjeev Kumar Prasad
Ajay Kumar Garg Engineering College, India

ABSTRACT

In this paper, the author presents the design of Superior Koch Curve with different scaling factor, which has wide applications in Fractals Graphics. The proposed curve has been designed using the technique of superior iteration. The Koch curve is the limiting curve obtained by applying the self similar divisions to infinite number of times but in Superior Koch Curve scaling factor is based on superior iteration.

1. NTRODUCTION

It is this similarity between the whole and its parts, even infinitesimal ones that make us consider this curve of von Koch as a line truly marvelous among all. If it were gifted with life, it would not be possible to destroy it without annihilating it whole, for it would be continually reborn from the depths of its triangles, just as life in the universe is. Koch Curve has wide application of Fractal Antenna (Fawwaz, 2008).

Recently, Rani and Kumar (2004) introduced superior iterations in the study of fractals and chaos and showed its power in a series of papers (Kumar et al., 2005, Negi et al., 2008, Rani et al., 2004, 2008). This paper was inspired by superior iterations (Rani et al., 2004) and created superior Koch Curves with different scaling factor.

2. PRELIMINARIES

Begin with a straight line (the blue segment in Figure 1). Divide it into three equal segments and replace the middle segment by the two sides of an equilateral triangle of the same length as the segment being removed (the two red segments in the figure). Now repeat, taking each of the four resulting segments, dividing them into three equal parts and replacing each of the middle segments

DOI: 10.4018/978-1-4666-3890-7.ch015

Figure 1. Koch curve

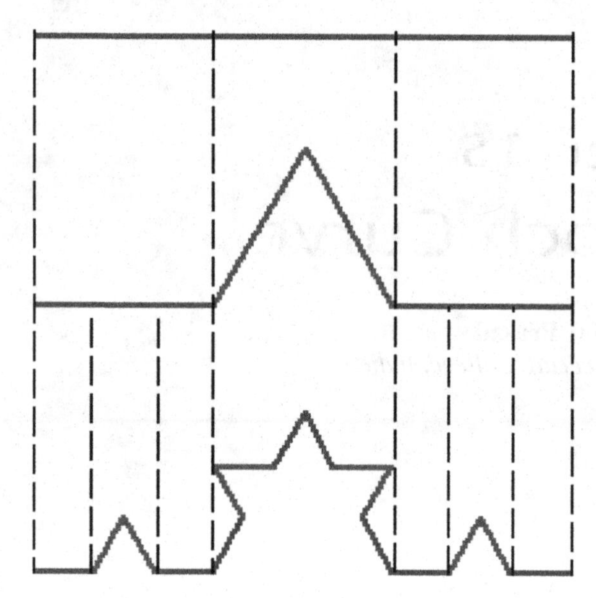

by two sides of an equilateral triangle (the red segments in the bottom figure). Continue this construction.

The Koch curve is the limiting curve obtained by applying this construction an infinite number of times. For a proof that this construction does produce a "limit" that is an actual curve, i.e., the continuous image of the unit interval, see the text by Edgar.

The first iteration for the Koch curve (Figure 2) consists of taking four copies of the original line segment, each scaled by **r** = 1/3. Two segments must be rotated by 60°, one counterclockwise and one clockwise. Along with the required translations, this yields the following Iterated Function System

$$f_1(x) = \begin{bmatrix} .333 & 0 \\ 0 & 0.333 \end{bmatrix} x$$

Scale by r

$$f_2(x) = \begin{bmatrix} .167 & -.289 \\ .289 & 0.167 \end{bmatrix} x + \begin{bmatrix} .333 \\ 0 \end{bmatrix}$$

Scale by r, rotation by 60°

$$f_3(x) = \begin{bmatrix} 0.167 & 0.289 \\ -.289 & 0.167 \end{bmatrix} x + \begin{bmatrix} .500 \\ .289 \end{bmatrix}$$

Scale by r, rotation by − 60°

$$f_4(x) = \begin{bmatrix} .333 & 0 \\ 0 & .333 \end{bmatrix} x + \begin{bmatrix} .667 \\ 0 \end{bmatrix}$$

Scale by r

The fixed invariant set of this IFS (Iterated Function System) is same as the Koch curve.

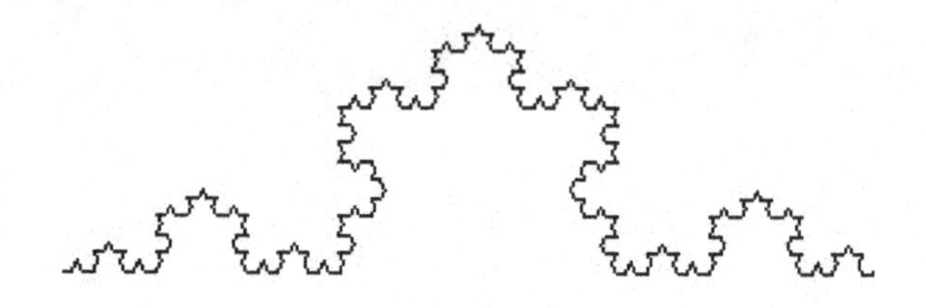

r = 0.29 n = 5 Redraw

2.1 Similarity Dimension

We have hyperbolic IFS (Iterated Function System) with each map being a similitude of ratio **r** < 1. Therefore the similarity dimension, **d**, of the unique invariant set of the IFS is the solution to

$$\sum_{k=1}^{4} r^d = 1 \Rightarrow d = \frac{\log(1/4)}{\log r} = \frac{\log 4}{\log 3} = 1.2619$$

Figure 2. First iteration of Koch curve

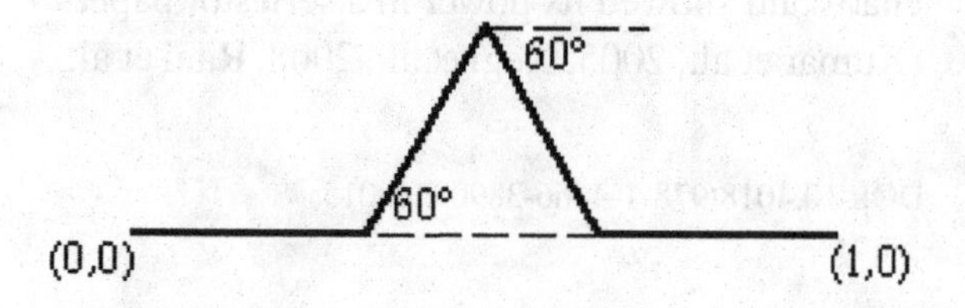

2.2 Special Properties

Koch constructed his curve in 1904 as an example of a non-differentiable curve, that is, a continuous curve that does not have a tangent at *any* of its points. Karl Weierstrass had first demonstrated the existence of such a curve in 1872. The article by Sime Ungar provides a simple geometric proof.

The length of the intermediate curve at the *n*th iteration of the construction is (4/3)^*n*, where *n* = 0 denotes the original straight line segment. Therefore the length of the Koch curve is infinite. Moreover, the length of the curve between any two points on the curve is also infinite since there is a copy of the Koch curve between any two points.

Three copies of the Koch curve placed around the three sides of an equilateral triangle form a simple closed curve that form the boundary of the Koch snowflake (Figure 3).

2.3 Superior Iteration

Let A be a subset of real or complex numbers and f: A → A. For $x_0 \in A$, construct a sequence $\{x_n\}$ in A in the following manner:

$x_n = \beta_n f(x_{n-1}) + (1-\beta_n) x_{n-1}$, where $0 < \beta_n \leq 1$ and $\{\beta_n\}$ is convergent to a non-zero number. Rani and Kumar (2005) called the sequence $\{x_n\}$ constructed above as superior orbit, which was, essentially, due to W. R. Mann, and denoted as $SO(f, x_0, \beta_n)$. Notice that $SO(f, x_0, \beta_n)$ with $\beta_n = 1$ is reduces to Picard orbit (Kumar et al., 2005; Negi et al., 2008; Rani et al., 2004a, 2004b, 2005, 2008, 2009).

3. SUPERIOR KOCH CURVE

Inspired by superior iterations (Rani et al., 2004), we compute superior Koch Curve. We classify it in four categories with their different scaling factor (Figures 4 through 7).

Figure 3. Closed Koch Curve

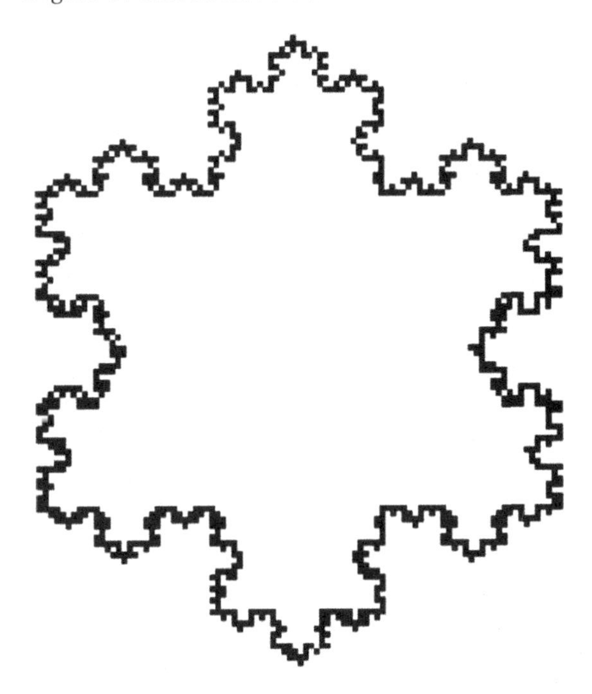

Category 1: Scaling factor of this Superior Koch Curve is 0.1.

Category 2: Scaling factor of this Superior Koch Curve is 0.15.

Category 3: Scaling factor of this Superior Koch Curve is 0 .2.

Category 4: Scaling factor of this Superior Koch Curve is 0.25.

4. CONCLUSION

Koch Curve, an example of classical fractals has wide applications in real word. This paper presents for different Superior Koch Curve with different scaling factor. Koch Curve has wide application of Fractal Antenna; this Superior Koch Curve can be use in Fractal Antenna.

Figure 4. Scaling factor 0.1

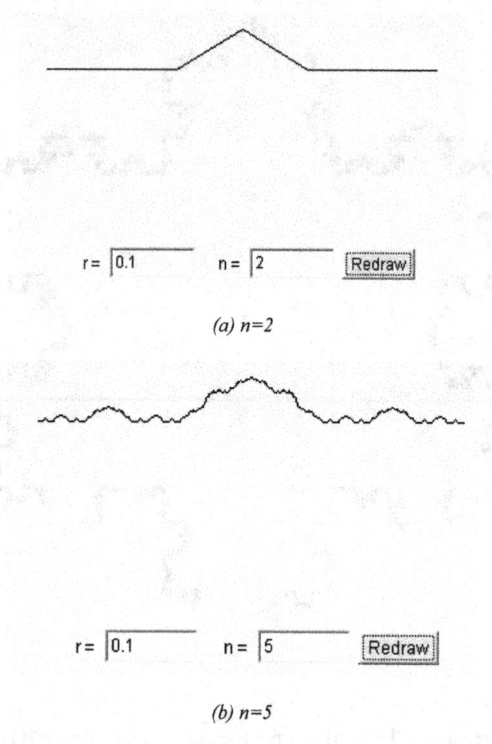

(a) n=2

(b) n=5

Figure 5. Scaling factor 0.15

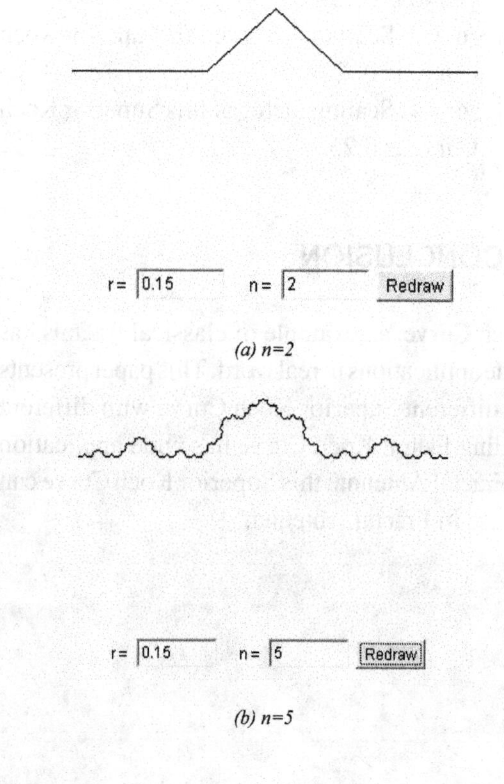

(a) n=2

(b) n=5

Figure 6. Scaling factor 0.2

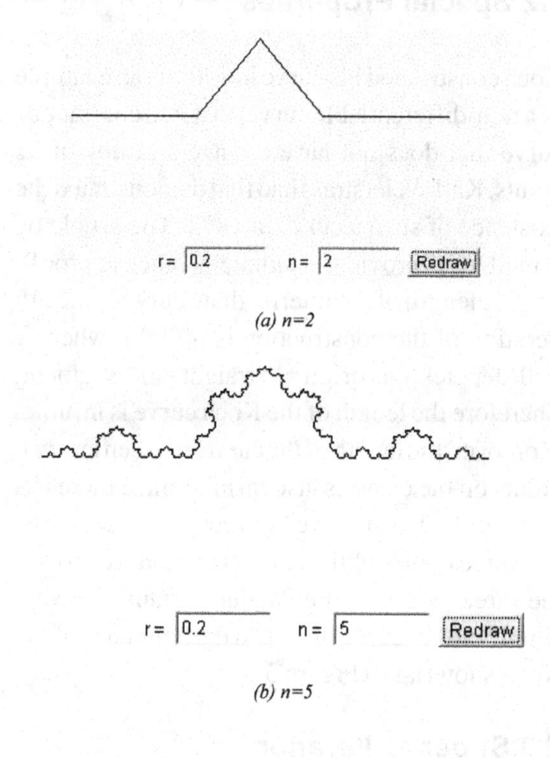

(a) n=2

(b) n=5

Figure 7. Scaling factor 0.25

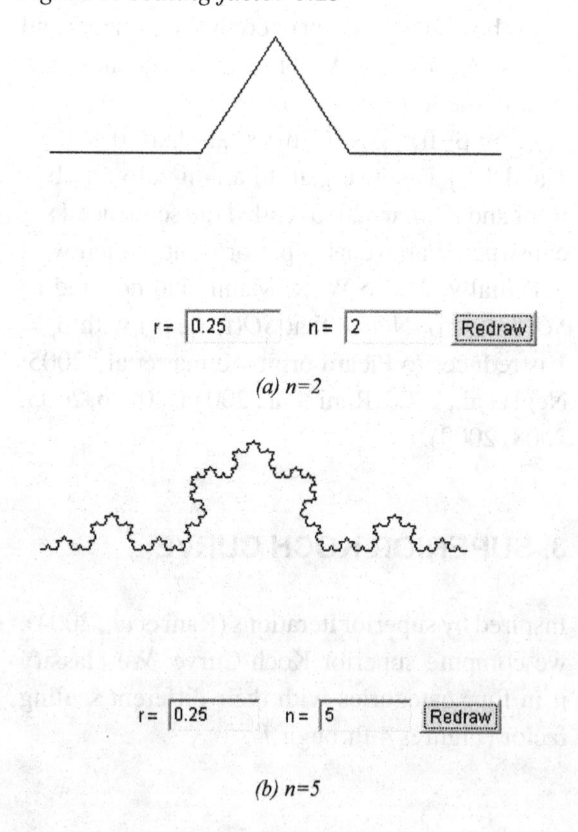

(a) n=2

(b) n=5

5. ALGORITHM

Here, presents the algorithm to generate Superior Koch Curve.

```
SuperiorKochCurve
Set         r = 0.1/0.15/0.2/0.25;
Set         n = 5;
SuperiorKochCurve(g, 1, 30, 190,
30+300, 190);
SuperiorKochCurve (Graphics g, int
level, double x1, double y1, double
x2, double y2)
{
if (level < n)
{
              Set         nx =
(2*x1+x2)/3;
                Set         ny =
(2*y1+y2)/3;
SuperiorKochCurve(g, level+1, x1, y1,
nx, ny);
                Set         ox = nx;
                Set         oy = ny;
                nx = (x1+x2)/2 -
r*(y1-y2);
                ny = (y1+y2)/2 +
r*(x1-x2);
SuperiorKochCurve (g, level+1, ox,
oy, nx, ny);
                ox = nx; oy = ny;
                nx = (x1+2*x2)/3;
                ny = (y1+2*y2)/3;
SuperiorKochCurve (g, level+1, ox,
oy, nx, ny);
SuperiorKochCurve (g, level+1, nx,
ny, x2, y2);           }
else          {
drawLine((int)x1, (int)y1, (int)x2,
(int)y2);
          }
      }
}
```

REFERENCES

Addison, P. (1997). *Fractals and Chaos: An Illustrated Course*. Bristol, UK: Institute of Physics. doi:10.1887/0750304006

Devaney, R. L. (1992). *A first course in chaotic dynamical systems: Theory and experiment*. Reading, MA: Addison-Wesley.

Dovgoshey, O., Martio, O., Ryazanov, V., & Vuorinen, M. (2006). The Cantor function. *Expositiones Mathematicae*, *24*(1), 1–37. doi:10.1016/j.exmath.2005.05.002

Falconer, K. (1990). *Fractal Geometry: Mathematical Foundation and Applications*. Chichester, UK: John Wiley & Sons.

Gianvitorio, J., & Rahmat, Y. (2002). Fractal Antennas: A Novel Antenna Miniaturization Technique and Applications. *IEEE Antennas and Propagation Magazine*, *44*(1), 20–36. doi:10.1109/74.997888

Horiguchi, T., & Morita, T. (1984). Fractal dimension related to devil's staircase for a family of piecewise linear mappings. *Physica A*, *128*(1-2), 289–295. doi:10.1016/0378-4371(84)90092-X

Jibrael, F. J. (2008). Miniature Dipole Antenna Based on Fractal Square Koch Curve. *European Journal of Scientific Research*, *21*(4), 700–706.

Kumar, M., & Rani, M. (2005). A new approach to superior Julia sets. *Journal of Natural and Physical Science*, *19*(2), 148–155.

Mann, W. R. (1953). Mean value methods in iteration. *Proceedings of the American Mathematical Society*, *4*, 506–510. doi:10.1090/S0002-9939-1953-0054846-3

Negi, A., & Rani, M. (2008). A new approach to dynamic noise on superior Mandelbrot set. *Chaos, Solitons, and Fractals*, *36*(4), 1089–1096. doi:10.1016/j.chaos.2006.07.026

Peitgen, H., Jürgens, H., & Saupe, D. (2004). *Chaos and Fractals: New frontiers of science* (2nd ed.). New York, NY: Springer.

Pickover, C. A., & McCarty, K. (1990). Visualizing Cantor cheese construction. *Computers & Graphics, 14*(2), 337–341. doi:10.1016/0097-8493(90)90046-Z

Rani, M., & Agarwal, R. (2008). A new experimental approach to study the stability of logistic map. *Chaos, Solitons, and Fractals, 41*(4), 2062–2066. doi:10.1016/j.chaos.2008.08.022

Rani, M., & Kumar, V. (2004). Superior Julia set. *Journal of the Korean Society of Mathematical Education Series D: Research in Mathematical Education, 8*(4), 261–277.

Rani, M., & Kumar, V. (2004). Superior Mandelbrot set. *Journal of the Korean Society of Mathematical Education Series D: Research in Mathematical Education, 8*(4), 279–291.

Rani, M., & Kumar, V. (2005). A new experiment with the logistic function. *Journal of the Indian Academy of Mathematics, 27*(1), 143–156.

Rani, M., & Kumar Prasad, S. (2010). Superior Cantor sets and superior devil staircases. In Koetsier, T., & Bergmans, L. (Eds.), *Mathematics and the divine: A historical study* (pp. 532–547). Amsterdam, The Netherlands: Elsevier. doi:10.4018/jalr.2010102106

Rani, M., & Negi, A. (2008). New Julia sets for complex Carotid-Kundalini function. *Chaos, Solitons & Fractals: The Interdisciplinary Journal of Nonlinear Science – Nano and Quantum Technology, 36*(2), 226-236.

Schroder, M. (1990). *Fractals, chaos, power laws: Minutes from an infinite paradise*. New York, NY: FreeMan.

Solomyak, B. (1997). On the measure of arithmetic sums of Cantor sets. *Indagationes Mathematicae, 8*(1), 133–141. doi:10.1016/S0019-3577(97)83357-5

Thiele, R. (2005). Georg Cantor and the divine. In Koetsier, T., & Bergmans, L. (Eds.), *Mathematics and the divine: A historical study*. Amsterdam, The Netherlands: Elsevier.

Tiehong, T., & Zheng, Z. (2003). A Novel Multiband Antenna: Fractal Antenna. In *Proceedings of the International Conference on Communication Technology* (pp. 1907-1910).

Troll, G. (1991). A devil's staircase into chaotic scattering. *Physica D. Nonlinear Phenomena, 50*(2), 276–296. doi:10.1016/0167-2789(91)90180-H

This work was previously published in the International Journal of Artificial Life Research, Volume 2, Issue 4, edited by E. Stanley Lee and Ping-Teng Chang, pp. 24-31, copyright 2011 by IGI Publishing (an imprint of IGI Global).

Section 3
Intelligent Information Processing and Applications

Chapter 16
Mitigation Strategies for Foot and Mouth Disease:
A Learning-Based Approach

Sohini Roy Chowdhury
Kansas State University, USA

Caterina Scoglio
Kansas State University, USA

William Hsu
Kansas State University, USA

ABSTRACT

Prediction of epidemics such as Foot and Mouth Disease (FMD) is a global necessity in addressing economic, political and ethical issues faced by the affected countries. In the absence of precise and accurate spatial information regarding disease dynamics, learning- based predictive models can be used to mimic latent spatial parameters so as to predict the spread of epidemics in time. This paper analyzes temporal predictions from four such learning-based models, namely: neural network, autoregressive, Bayesian network, and Monte-Carlo simulation models. The prediction qualities of these models have been validated using FMD incidence reports in Turkey. Additionally, the authors perform simulations of mitigation strategies based on the predictive models to curb the impact of the epidemic. This paper also analyzes the cost-effectiveness of these mitigation strategies to conclude that vaccinations and movement ban strategies are more cost-effective than premise culls before the onset of an epidemic outbreak; however, in the event of existing epidemic outbreaks, premise culling is more effective at controlling FMD.

DOI: 10.4018/978-1-4666-3890-7.ch016

1. INTRODUCTION

Predictive epidemiology refers to the analytical study of disease dynamics to predict future outbreaks in space and time so that effective mitigation strategies can be implemented to curb the recurrence of epidemics. Since epizootic diseases like the Foot and Mouth Disease (FMD) raise several political, administrative, economic and welfare issues, it is imperative to analyze the disease dynamics to facilitate adequate preventive measures, especially in countries that report recurring epidemic outbreaks instances. Since the FMD outbreak in the United Kingdom in 2001, several analytical spatio-temporal models have been developed to spatially locate such epidemic outbreaks in time (Morris et al., 2001 ; Bates et al., 2003a, 2003b; Carpenter et al., 2004; Ferguson et al., 2001; Keeling et al., 2001). However, it is important to address that spatio-temporal models have parameters of a possibly global structure. Such structures allow region-independence and adaptability of the models by taking information regarding the environment and neighborhood of geographical locations expressed in terms of model parameters. But, in the absence of the sensitive spatial parameters, we attempt to train a learning-based model on a certain regional data with latent parameters to mimic the predictive performance of spatial predictive models. The novel contribution of this paper is that we study local information regarding the temporal evolution of infection that is hard-coded in geographical regions, by using different learning- based models. Additionally, we simulate instances of mitigations strategies to study the cost-effectiveness of culling, vaccination and movement strategies to reduce the total number of infected livestock at the end of a period under study. Also, the utility function to assess the cost-effectiveness of mitigation strategies is defined in terms of the percentage reduction in the total number of infected livestock to the total cost incurred in million US dollars.

Numerous learning-based models have been developed so far to achieve temporal epidemic predictions. For example, neural network models have been argued to effectively model the dynamics of temporal data (Abidil & Goh, 2006), while time series models have been applied for forecasting the incidences of influenza-like illnesses (ILI) in France (Hawksworth et al., 2003). Also, Bayesian networks are useful for reasoning under uncertainty in artificial intelligence which not only detects an outbreak, but also estimates how acute the epidemic is (Lagazio et al., 2001; Jiang & Wallstrom, 2006). Regressive models have also been implemented to fit and predict outbreak related data (Kobayashi et al., 2007a, 2007b). Additionally, learning-based prediction models have found their importance in predicting wheat leaf wetness (Francl & Panigrahi, 1997; Chtioui et al., 1999), soy- rust in plants (Alexandersen et al., 1997) and critical diseases like influenza (Viboud et al., 2003), malaria (Krishnamurti et al., 2007; Brit et al., 2008) and SARS (Lai, 2005) in humans. However, such models have not found any application in prediction of global epizootic epidemics like FMD so far. Our study aims at analyzing the temporal prediction capability of various temporal prediction models and applying them for spatial predictions of FMD epidemic outbreaks in time.

Learning-based predictors are proactive methods for the classification of epidemic severity and for the development of preemptive disease mitigation strategies. They are good tools to analyze infection spread patterns without relying on background spatial information which is generally unknown or estimated. However, it is noteworthy that learning-based models suffer from higher prediction errors than spatio-temporal predictive models in the absence of a high volume of well correlated data (Chowdhury et al., 2009). This is because learning-based models require a considerable training, validation and testing data in order to predict well. Evidently, while an under-trained model produces higher prediction errors, an over-trained model will generate a high variance in its predictions. Thus, due to the generally sparse nature of sensitive data regarding epidemic outbreaks, learning-based models can be trained to estimate logical bounds to the rate of spread of disease infection with time (Viboud et al., 2003; Abeku et al., 2004). Such bounds can eventually be used for development of mitigation strategies to curb the impacts of the epidemic thus predicted.

In this paper, we propose a few learning-based prediction models that can be used to study the temporal evolution of FMD infection and susceptibility at different administrative districts in Turkey. The predictive models when trained for each administrative district separately can be used to predict the probability of infection and the probability of susceptibility to the FMD Virus (FMDV) in future time instants. The different temporal prediction models are neural networks, autoregressive models and Bayesian networks backed up with Monte-Carlo simulation models. Neural networks are non-linear models to detect sudden fluctuations in infection incidence data and respond accordingly. Contrarily, autoregressive models map the randomness in infection incidence data correlated in space, by trying to estimate a probability distribution function to fit the time series data. Bayesian networks can be applied in instances where no good approximations regarding the probability distribution functions are feasible so that Bayesian estimators based on historical data can be used to predict future occurrences of infection. Also, Monte Carlo simulations help to achieve the gold standards regarding the 95% confidence intervals in the infection and susceptibility probability distribution. We evaluate the performance of each prediction model in simulating mitigation strategies, and we eventually justify their predictive performances in terms of the effectiveness of the various mitigation strategies proposed.

We have simulated and analyzed the effectiveness of mitigation strategies for FMD based on the temporal infection predictions in space. From these simulations, we infer that vaccination and movement ban strategies are effective in impeding the spread of FMDV before the onset of an epidemic outbreak, whereas premise culls are important to severely impede the spread of FMDV after an outbreak has set in. In countries which report recurrences of FMD such as the example of Turkey, it can be effective to develop adequate infrastructure to facilitate mass vaccinations for long term disease mitigation.

In this paper, section 2 describes the four learning-based temporal models while section 3 analyzes the predictive performances of each model. Section 4 describes the practical simulative mitigation strategies followed by the analysis of the mitigation strategies in section 5. Finally, concluding remarks and discussion are presented in section 6.

2. MODEL DESCRIPTION

This section describes the application of neural networks, autoregressive models, Bayesian networks and Monte-Carlo simulations for temporal epidemic predictions. The method for parameter estimation is described, and the predictive performance of each model is also analyzed. Here, we primarily categorize the period under study into two categories. The period wherein the number of reported outbreaks and infected livestock remains at a steady low value is defined as the pre-outbreak period. Contrastingly, the outbreak period is defined as the period when the number of infected livestock increases rapidly to a very large value. The transition period between the pre-outbreak and the outbreak period, is the time when the epidemic sets in. Thus, a well trained model is one which can predict the transition period, and also estimate the total number of infected livestock in the epidemic period under study.

3. EXPERIMENTAL DATA

Due to the lack of a high volume of sensitive data regarding the impact of an FMD outbreak, we study the infection incidence data in Turkey from January 2005 through December 2006. Here, the first 12 months indicate the pre-outbreak period, while the later 12 months indicate the outbreak period with a high number of livestock being reported infected per monthly time period. It is important to assess the impact of the peak seasons and the off-peak seasons on the spread of the FMDV so as to extract information regarding the seasonality and trends followed by rate of infection, thus enabling predictions in future time instants. Although, the study of a 24 month time cycle may seem insufficient to provide substantial conclusions regarding the duration of future outbreaks from a machine learning perspective, it is important to note that we are interested in studying a complete cycle of infection comprising of the pre-outbreak and the outbreak period rather than a yearly variation in the rate of infection. It may therefore be possible that in a certain year in the future time instants, the pre-outbreak and outbreak periods occur within a 12 month time period.

The 24 month period under study here, is divided into two segments, the training data set, which corresponds to the first 15 months (60% of entire data), and the validation data set, which corresponds to the last 9 months (40% of the entire data set). We study the probability of infection incidence, and the probability of susceptibility to infection at 79 administrative districts in Turkey denoted by nodes in Figure 1, assuming that each node is independent and identical in nature. Each node represents a region representing breeding farms, grazing lands, feedlots and meat markets which house livestock that are susceptible to infection and that may contribute to the spread of the FMDV.

We assess the performance of the prediction models on two data sets namely data set S that represents the population of susceptible livestock, and data set I that represents the population of infected livestock at a particular time instant. At each monthly time instant, each node houses a certain amount of livestock that are susceptible to infection and thus, they belong to data set S. It is important to note here that the susceptible population at each node changes at every time instant and, that only a portion of this susceptible population may actually get infected. Due to the presence of infection in neighboring nodes, a fraction of the susceptible livestock become infected and thus, this fraction of infected livestock transition from the data set S to the data set I . Once infected, livestock belonging to data set I, may be removed from the cycle of infection in future time steps by culling. It is noteworthy that at a particular time instant in a particular node, livestock can belong to either the data set S or the data set I and, that transition is only possible from data set S to data set I and not in the reverse direction. So, at each node we compute the probability of infection on data set I as a ratio between the number of infected animals at that node to the total number of infected animals, and probability of susceptibility on data set S as a ratio between the number of susceptible animals at that node to the total number of susceptible animals at a particular monthly time instant . For the infection data, the training data set is I_t, while the validation data set is I_v . Similarly, for the susceptibility data,

Figure 1. Map of Turkey depicting 79 administrative districts represented by circular nodes. The temporal evolution of FMDV in each node is studied separately.

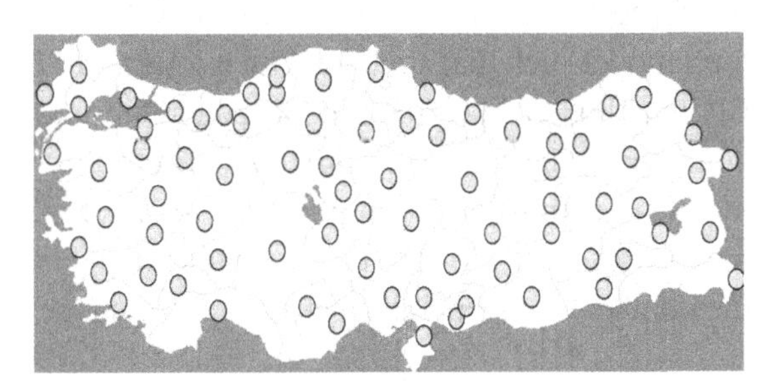

the training data set is S_t, while the validation data set is S_v. All data sets are normalized so that they represent the fraction out of the total number of susceptible or infected population of livestock over the 24 month period at each monthly time instant for every node.

4. STATISTICS FOR MODEL PARAMETERIZATION

We compute spatio-temporal variation in infectivity and susceptibility by studying the temporal evolution in the probability of infection and probability of susceptibility at different regions in space (nodes). Thus, the predictive performance of the learning-based models is analyzed with respect to the actual data such that all prediction errors are probabilistic. The models predict the probability of infection and the probability of susceptibility at each node in progressive time steps of one month, which is compared to the actual probability of infection/susceptibility calculated by the ratio of infected/susceptible animals at a node at a certain time step over the total number of infected/susceptible animals all over Turkey in that time step.

Predictive performance of each model is analyzed in terms of the prediction error between the actual and predicted probabilities expressed in terms of Mean Squared Error (MSE), and Root Mean Squared Error (RMSE). The goodness of model fit is defined in terms of symmetric Maximum Absolute Percentage Error (sMAPE) and the coefficient of determination (R^2). Besides, Akaike Information Criterion (AIC), Bayesian Information Criterion (BIC) are the measure of goodness for a model such that a model with a lower AIC/BIC has higher likelihood to fit the actual data while incurring a lower computational complexity due to parameterization. If k is the number of parameters to be estimated by the model and n is the number of data points, then if we assume the observations to be identically normally distributed, AIC is given below.

$$AIC = 2k + n\ln\left(MSE\right) \quad (1)$$

Similarly, BIC is defined below for normally distributed error with error variance (σ^2).

$$BIC = \ln\left(\sigma_e^2\right) + \frac{k}{n}\ln\left(n\right) \quad (2)$$

Kullback Leibler (KL) divergence distance is a method used to measure the goodness of fit by estimating a non-symmetric measure of distance between two distributions, the actual probability distribution (P), and the probability distribution predicted by a model (Q). This method relies on the concept of relative entropy given in the following equation, such that the model with the lowest KL divergence distance is the best fit model.

$$D_{KL}(P \,||\, Q) = \sum_i P\left(i\right)\frac{P\left(i\right)}{Q\left(i\right)} \quad (3)$$

5. NEURAL-NETWORK MODEL

We implement a three layered feed-forward neural network (NN) with an input layer, hidden layer and output layer using a sigmoidal activation function depicted in Figure 2. The reason for choosing the sigmoidal function is that it is a continuous function that simplifies the backprogation process. Also, the derivative of a sigmoidal function can be easily calculated for the purpose of weight updates using the chains rule.

The first important task in building a NN is the selection of an optimal structure for the multi-layered perceptron. By experience from all the applications of NN, preference goes to the structure in which there are fewer neurons in the hidden layer than neurons in the input layer, so as to

Figure 2. Showing a multi-layered neural network

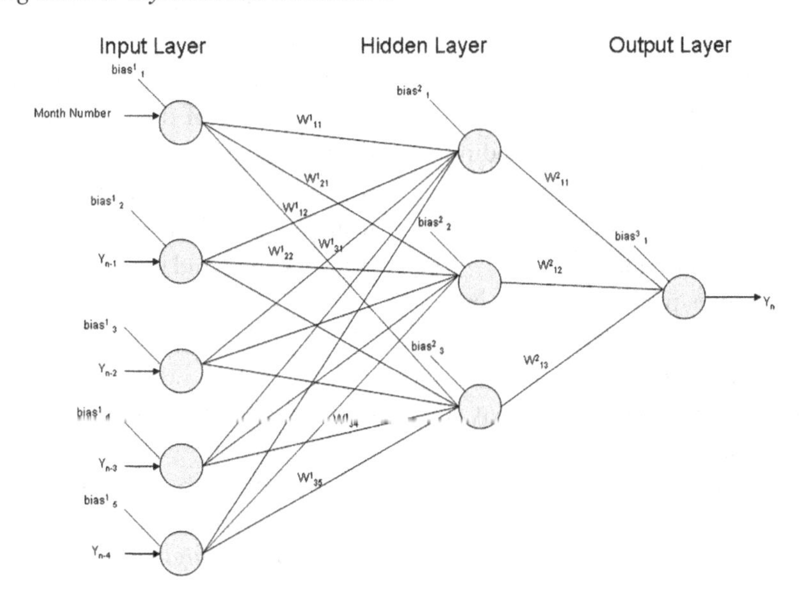

minimize the final size of the NN. Also, according to Ockham's razor principle, it is more likely that a smaller and simpler computing model would have better generalization abilities. There exists a problem known as the bias/variance dilemma which says that when the model is too small it will generally be biased whereas when the model is too large its parameter estimates can have large variance leading to a large variance in the output. Thus, we train our NN to satisfy the conditions of having a minimal average and variance of prediction error using the data sets I_t, S_t, and we validate the neural network using I_v, S_v respectively. The errors for all these 4 data sets are studied separately to find the best NN structure that minimizes RMSE variation in each data set.

The procedure to estimate the best size of the NN is as follows: The output layer shall have one neuron since we are interested in single step predictions only. Next, it is established that the number of neurons in the hidden layer are at most the number of neurons in the input layer for a minimized network structure. We may vary the number of neurons in the input layer from 2 through 11, since the pre-outbreak period for the data set of Turkey shows the onset of an epidemic at the end of the twelfth month and hence, to predict an outbreak in time, the pre-outbreak data (comprising of 11 months) must be sufficient to predict the time of onset of an epidemic in the following time step. Also, the neurons in the input layer correspond to the month number and scaled probability of infection/susceptibility in previous time instants. Due to the lack of data, the size of input layer must be kept minimally small although the input layer cannot have only 1 neuron since a 1x1x1 neural network would intuitively be biased. Thus, we implement all combinations of NN from 2x1x1 through 11x11x1 and we evaluate the RMSE of each network using 200 training iterations for 10 separate generations for each network. Each NN has a learning rate of 0.5 and momentum of 0.18. The ideal NN is the one with the lowest average RMSE and the lowest standard deviation in RMSE. Thus from Figures 15 and 16 in the Appendix, we observe that the 4x4x1 NN has a heuristically optimal size.

Having determined the best size of the predictive neural network, we examine the predictive performance of the optimal NN to predict the probability of infection and probability of susceptibility at each node. For computation of AIC and BIC at each node, it is important to note that the number of parameters to be estimated is equal to the size of $S_{t,}$ I_t times the number of weights and biases that are updated by each backpropagation operation. Thus, Tables 7, 8, and 9 depict BIC corresponding to the prediction of infection at each node using the optimal NN. In our case, for a 4x4x1 network, we have 29 such weights and biases.

Finally, the predictive performance of NN on data set *I*, S are shown in Figure 3. Here the input refers to the actual probability in the training and validation data sets while, the output is the predicted probability for the respective data set. A sample fit for node ID 52, representing the district of Gazaiantep is shown in Figure 4 and it depicts the under-prediction errors introduced by the NN. Since Gaziantep represents a lowly populated node, good prediction performance at this node works against over-fitting of the prediction models.

6. AUTOREGRESSIVE MODEL

The second temporal prediction approach is based on time-series analysis of FMD outbreak. An autoregressive (AR) model is a random process that models the randomness in natural phenomena such as epidemic spread. These models predict the output at a time instant *t* based on the outputs of previous time instants by modeling the randomness in correlated data. To ensure correlation of data, we consider the evolution of probability of infection and probability of susceptibility at each node separately, thus ensuring the data to be correlated in space. The first step towards time-series modeling is to subtract the mean value of data followed by eliminating the seasonality and trend in data. Next, the error in the probability is modeled and hence it becomes possible to fit a probability distribution function at each node corresponding to probability of infection and probability of susceptibility. In our analysis, since the data is sparse and having irregular trends, we analyze sliding windows of varying sizes for prediction. For example, in the training phase, a sliding window size of 6 would imply the use of data regarding probabilities at monthly time instants of 1 through 6 to predict the probability at monthly time instant 7, data of time instants 2 through 7 to predict the probability at time instant 8 and so on. In the validation phase,

Figure 3. Output versus input characteristic feature for NN towards data set I and S. The input is the actual probability of infection (I) or susceptibility (S) while the output is the probability predicted by the NN. The straight line is the ideal input-output curve such that data points lying along this ideal curve depict accurate predictions.(a) Data set I (b) Data set S.

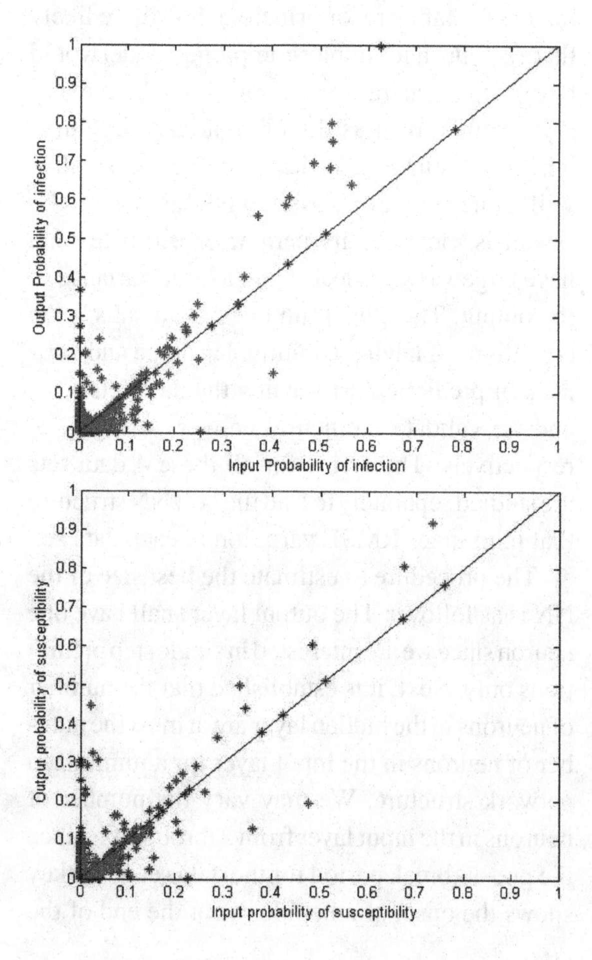

Figure 4. Neural network predicted probability of infection or susceptibility fitted against actual data on node 52, Gaziantep district, in Turkey. Both data sets S and I depict a high probability of susceptibility and infection in month 12, i.e. December 2005, thus depicting the onset of an FMD outbreak. Months following December 2005 have a low variation in infection and susceptibility owing to lowly populated farmlands and meat markets Gaziantep after the onset of the outbreak. (a) Data set I (b) Data set S

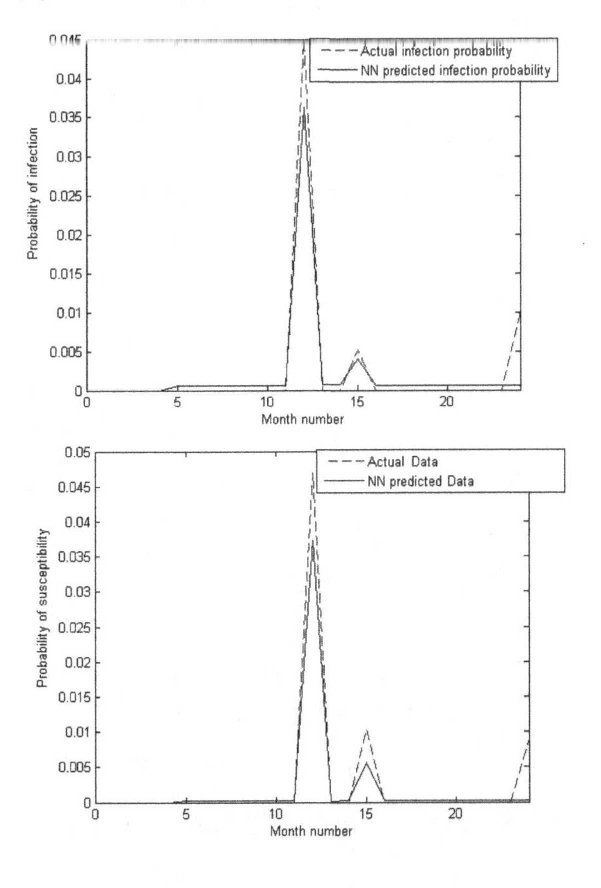

the predictions from previous iterative steps are used as data for future predictions.

The window size and order of autoregressive models with the lowest AIC and lowest MSE shall be deemed to be heuristically optimal, and that combination of window size and order shall be used for future predictions for any particular node.

Since we wish to minimize parameter estimation, we consider only autoregressive models without the moving average parameters. Autoregressive models are defined using the following equation.

$$X(t) = \varnothing_0 + \sum_{i=1}^{i=p} \varnothing_i X(t-i) + \varepsilon_t \quad (4)$$

Here, ε_t is additive white Gaussian noise which models the randomness part in the data while $X(t)$, $X(t-i)$ are the probability of infection/susceptibility at time instants t, $t-i$ respectively. $\varnothing_0$ is a constant and p is the order or the autoregressive model. Also, the order of the autoregressive model can be estimated using the Yule-Walker Equations. To find the best data windows size and order of an AR model pertaining to each node separately, we first analyze the seasonality of infection incidence. From Figure 5 we see that the seasonality of infection incidence is six months, i.e. the peak season with high probability of infection lasts for six months. Hence, the order and window size of the AR models must be at most equal to 6 since we wish to reduce the error that will creep into the predicted probability of infection during the transition from a peak

Figure 5. Seasonality of Infection Incidence using the OIE Incidence reports in Turkey

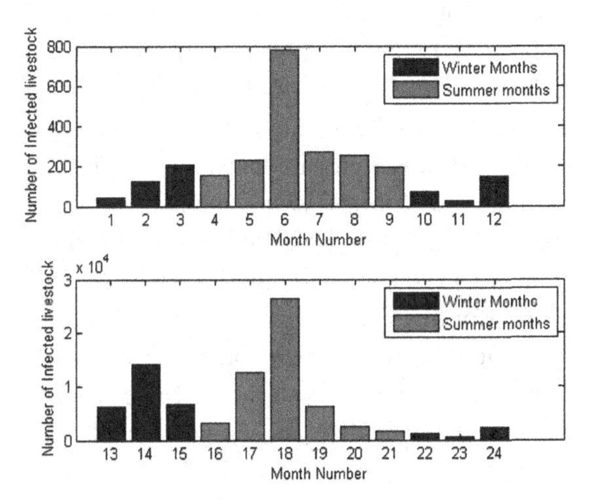

season to an off-peak season of infection in Turkey.

We classify the 12 months of each year according to conditions conducive to the growth and rapid spread of FMDV. The months of April through September have a higher incidence of infection reports than the other months and hence they are the peak months. The rest of the months represent the off-peak seasons for an epidemic outbreak. The top figure represents infection incidence reports in the year 2005 starting with January as month number 1. The figure in the bottom represents infection incidence in the year 2006 starting with January as month number 13 for the 24 month simulation period. This figure aggregates over all nodes, the seasonality component with regards to the peak and off-peak seasons for the spread of FMDV. It is noteworthy that certain nodes may transition from the pre-outbreak to the outbreak period slightly before or after January 2006 (month 13) with an example of early transitions visible in the Gaziantep district. However, the aggregated trend remains consistent as depicted here.

We estimate the best data window size and order of AR models (p) at each node separately, by evaluating AIC/BIC for all combinations of window size from 2 through 6 order p from 1 through 5 on data sets S_t, I_t . The combination of window size and order (p) that yields the lowest AIC/BIC is selected as optimal. Next, the goodness of fit at node ID 52 of the Gaziantep district is shown in Figure 6, where it is evident that this node transits from the pre-outbreak to the outbreak period a before January 2006 (month 13), which is shown to be the aggregate transition time instant from the pre-outbreak to the outbreak period in Figure 5. Also, the predicted versus the input probability is depicted in Figure 7. However, it is noteworthy that since the number of parameters for estimation is quite low for autoregressive models, they incur a low BIC as shown in Table 7, 8 and 9. Thus, these models are computationally less

Figure 6. AR predicted probability of infection or susceptibility fitted against actual data on node 52, Gaziantep district, in Turkey. (a) Data set I (b) Data set S

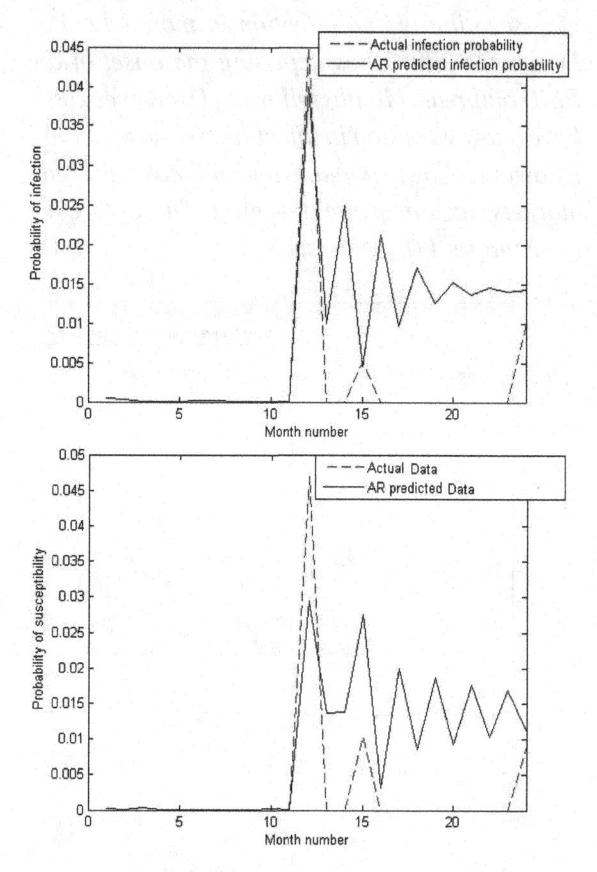

expensive and find feasibility when the number of nodes under consideration is high.

7. BAYESIAN NETWORK MODEL

A Bayesian network is a directed acyclic graph representing a set of random variables and their conditional or probabilistic relationships. Since FMD outbreaks may be modeled as random events, in the absence of a specific probability distribution function, they can be successfully modeled using probabilistic methods like Bayesian Networks for temporal predictions regarding the number of infected livestock at any time instant. Each random variable in the network is represented by

Figure 7. The input is the actual probability of infection (I) or susceptibility (S) while the output is the probability predicted by the autoregressive model. The straight line is the ideal input-output curve. (a) Response in I (b) Response in S

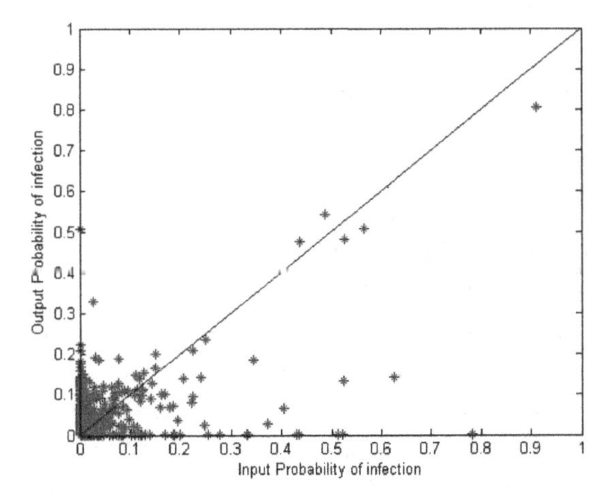

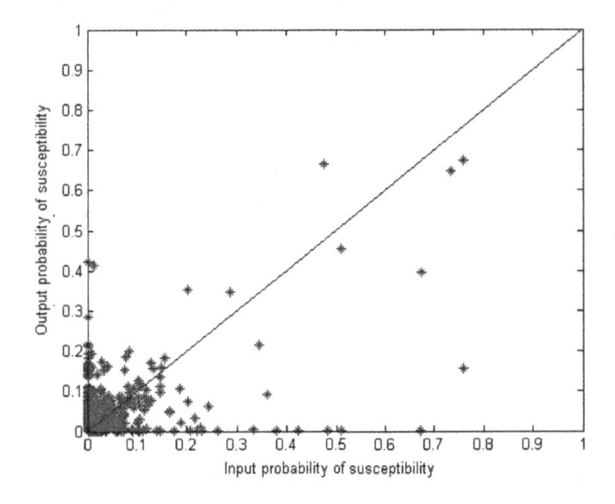

a node in the graph with links between the parent and offspring nodes, considering input random variables to be parent nodes and the predicted output to be the offspring node respectively. A single link leading a parent node to an offspring node depicts the conditional dependence between the offspring and parent nodes. However, if there is no connection between two nodes, it indicates conditional independence. There is a conditional probability table for each offspring node, which can be computed by the prior probabilities of the parent nodes (Gui, 2009).

A single-layer discrete Bayesian network shown in Figure 8, is constructed with two input parent nodes representing the month type and the probability of infection/susceptibility in the previous time step, and one offspring output node representing the present probability of infection /susceptibility, such that inputs and outputs are classified into discrete levels. Based on the OIE Incidence reports regarding FMD outbreaks, a conditional probability table and causal relationships are derived for each data set. Bayesian parameter estimation is carried out using Maximum Likelihood Estimation (MLE) and the mean expected output probability is compared with the actual data set. In our analysis, we consider 10 discrete input levels and 10 discrete output levels. Since we aim at magnifying the impact of input probability in the previous time instants on the output probability in the present time instant, thus we consider 5 levels of input probability and 2 types of month classifications (5x2=10 input levels). The output probability, for the sake of evenness has 10 discrete probability levels, each corresponding to a mean expected output probability level.

Figure 8. A single layer Bayesian network model for predicting the probability of infection/susceptibility. The month type is classified into 2 types according to the peak and off-peak season for the virulence of FMDV. The 'previous record level' input is the probability in the previous time step. The 'present record level' output is the probability in the current time instant that is predicted by the Bayesian network.

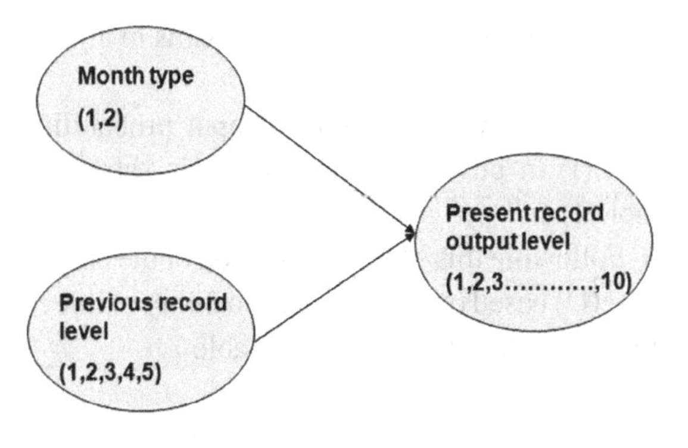

It is necessary that the classification for the input parent nodes should be done in a way such that all the data points which have similar influences on the offspring output nodes are grouped into the same level. For the offspring node, the classification should be done so as to have as many levels as possible, with relevant number of data entries in each level (Gui, 2009). Classification of the input month type is based on the number of infections reported each month. Survival conditions of FMDV depend on the initial concentration of virus in the material, the strain of virus, the humidity, the pH and ambient temperature. Considering the seasonal average number of FMD outbreaks annually in Turkey, we find that reported foot and mouth disease outbreaks in summer (60%) is higher than in winter (40%) (Tufan, 2003). All regions in Turkey have lower incidence in December-February (winter) and September-October (autumn) with incidence rising over summer to mainly a peak in March to July (Tufan, 2003). Thus, we categorize the months for Turkey as the following:

- **Month Type 1:** Months of January, February, March, October, November, and December in each year. From the Figure 5 we observe that during these months, the number of infected livestock is comparatively lower than the rest of the months in that year cycle.
- **Month Type 2:** Months of April, May, June, July, August, and September in each year. From the Figure 5 it is evident that these months have a higher number of infected than the rest of the months in a year.

Next, the classification of input probability level (I) in previous time instants is shown in Table 1.

Following this, the classification of the output level (O) based on the output probability of infection/susceptibility is given by Table 2.

The conditional probability table (CPT) which gives the probability of occurrence of each output level given the probability of input in the previous time instant and the month classification type is given as follows.

$$P\,(\text{Output level} = \text{i} \mid \text{Input state} = \text{m}) = \frac{N_i}{T_m}$$

Where, N_i is the number of occurrences in output level i and T_m is the total number of occurrences in input state m.

It is important to note at this stage, that we construct a single CPT using the S_t, I_t of all 79 nodes. This is done since we want the CPT to be as non-sparse as possible and it would not be possible to populate such a CPT for each node individually, due to the lack of data. Thus, we

Table 1. Input probability level classification

Input Level(I)	Probability at time (t — 1)
1	< 0.0001
2	0.0001-0.007
3	0.0071-0.04
4	0.041-0.1
5	> 0.1

Table 2. Output probability level classification

Output Level(O)	Probability at time instant (t)
1	0-0.00002
2	0.000021-0.00005
3	0.000051-0.0014
4	0.00141-0.0035
5	0.00351-0.0080
6	0.0081-0.0133
7	0.01331-0.05
8	0.051-0.1
9	0.1-0.3
10	> 0.3

construct the CPT using the training data sets and we validate the performance of the Bayesian network using the validation data sets S_v, I_v. Next, we evaluate the expected output probability level as the prediction obtained from the CPT, which is based on the expected output for each input level given below.

$$E(N_i|Input\,State = j) = \sum_{k=1}^{10} P\left(O_k|I_j\right) * Average\left(O_k\right) \quad (5)$$

where, N_i: Output Probability

O_k: Output level=k.
I_j: Input state=j.

The predictive performance of the Bayesian network for a node ID 52, Gaziantep district, is shown in Figure 9. The input probability versus the predicted probability of infection/susceptibility against the ideal prediction curve is shown in Figure 10.

8. MONTE-CARLO SIMULATIONS

Monte-Carlo simulations (MCS) refer to a particular class of algorithms that rely on repeated random sampling to compute results. Since these simulations rely on repeated computation of random events, these methods are most suited to calculation by a computer and tend to be used when it is unfeasible or impossible to compute an exact result with a deterministic algorithm or to simulatively verify the results obtained by other deterministic algorithms. In the Bayesian network model, we assume that the predicted value for each output state is the expected value, which represents a point estimate for the output probability. But since a probability input level is itself composed of a number of unevenly distributed entities, it is only a rough classification of the effects of month

Figure 9. Bayesian network predicted probability of infection or susceptibility fitted against actual data on node ID 52, Gaziantep district, in Turkey. (a) Data set I (b) Data set S.

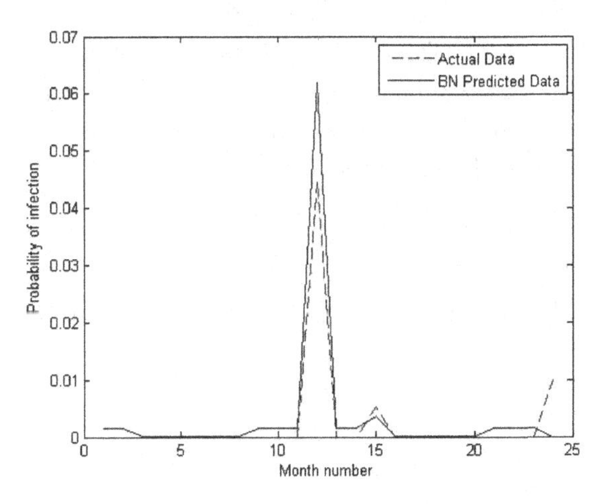

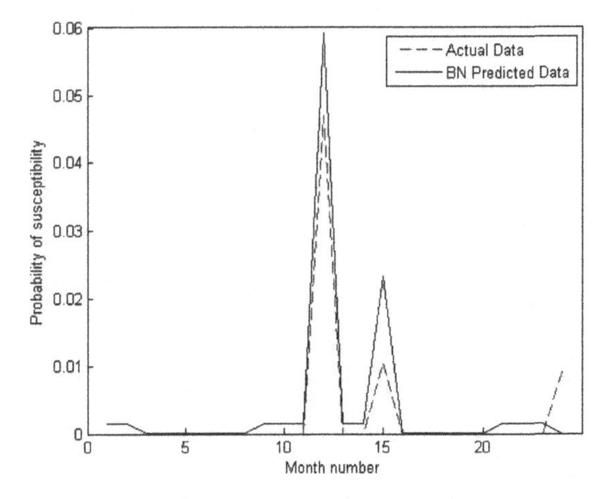

and input probability level on the predicted output probability. Thus, the model is expected to have errors in prediction and we should simulate and find a range of values within which the output probability is expected to lie. MCS is a common method to find out the confidence intervals and to simulate and calculate the expected output probability of infection/susceptibility (Gui, 2009) by averaging several simulated instances of an output probability.

In our experiments, MCS predictions not only indicate that they are in accordance with the Bayesian Network predictions but they also set

Figure 10. Output versus input characteristic feature for Bayesian network towards data set I and S. The input is the actual probability of infection (I) or susceptibility (S) while the output is the probability predicted by the Bayesian network. The straight line is the ideal input-output curve. Also, the observed data points assume discrete levels due to the discrete Bayesian estimates. (a) Data set I (b) Data set S.

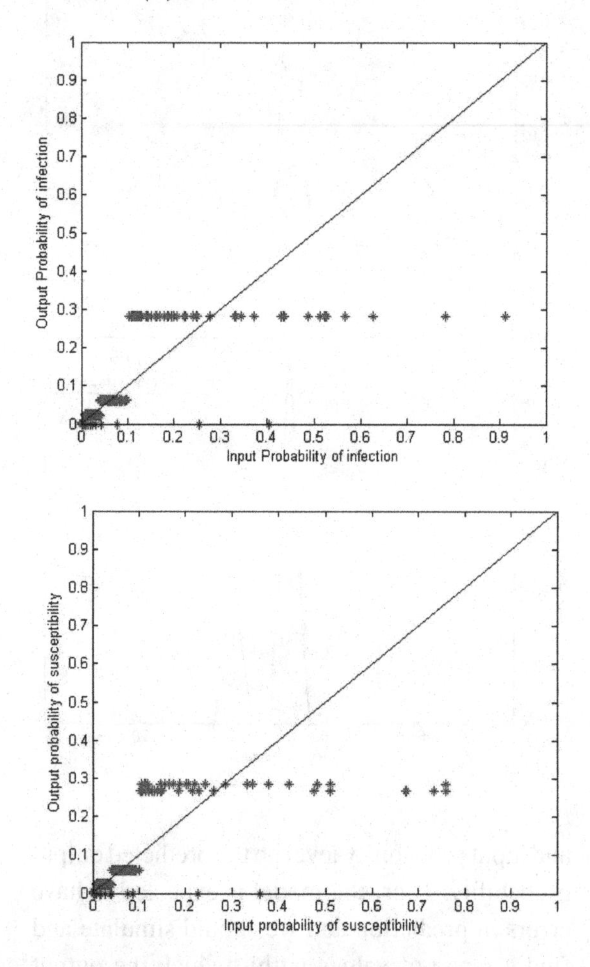

the 95% confidence intervals regarding the probability of infection and susceptibility with respect to the actual data (Roy Chowdhury, 2010). This aids comparative evaluation of each prediction model with respect to the gold standards in the following section. The goodness of fit of the MCS predictions for a sample node ID 52 are shown in Figure 11, and also the input probability versus the output probability comparison plots against the ideal input-output curve are shown in Figure 12.

9. MODEL ANALYSIS

Having trained the learning-based models, the predictions regarding the probability of infection and the probability of susceptibility obtained using NN, AR, BN and MCS models are comparatively analyzed to evaluate the predictive performance of each model. We can now assume that so far we have performed spatio-temporal predictions regarding the evolution of FMD, since we consider temporal predictions at different administrative districts in space, separately. At any time step, the product of the probability of infection at a particular node,

Figure 11. MCS predicted probability of infection or susceptibility fitted against actual data on node 52, Gaziantep district, in Turkey. (a) Data set I (b) Data set S.

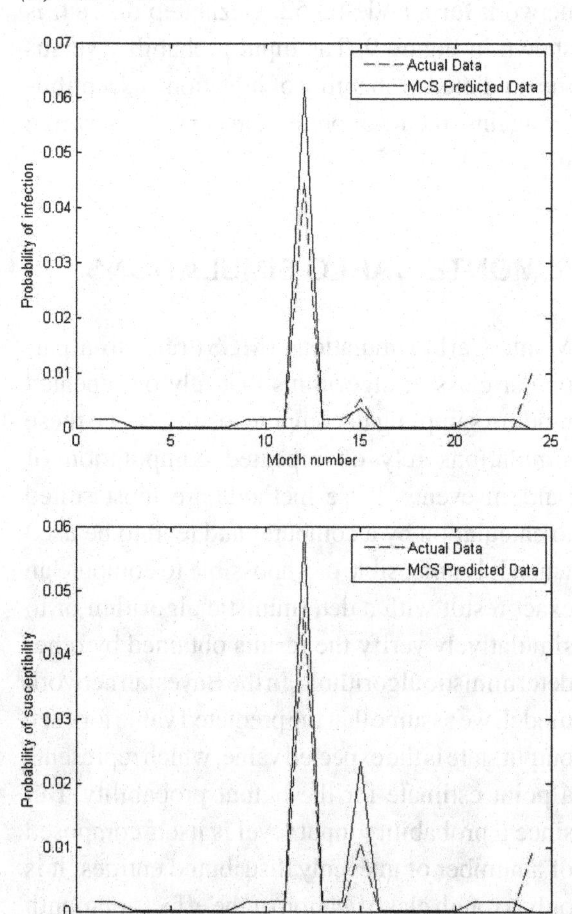

Figure 12. Output versus input characteristic feature for MCS towards data set I and S. The input is the actual probability of infection (I) or susceptibility (S) while the output is the probability predicted by the Monte-Carlo simulations. The straight line is the ideal input-output curve. The observed data points are scattered in discrete levels parallel to the Bayesian network output. This performance is similar to the performance of the Bayesian Network. (a) Data set I (b) Data set S.

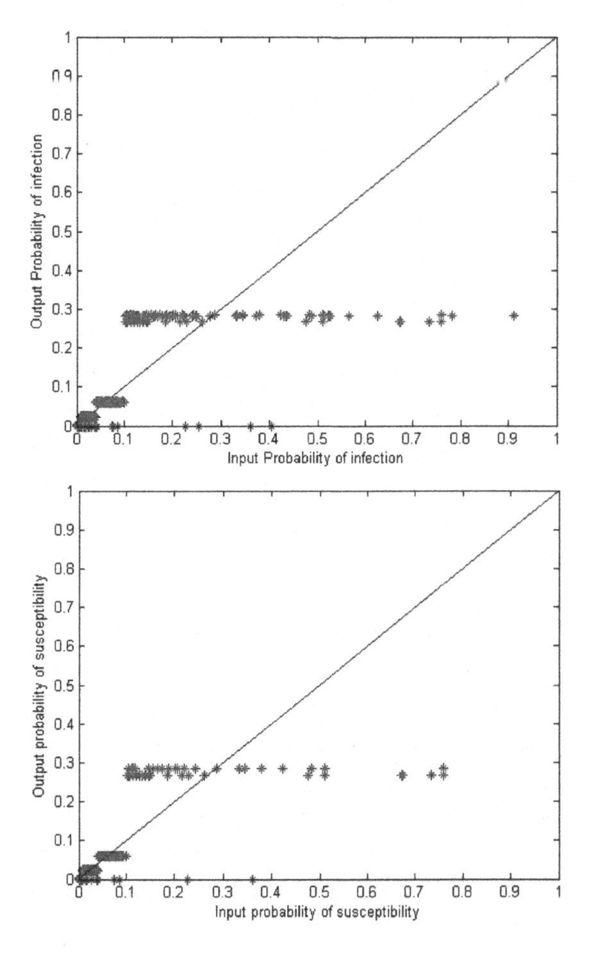

and the total number of animals that can be infected at that node throughout the period of study, yields the number of infected animals at that time step at that particular node. Thus, it is possible to estimate the total number of infected animals all over Turkey at a particular time step, by aggregating the number of infected animals at each node. The pre- diction errors between the predicted and actual number of infected animals all over Turkey in progressive time steps is elucidated in terms of sMAPE, R^2 and D_{KL} in Table 3, by using the four learning-based prediction models.

On comparing the KL divergence distance of spatio-temporal predictions for the data set S, we see that the divergence distance for NN is 0.0061, for autoregressive models it is 0.00794, for Bayesian networks it is 0.0122 and for MCS it is 0.0120 respectively. As an inference, NN has the lowest KL divergence distance, lowest error and best R^2 statistic. Hence, it provides best estimates spatially and temporally regarding the evolution of FMDV. Autoregressive models on the other hand incur higher prediction errors than neural networks, yet they fit the data aggregately and thus, they incur a considerable R^2 statistic of 0.9535. Bayesian network on the other hand have a consistent error with respect to under-predictions and over-predictions incurred spatio-temporally and hence, they set good bounds to the extent of infection spread. Also, each learning-based prediction model has an acceptable prediction standard within the 95% confidence interval generated using the Monte-Carlo Simulations.

Thus, when computational complexity is not a constraint, NN models provide better temporal predictions with lower prediction errors and better coefficient of determination (R^2) than rest of the models . The BIC of the NN and AR models is summarized in Table 7,

Table 3. Comparative analysis of spatio-temporal predictions for data set I.

Model	sMAPE	R^2	^{D}K L
NN	11.2164	0.9816	0.0056
BN	16.3127	0.9371	0.0095
MCS	18.164	0.9387	0.0094
AR	22.2083	0.9535	0.00833
MCS upper bound	44.925	0.5071	-
MCS lower bound	30.5148	0.4398	-

8, 9 for each node separately, and from these tables we infer that autoregressive models incur a lower AIC/BIC since they require lower parameter estimation than neural network models. Hence, if the number of nodes is large and computational complexity is a constraint, AR models can be preferred over NN models in spite of the fact that AR models introduce larger prediction errors than NN models. Otherwise, for relatively small number of nodes, NN models provide a much more accurate prediction as compared to AR models. Finally, we observe that BIC_{score} for spatio-temporal Bayesian network predictions is -185.3445. This is indicative that Bayesian networks incur a very low computational complexity and thus, they are useful for mainly for setting prediction bounds since the prediction performance of a BN model is worse than the AR and NN models in this case.

Having analyzed the predictive performance of the learning-based models, we move on to analyze the performance of various mitigation strategies that may be simulated using the probabilistic results from these predictive models.

10. PRACTICAL MITIGATION

Mitigation of epidemics is necessary to alleviate the devastating impacts of FMD. Further, theoretical mitigation strategies based on isolation of nodes with high probability of infection (Chowdhury et al., 2010) are practically infeasible since complete isolation of nodes is impossible in real-world situations. Also, there is a difference in the impedance to the spread of FMD due to the various practical control policies that are adopted such as: movement bans, vaccination, infected premise (IP) culls or dangerous contact (DC) culls. Though IP and DC culls are very successful in immediately retarding infection spread, they are highly unethical (Anderson, 2002). Conversely, vaccination and movement bans are ethical but they may be costly policies, and they may not retard FMDV spread as much as the culling strategies. Thus, it is absolutely imperative to formulate mitigation strategies that are practically feasible, ethical and cost-effective too (Chowdhury, 2010).

As practically feasible strategies, we simulate multiple instances of mitigation strategies which combine the implementation of several mitigation tasks to curb the impacts of FMD epidemic outbreak. A particular instance of any mitigation strategy is realized by randomly selecting a threshold probability (p_{th}) in [0, 1] for each node such that, if the probability of infection ($p_{t,i}$) at any node i at time step t, is greater than p_{th}, the node is deemed to be infected, or else the node remains susceptible. In the simulations we have restricted p_{th} in [0.01, 0.5] to reduce the range of variation in the mitigation strategies, since a smaller range of values eases analysis. Various mitigation tasks are implemented by introducing a multiplicative constant $c_{t,i}$ for each $p_{t,i}$. Since these mitigation tasks impede the spread of FMDV, $c_{t,i}$ represents the extent of impedance by dampening the probability of infection. When no mitigation tasks are enforced, $c_{t,i} = 1$, and when mitigation tasks such as premise culls, vaccination and movement bans are enforced, $c_{t,i} = c$, where c is a constant dependent on the mitigation task.

In the simulation *Box 1*, *sus(i)* is the total number of animals that can be infected in node *i* throughout the period of study, and *Inf*(t) is the evaluated total number of infected animals across all nodes in time *t*. The mitigations strategies are applied from March 2005 and onwards, since the first two months are considered for sample run, and they also indicate the delay in detection of FMD outbreak. It is noteworthy that the simulations of the mitigation strategies do not involve re-evaluation of the probability of infection after each time step as done in the case of spatio-temporal models (Chowdhury et al., 2010). However, the conclusions regarding the effectiveness of the mitigation strategies thus obtained are parallel to that using the spatio-temporal analysis. The various mitigation tasks applied to impede FMDV spread are as follows.

Box 1. Simulative practical mitigation strategies

Randomly generate p_{th} in $[0.01,0.5]$. Given $c_{1,i} = 1$, $c_{2,i} = 1$.
for *nodes* **i** $= 1$ **to** *N umber of nodes* **do for** *time* **t** $= 3$ **to** *N umber of months* **do**
Evaluate $p_{th1}, p_{th2}, p_{th3}, p_{th4}$ such that
$pth_1 = 0.95 * p_{th}$ $pth_2 = 0.9 * p_{th}$ $pth_3 = 0.85 * p_{th}$ $pth_4 = 0.8 * p_{th}$
if $c_{t,i} * p_{t,i} > p_{th}$ **then**
$Inf(t) = Inf(t) + c_{t,i} * p_{t,i} * sus(i)$

end if

if $c_{t,i} * p_{t,i} > p_{th1}$ **then**

$c_{t+1,i} = c$ **for** *Task 1 due to high impeding infection*
If *Task 1 is culling or vaccination,* $cull_i$ *or* $vacc_i = c_{t,i} * p_{t,i} * sus(i)$
else if $p_{th1} \geq c_{t,i} * p_{t,i} \geq p_{th2}$ **then**

$c_{t+1,i} = c$ *for Task 2 due to considerably high impending infection*
If *Task 2 is culling or vaccination,* $cull_i$ *or* $vacc_i = c_{t,i} * p_{t,i} * sus(i)$
else if $p_{th2} \geq c_{t,i} * p_{t,i} \geq p_{th3}$ **then**

$c_{t+1,i} = c$ *for Task 3 due to considerable impending infection*
If *Task 3 is culling or vaccination,* $cull_i$ *or* $vacc_i = c_{t,i} * p_{t,i} * sus(i)$
else if $p_{th3} \geq c_{t,i} * p_{t,i} \geq p_{th4}$ **then**

$c_{t+1,i} = c$ *for Task 4 due to impending infection in near future*
If *Task 4 is culling or vaccination,* $cull_i$ *or* $vacc_i = c_{t,i} * p_{t,i} * sus(i)$

else

$c_{t+1,i} = 1$, *No mitigation task performed due to low impending infection.*

end if end for
end for

1. **Infected Premise Cull (IP):** This strategy is adopted at nodes with very high probability of incident infection. Accordingly, all livestock in the particular premise (node) are culled within next 24-48hours. Thus c = 0.73, since IP culls induce almost 27% impedance to the infection spread (Tildesley et al., 2005). IP culls provide the most impedance to the spread of infection.
2. **Dangerous Contact Cull (DC):** This strategy is adopted at nodes that are not yet infected but they have an infected node in their neighborhood and hence, there is a high probability of infection in future time steps. Accordingly, all livestock in such a premise (node) are culled within the next 4-10 days, and thus c = 0.87, since DC culls can be estimated to yield 13% impedance to the virus spread (Tildesley et al., 2005).
3. **Vaccination (Vacc):** If the vaccine to the particular strain of FMDV is available, its administration will reduce the probability of future infections. The impedance to the spread of infection is represented by c = 0.77 (Wallace et al., 2006), due to almost 23% impedance by a potent oil-based vaccine.
4. **No Movement Bans (NM):** For nodes not having very high probability of incident infection, but with a potential of getting infected in the near future, the grazing movements of animals are banned. Human movements are also banned from such territories to prevent the spread of the virus. However, a high cost per day is incurred due to these movement restrictions and thus, movement bans are implemented by varying c = 0.80, to incur 20% impedance to infection spread. Although (Ferguson et al., 2001) states that movement restrictions could lead to as much as 50% reduction in the rate of disease transmission, we dampen this effect in our data set considering the granularity of the data set, and considering the delays in implementation of such movement restrictions.

We evaluate the impact of six different mitigation strategies on reducing the total number of infected animals at the end of the 24 month study period in Turkey. These mitigation strategies are identified by a set of mitigation tasks defined in Table 4. These mitigation strategies include a logical sequence of mitigation tasks that may be performed.

To devise economic mitigation strategies, it is imperative to understand the cost-effectiveness of the mitigation strategies. We compute the direct costs of implementation of mitigation tasks by assuming that the cost per head to breed cattle is 1133.5 US dollars and that of sheep is 121.0 US dollars (Kobayashi et al., 2007a). The cost of vaccinating cattle or sheep is 6.0 US dollars per head. However, vaccinations must be well planned by training vaccination teams, providing proper equipment, and preparing for quick transportation of vaccines to affected sites. All these costs are indirect costs and hence, they are not included in our simulations. The cost to administer euthanasia or cull cattle is 16.5 US dollars per head and 2.31 US dollars per head for sheep (Kobayashi et al., 2007a). Premise culling costs would thus include the breeding costs for livestock, the cost for the safe disposal of carcass, for administering euthanasia and decontamination costs. In case of movement restrictions, a cost of 157,968 US dollars is incurred per day for every geographic location (Kobayashi et al., 2007b). Although the mitigation strategies suggested above are effective in reducing unnecessary and unethical livestock culls, it is imperative to understand their cost-effectiveness as well. These costs help in formulating the different mitigation strategies, wherein the priority of each of the four mitigation tasks is different. The cost-effectiveness of a mitigation strategy is defined as following.

effectiveness =% Reduction in number of infected livestock (6)

Cost incurred (million U S $)

11. RESULTS

We analyze the impact of the 6 mitigation strategies using the learning-based models on the data set of Turkey. Since the learning-based models incur a higher error in predictions than spatio-temporal models in (Chowdhury et al., 2010), the impacts of mitigation using these models are not very quantitatively accurate. However, the significances of the mitigation strategies in terms of their cost-effectiveness, and in the impedance to the spread of FMDV, remain unchanged. Learning-based prediction models can be used to realize the possible range of *effectiveness* of these mitigation strategies, and to understand the importance of each strategy. Thus, we wish to analyze the impact of each mitigation strategy towards reducing the total number of infected animals, while keeping the implementation costs to a minimum. We must consider the facts that the development and deployment of vaccines could incur a time delay, and that massive culling is unethical, thus we wish to analyze the mitigation strategies that minimize the number of livestock vaccinated and culled as well. Figures 13 and 14 depict the cost of implementation of mitigation strategies and their cost-effectiveness respectively.

Table 4. Sequence of mitigation tasks in different mitigation strategies

Strategy	Mitigation Task 1	Mitigation Task 2	Mitigation Task 3	Mitigation Task 4
1	IP	DC	Vacc	NM
2	Vacc	NM	-	-
3	IP	DC	NM	
4	IP	DC	NM	Vacc
5	IP	Vacc	DC	-
6	IP	NM	-	-

Analysis of the impacts of the 6 mitigation strategies in terms of the percentage reduction in the total number of infected livestock at the end of the period of study, the cost of implementation of the mitigation strategies in million US dollars and the total number of culled and vaccinated livestock are presented in Table 5. Thus, we infer that strategy 5 incurs the mostly culling and vaccinations, because of which, this strategy results in the maximum reduction in the total number of infected animals. Strategy 5 is thus very effective in impeding the rate of spread of FMDV when an epidemic has already set in. However, strategies 1 and 4 which rely on all 4 mitigation tasks have cost-effectiveness similar to each other, and they both incur a high cost of implementation.

This shows that implementation of all 4 mitigation task merely increases the total cost of implementation without improving on the percentage reduction of the number of infected animals or the strategy *effectiveness*. Thus, it is important to plan the order of mitigation tasks instead of considering each task to be equally important. Also, we observe that the reduction in the total number of infected livestock is almost double for culling based strategies 1, 3, 4, 5, 6 when compared to vaccination based strategy 2; however, the *effectiveness* of strategy 2 is the highest. Thus, vaccinations must be preferred to culling strategies in the pre-outbreak period. Finally, we observe that strategy 2 and 5 which rely on vaccination strategies have the highest range of *effectiveness*. This shows that vaccination strategies need to be well planned, and a good vaccination program needs to be developed, to obtain high cost-effectiveness of mitigation strategy 2, otherwise vaccination will not be effective in impeding the spread of FMDV. Also, potency of vaccines plays an important role in determining the cost-effectiveness of any vaccination strategy, since oil-based vaccines are considered to be more potent and effective than the water-based ones.

When we compare the predictive performance of the learning-based models in determining the

Figure 13. Total number of infected animals presented against the total cost incurred for the 6 mitigation strategies using learning-based predictive models. (a) Neural Network model (b) Auto Regressive model (c) Bayesian Network model (d) Monte Carlo Simulations.

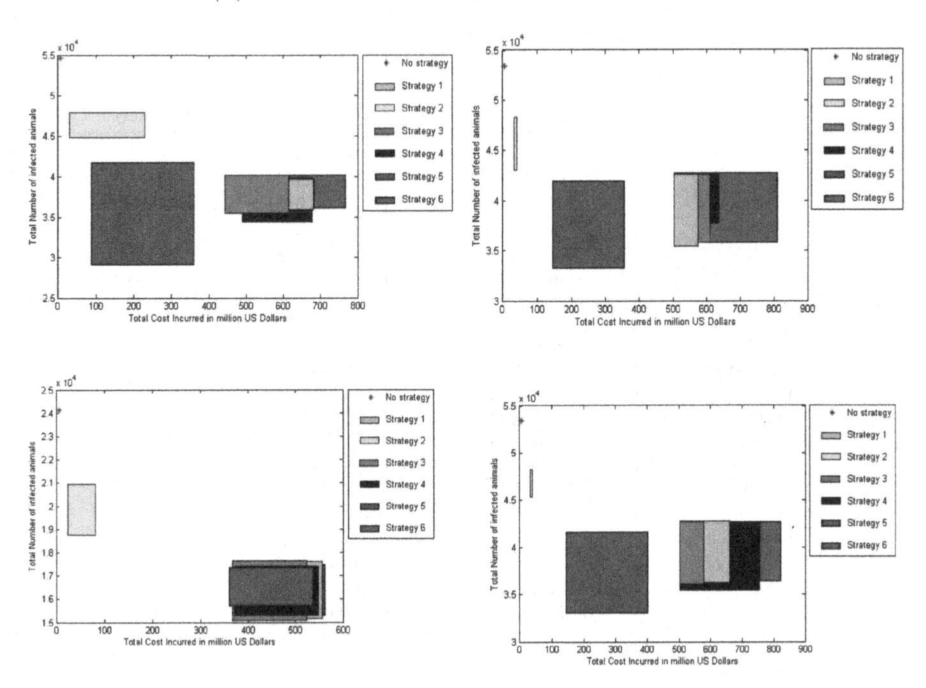

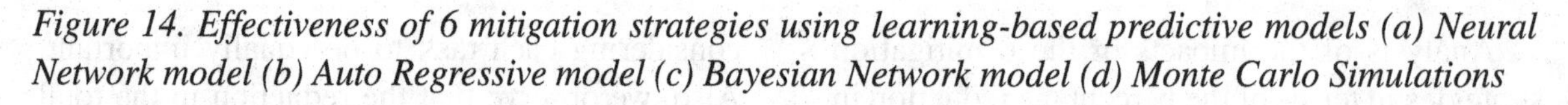

Figure 14. Effectiveness of 6 mitigation strategies using learning-based predictive models (a) Neural Network model (b) Auto Regressive model (c) Bayesian Network model (d) Monte Carlo Simulations

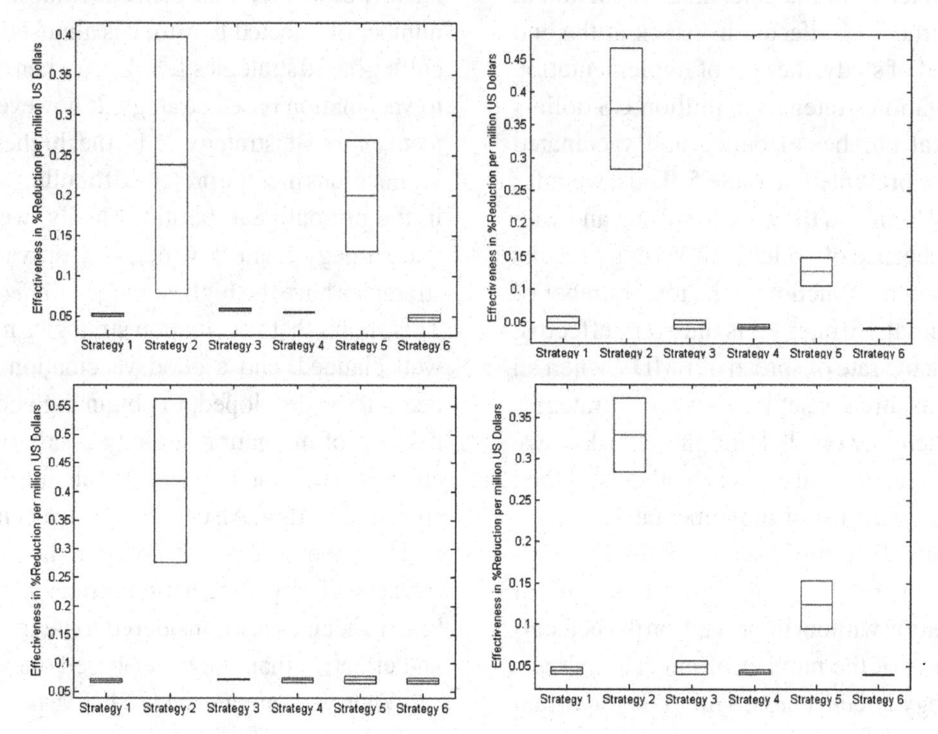

Table 5. Cost-effectiveness of mitigation strategies based on learning-based predictive models

Model	Strategy	%Reduction	Cost(million USD)	effectiveness	Culled	Vaccinated
NN	1	27.338-34.253	514.514-680.597	0.05-0.053	154 -1108	0- 207
	2	12.419-18.019	31.244-229.052	0.079-0.397	0	195-373
	3	26.488-35.023	443.186-614.489	0.057-0.06	154-905	0-199
	4	27.134-36.896	490.215- 677.393	0.054-0.055	154-903	0-222
	5	23.663-46.794	87.892-360.019	0.13-0.269	2217-15295	0- 1310
	6	26.54-33.891	507.738-767.864	0.044-0.052	154-1064	0
AR	1	27.055-37.178	371.233-557.004	0.067-0.073	110-1388	0-405
	2	13.23-22.299	23.511-80.703	0.276-0.563	0	140-1761
	3	26.961-37.552	367.508 -523.402	0.072-0.073	110-1388	0 -372
	4	27.879-36.669	371.624 - 548.711	0.067-0.075	110-1388	0-225
	5	28.183-34.972	360.779 - 536.617	0.065-0.078	110-1388	0-207
	6	27.606-36.569	372.846-561.948	0.065-0.074	110-1388	0
BN	1	20.177-31.029	501.88-768.027	0.04- 0.042	483-965	0-21
	2	9.627-17.408	33.88-42.338	0.284-0.411	0	770-1224
	3	20.006-31.644	501.88-584.517	0.04-0.054	483-965	0
	4	20.002-31.872	501.88-906.258	0.035-0.04	606-965	0
	5	20.608-34.468	143.742-384.179	0.09-0.143	3318-8888	0-437
	6	20.278-31.675	501.88-753.125	0.04-0.042	585-965	0
MCS	1	20-32.19	501.88-660.437	0.04-0.049	483.004-965	0-21
	2	9.627-15.07	33.88-0.377	0.284-0.373	0	743-1224
	3	20.014-32.388	501.88-579.315	0.04-0.056	800-965	0-21
	4	20.107-33.666	501.88-753.479	0.04-0.045	610-965	0
	5	22.084-38.085	144.094-402.147	0.095-0.153	3976-8888	0-437
	6	20.087-31.84	501.88-821.285	0.039-0.04	585-965	0

cost-effectiveness of mitigation strategies, we observe that the range of variation in*effectiveness* of mitigation strategies is higher using NN and autoregressive models since they are more sensitive to the actual data. Bayesian networks and MCS on the other hand have smaller range of strategy *effectiveness*. Finally, we compare the downstream impact of prediction errors on the effectiveness of the mitigation strategies as follows.

12. IMPACT OF PREDICTION ERRORS ON MITIGATION

Having quantified the impact of mitigation strategies using their cost-effectiveness, we observe that prediction errors cause non-linear variations in the effectiveness of these mitigation strategies. Hence, it is imperative to analyze the variations in the *effectiveness* of fixed mitigation strategies caused due to prediction errors, which in turn implies the robustness of mitigation strategies against prediction errors. Table 6 shows the range of variation in *effectiveness* of the 6 mitigation strategies ($\Delta eff\ 1$ through $\Delta eff\ 6$) with variation in prediction errors ($\Delta sMAPE$).

Here, we analyze the variation in prediction error, and variation in cost-effectiveness of the mitigation strategies between the learning-based models and the spatio-temporal predictive model in (Chowdhury et al., 2010). Since this spatio-temporal model incurs lower prediction errors due to its global parameterization, we assess the impact of increased prediction errors, by the learning-based models, on the downstream utility function defined in terms of *effectiveness* of the mitigation strategies. We observe that for NN models almost 14% change in prediction error causes about 30% variation in *effectiveness*, while AR models show 24% variation in *effectiveness* for 56% prediction error. BN and MCS show around 18% variation in *effectiveness* due to a prediction error of 48%. Thus, we are able to assess the non-linear variation and robustness in *effectiveness* of mitigation strategies for prediction errors from the learning-based prediction models since, as much as 50% prediction errors incur as much as 30% variation in cost-effectiveness of mitigation strategies. Further, we find that the variation in prediction errors for NN and AR with respect to the spatio-temporal model in (Chowdhury et al., 2010), which requires parameters such as wind, human movement and grazing movement of animals, is lesser than that of other spatio-temporal models with lower parameterization. Hence, we infer that the predictive performance of learning-based models can be better than under-parameterized spatio-temporal models.

13. CONCLUSION AND DISCUSSION

In this paper we apply learning-based models to mimic the spatial characteristics of net- work based spread models (Chowdhury et al., 2010). Our study shows that it is possible to generalize local learning-based models such as neural network, autoregressive, Bayesian network and Monte-Carlo simulation models to recover the latent spatial parameters in it. Temporal predictions regarding the time of onset of an epidemic can be achieved, and certain bounds can be established with regards to the number of infected livestock at different geographical regions in different time instants using these learning-based models. In our case study, we find that neural network models perform significantly better than autoregressive, Bayesian network and MCS models in predicting the probability of infection and susceptibility at each node representing a particular administrative district in Turkey. However, this observation is specific to our particular case since the non-linear neural network is able to respond to the sudden transition from the pre-outbreak to the outbreak period. Here, Bayesian network and MCS models are mostly important in determining

Table 6. Impact of prediction errors using learning-based models

Statistic	NN	AR	BN	MCS
min $\Delta sMAPE$ max $\Delta sMAPE$	1.5469 2.6581	12.5388 13.6500	6.6432 7.7544	8.4945 9.6057
min $\Delta eff1$ max $\Delta eff1$	0.0352 0.0440	0.0188 0.0242	0.0452 0.0569	0.0368 0.0572
min $\Delta eff2$ max $\Delta eff2$	0.0701 0.2638	0.1276 0.4291	0.1505 0.2624	0.1505 0.2245
min $\Delta eff3$ max $\Delta eff3$	0.0661 0.0716	0.0513 0.0580	0.0689 0.0915	0.0671 0.0915
min $\Delta eff4$ max $\Delta eff4$	0.0420 0.0465	0.0268 0.0296	0.0613 0.0620	0.0518 0.0618
min $\Delta eff5$ max $\Delta eff5$	0.0006 0.1387	0.0524 0.0654	0.0129 0.0409	0.0228 0.0359
min $\Delta eff6$ max $\Delta eff6$	0.0415 0.0583	0.0197 0.0373	0.0533 0.0604	0.0537 0.0636

the 95% confidence bounds for prediction. Thus, we infer that learning-based models can extract the seasonality and trends in the rate of spread of FMDV, and they can be used to predict the number of infected livestock in future time instants with a significant confidence. These predictions are useful in simulating cost-effective mitigation strategies that can effectively impede the impacts of future occurrences of FMD outbreaks.

Predictions regarding the number of infected livestock in future time instants obtained from the learning-based models can be used to assess the importance of various mitigations strategies by simulations. Different mitigation tasks such as Infected Premise (IP) culls, Dangerous Contact (DC) culls, and Vaccination (V) and No Movement (NM) restrictions are simulated and their importance in impeding the spread of FMDV is analyzed. Conclusively, we observe that IP and DC culls followed up by vaccinations result in a cost-effective reduction of the total number of infected livestock, and this strategy retards the rapid spread of FMDV. However, the cost incurred in this process is quite high. Hence, this mitigation strategy may be preferred after the onset of an epidemic outbreak when immediate reduction in the number of infected livestock is mandatory. Although a potent vaccination strategy, followed by movement restrictions, incurs a lower reduction in the total number of infected livestock, this strategy results in the most cost-effective control of the epidemic. Thus, vaccinations and movement ban strategies may be adopted before the onset of an epidemic outbreak to control the rapid spread of FMDV.

The variation in effectiveness of mitigation strategies with prediction errors provides insights into the robustness of our simulative mitigation strategies. Such downstream utility impacts of prediction errors in mitigation strategies may be studied in the future to analyze the non linear relation between prediction error and variations in the cost-effectiveness of mitigation strategies.

Future work may involve analyzing the predictive performance of other learning-based local models such as recurrent networks, moving average and wavelet-based prediction models to name a few. The impact of delays in implementation of mitigation tasks, may provide better insights into the critical nature of the mitigation strategies. Also, since we observe a notably high cost-effectiveness of mitigation strategies involving vaccination and movement bans in Turkey, we may conclude that training of personnel, providing a good transportation infrastructure, and enhancing research facilities for quicker development of vaccines may be beneficial for long term eradication of FMD in such cases.

ACKNOWLEDGMENT

This research was supported by National Agricultural and Biosecurity Center (NABC) at Kansas State University.

REFERENCES

Abeku, T., Hay, S., Ochola, S., Langi, P., Beard, B., & De Vlas, S. (2004). Malaria epidemic early warning and detection in African highlands. *Trends in Parasitology*, *20*, 400–405. doi:10.1016/j.pt.2004.07.005

Abidil, S. S. R., & Goh, A. (2006). *Applying knowledge discovery to predict infectious disease epidemics* . Berlin, Germany: Springer-Verlag.

Alexandersen, S., Zhang, Z., Donaldson, A. I., & Garland, A. J. M. (1997). The pathogenesis and diagnosis of foot-and-mouth disease. *Transactions of the ASAE. American Society of Agricultural Engineers*, *40*, 247–252.

Anderson, I. (2002). Foot and mouth disease 2001: Lessons to be learned inquiry. *The Stationary Office, 187.*

Bates, T. W., Carpenter, T. E., & Thurmond, M. C. (2003a). Description of an epidemic simulation model for use in evaluating strategies to control an outbreak of foot-and-mouth disease. *American Journal of Veterinary Research*, *64*, 195–204. doi:10.2460/ajvr.2003.64.195

Bates, T. W., Carpenter, T. E., & Thurmond, M. C. (2003b). Results of epidemic simulation modeling to evaluate strategies to control an outbreak of foot-and-mouth disease. *American Journal of Veterinary Research*, *64*, 205–210. doi:10.2460/ajvr.2003.64.205

Brit, O., Vounatsou, P., Gunawardena, D., Galappaththy, G., & Amerasinghe, P. (2008). Models for short term malaria prediction in Sri Lanka. *Malaria Journal*, *7*, 1475–2875.

Carpenter, T., Thurmond, M., & Bates, T. (2004). A simulation model of intraherd trans- mission of foot and mouth disease with reference to disease spread before and after clinical diagnosis. *Journal of Veterinarian Diagnostic Investigation*, *16*, 11–16. doi:10.1177/104063870401600103

Chowdhury, R. S. (2010). Mathematical models for prediction and optimal mitigation of epidemics (Doctoral dissertation, University of Minnesota). *K-State Electronic Theses, Dissertations, and Reports.*

Chowdhury, R. S., Scoglio, C., & Hsu, W. (2009). Evolution and control strategies of the foot and mouth disease epidemic on a weighted contact network. In *Proceedings of the Second International Conference on Infectious Diseases Dynamics*, Athens, Greece (p. 2.01).

Chowdhury, S. R., Scoglio, C., & Hsu, W. (2010). Simulative modeling to control the foot and mouth disease epidemic. *Procedia Computer Science*, *1*(1), 2261–2270. doi:10.1016/j.procs.2010.04.253

Chtioui, Y., Panigrahi, S., & Francl, L. (1999). A generalized regression neural network and its application for leaf wetness prediction to forecast plant disease. *Chemometrics and Intelligent Laboratory Systems*, *48*, 47–58. doi:10.1016/S0169-7439(99)00006-4

Ferguson, N., Donnelly, C., & Anderson, R. (2001). The foot-and-mouth epidemic in Great Britain: pattern of spread and impact of interventions. *Science*, *292*, 1155–1160. doi:10.1126/science.1061020

Francl, L., & Panigrahi, S. (1997). Artificial neural network models of wheat leaf wetness. *Agricultural and Forest Meteorology*, *88*, 57–65. doi:10.1016/S0168-1923(97)00051-8

Gui, M. (2009). Advanced methods for prediction of animal-related outages in overhead distribution systems (Doctoral dissertation, Kansas State University). *K-State Electronic Theses, Dissertations, and Reports.*

Hawksworth, A., Hansen, C., Good, P., Ryan, M., Russell, K., & Kelley, P. (2003). Using autoregressive epidemic modeling to augment the existing department of defense (dod) febrile respiratory illness surveillance system at military training centers. *Journal of Urban Health*, *80*, 38. doi:10.1007/BF02416933

Jiang, X., & Wallstrom, G. (2006). A bayesian network for outbreak detection and prediction. In *Proceedings of the 21st National Conference on Artificial Intelligence* (Vol. 2, pp. 1155-1160).

Keeling, M. J., Woolhouse, M. E. J., Shaw, D. J., Matthews, L., Chase-Topping, M., & Haydon, D. T. (2001). Dynamics of the 2001 UK foot and mouth epidemic: Stochastic dispersal in a heterogeneous landscape. *Science*, *294*, 813–817. doi:10.1126/science.1065973

Kobayashi, M., Carpenter, T., Dickey, B., & Howitt, R. E. (2007a). A dynamic, optimal disease control model for foot-and-mouth disease: Model description. *Preventive Veterinary Medicine*, *79*, 257–273. doi:10.1016/j.prevetmed.2007.01.002

Kobayashi, M., Dickey, B., Carpenter, T., & Howitt, R. E. (2007b). A dynamic optimal disease control model for foot-and-mouth disease: Model results and policy implications. *Preventive Veterinary Medicine*, *79*, 274–286. doi:10.1016/j.prevetmed.2007.01.001

Krishnamurti, T., & Chakraborti, V. A.and Mehta, & A.V., M. (2007). *Experimental prediction of climate-related malaria incidence.* Retrieved from www.pitt.edu/~super7/29011-30001/29021.ppt

Lagazio, C., Dreassi, E., & Biggeri, A. (2001). A hierarchical Bayesian model for spacetime variation of disease risk. *Statistical Modelling*, *1*, 1729. doi:10.1191/147108201128069

Lai, D. (2005). Monitoring the sars epidemic in China: Time series analysis. *Journal of Data Science*, *3*, 279–293.

Morris, R., Wilesmith, J., Stern, M., Sanson, R., & Stevenson, M. A. (2001). Predictive spatial modeling of alternative control strategies for the foot-and-mouth disease epidemic in Great Britain. *The Veterinary Record*, 137–144. doi:10.1136/vr.149.5.137

Tildesley, M. J., Savill, N. J., Shaw, D. J., Deardon, R., Brooks, S. P., & Woolhouse, M. E. J. (2005). Optimal reactive vaccination strategies for a foot-and-mouth outbreak in the UK. *Nature*, *440*, 83–86. doi:10.1038/nature04324

Tufan, M. (2003). Report of the foot-and-mouth situation in Turkey from 1990 to 2002 . In *Consultant's report to the food and agriculture organisation of the United Nations* (p. 44). New York, NY: United Nations.

Viboud, C., Pierre-Yves, B., Alain-Jacques, F. C. V., & Flahault, A. (2003). Prediction of the spread of influenza epidemics by the method of analogues. *American Journal of Epidemiology*, *158*, 996–1006. doi:10.1093/aje/kwg239

Wallace, S., Maki-Petaja, K., Cheriyan, J., Davidson, E., McEniery, C., & Wilkinson, I. (2006). Simvastatin prevents acute inflammation-induced aortic stiffening and endothelial dysfunction in healthy volunteers. *British Journal of Clinical Pharmacology*, *70*(6), 799–806. doi:10.1111/j.1365-2125.2010.03745.x

APPENDIX

Table 7. Statistics of model parameterization specific to each administrative district (node). The latitude and longitudinal information of each node along with the Bayesian Information Criterion obtained by implementing a 4x4x1 NN and an AR model with an optimal window size and order is presented.

Longitude	Latitude	Node ID	BI CN N	BI CAR
26.6667	41.25	1	31.587294	-13.768624
26.8333	40.0833	2	30.855536	-15.717734
27.5	38.25	3	34.714106	-6.121068
27.5	41	4	30.927472	-10.98562
27.5	41.6667	5	33.589435	-8.424702
28	37.75	6	40.329401	-3.161489
28	39.75	7	33.102374	-8.826055
28.1667	38.8333	8	32.766836	-9.886397
28.5	37.1667	9	32.113432	-10.414174
28.75	41.1667	10	31.69056	-11.606907
29.0833	40.1667	11	32.632659	-5.974234
29.1667	40.5833	12	31.700809	-15.017764
29.25	37.7	13	34.713502	-5.692222
29.4167	38.5	14	40.152734	-3.813143
29.5	39.25	15	35.553555	-3.042159
29.9167	40.9167	16	32.16367	-5.781231
30	37.5	17	33.496308	-8.844905
30.1667	40	18	34.616385	-4.575703
30.5833	40.75	19	32.803956	-8.41158
30.6667	38.75	20	37.454428	-4.252727
31	37	21	32.124753	-10.102669
31	38	22	34.922849	-6.867442
31.1667	39.6667	23	35.3125	-5.594083
31.1667	40.8333	24	30.159604	-5.71026
31.5833	40.6667	25	34.83023	-5.6856
31.8333	41.25	26	33.161459	-9.976079

Table 8. Statistics of model parameterization specific to each administrative district (node) obtained by implementing a 4x4x1 NN and an AR model with an optimal window size continued

Longitude	Latitude	Node ID	BI CN N	BI CAR
32.5	38.1667	27	40.320166	-2.835948
32.5	41.25	28	32.003598	-11.330234
32.5	41.5833	29	31.285472	-12.587395
32.8333	39.9167	30	37.790066	-1.748802
33.25	37.0833	31	32.248835	-10.263417
33.4167	40.6667	32	39.59402	-3.380626
33.6667	41.5	33	39.102188	-4.434369
33.75	39.8333	34	33.476782	-6.969537
33.8333	38.5	35	34.09141	-7.629165
34	36.75	36	33.481779	-8.639217
34.1667	39.3333	37	37.63204	-4.280764
34.6667	38.9167	38	39.940679	-3.451039
34.75	37.8333	39	37.178305	-2.825407
34.75	40.5	40	34.708288	-4.735314
35	41.6667	41	32.739352	-4.041312

continued on following page

Table 8. Continued

Longitude	Latitude	Node ID	BI CN N	BI CAR
35.3333	39.5833	42	34.234868	-7.686428
35.4167	37.25	43	34.039976	-7.320137
35.8333	40.6667	44	40.368819	-2.28734
35.9167	38.75	45	42.021822	-1.260702
36.25	36.5	46	36.537339	-4.191847
36.25	37.25	47	35.014555	-7.502909
36.3333	41.25	48	37.027359	-2.514365
36.5833	40.4167	49	36.804883	-4.57921
36.9711	37.8983	50	33.37184	-5.344075
37.0833	36.8333	51	31.681558	-12.951114
37.3333	37.0833	52	34.242086	-6.785186

Table 9. Statistics of model parameterization specific to each administrative district (node) obtained by implementing a 4x4x1 NN and an AR model with an optimal window size continued

Longitude	Latitude	Node ID	BI CN N	BI CAR
37.4167	39.5	53	35.611263	-5.300083
37.5	40.8333	54	36.627991	-6.630713
38	38.5	55	39.911486	-2.264711
38.25	37.75	56	38.590554	-4.370979
38.5	40.5	57	32.191497	-10.944446
39	37.25	58	31.55844	-12.573825
39.5	38.5	59	40.934662	-2.390237
39.5	39.0833	60	32.237379	-5.52672
39.5	39.75	61	38.419929	-3.612209
39.5833	40.25	62	33.657435	-5.687787
39.8333	40.9167	63	34.281478	-5.539816
40.25	40.25	64	33.739998	-5.699851
40.5	38	65	36.023801	-5.698815
40.8333	37.4167	66	31.059333	-16.49936
40.8333	39.0833	67	32.978381	-9.155027
41	41	68	34.474717	-5.081739
41.3333	38	69	30.461939	-16.509473
41.5	40	70	40.923716	-2.020223
41.75	39	71	38.076008	-3.789508
41.8333	41.1667	72	33.95905	-7.270599
42.25	38.5	73	33.881279	-7.287346
42.8333	41.0833	74	39.939539	-3.032314
43.0833	40.4167	75	39.991685	-2.61949
43.1667	39.6667	76	32.534128	-6.859712
43.5	38.5	77	34.252427	-4.818717
44	39.9167	78	31.063646	-5.768835
44.1667	37.5833	79	36.483662	-4.88638

Figure 15.

(a) Standard deviation of RMSE for I_t.

Layer	Hidden	1	2	3	4	5	6	7	8	9	10	11
Input	2	6.310552	0.947369	0	0	0	0	0	0	0	0	0
*10^-4	3	3.771548	1.780645	1.040586	0	0	0	0	0	0	0	0
	4	3.843783	0.950913	1.374315	0.762992	0	0	0	0	0	0	0
	5	4.484188	1.220665	1.173147	0.800148	0.822272	0	0	0	0	0	0
	6	2.433716	3.388474	1.808123	2.521226	2.273526	1.195965	0	0	0	0	0
	7	5.401291	0.921287	1.11247	1.74463	1.115324	2.284228	1.405691	0	0	0	0
	8	7.431936	1.553089	1.554445	1.295002	0.81395	0.881021	0.897449	0.70846	0	0	0
	9	4.035966	0.667553	2.0221	1.129283	1.020675	0.83553	0.73364	0.726408	0.511091	0	0
	10	3.822984	1.926419	0.906339	1.035606	1.396592	0.807239	0.838998	3996.012	0.891423	2998.165	0
	11	1.698598	1.816232	0.791196	0.678329	0.69399	0.634915	0.721658	0.580931	0.92904	4580.282	3998.067

(b) Average RMSE for I_t

Layer	Hidden	1	2	3	4	5	6	7	8	9	10	11
Input	2	0.062173	0.062413	0	0	0	0	0	0	0	0	0
*10^-4	3	0.066589	0.066884	0.067132	0	0	0	0	0	0	0	0
	4	0.062278	0.062401	0.062444	0.062359	0	0	0	0	0	0	0
	5	0.062423	0.062355	0.062373	0.06229	0.062183	0	0	0	0	0	0
	6	0.065241	0.065498	0.065429	0.065314	0.06503	0.064803	0	0	0	0	0
	7	0.065512	0.065376	0.06527	0.065101	0.064964	0.064771	0.064478	0	0	0	0
	8	0.051403	0.049826	0.049699	0.049693	0.049584	0.049464	0.049441	0.049339	0	0	0
	9	0.051096	0.049931	0.049787	0.049651	0.049592	0.049467	0.049413	0.049352	0.049233	0	0
	10	0.053941	0.052719	0.052587	0.052439	0.05236	0.052242	0.052066	0.241486	0.051873	0.146603	0
	11	0.053802	0.052762	0.052484	0.052428	0.052333	0.05219	0.05206	0.051911	0.051819	0.336175	0.241253

(c) Standard deviation of RMSE for I_v.

Layer	Hidden	1	2	3	4	5	6	7	8	9	10	11
Input	2	5.948035	0.729169	0	0	0	0	0	0	0	0	0
*10^-4	3	4.48631	2.498337	0.873248	0	0	0	0	0	0	0	0
	4	3.369987	0.836054	1.291679	0.902712	0	0	0	0	0	0	0
	5	3.889623	1.671172	1.433332	1.216624	0.950433	0	0	0	0	0	0
	6	2.760118	4.260988	2.271009	3.295078	2.941242	1.748568	0	0	0	0	0
	7	5.841709	1.5541	1.370496	2.086569	1.622629	3.174777	2.137844	0	0	0	0
	8	7.748699	1.847479	2.164124	1.704097	1.123234	1.133637	1.223308	0.956703	0	0	0
	9	4.676097	0.963664	2.526879	1.635854	1.340276	1.096209	1.063275	0.932634	0.681954	0	0
	10	3.94085	2.483713	1.276139	1.387756	2.005633	1.067103	1.360633	3996.594	1.266113	2998.908	0
	11	1.55022	2.06739	1.116844	0.837765	0.977891	0.698742	0.923908	0.678229	1.174378	4581.556	3999.21

(d) Average RMSE for I_v.

Layer	Hidden	1	2	3	4	5	6	7	8	9	10	11
Input	2	0.062029	0.06238	0	0	0	0	0	0	0	0	0
	3	0.066233	0.066538	0.066934	0	0	0	0	0	0	0	0
	4	0.062183	0.062363	0.062401	0.062418	0	0	0	0	0	0	0
	5	0.06229	0.062262	0.062311	0.062249	0.062218	0	0	0	0	0	0
	6	0.065266	0.065353	0.06518	0.064931	0.064542	0.064205	0	0	0	0	0
	7	0.065449	0.065107	0.064959	0.06465	0.06443	0.064099	0.063644	0	0	0	0
	8	0.051505	0.049877	0.049771	0.049795	0.049681	0.049563	0.049573	0.049458	0	0	0
	9	0.051193	0.050035	0.049889	0.049748	0.049711	0.049575	0.049546	0.049487	0.04938	0	0
	10	0.053986	0.052741	0.052608	0.052409	0.052314	0.052162	0.051962	0.241365	0.051702	0.146391	0
	11	0.053771	0.052722	0.052429	0.052343	0.052232	0.052052	0.051907	0.051718	0.051604	0.33599	0.241033

Figure 16.

(a) Standard deviation of RMSE for S_t.

Layer	Hidden	1	2	3	4	5	6	7	8	9	10	11
Input	2	1.713061	0.400871	0	0	0	0	0	0	0	0	0
*10^-3	3	2.806966	0.599809	0.292405	0	0	0	0	.0	0	0	0
	4	1.600388	0.489289	0.195729	0.05471	0	0	0	0	0	0	0
	5	1.324842	0.5833	0.185242	0.05588	0.034803	0	0	0	0	0	0
	6	1.98665	0.493961	0.241961	0.070448	0.046309	0.060303	0	0	0	0	0
	7	0.854338	0.287385	0.095285	0.069083	0.058943	0.085812	0.066988	0	0	0	0
	8	0.962515	0.654814	0.398105	0.375922	0.211389	0.198211	0.13068	0.226303	0	0	0
	9	0.550287	0.630244	0.493983	0.238393	0.166391	0.190363	0.096626	0.104979	0.084559	0	0
	10	0.812954	0.399734	0.366642	0.346248	0.237842	0.058644	0.043083	0.252233	0.052933	0.626025	0
	11	0.781274	0.47561	0.313265	0.303514	0.172357	0.051755	0.16418	0.095505	0.049768	0.430119	413.2222

(b) Average RMSE for S_t.

Layer	Hidden	1	2	3	4	5	6	7	8	9	10	11
Input	2	0.042772	0.037386	0	0	0	0	0	0	0	0	0
	3	0.043738	0.039488	0.039082	0	0	0	0	0	0	0	0
	4	0.039551	0.03605	0.034869	0.034688	0	0	0	0	0	0	0
	5	0.038238	0.035644	0.034912	0.034757	0.034701	0	0	0	0	0	0
	6	0.037893	0.034361	0.033469	0.033189	0.033123	0.033051	0	0	0	0	0
	7	0.037156	0.034042	0.033435	0.033227	0.03317	0.033103	0.033083	0	0	0	0
	8	0.028614	0.023784	0.020721	0.019397	0.018657	0.01831	0.018047	0.017866	0	0	0
	9	0.028194	0.023567	0.020752	0.019267	0.018538	0.018359	0.017903	0.017838	0.017677	0	0
	10	0.03111	0.025245	0.022558	0.021452	0.020699	0.020207	0.019983	0.01998	0.019861	0.020081	0
	11	0.030857	0.025084	0.022463	0.021383	0.020604	0.020155	0.020064	0.019902	0.019834	0.020054	0.215952

(c) Standard deviation of RMSE for S_v .

Layer	Hidden ne	1	2	3	4	5	6	7	8	9	10	11
Input neu	2	1.498971	0.596276	0	0	0	0	0	0	0	0	0
*10^-3	3	2.193729	0.608541	0.267397	0	0	0	0	0	0	0	0
	4	1.862968	0.578421	0.179852	0.071873	0	0	0	0	0	0	0
	5	1.693829	0.268808	0.167048	0.096516	0.096606	0	0	0	0	0	0
	6	1.028045	0.334211	0.21463	0.137801	0.073177	0.085833	0	0	0	0	0
	7	1.720591	0.386386	0.089197	0.104419	0.092843	0.130227	0.137562	0	0	0	0
	8	0.824314	0.730889	0.629799	0.508851	0.212442	0.161741	0.212383	0.134998	0	0	0
	9	0.88074	0.440748	0.326809	0.36627	0.186122	0.213512	0.197475	0.207808	0.19503	0	0
	10	0.543959	0.451378	0.212871	0.387977	0.143649	0.115	0.17336	0.111604	0.098648	309.1754	0
	11	0.460607	0.364565	0.270063	0.130202	0.183241	0.137091	0.053968	0.088344	309.7613	1.215907	413.1375

(d) Average RMSE for S_v .

Layer	Hidden	1	2	3	4	5	6	7	8	9	10	11
Input	2	0.041605	0.037484	0	0	0	0	0	0	0	0	0
	3	0.042563	0.039249	0.038832	0	0	0	0	0	0	0	0
	4	0.039436	0.036132	0.034661	0.034444	0	0	0	0	0	0	0
	5	0.039248	0.035436	0.034734	0.034591	0.034454	0	0	0	0	0	0
	6	0.037464	0.034351	0.033331	0.032992	0.03284	0.032798	0	0	0	0	0
	7	0.03693	0.034243	0.033273	0.033054	0.033005	0.032898	0.032855	0	0	0	0
	8	0.028973	0.02388	0.021138	0.019671	0.018648	0.018313	0.018147	0.017891	0	0	0
	9	0.028317	0.023185	0.020704	0.019378	0.01869	0.018338	0.018078	0.017886	0.017881	0	0
	10	0.030508	0.025244	0.022643	0.021403	0.020516	0.020301	0.020138	0.019976	0.019958	0.119827	0
	11	0.030417	0.025062	0.022412	0.021048	0.020689	0.020211	0.01999	0.019985	0.118111	0.020401	0.21606

This work was previously published in the International Journal of Artificial Life Research, Volume 2, Issue 2, edited by E. Stanley Lee and Ping-Teng Chang, pp. 42-76, copyright 2011 by IGI Publishing (an imprint of IGI Global).

Chapter 17
Considerations on Strategies to Improve EOG Signal Analysis

Tobias Wissel
Otto von Guericke University, Germany & University of Essex, UK

Ramaswamy Palaniappan
University of Essex, UK

ABSTRACT

Electrooculogram (EOG) signals have been used in designing Human-Computer Interfaces, though not as popularly as electroencephalogram (EEG) or electromyogram (EMG) signals. This paper explores several strategies for improving the analysis of EOG signals. This article explores its utilization for the extraction of features from EOG signals compared with parametric, frequency-based approach using an autoregressive (AR) model as well as template matching as a time based method. The results indicate that parametric AR modeling using the Burg method, which does not retain the phase information, gives poor class separation. Conversely, the projection on the approximation space of the fourth level of Haar wavelet decomposition yields feature sets that enhance the class separation. Furthermore, for this method the number of dimensions in the feature space is much reduced as compared to template matching, which makes it much more efficient in terms of computation. This paper also reports on an example application utilizing wavelet decomposition and the Linear Discriminant Analysis (LDA) for classification, which was implemented and evaluated successfully. In this application, a virtual keyboard acts as the front-end for user interactions.

1. INTRODUCTION

As part of the Human-Computer Interface (HCI) research field, the electrooculo-graphy (EOG) has been established as a promising modality among other sources like electroencephalography (EEG) or electromyography (EMG). In this context various eye-based applications have been proposed, acting as hands-free interfaces to a general computer system (Estrany, Fuster, Garcia, & Luo, 2009) or even wearable embedded system, in order to aid the context awareness of machines (Bulling, Roggen, & Tröster, 2008).

DOI: 10.4018/978-1-4666-3890-7.ch017

Applying a combined approach of Continuous Wavelet Transform (CWT) and threshold comparisons for feature extraction, Bulling et al. has shown that it is possible to detect saccade movements revealing clues about the context and in particular the reading activity of a subject (Bulling, Roggen, & Tröster, 2008).

Apart from those applications, parts of the research have been motivated by the idea to enhance the quality of life especially for disabled people who suffer from severe motor impairments (Barea, Boquete, Mazo, & López, 2002; Teccea, Gips, Olivieri, Pok, & Consiglio, 1998).

People suffering from neurodegenerative diseases like amytropic lateral sclerosis (ALS) are often locked into their own body, whereas the oculomotor system, being one of the last capabilities remaining, has a certain resistance to the derogating process (Dhillon, Singla, Rekhi, & Jha, 2009).

With respect to that motivation, several eye-based spelling devices using a virtual keyboard on a screen have been designed to provide a means of communication for that particular target group (Dhillon, Singla, Rekhi, & Jha, 2009; Hori, Sakano, Miyakawa, & Saitoh, 2006; Usakli & Gurkan, 2010).

Those interfaces were proposed as an alternative to already existing, commercial applications as described by Krolak et al. (2009), which are vision-based and entail the necessity of a bulky headgear.

The work presented here embraces the idea of a virtual keyboard to form the context for the investigation of different methods of feature extraction, which is the main focus of this study.

Further, most keyboard interfaces use the eye blink, for the selection command. However, eye blinks can also occur involuntarily, wherefore a long blink is proposed here as an alternative as it is more distinguishable from involuntary ones which tend to be shorter.

The methods underlying most of the approaches mentioned earlier are time-based and some are even designed without significant feature extraction stage in the reasoning that the interface should be kept simple and without complicated algorithms (Usakli & Gurkan, 2010).

Thus, the investigation being done here attempts to answer the question, whether simplicity does always correlate with efficiency. It aims at a comparative study of time, frequency-based methods as well as hybrid approaches, which combine both domains.

With regards to spectral analysis, only parametric methods are applicable to EOG signals, the amplitudes of which range between 50 to 3500 μV, with its major components below 35 Hz (Barea, Boquete, Mazo, & López, 2002). A non-parametric FFT would hence result in a very poor frequency resolution due to a small amount of samples available for a particular time interval.

Parametric approaches, on the contrary, attempt to model the system based on a set of coefficients. Those models, which are mostly autoregressive (AR) systems due to their good prediction results, have successfully been utilized for the analysis of EEG characteristics (Faust, Acharyaa, Allen, & Lin, 2008; Tseng, Chen, Chong, & Kuo, 1995), but have not sufficiently been taken into account for eye-based interfaces.

Further on, investigations have revealed that wavelet-based procedures are superior to frequency-based methods for certain applications (Fargues & Bennett, 1995) and are even found to be promising for biomedical systems (Unser, 1996).

The multi-resolution approach, considering the given signal at several distinct scales, was employed for signal compression (Bhandari, Khare, Santhosh, & Anand, 2007), which is noise reduction in more general terms, as well as for the extraction of specific information.

The latter aspect led Magosso et al. to propose solutions capable of detecting slow eye movements (SEM) during certain sleep phases in the EOG signals (Magosso, Ursino, Zaniboni, & Gardella, 2006) and moreover of identifying patterns in the brain activity during an epileptic seizure (Magosso, Ursino, Zaniboni, & Gardella, 2009).

In contrast to the examination of energy distributions across all scales, as studied in the articles mentioned, this study utilizes the discrete wavelet transform (DWT) to obtain an efficient representation of the data within the chosen feature space.

The results are compared with AR-modeling as well as straightforward time-based template matching used by Usakli and Gurkan (2010) for instance.

Efficiency is hence studied here as a combination of sufficient performance, which is necessarily accompanied by an optimal class separation, and small computation overhead.

After describing the experimental setup and data acquisition, the paper proceeds with a general elaboration of the methodology used. Finally the results are appraised and concluded with the online EOG-based virtual keyboard implementation.

2. METHODOLOGY

The project scope involves the acquisition of EOG data, which is used for a subsequent off-line analysis to justify the methods for feature extraction and classification stage. Finally an on-line system is implemented and connected to the front-end speller.

2.1 Data Acquisition

The EOG data was recorded from the five participants listed in Table 1 (all had normal or corrected to normal vision), whereas subjects 1 to 3 participated in two sessions, one of which was the basis for the off-line investigations. The other session, which was also recorded from the remaining two subjects, provided the training sets for the evaluation of the on-line implementation later on.

Active Ag/AgCl electrodes were placed above the right eyebrow and in a vertical plane on the malar prominence below the eye (Denney & Denney, 1984) to form the vertical pair and on the outer canthi of both eyes near the temples for the horizontal electrode pair.

Table 1. Participant listing

Subject	*Age*	*Gender*	*Vision*
1	*24*	*Male*	*Glasses*
2	*23*	*Male*	*Normal*
3	*27*	*Female*	*Normal*
4	*24*	*Male*	*LASIK*
5	*26*	*Male*	*Glasses*

Since these were used in terms of differential pairs, referenced to each other, a fifth reference electrode was not necessary. Only a less intrusive ground reference for the ADC hardware was attached to the left hand of the subject.

The participants were required to perform eye movements in four directions and long blinks as requested in separate sections on a computer monitor. The sessions, which were recorded for the investigation, also involved random sections, where the individual produced EOG objects (i.e. eye movements and blinks) following the order which the subject determined earlier. To prevent fatigue during the session, subjects were instructed occasionally to perform short blinks during the collection of ten EOG objects in each session.

2.2 Pre-Processing

Using the EOG signals that were sampled with 256 Hz, a vertical and a horizontal channel were computed by referencing, in order to eliminate common noise and drift effects.

Subsequently, they were passed to a cascade of two digital, low-order Bessel filters acting as a bandpass with the lower cut-off frequency at 1 Hz and the higher at 5 Hz. The filter type was chosen due to its linear phase characteristic in the passband without any ripples in the amplitude response. The lower cut-off frequency eliminates parts of the EOG signal for the benefit of a very stable baseline, the components of which overlap the information inseparable in the lower bands. Preliminary investigation found this to be optimal for the following stages of feature extraction and classification.

Finally, avoiding aliasing effects, down-sampling by a factor of four was carried out, which resulted in a Nyquist frequency of 32 Hz.

Further operations for on-line setup are described as follows. First, blocks of 112 data samples were obtained from a sliding window, which were moved in steps of 16 samples through the signal. Hence, the buffer will be large enough to cover the duration of a long blink and will be updated every 250 ms in the final on-line solution.

2.3 Feature Extraction

The features are extracted with three different approaches.

First, a straightforward time-based approach takes the whole buffer content without any further operations, resulting in a 112 dimensional feature space. The classification in this case is exclusively done using a nearest neighbor classifier (with single neighbor, 1NN) similar to Usakli and Gurkan (2010).

This classifier computes the Euclidean distance as a measure of similarity to all sets of stored class templates for each EOG object type and assigns the nearest class to the test feature vector. Thus, this approach is also referred to as template matching.

Second, due to the obvious limitations of FFT based methods, a parametric, frequency-based approach using AR-modeling was chosen. As a result, a set of coefficients was estimated approximating the physical system with the following equation, where m is the chosen model order:

$$H\left(z\right) = \frac{1}{\sum_{j=1}^{m} a_j z^{-j}} \quad (1)$$

In this context, the method proposed by Burg (1975) is used in its order-recursive implementation to obtain the parameters. The algorithm calculates the coefficients a_j, being characteristic of the data, based on the optimization criterion:

$$\epsilon_r = \sum_{i=m}^{N-1}\left(\left|e_{fr}\right|^2 + \left|e_{br}\right|^2\right) \rightarrow Min \quad (2)$$

where N is the buffer size, r the current order with regards to the recursion and e_{fr} and e_{br} the forward and backward prediction errors, respectively.

Using the Levinson-Durbin recursion, the method finally provides m coefficients, which are used as features stretching an m dimensional feature space. Due to inaccuracies and computations related to the autocorrelation function, this method was preferred to the Yule-Walker approach (Hoon, Hagen, Schoonewelle, & Dam, 1996).

By minimizing the Akaike Information Criterion (AIC), the necessary order was found and fixed to be nine corresponding to a sufficient estimation of the data block contents. This leads to ten distinct features separating the classes. However, the first coefficient is normally chosen to be 1 and hence can be discarded when used as features.

Finally, wavelet decomposition based on DWT is employed to obtain specific features in a particular time and frequency resolution.

In each stage, the signal is decomposed into an approximation space V_j stretched by scaling functions $\varphi_{j,n} = 2^{-j/2} \cdot \varphi(2^j x - n)$ and a detail space W_j stretched by wavelet functions $\psi_{j,k} = 2^{-j/2} \cdot \psi(2^j x - k)$, with the following conditions fulfilled:

$$V_j = V_{j-1} \cup W_{j-1} \quad (3)$$

Hence

$$V_0 \subset V_1 \ldots \subset V_j \quad (4)$$

For feature extraction, only the projection on the approximation space is kept, which is needed to compute the next approximations of the subsequent, coarser decomposition stage. For this projection, the Haar basis, being piecewise constants on unit intervals, was used with its coefficients $h_0 = 1/\sqrt{2}$ and $h_1 = 1/\sqrt{2}$, where according to Mallat (1989):

$$\forall n \in \mathbb{Z},\ h_n = \varphi_{j,0},\ \varphi_{j-1,n} \tag{5}$$

As proposed, based on the relationship between the coefficients of the refinement equations and FIR filters found by Mallat (1989), the implementation was realized with a bank of quadrature mirror filters, which are followed by a downsampling operator of factor two in each stage. Thus, after four stages of decomposition the input buffer results in seven approximation coefficients to be used as the representative feature set.

For an objective evaluation of the class separation capabilities for each proposed method, a measure R_{ij} from the Davies Bouldin Index (Davies & Bouldin, 1979) is applied to the feature sets.

This measure computes the ratio between the compactness and distance of two classes i and j in the chosen feature space:

$$R_{ij} = \frac{d_i + d_j}{\mu_i + \mu_j} \quad \mathrm{i,j} \in \{1,2\ldots,q\} \tag{6}$$

Assuming a number of Q labeled reference vectors for each of the q classes, the centroid μ_i and the class scatter d_i are obtained as follows:

$$\mu_i = \frac{1}{Q}\sum_{\mathrm{r} \in \mathrm{C_i}} \mathrm{r} \quad \mathrm{i} \in \{1,2\ldots,q\} \tag{7}$$

$$d_i = \frac{1}{Q}\sum_{\mathrm{r} \in \mathrm{C_i}} \mathrm{r} - \mu_i \quad \mathrm{i} \in \{1,2\ldots,q\} \tag{8}$$

For all operations, the Euclidean distance was used to measure the norm for the separation of two objects in the feature space.

2.4 Classification

Taking the extracted feature set into consideration, the classification stage was to decide to which of the defined classes the current signal sequence was to be assigned. In order to distinguish between all EOG objects and to discard involuntary short blinks, six classes have been set up:

$C_1 \rightarrow$ Eyes move down
$C_2 \rightarrow$ Eyes move up
$C_3 \rightarrow$ Long blink
$C_4 \rightarrow$ Eyes move left
$C_5 \rightarrow$ Eyes move right
$C_6 \rightarrow$ Short blink

Apart from the 1NN classifier, which was the only classifier used with template matching, two other classification methods have been employed here: Artificial Neural Network (ANN) and Linear Discriminant Analysis (LDA).

The two stages related to the pre-processing were only triggered in case of a prominent activity in one of the channels. This reduced unnecessary computation and minimized wrong decisions due to a random sequence in the buffer, which would not represent any of the defined classes.

The detection was done by applying a threshold to the absolute value of the signal, which corresponds to the square root of the signal energy. If the threshold, which was adjusted for each subject, was exceeded, the activity was evaluated as containing significant eye movement or blink data.

Moreover, based on preliminary investigations, the number of classes to be separated is reduced to four for the vertical channel and to two for the horizontal channel, since left and right movements of the eye will only influence the potential induced to the electrodes on the outer canthi, whereas the other four EOG objects affect the vertical channel only.

For each channel, a neural network using the multilayer perceptron (*MLP*) structure was trained with reference feature vectors using the backpropagation algorithm along with a sigmoid transfer function for each neuron. The hidden unit size was fixed at 30.

The pooled covariance matrices and class means were computed based on the training data for both channels and passed to the linear statistical classifier (i.e. LDA).

The computation costs for both ANN and LDA were not significantly different simple because the training is done off-line and during on-line testing, both classifiers perform similarly in speed. In this context, adjustments that will change the amount of training data will not affect the execution time.

3. RESULTS AND DISCUSSION

In order to explore the different vector spaces and the proportion of the overall information they contain, recorded EOG data have been decomposed using the 112 samples buffer content.

As an example, consider the class 'Eyes Down' (Figure 1). As it is apparent, the decomposition results in seven detail spaces W_j and one approximation space V_0. In each level of decomposition, the algorithm takes every sample after convolving with the orthogonal filters accordingly.

By examining the different resolutions, it is obvious that the coefficients significantly exceed 500 µV from space W_2 onwards, which corresponds to the fact that its major components are located in the subsequent spaces.

Consequently, a decomposition using four stages has been chosen as appropriate, which stops with the projection indicated below:

$$V_4 = V_3 \bigcup W_3 \quad (9)$$

After discarding the details in W_3, the algorithm results in seven order approximation coefficients, which constitute the new feature set to be sent to the

Figure 1. Haar wavelet decomposition for the class 'Eyes Down'

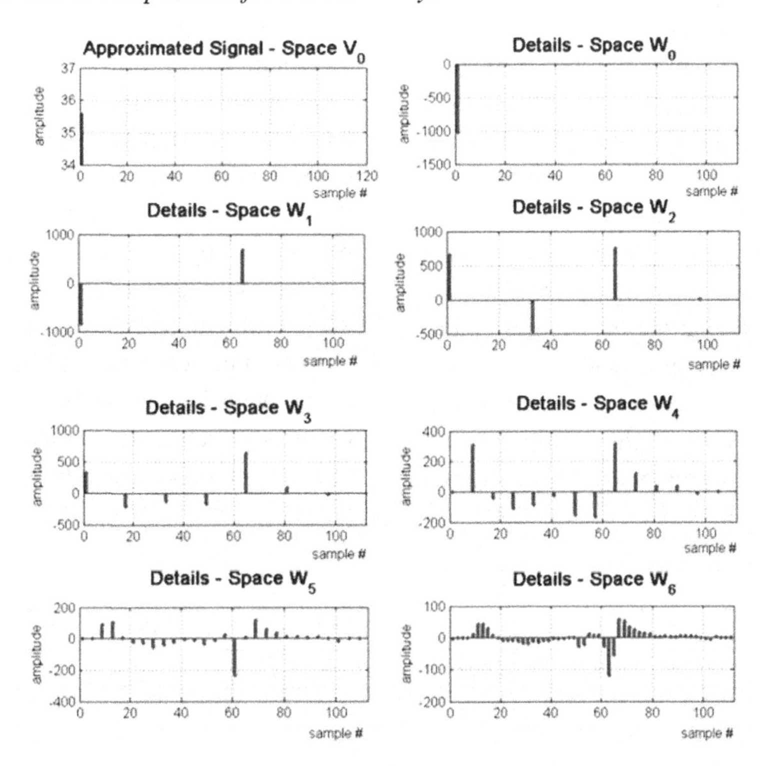

classifier. Apart from this feature representation, nine coefficients from a ninth order AR-system have been computed using the Burg method in its recursive implementation.

Finally, the time sequences being used as feature sets by the template matching approach were also taken into consideration for evaluating the class separation capabilities by means of the Davies-Bouldin Index. The results comparing the scores for each of the three methods are listed in Table 2 and Figure 2.

First, it is apparent that the frequency based algorithm performed much worse compared to the others. For that method, the index values denote results that were always above 1.5 and significantly higher compared to the other feature sets. In this context a lower index indicates better separability. The results were confirmed by applying this feature extraction to the offline data in order to obtain the rate of correct detections. As expected, it was found that the classes were hardly distinguishable which led to unreliable detections with less than 50% success rate.

Looking at the theory behind this approach, one will notice that the criterion for deriving the AR-system is based on a minimal sum of squared prediction errors. Consequently the algorithm is not sensitive with regard to phase information, which is however an essential requirement for a good class separation for this data.

Considering the time characteristics of the EOG data, it could be seen that a considerable part of the information about the eye movement is carried by the phase component.

In rough terms, the class 'Eyes Left' could be seen as the class 'Eyes Right' multiplied with minus one. This corresponds to a phase shift of

Figure 2. Illustrated separation indices

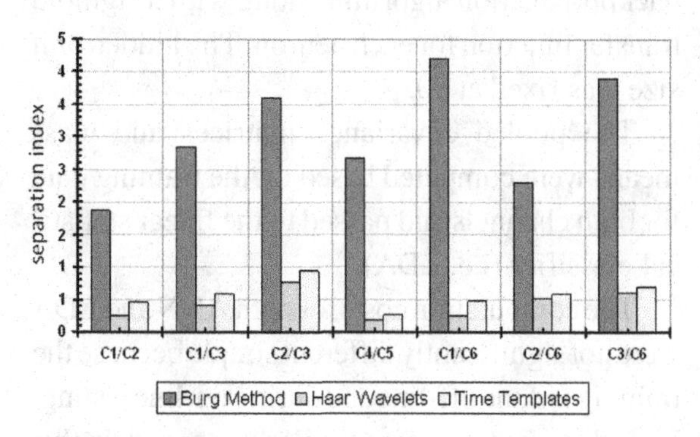

180° (though this is somewhat simplified explanation); the information about which is however eliminated by squaring the prediction error. As a conclusion the optimization criterion provides no clue about a possibly different phase. Both eye movements could have the same representation with respect to their AR coefficients.

This reasoning also applies for the other classes, which has been confirmed by the indices in Figure 2.

In the optimal case, the AR predictor generates white noise as its output, which means that it entirely compensates the physical system being modeled. The predictor filter hence models the amplitude of all components, whereas it suppresses the phase information. A particular phase shift in the random components of the white noise at the output has no influence on the noise energy represented by the prediction error.

Therefore this frequency-based method is not useful with regards to the separation of EOG objects. Furthermore, all the major frequency components for all objects are located in the same frequency bands, as confirmed by the wavelet decomposition earlier. Thus, a frequency-based

Table 2. Class separation indices

	$C1 \leftrightarrow C2$	$C1 \leftrightarrow C3$	$C2 \leftrightarrow C3$	$C4 \leftrightarrow C5$	$C1 \leftrightarrow C6$	$C2 \leftrightarrow C6$	$C3 \leftrightarrow C6$
Burg	1.866	2.845	3.600	2.664	4.208	2.295	3.902
Haar	0.241	0.397	0.756	0.169	0.248	0.507	0.591
Templates	0.456	0.578	0.927	0.259	0.480	0.574	0.677

algorithm, which aims at being successful, *must* retain the phase information, in order to keep the information necessary to distinguish between the classes.

The separation index also reveals an improvement in case of employing the wavelet based approach compared to template matching and the unprocessed time sequence.

Consequently, apart from reducing the number of dimensions significantly from 112 to 7, which will result in less computation requirements, the decomposition into the approximation space V_3 will also simplify the problem for the classifier, which could lead to higher performance.

Due to its localization in the time domain, the features obtained from the DWT retain the phase information in particular related to the frequency range extracted by the decomposition procedure. An example is shown for the movement 'Eyes Down' from the data of participant 3 in Figure 3.

The features contain the information located in the vector space V_3, which corresponds to

$$V_3 = V_0 \cup \left(\bigcup_{j=0}^{2} W_j \right). \tag{10}$$

For the classification step, template matching was combined with a 1NN classifier as suggested by Usakli and Gurkan (2010). Likewise, the proposed wavelet approach here was utilized along the LDA, ANN as well as 1NN.

Each of those four combinations (i.e. 1NN (template matching), DWT+LDA, DWT+ANN and DWT+1NN) was applied offline to the random sequences of the EOG data for each of the first three participants P1, P2 and P3. The performance in terms of correctly detected EOG objects is summarized in Table 3. The results in Table 3 shows that all methods worked well espe-

Figure 3. Features for 'Eyes Down'

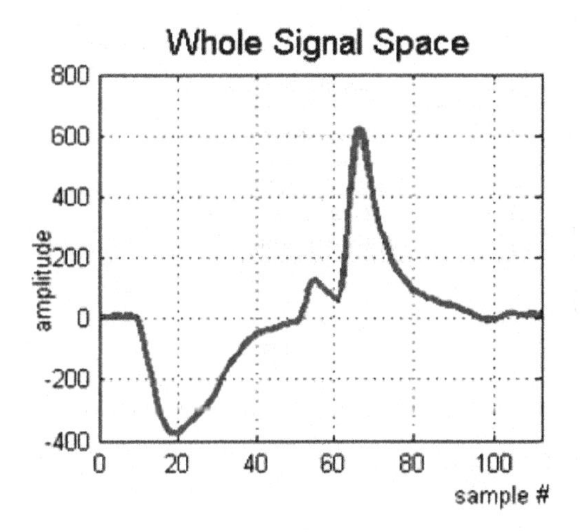

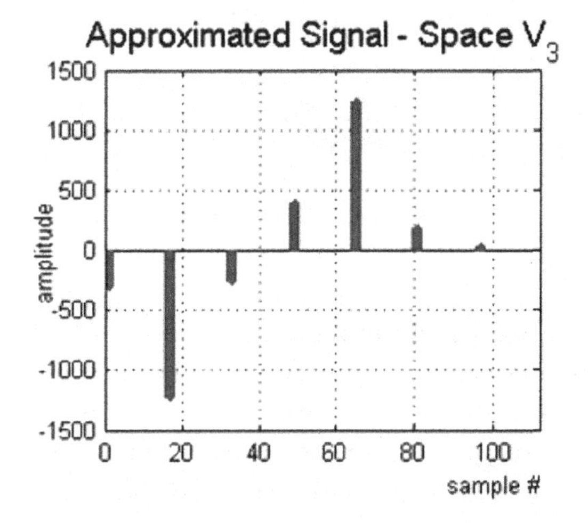

Table 3. Performance results

Method	*P1*	*P2*	*P3*	**Average**
DWT+1NN	98.10%	93.30%	80.00%	90.50%
DWT+ANN	90.60%	86.70%	80.00%	85.80%
templates	96.20%	93.30%	80.00%	89.80%
DWT+LDA	98.10%	93.30%	93.30%	94.90%

Figure 4. Number of operations versus buffer size for the vertical channel having four classes

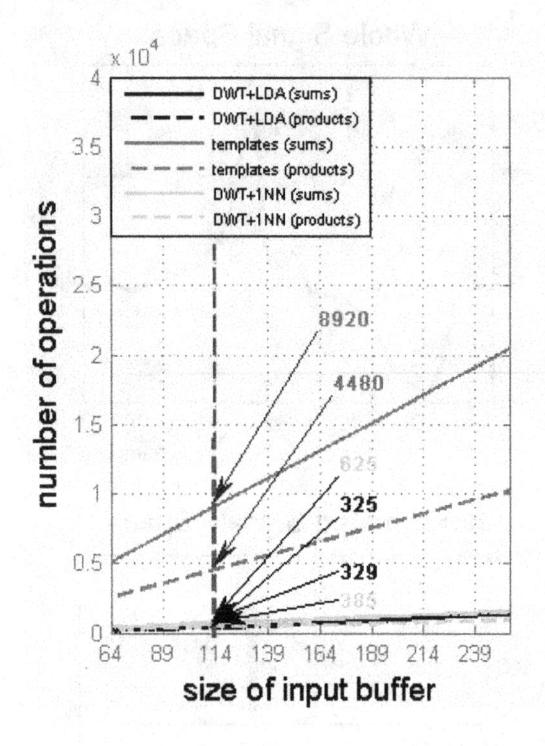

Figure 5. Number of operations versus amount of training data for four classes

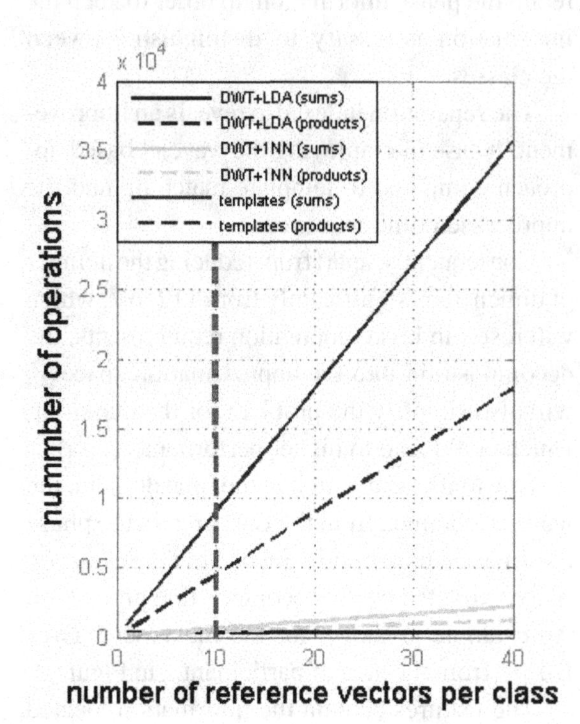

cially for participant 1 and slightly less well with participant 3; this could be caused by the 'Long Blink' and 'Eyes Up' signal characteristics which were very similar for participant 3. In this case, the combination of DWT and LDA performed better than the others, although it resulted in a similar outcome compared to DWT and 1NN for the other two subjects.

Moreover, the ANN classifier achieved hardly more than 90% of correct detections, which is in any case inferior to the performance of the other methods.

Considering the computation costs as a more practical aspect for both the feature extraction and classification stages, yields the following reasoning. Figures 4 and 5 illustrate the costs with respect to two different independent variables: the size of the input buffer and the number of labeled reference vectors (i.e. training data) per class. The diagrams indicate the number of multiplications and summations (being primitive operations), in which the algorithms can be decomposed.

In this context, it has to be mentioned, that the square root function has not been counted for the kNN classifier, since there are different ways for implementing it depending on the aspired accuracy or exact realization. Referring to the latter point, a hardware implementation might use a LUT or the Taylor series, for instance.

In any case, it will add a certain number of operations to the combinations utilizing kNN. Depending on the application, a tradeoff will always arise between hardware resources and computation time in order to cover those costs.

For the results, it can be deduced that template matching without any feature extraction turns out to be very inefficient. Applying this method entails more computation, whereas the computation cost of the other combinations after extracting the wavelet approximation coefficients is very close to each other.

Increasing the input buffer size seems not to be very useful from the practical point of view, since it will increase the latency and the hardware requirements of the system. Thus an implementation should aim at a reduction rather than at increased buffer size.

On the other hand, it is more likely that the training data per class might vary, since it is directly related to the performance of the system. In this case, the LDA and ANN classifiers are superior compared 1NN due to the independence on the amount of training data.

Conversely, the kNN classifier would have to compute more distance measures, if the number of reference vectors, i.e. training data, per class is increased.

Thus, its performance and the fact that the combination of DWT and LDA exhibits least cost for the chosen parameter set, leads to it being the chosen approach with regards to the implementation stage.

4. IMPLEMENTATION

The algorithm explained earlier was embedded in a LabVIEW application running on a common desktop computer. A BioSemi recording device along with four active electrodes was used to açquire the EOG data from a subject interacting with the computer.

The implementation, the block diagram of which is shown in Figure 6, was placed into a general state machine which was working synchronously with one iteration requiring a time of 250 ms.

Taking the initial sampling rate of 256 Hz into account, 64 samples per iteration were passed as an input to the application illustrated in Figure 6.

The pre-processing step included the cascade of Bessel filters for each of the two channels, which were computed from the electrodes by referencing.

Since aliasing was avoided due to the elimination of higher frequency components by the cascaded filter, the signal was down-sampled again by a factor of four resulting in 16 samples per iteration for each channel.

This data was shifted into a register, which contained 2x112 samples by discarding the 16 oldest samples from the register. Based on the buffer size of 112 samples, the features were extracted by calculating the seven approximation coefficients obtained from the fourth decomposition level of the Haar wavelet transform.

This decomposition was efficiently implemented by a cascade of downsampling FIR filters as shown in Figure 7.

Next, the feature extraction involved a threshold based onset detection for both channels, which triggered the extraction only if a prominent activ-

Figure 6. Flow diagram of the EOG interface

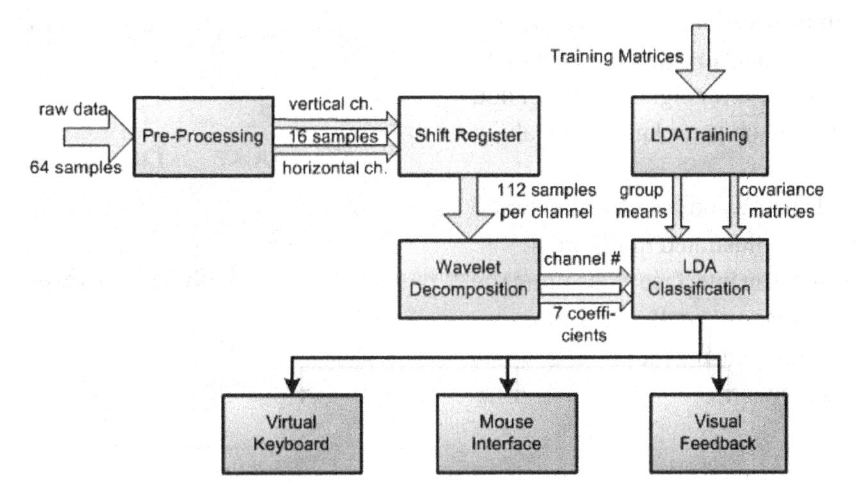

Figure 7. First level of wavelet decomposition

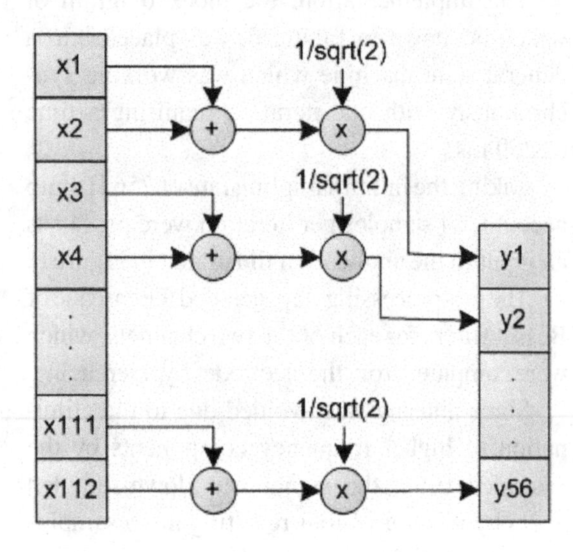

ity was detected in one of the channels. Thus, the data block in which the activity was detected and the number of the channel namely horizontal or vertical is passed to the LDA classifier.

This stage decided on the class assignment based on the covariance matrix and the class means obtained from the offline training data. Strictly speaking, the application worked with two classifiers, one for each channel, which were selected according to the output of the feature extraction.

If there was no detected activity in any channel, both the wavelet decomposition and the classification steps will not be executed, which reduces unnecessary computation.

The threshold applied to the data stream corresponds to a simple decision between two cases. Since its purpose was only to detect whether there is a significant increase in the signal energy or not, it does not overly rely on prior knowledge about the EOG characteristics.

The front-end focuses on a virtual keyboard application, which is illustrated in Figure 8.

It can also provide an interface controlling the Windows mouse cursor as well as a simple visual and textual feedback using the processed data from the buffer's content.

After receiving the classification results from the statistical classifier, they could be connected to any front-end by mapping the decisions to actual commands, which are interpreted by the machine.

In this particular keyboard, short blinks were not used as commands and hence were ignored. Consequently, the keyboard used only the four directions of the eye movements to move the cursor in discrete steps from one key to next one and the long blink was used as the selection command.

Differing from other virtual keyboard applications, this front-end intentionally utilized a mobile phone layout, which provided three to four letters per key (Dhillon, Singla, Rekhi, & Jha, 2009; Hori, Sakano, Miyakawa, & Saitoh, 2006; Usakli & Gurkan, 2010). This was seen as a promising alternative, since eye movements are expected to be much more tiring compared to blinks. Consequently, the layout would decrease the number of movements for the benefit of blinks,

Figure 8. Layout for the virtual keyboard

since the subject is required to select the correct letter by browsing through the options provided by each key.

As a structured way of designing the keyboard, a Mealy state machine was used, the output of which is generated depending on the current state of the machine and its input.

The basic layout consists of a 3x4 key array that covers the whole alphabet and a few special characters including 'space' as well as a return key to make corrections if necessary.

Additionally, it should be mentioned that after a completed classification, the classifier output is blocked until the buffer is emptied. This is to avoid a double classification of the same EOG data.

5. EVALUATION

The interface was evaluated with five subjects as listed in Table 1. Before interacting with the system, training data was collected, i.e. ten reference vectors for each of the six classes.

Those were pre-processed and arranged in a matrix representation to be fed into as inputs to the LabVIEW online system. The actual testing and the training were done in two different sessions, since it was found that the algorithm works robustly across sessions without significant decrease in performance.

However, the system does not work well across subjects, i.e. it is subject dependent, since appearance of the EOG objects might differ due to the way a subject performs them. For example, the speed and angle in which a subject moves his or her eyes would be different for each subject. Furthermore, the signals might be weaker or stronger depending on the ability of the subject.

Each participant was asked to carry out the defined EOG objects, whereas the amount of correct classifications was used as a measure of the performance for the online system. Comparable to the offline measurements, the number of right detections was between 96% and 100%. The performance is slightly better compared to the offline, possibly as the subject could unconsciously adapt to the system.

The second task (as also used for evaluation purposes by Usakli & Gurkan, 2010) was to write the word 'WATER' on the virtual keyboard. For this task, the number of corrections and the elapsed time were recorded as a measure of the performance for the virtual keyboard.

All the subjects were able to produce the word as the text output and none of them needed more than one correction. In total, each participant had two trials, in which he was required to write the specified word. The time needed to complete the five letters was used to calculate the average rate indicating how many letters could be written in one minute. For all subjects, the rate was between 3 and 5 letters per minute even though all subjects except subject 1 interacted with the system for the first time. A trained individual as participant 1 was able to finish the task within one minute. Movies on the operation of the virtual keyboard could be found at http://csee.essex.ac.uk/staff/palaniappan.

It should be mentioned that the fact of using long blinks as a class prevented the system from being faster, since the buffer size had to be chosen accordingly to cover the duration of a long blink, which could be more than one second.

Similar applications described in other articles were exclusively employing short blinks for selection, which would increase the writing performance in terms of time, but it will not be robust against involuntarily performed blinks.

Finally the participants were asked to fill in a questionnaire, in order to receive a subjective feedback about the system. All participants were of the opinion that blinks were less tiring compared to eye movements and hence supporting the phone layout rather than the QWERTY layout. On the whole, they agreed that the EOG interface was intuitive and easy to use.

6. CONCLUSION

This paper has considered several strategies to improve the analysis and processing of EOG signals. It was shown that frequency based approaches, in particular the AR parametric approach using Burg's method, is not a good feature extraction approach in order to achieve a robust class separation.

The introduced method of Haar wavelet decomposition was found to be superior to other approaches with regard to class separation and especially in terms of computation overhead due to the reduction of the dimension size in the feature space.

The combination of this feature extraction stage and the statistical LDA classifier was implemented with a virtual keyboard acting as the user front-end and evaluated to give promising detection rates.

In particular, real-time or embedded systems would benefit from improved feature extraction since hardware resources and processing power are very limited for such systems. Moreover this would allow using more sophisticated classifiers instead of the computationally intense kNN approach as used by others. Systems, which have to meet strict timing and have to be predictable, would benefit from the independence on the amount of training data as shown for the case of using LDA.

REFERENCES

Barea, R., Boquete, L., Mazo, M., & López, E. (2002). System for Assisted Mobility Using Eye Movements Based on Electrooculography. *Neural Systems and Rehabilitation Engineering. IEEE Transactions on, 10*(4), 209–218.

Bhandari, A., Khare, V., Santhosh, J., & Anand, S. (2007). Wavelet based compression technique of Electro-oculogram signals. *3rd Kuala Lumpur International Conference on Biomedical Engineering 2006. 15*, pp. 440-443. Kuala Lumpur: Springer.

Bulling, A., Roggen, D., & Tröster, G. (2008). It's in Your Eyes - Towards Context-Awareness and Mobile HCI Using Wearable EOG Goggles. *ACM Proceedings of the 10th International Conference on Ubiquitous Computing (UbiComp 2008)*, (pp. 84-93). Seoul, Korea.

Bulling, A., Roggen, D., & Tröster, G. (2008). Robust Recognition of Reading Activity in Transit Using Wearable Electrooculography. *Proceedings of the 6th International Conference on Pervasive Computing (Pervasive 2008)* (pp. 19-37). Sydney, Australia: Springer.

Burg, J. P. (1975). *Maximal Entropy Spectral Analysis, PhD Thesis.* Stanford University, Department of Geophysics.

Davies, D. L., & Bouldin, D. W. (1979). A cluster separation measure. *Pattern Analysis and Machine Intelligence. IEEE Transactions on, PAMI-1*(2), 224–227.

Denney, D., & Denney, C. (1984). The eye blink electro-oculogram. *The British Journal of Ophthalmology, 64*, 225–228. doi:10.1136/bjo.68.4.225

Dhillon, H. S., Singla, R., Rekhi, N. S., & Jha, R. (2009). EOG and EMG Based Virtual Keyboard: A Brain-Computer Interface. *Computer Science and Information Technology, 2nd International Conference on* (pp. 259-262). Beijing, China: IEEE.

Estrany, B., Fuster, P., Garcia, A., & Luo, Y. (2009). EOG signal processing and analysis for controlling computer by eye movements. *Pervasive Technologies Related to Assistive Environments, Proceedings of the 2nd International Conference on.* Corfu, Greece: ACM.

Fargues, M. P., & Bennett, R. (1995). Comparing Wavelet Transforms and AR Modeling as Feature Extraction Tools for Underwater Signal Classification. *Proceedings of the Conference Record of the Twenty-Ninth Asilomar Conference on Signals, Systems and Computers (ASILOMAR '95)* (pp. 915 - 919). Pacific Grove, CA, USA: IEEE.

Faust, O., Acharyaa, R., Allen, A., & Lin, C. (2008). Analysis of EEG signals during epileptic and alcoholic states using AR modeling techniques. *IRBM*, *29*(1), 44–52. doi:10.1016/j.rbmret.2007.11.003

Hoon, M. J., Hagen, T. H., Schoonewelle, H., & Dam, H. v. (1996). Why Yule-Walker should not be used for autoregressive modelling. *Annals of Nuclear Energy*, *23*(15), 1219–1228. doi:10.1016/0306-4549(95)00126-3

Hori, J., Sakano, K., Miyakawa, M., & Saitoh, Y. (2006). Eye Movement Communication Control System Based on EOG and Voluntary Eye Blink. *Computers Helping People with Special Needs. Lecture Notes in Computer Science*, *4061*, 950–953. doi:10.1007/11788713_138

Kiamini, M., Alirezaee, S., Perseh, B., & Ahmadi, M. (2008). A wavelet based algorithm for Ocular Artifact detection In the EEG signals. *Multitopic Conference, 2008. INMIC 2008. IEEE International*, (pp. 165-168). Karachi.

Krolak, A., & Strumillo, P. (2009). Eye-Blink Controlled Human-Computer Interface for the Disabled. *Human-Computer Systems Interaction. Advances in Soft Computing*, *60*, 123–133.

Magosso, E., Ursino, M., Zaniboni, A., & Gardella, E. (2006). A wavelet based method for automatic detection of slow eye movements: A pilot study. *Medical Engineering & Physics*, *28*(9), 860–875. doi:10.1016/j.medengphy.2006.01.002

Magosso, E., Ursino, M., Zaniboni, A., & Gardella, E. (2009). A wavelet-based energetic approach for the analysis of biomedical signals: Application to the electroencephalogram and electro-oculogram. *Applied Mathematics and Computation*, *207*(1), 42–62. doi:10.1016/j.amc.2007.10.069

Mallat, S. G. (1989). A Theory for Multiresolution Signal Decomposition: The Wavelet Representation. *IEEE Transactions on Pattern Analysis and Machine Intelligence*, *11*(7), 674–693. doi:10.1109/34.192463

Teccea, J. J., Gips, J., Olivieri, C. P., Pok, L. J., & Consiglio, M. R. (1998). Eye movement control of computer functions. *International Journal of Psychophysiology*, *29*(3), 319–325. doi:10.1016/S0167-8760(98)00020-8

Tseng, S.-Y., Chen, R.-C., Chong, F.-C., & Kuo, T.-S. (1995). Evaluation of parametric methods in EEG signal analysis. *Medical Engineering & Physics*, *17*(1), 71–78. doi:10.1016/1350-4533(95)90380-T

Unser, M. (1996). Wavelets, Statistics, and Biomedical Applications. *Proceedings of the 8th IEEE Signal Processing Workshop on Statistical Signal and Array Processing (SSAP '96)*, (pp. 244 - 249). Corfu, Greece.

Usakli, A. B., & Gurkan, S. (2010). Design of a Novel Efficient Human–Computer Interface: An Electrooculagram Based Virtual Keyboard. *Instrumentation and Measurement. IEEE Transactions on*, *59*(8), 2099–2108.

This work was previously published in the International Journal of Artificial Life Research, Volume 2, Issue 3, edited by E. Stanley Lee and Ping-Teng Chang, pp. 6-21, copyright 2011 by IGI Publishing (an imprint of IGI Global).

Chapter 18
An Autonomous Multi-Agent Simulation Model for Acute Inflammatory Response

John Wu
Kansas State University, USA

David Ben-Arieh
Kansas State University, USA

Zhenzhen Shi
Kansas State University, USA

ABSTRACT

This research proposes an agent-based simulation model combined with the strength of systemic dynamic mathematical model, providing a new modeling and simulation approach of the pathogenesis of AIR. AIR is the initial stage of a typical sepsis episode, often leading to severe sepsis or septic shocks. The process of AIR has been in the focal point affecting more than 750,000 patients annually in the United State alone. Based on the agent-based model presented herein, clinicians can predict the sepsis pathogenesis for patients using the prognostic indicators from the simulation results, planning the proper therapeutic interventions accordingly. Impressively, the modeling approach presented creates a friendly user-interface allowing physicians to visualize and capture the potential AIR progression patterns. Based on the computational studies, the simulated behavior of the agent–based model conforms to the mechanisms described by the system dynamics mathematical models established in previous research.

1. INTRODUCTION

The function of the human immune system is to respond to intruding pathogens or damage tissues (e.g., trauma) and to prevent them from spreading to the entire body by producing warning chemical signals, activating relevant immune cells in the blood circulation system near the infected area, and then killing the intruded pathogens or microbial organisms. The process to protect the human body from further infection by harmful stimuli is commonly referred as the immune responses

DOI: 10.4018/978-1-4666-3890-7.ch018

or acute inflammatory responses. However, an uncontrolled series of Acute Immune Responses (AIR) may lead to possible sepsis, severe sepsis or sepsis shocks since the immune cells and their released cytokines eliminate pathogens and microbial organisms but which also kill neighboring healthy cells. Recent census found that more than 750,000 severe sepsis or sepsis shock cases developed from sepsis in the US (Angus, 2001) with mortality rates between 20% and 80% (Zeni, 1997). In the United States alone, almost $17 billion is spent each year, treating patients with sepsis (Angus, 2001). Therefore, it is necessary to find an effective methodology that can help physicians predict the outcomes of an AIR, prevent possible severe sepsis or septic shocks, and control the involved risks for patients, which is the focus of this research.

This article presents a new modeling approach to predict the evolution of the Acute Inflammatory Response (AIR) which is the initial stage of sepsis pathogenesis. This predictive agent-based model (ABM) uses the system dynamics model developed by Reynolds et al. (2006) as a benchmark.

The organization of this paper is as follows: first we present the basic biological process of AIR, using a system dynamics model developed in previous research. Next, the agent-based model embedded with an existing system dynamics model is presented while its implementation detail is discussed. Outcomes of the agent based simulation are demonstrated and a sensitivity analysis is presented. Finally, conclusions and potential applications of the proposed model are discussed.

2. BIOLOGICAL MECHANISM OF ACUTE INFLAMMATORY RESPONSE

2.1 Process Description

The Acute Inflammatory Response, which can be the initial stage of sepsis, usually occurs when the human immune system detects intruding pathogens or existing tissue damages and sends out a signal (e.g., Interleukin-8 (IL-8) and C5a, the process is referred to as the chemotaxis) to the resting phagocyte cells such as the neutrophils initially and followed by the monocytes (two typical immune cells in the human body) in the blood vessel near the infected tissue. The resting phagocyte cells are activated and start to migrate towards the pathogens or damaged tissue whose recognizable protein on the surface is similar to those of the immune cells. Once the activated phagocyte cells reach the infection site, they start to engulf and consume the pathogens. Meanwhile, these activated phagocyte cells release pro-inflammatory cytokines such as Tumor Necrosis Factor (TNF), Interleukins (IL-1), IL-6, IL-8 and High Mortality Group Box-1 (HMGB-1) that activate more phagocyte cells and recruit them to the infection site. All those activated phagocyte cells not only eliminate the pathogens but also secrete substances which contribute to killing healthy cells and induce more inflammation in the initial progression of sepsis. Almost at the same time, several types of anti-inflammatory mediators such as IL-6, IL-10, soluble TNF receptors (sTNFRs) and IL-1 receptor antagonist (IL-1ra) are also released by the activated phagocyte cells in this stage. These anti-inflammatory mediators inhibit the production of pro-inflammatory mediators and therefore inhibit recruiting more phagocyte cells (Gogos, 2000).

2.2 System Dynamics Modeling of AIR

Undoubtedly, the complex mechanism of the AIR allows various possibilities of sepsis progression which may lead to a healthy response or a septic shock. Thus, based on insights into the biological mechanism of AIR a three equation system dynamics model was developed by Kumar et al. (2004). In the three equations model, pathogen level, early pro-inflammatory mediator, and late pro-inflammatory mediators were defined respectively. Moreover, those three essential indicators in AIR were measured by three individual equations. However, considering many other important indicators involved in AIR, a more complete system dynamics model based on five equations was developed by Reynolds et al. (2006). This model is shown next:

$$\frac{dP}{dt} = K_{pg} P\left(1 - \frac{P}{P_{\infty}}\right) - \frac{K_{pm} S_m P}{\text{ì}_m + K_{mp} P} - K_{pn} f\left(N^*\right) P \quad (1)$$

$$\frac{dN_R}{dt} = S_{nr} - R1 N_R - \text{ì}_{nr} N_R \quad (2)$$

$$\frac{dN^*}{dt} = R1 N_R - \text{ì}_n N^* \quad (3)$$

$$\frac{dD}{dt} = K_{dn} f_s\left(f\left(N^*\right)\right) - \text{ì}_d D \quad (4)$$

$$\frac{dC_A}{dt} = s_c + \frac{K_{cn} f\left(N^* + K_{cnd} D\right)}{1 + f\left(N^* + K_{cnd} D\right)} - \text{ì}_C C_A \quad (5)$$

where

$$R_1 = f\left(K_{nn} N^* + K_{np} P + K_{nd} D\right)$$

$$f\left(V\right) = V / \left(1 + \left(C_A / C_{\infty}\right)^2\right)$$

$$f_s\left(V\right) = V^6 / \left(x_{dn}^6 + V^6\right)$$

In this five equations model, the variables P, N_R, N^*, D, and C_A represent pathogen level, resting phagocyte cells, activated phagocyte cells, damaged tissue and anti-inflammatory mediators respectively. All the system parameters (K_{nn} et al.) reflect the strength of the immune system but their detailed description is beyond the scope of this paper.

The measurements in both mathematical models are based on time-based rates, so their function is to calibrate the change in number of indicators in AIR with the progress of time. Since the change in number of indicators in the next time unit depends on the previous number of indicators, the mathematical models alone are not convenient to automatically calibrate the indicators during the development of acute inflammatory response. Thus, two Simulink models in MatLab were implemented to separately simulate both mathematical models. The results from the simulation reflect the progression of acute inflammatory responses since it measures the corresponding changes in the number of pathological and physiological markers during the acute inflammatory response. One of the results from the Simulink model of five equations system dynamics model is shown in Figure 1 to demonstrate the feasibility of measurement.

In Figure 1 the Y axis represents the change of pathological indicators in AIR, and the X axis represents the simulation time corresponding to the progression of AIR. Thus the models established are capable of reflecting the progression of AIR.

Both mathematical models provide a high level view of the AIR progression by defining the changes in the number of indicators which react with the cells or in the blood circulation system. However, the actual immune system response is much more complicated and is highly stochastic in nature. For example, the strength of immune response differs among individual patients and even is different among the organs of the patient. Thus, the system dynamics models presented so far have limited ability of capturing these variations since they use deterministic scalar parameters. In order to model the important stochastic nature of

Figure 1. Simulation results of the five equations model

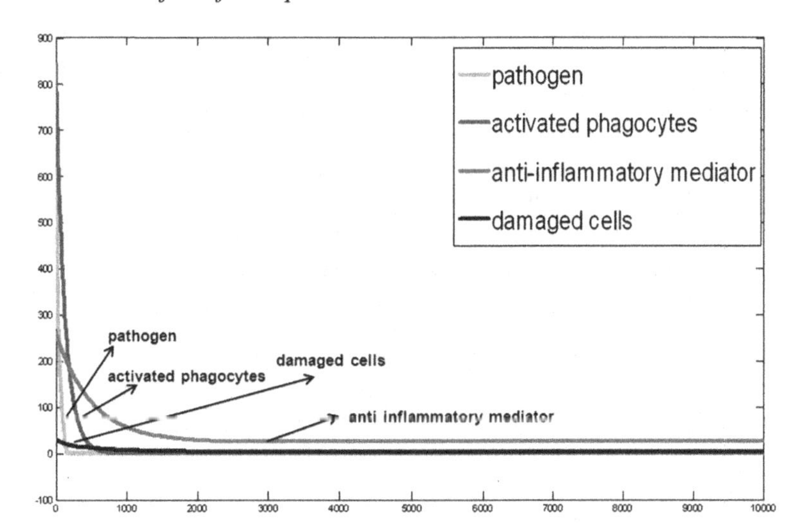

the biological system in focus and allow the ability to include the correct boundary conditions we expand the modeling method to an agent based modeling approach as shown next.

3. AN AGENT-BASED MODEL (ABM) EMBEDDED WITH SYSTEM DYNAMICS MATHEMATICAL MODEL

The agent-based model employs agents representing the various indicators in the progression of AIR. With each type of agent defined based on the variables described in the system dynamics model, it constructs the link between the system dynamics models and a real AIR environment. Moreover, agents can have autonomous and probabilistic behaviors, and therefore, provide improved modeling capability capturing the stochastic nature of the AIR progression episode. Compared to system dynamics models, the agent-based modeling approach is more flexible and more realistic. Furthermore, our ABM embedded with system dynamics models is applied at the intercellular level and expands to the tissue level, an improvement over the agent-based models previous established (An, 2000, 2004; Wakeland, Macovsky, & An, 2000). Thus, this ABM embedded with system dynamics mathematical model has the advantages of both system dynamics and traditional ABM models.

In this modeling approach, there are nine agent types: (1) *pathogen,* (2) *resting phagocytes,* (3) *activated phagocytes,* (4) *damaged cells,* (5) *IL-10s,* (6) TNF-as, (7) IL-1s, (8) HMGB-1s, and (9) IL-6s. Since the damaged tissue in the system dynamics model is hard to calibrate, we use *damaged cells* agent in our agent-based model as a representation of the extent of tissue damaged. The agent types and their description are shown in Table 1.

Each type of agent is defined as a homogeneous set of agents that may contain thousands of body cells, microorganisms or microbial antagonisms. For instance, the *pathogen* agent set contains pathogen agents, each consisting of thousands of pathogens. Figure 2 shows the logical structure of the agent set in the agent-based model.

An increase or decrease (e.g., creation, death or transformation) in the number of cells for a particular type of agent is executed by changing the current states of that agent in the agent set, and the current state of any agent is tracked by a set of pre-defined state variables. Since the ABM is implemented as loops and each loop is corresponding to one simulation time unit, the amount

Table 1. Agent type and its description

Agent Type	Description
Pathogen	Instigator of AIR; AIR starts when pathogen intrude into body
Resting phagocytes	Inactive immune cells such as neutrophils and macrophages existing in the blood vessel
Activated phagocytes	Activate immune cells that respond with intruding pathogen and secrete corresponding cytokines such as TNF-as, IL-1s etc.
Damaged cells	Normal cells damaged by intruding pathogen or cytokines released by the activated phagocytes
IL-10s	Anti-inflammatory mediator released by the activated phagocytes
TNFs	Early pro-inflammatory mediator released by the activated phagocytes
IL-1s	Early pro-inflammatory mediator released by the activated phagocytes
HMGB-1s	Late pro-inflammatory mediator released by the activated phagocytes
IL-6s	Late pro-inflammatory mediator released by the activated phagocytes

Figure 2. Basic structure of agents changing

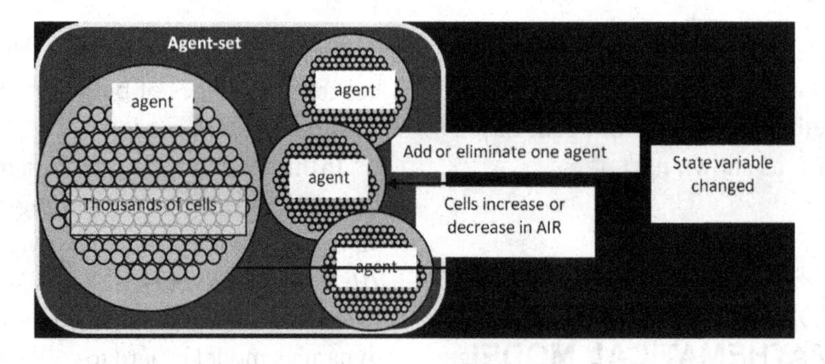

of changes in state variables follows a deterministic behavior similar to the systemic dynamic equations (1)-(5). In our agent based model, *pathogen*, *resting phagocytes*, *activated phagocytes*, *damaged tissue* and *IL-10s* are the five main agents whose state changes are defined by the system dynamics model (i.e., eqns. (1)-(5)). Each positive mathematical term in the equations such as $K_{pg}P\left(1-\frac{P}{P_{\infty}}\right)$ contributes to define the increase of cells in one agent while negative terms are used to define the decrease process. Figure 3 shows a comprehensive structure of the agent based model.

As an example, one of the state variables defined in the agent based model is called "increased-number-of-pathogen-cells-in-one-agent". This state variable measures the increased number of pathogen cells in one *pathogen* agent and it will keep increasing by $+k_{pg}P\left(1-\frac{P}{P_{\infty}}\right)$. per loop when the agent based model evolves. When the cell population in a *pathogen* agent grows beyond a predefined boundary condition, (maximal number of cells in a *pathogen* agent), the total number of *pathogen* agents will increase by one and the value of this state variable in the newly generated *pathogen* agent is set to its proper level. Also often, one agent will turn into another agent or generate other type of agents depending on the real physiological behaviors patterns modeled in the ABM. Therefore, the proposed ABM is quite flexible and functional in modeling the microbial pathogenesis of the AIR.

Figure 3. Comprehensive structure of the agent based model

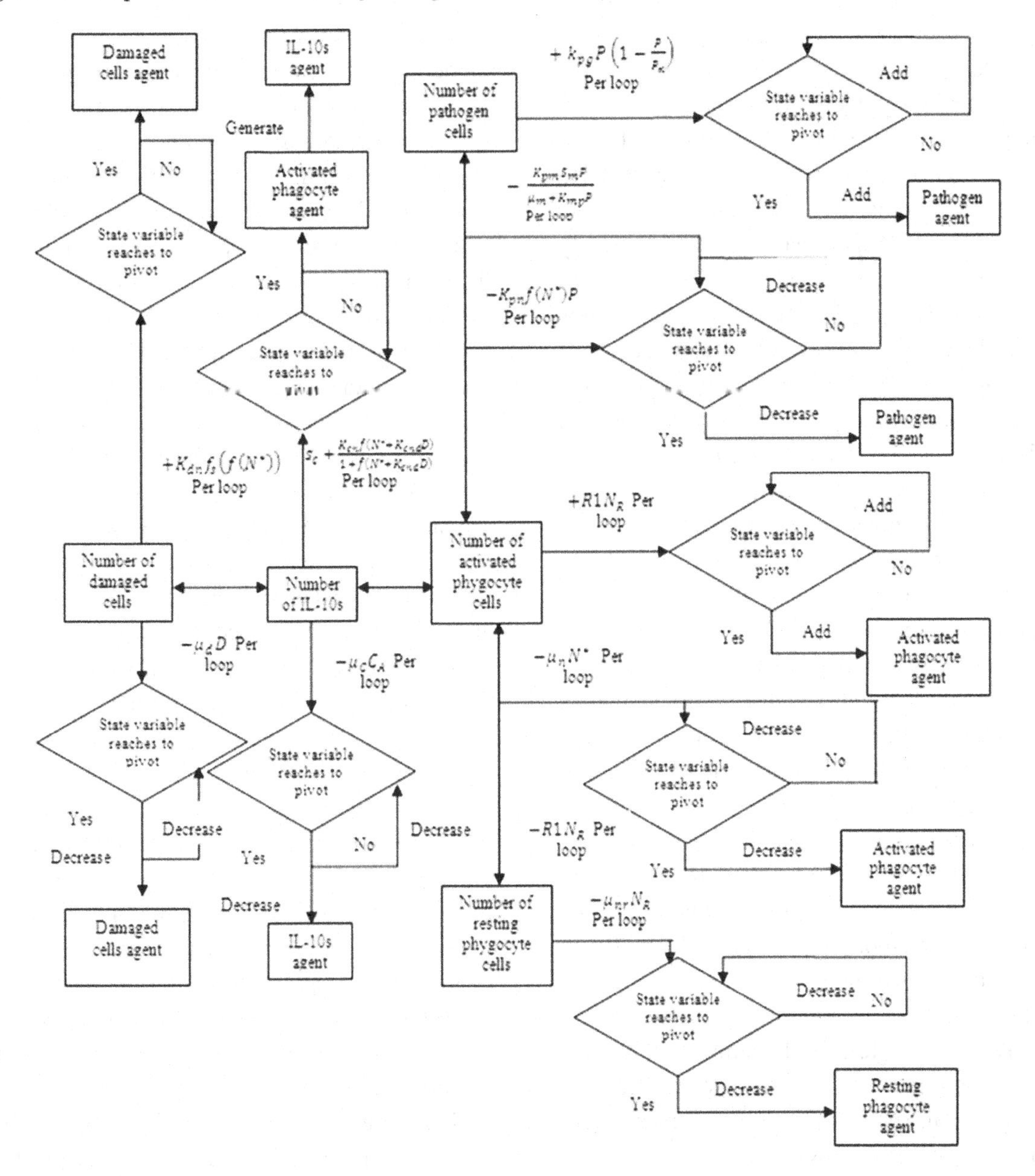

In addition, the stochastic behaviors of the agents are included in the ABM to enhance the probabilistic nature of agents' creations, deaths and transformations. Once the state of the agent reaches a point of change (defined as the boundary conditions of the system), the agent will choose one of four possibilities: keep the original agents population, increase its population, be eliminated, or transform into another type of agent. For example, *pathogen* agent will be eliminated when the pathogen dosages in the agent turns to 0. The *resting phagocytes* will transform into *activated phagocytes* during the chemotaxis process when pathogens exist in the surrounding environment. The states of one type of agents can be changed by other agents either of the same type or of different types, allowing interactions between various types of agents. Moreover, the possibility of those creations, deaths and transformation is easier to define as the probability distribution in the ABM. For instance, the growth rate of pathogens could be assigned as a normal distribution with mean equal to 0.1 and deviation equal to 0.01. Thus, the proposed autonomous multiagent-based model effectively describes the processes of acute inflammatory response, quantitatively defining the relationships among the various indicators (e.g. pathogen, resting phagocyte cells, activated phagocyte cells, damaged tissue, pro-inflammatory mediators and anti-inflammatory mediators) and capturing the complex and stochastic interactions among the pathological or physiological indicators.

Unlike the mathematical model, the ABM computer simulation allows the implementers and users to simulate and observe the interactions among different agents, thus it is more intuitive and flexible than the traditional mathematical models (this will be further demonstrated in the next section). However, the ABM approaches generally require large amount of computing resources for a precise simulation of complex behaviors. The use of large agent populations results in more accurate system behaviors but could require large amount of computer memory, and a detailed simulation may be time consuming.

The novel ABM approach presented above permits us to implement the simulation models on microcomputers platforms with moderate configuration. Such simulation experiments were completed within few minutes.

4. IMPLEMENTATION OF THE AGENT-BASED MODEL

The ABM presented in the previous section, as a methodology, is well-suited for modeling the complicated relationships and behaviors in the progression of acute inflammatory response. However, the implementation needs a certain type of computer simulation tool. In this research, we used Netlogo 4.0.4 (Netlogo, 2005), a Java based modeling platform for implementing the proposed agent-based model. This tool allows modelers to specify the behaviors of hundreds or thousands of "agents", which makes it possible to explore the connections between the microbial agent behaviors and the macro-level patterns that emerge from the interactions of multiple autonomous agents.

The main user interface of Netlogo is made up of two-dimensional grids. The agents can be divided into two categories: "patches" and "turtles". The "patches" are fixed agents placed on the background grids in the model workspace. The "turtles" are mobile agents who could occupy a position or move freely on the surface of patches and execute certain functions or actions. In the AIR agent-based model, the *damaged cell* agents are defined as patches because they simulate the tissue or artery cells, which are not movable. All the other eight agent types are defined as "turtle". Moreover, Netlogo offers a way to define agent set as "breed" which means agent types whose behaviors are similar or controlled by the same mechanisms. This allows the modeler to define a class of agents with a set of common state variables and establish various functions or actions (autonomous behaviors) for agent types. Also, the modeler can generate the output of a simulation and set parameters in a special area of the Netlogo interface. This interface as implemented in our agent-based model is shown in Figure 4.

In Figure 4, the upper left side of the interface window consists of two-dimensional grids. It allows the modeler to observe several different ways of interactions between agents: one agent could turn into another agent or release another type of agent under certain predefined conditions, and the population of a breed of agents could increase or decrease as well. The main user interface grid shows the process of activated phagocytes responding with the pathogens, releasing anti-inflammatory cytokines. The right hand side of the user interface window shows the user-defined

Figure 4. The whole user interface of the proposed agent based model

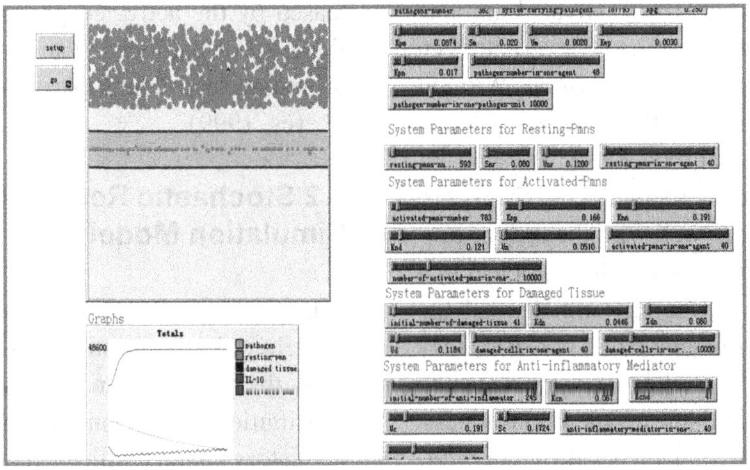

system parameters and initial populations of the agents. In this model all the system parameters correspond to the ones defined by equations (1)-(5). The picture on upper-left corner of Figure 4 animates the progression of an AIR episode, including initial pathogen intrusion, phagocyte activation and migration, pathogen elimination, and tissue damage/healing, which provides a real-time visualization for the modeler. Furthermore, the graph on the lower-left allows the modeler to easily track the real-time results of the simulation. Thus, once the simulation starts the population changes of different agents over the course of the simulation can be easily tracked and visualized.

5. PREDICTIVE RESULTS OF THE SIMULATION MODEL

5.1 Deterministic Results of the Simulation Model

In this section, the simulation tool developed is used as a predictive model for the prognosis of AIR. The predictive model is a useful tool to assess patients with different initial pathogens load levels or physiological conditions. To predict the pathogenesis of acute inflammatory response, the simulation model is run with the corresponding initial profiles of the patients of interest and a combination of adjustable system parameters, resulting is the behavior shown, for example, in Figure 5.

In Figure 5, the trajectories of all the indicators show that a patient with a low-level initial pathogen load is most likely to recover from an acute inflammatory response episode. Here all the indicators (pathogens, resting phagocyte cells, activated phagocyte cells, damaged tissue and anti-inflammatory mediators) return to a relatively low level and stay in stable state after a moderately long period of simulation. Furthermore, the mechanism of acute inflammatory response could

Figure 5. Healthy response with low pathogen load

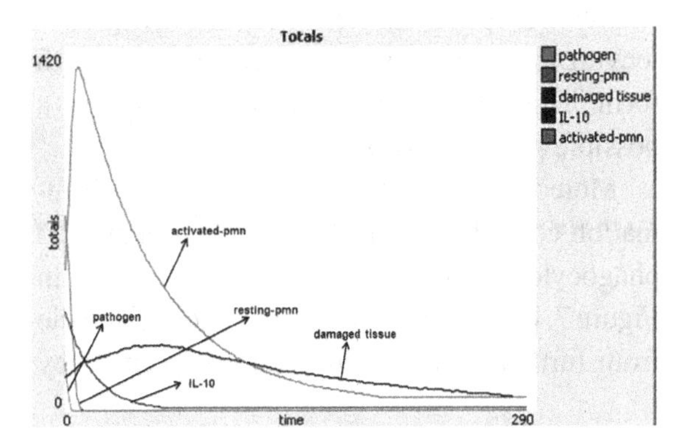

be explained by the predicted outcomes from the ABM simulation. That is, the activated immune cells could quickly eliminate a low-level pathogen load at the early stage of the episode. Without more pathogens recruited, the activated immune cells will then gradually decrease. The anti-inflammatory mediators (depicted as the IL-10 level) will decrease as well with the lack of production of activated immune cells. It is clear that the damaged tissue will increase initially with the effect of activated immune cells; however, they smoothly recover from the damaged status to normal cells under the tissue regeneration process.

In contrast, Figure 6 depicts a different prognosis of AIR. The predictive ABM demonstrates that the pathogens will initially decrease rapidly and then sharply increase to saturation level if the initial load of pathogens is elevated. Meanwhile, the value of system parameter K_{pn} (Branwood, 1991) of the predictive ABM decreases from 0.01 to 0.005 responding to the relatively-high initial load of pathogens. The system parameter K_{pn} represents the efficiency of pathogens elimination by activated phagocyte cells ($K_{pn} = 0.01$ means 1 percent of pathogens will be consumed by 1 measure unit of activated phagocytes cells per hour). The change of its value shows that the effectiveness of the immune system of a hypothetical patient who had a high initial load of pathogens decreases with time. Thus, the pathogens could not be entirely eliminated at the acute initial stage of the episode. After the number of activated immune cells decrease to a relatively low level (depicted as the activated-PMN granulocyte), the pathogens elevate in a logistic growth form. This situation is frequently referred as a possible prognosis of a septic shock.

Moreover, the persistent non-infectious inflammation could happen if the number of activated phagocyte goes to saturation, which is shown in Figure 7. Under this condition, a patient could die from further inflammation reactions caused by the persistent pro-inflammatory cytokines released by the activated phagocyte even though the level of pathogens vanishes. Thus, it would eventually lead to multiple organ failure and death (Reyes, 1999).

5.2 Stochastic Results of the Simulation Model

Our agent-based model has the special ability of modeling the stochastic process of AIR. In this case, the change of indicators follows a certain distribution since randomization exists when indicators interact with each other. For example, the growth rate of pathogen is not deterministic but having a normal distribution with mean and

Figure 6. Severe sepsis

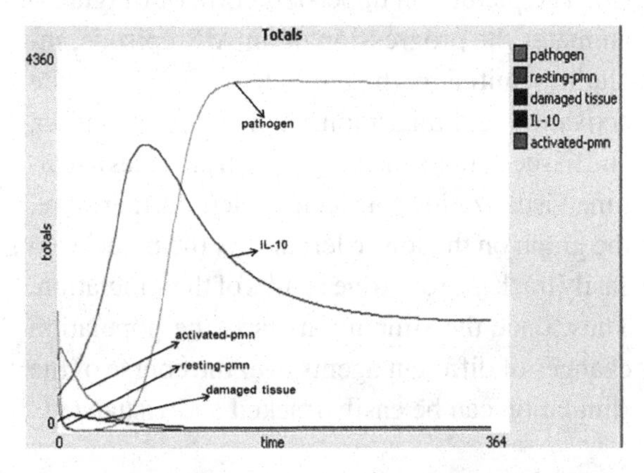

Figure 7. Persistent non-infectious inflammations

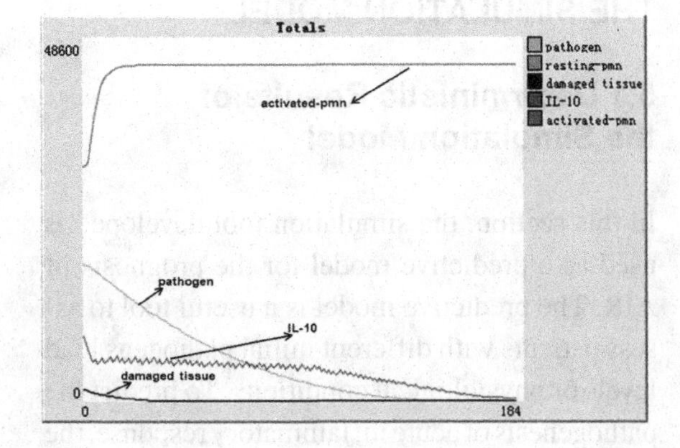

variance. This situation is common especially when the inner environment of cells varies corresponding to different individual patients.

When the pathogen growth has a normal distribution of (0.25, 0.04) for mean and variance, the progression of the episode can be seen in Figure 8. After running the simulation 20 times, it is found that the curve of those indicators such as activated-pmns, IL-10s have a narrower gap between maximum number and minimum number (refer to part (b) and part (c) of Figure 8). While, the pathogen with probabilistic growth rate has the bigger deviation over multiple replications (refer to part (a) of Figure 8).

6. SENSITIVITY ANALYSIS OF THE AGENT-BASED MODEL

Predictive results show that the pathogenesis of AIR did have a strong relationship with the threshold of pathogen level. However, the range of initial pathogen loads leading to a healthy response or towards an undesired acute inflammatory response is not clear. Thus, this section explores the sensitivity of the AIR episode to different initial loads of pathogen dosage. Figure 9 illustrates the impacts of the initial pathogen loads and shows that the model is quite sensitive to the pathogenesis of AIR while the influence of anti-inflammatory mediators is kept at a relatively low level.

In Figure 9, the initial load of pathogens (the X-axis) starts from 0 and is increased by 120 for each simulation run. The Y-axis ranges from 0 to 1 representing the state of AIR, with a value of 0 presenting a healthy acute inflammatory response and 1 a possible severe sepsis outcome. It is clearly seen that under an initial pathogen load below 480 the progression of AIR leads to a healthy response. Statistically, the initial value of pathogens which cause the different prognosis of AIR is 480 $\pm$ 2.56 with 95% confidence interval after multiple simulation experiments.

This result is achieved under a low anti-inflammatory mediator environment. Our simulated prognosis experiments suggest that the initial anti-inflammatory mediator or the system parameters controlling the change of anti-inflammatory mediator are sensitive to the AIR pathogenesis. Thus, next we tested the influences of anti-inflammatory mediator on the progression of AIR. A set of simulation runs using different combinations of initial loads of anti-inflammatory mediator (IL-10s), and initial loads of activated phagocytes, under the condition of low initial load of pathogen (range from 0 to 480), were executed. The results are shown in Figure 10.

This experiment shows that there are two essential pivot points for a change of the AIR prognosis. One relates to the initial number of anti-inflammatory mediator, and the other to the initial level of activated phagocytes. When the initial load of anti-inflammatory mediator ranges from 0 to 600 the AIR episode is stable and healthy regardless the initial value of activated phagocytes. An initial load of activated phagocyte higher than 600 cells will expose patients to an elevated risk for severe sepsis when initial load of anti-inflammatory mediator is lower than 200. This means that an uncontrolled AIR episode in which overly activated immune cells lacking proper self-generated or medicated anti-inflammatory controls could develop, even with a low initial pathogen load.

Based on the analysis, one can realize that patients still experience different AIR outcomes due to the variation in their immune systems, even though they have the same threshold of pathogen induction. Since the initial value of system parameters in ABM are associated with the initial conditions of patients, it is necessary to consider the relationships between the initial values of physiological indicators and the system parameters as defined in the system dynamic mathematical model.

Figure 8. AIR progression under multiple replications with pathogen growth rate having a normal distribution (0.25, 0.04)

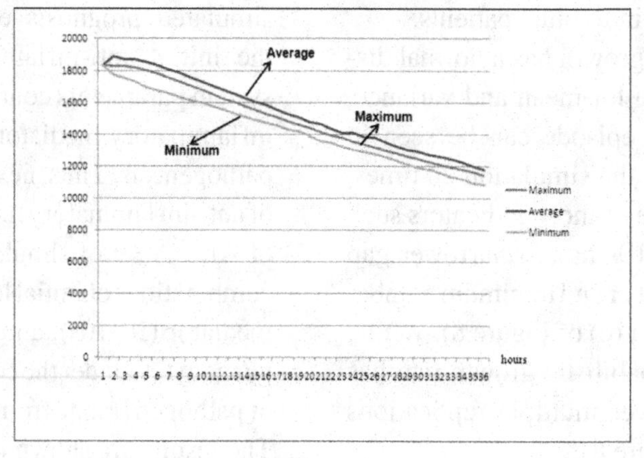

a

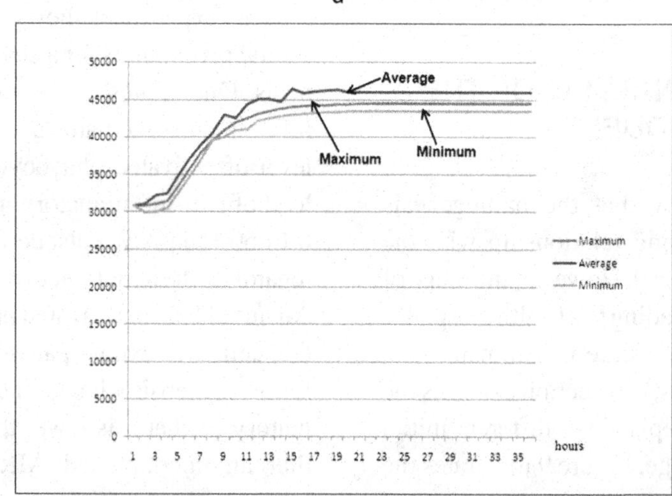

b

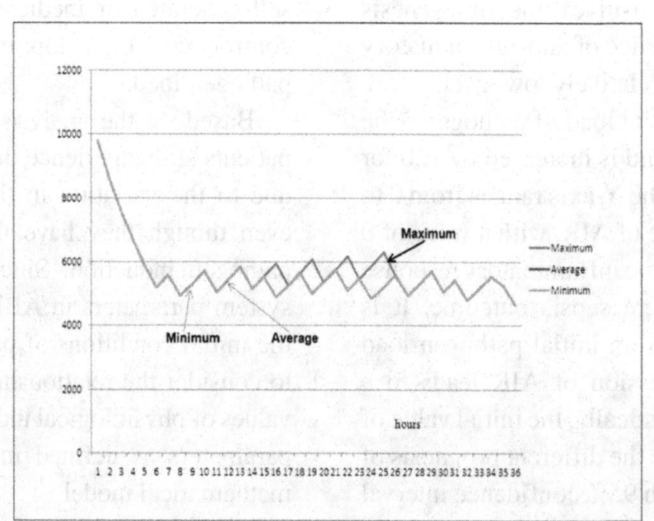

c

Figure 9. AIR response to the range of the initial pathogen load

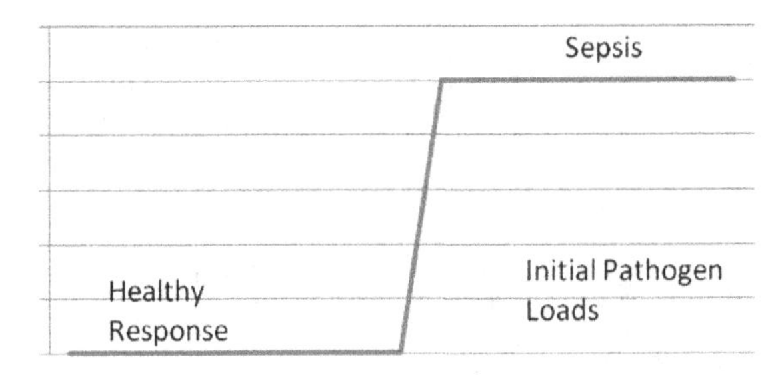

Figure 10. The influence of initial load of anti-inflammatory mediator and activated phagocytes

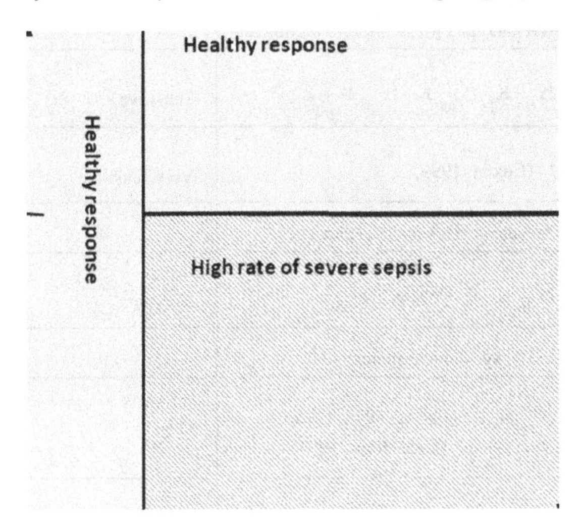

First we consider the influence of system parameters K_{nn} and K_{np} on the AIR in equation (1). The reduction of the pathogens is assumed to follow the standard competitive enzyme inhibition effects among the three agents namely, the *activated phagocytes*, *current pathogen load*, and *damaged tissue* (Branger, 2002). The K_{nn} represents the ability of activated phagocyte cells to recruit more resting phagocytes to the infectious location, and the K_{np} calibrates the number of resting phagocyte cells to be activated in order to eliminate the existing pathogens. Based on three different initial pathogen loads, our simulation results have shown that there is a certain relationship between these two system parameters contributing to a stable ultimate state of AIR. That is, the numbers of activated phagocytes needs to be balanced in order to achieve a healthy AIR response. Otherwise, the acute inflammatory response will turn out to be a persistent non infectious or septic shocks episode. This relationship is shown in Figure 11.

From the results shown in Figure 11, one can see that if K_{np} decreased and K_{nn} increased (the number of activated phagocytes induced by pathogens could be supplemented by previously activated phagocytes), the patient can still be on track for a healthy response. However, since the boundary of activation rate of resting phagocytes by previously activated phagocytes is much lower than activation rate of resting phagocytes by pathogens, the patients with a relatively low activation rate of resting phagocytes by pathogens under the condition of previously suffering infection or trauma still could not recovery from sepsis by increasing the value of K_{nn}. This explains why patient populations such as HIV infected patients or those with organ transplants are more likely to suffer immune-deficiency after AIR (Kumar et al., 2004).

To conclude, we summarize the sensitivity of the other system parameters in our ABM simulation. The test of sensitivity of our system parameters is based on the observation of curve-changing of indicators. It is believed that one parameter is sensitive to the AIR progression if the curve of indicators goes to different directions when this specific parameter interested changes and others fix. The results are shown in Tables 2 and 3.

Figure 11. Activated phagocyte affect on the evolution of AIR

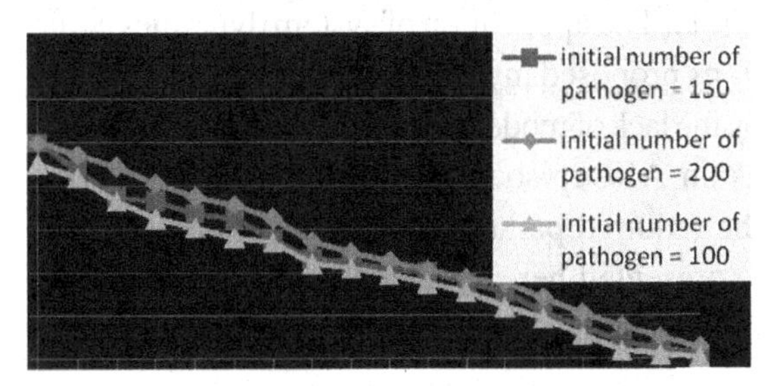

Table 2. The sensitivity of AIR progression to the initial load of indicators

Initial value of indicators in the AIR progression	If it is sensitive to the AIR progression
Pathogen	Very sensitive
Activated Phagocytes	Sensitive under certain condition
Damaged cells	Not quite sensitive
Anti inflammatory mediator	Very sensitive specially when the initial load of pathogen is low

These tables provide insights for various indicators and system parameters which are sensitive to the AIR progression and are essential for a better understanding of the pathogenesis of the AIR. The potential benefits of this research include helping physicians plan proper medication interventions for patients who develop AIR. Moreover, a greater benefit to clinicians will be provided if some of the essential system parameters could be converted or translated into the patient's physiological indicators, which will create an individualized prediction of the AIR progression.

7. DISCUSSION AND CONCLUSION

Systemic inflammation and multiple organ dysfunctions are two of the major causes of mortality today (Wang, 2001; Eiseman, 1977). With the advent and improvement of antibiotics and organ support therapy, these conditions have become increasingly relevant (Bone, 1992). The incidences of systemic inflammation are also expected to increase with further advancement of medical technology and the aging of our population (Angus, 2001). The application of system dynamics equations proposed in the existing literature is limited by the lack of modeling stochastic phenomena such as an AIR episode and difficulty in measuring the required parameters. An agent-based model is presented herein, capable of synthesizing the information acquired from the biological process

Table 3. The sensitivity of the AIR progression to various system parameters

Pathogen system parameters	If it is sensitive to the AIR progression
K_{pg} (Spector, 1956)	Very sensitive
K_{pm} K, S_m S, $\grave{\imath}_m$ (Janeway, 2001; Zouali, 2001), K_{mp} K	Not quite
K_{pn} *K*	Sensitive
Activated phagocyte system parameters	
K_{nn} K, K_{np} K, K_{nd} K	Sensitive
$\grave{\imath}_n$ (Coxon, 1999)	Very sensitive
Damaged cells system parameters	
K_{dn} K, $\grave{\imath}_d$ (Wang, 1999)	Not quite
IL-10s system parameters	
K_{cn} *K*, $\grave{\imath}_c$ (Bacon, 1973; Bocci, 1991; Fuchs, 1996; Huhn, 1997)	Sensitive
K_{cnd} *K*, S_c (Tsukaguchi et al., 1999)	Not quite

(interaction between indicators) into a modeling/calibrating process while preserving the complexity of the acute inflammatory response process.

Thus, the proposed agent-based model combined the strength of the system dynamics models with that of the simpler agent-based model, decomposing the dynamics model into multiple autonomous agents and capturing the stochastic nature of a biological system.

In order to validate the ABM simulation we compared results from the DNA-Neumococo Study Group (2009) with our simulation. This group sampled 353 patients with community-acquired pneumonia and found that bacterial load is highly correlated with the outcomes in patients with pneumonia. The bacterial load of $\geq 10^3$ copies per milliliter occurred in 29.0% of patients (27 of 93 patients; 95% CI, 20.8 to 38.9%) being associated with a statistically significant higher risk for septic shocks, the need for mechanical ventilation, and hospital mortality (Rello, 2009). Our simulated prognosis experiments also have shown that the initial pathogen load is highly associated with prognosis of AIR pathogenesis.

In addition, maintenance of elevated levels of the anti-inflammatory mediators or the transient administration of the anti inflammatory mediator in patients who would otherwise survive or evolve to septic death is needed (Reynolds, 2006). This finding corresponds to the simulated results from our model that either AIR turns to healthy response or high rate of severe sepsis when the initial load of anti-inflammatory mediator is elevated. Thus, the proper management of anti-inflammatory mediators plays an important role in the acute phase of infection. The fact that a significant body of recent clinical evidence suggests that low-dose immune-suppression with an anti-inflammatory mediator may in fact improve outcomes in patients with severe infection, particularly in patients with an insufficient anti-inflammatory response (Annane, 2002) is also illustrated by the outcome of our model.

The major strength of the proposed agent-based modeling and simulation approach is that it can help to predict the possible pathogenesis of acute inflammatory response based on the patients' initial physiological conditions. Furthermore, the new approach it is more flexible, visible and more accurate than existing of mathematical models. Even with the complex non-deterministic system presented herein, the proposed ABM uses only few types of agents in the simulation, and it is modular, more flexible and applicable for the development of more complex and accurate models for simulating disease progression.

For further research, we expect to use real clinical data such as the measurement of pathogen load, activated immune cells, resting immune cells as well as the measurement representing damaged tissue as input to this model. This will allow applying this predictive agent-based model towards real clinical environments after a proper calibration. Furthermore, even though the system's parameters could be predicted by the model, these parameters are difficult to measure clinically in practice. Thus our next step is to convert the current system parameters into physiological indicators such as temperature, blood pressure, CD14 markers, etc. that are easier to measure. These parameters will be used then to model the progression of a sepsis pathogenesis for different patients using patient specific physiological markers.

REFERENCES

An, G. (2000). Agent-based computer simulation and SIRS: Building a bridge between basic science and clinical trials. *Shock (Augusta, Ga.)*, *16*(4), 266–273. doi:10.1097/00024382-200116040-00006

An, G. (2004). In silico experiments of existing and hypothetical cytokine-directed clinical trials using agent-based modeling. *Critical Care Medicine*, *32*(10), 2050–2060. doi:10.1097/01.CCM.0000139707.13729.7D

Angus, D. C., Linde-Zwirble, W. T., Lidicker, J., Clermont, G., Carcillo, J., & Pinsky, M. R. (2001). Epidemiology of severe sepsis in the United States: Analysis of incidence, outcome, and associated costs of care. *Critical Care Medicine*, *29*(7), 1303–1310. doi:10.1097/00003246-200107000-00002

Annane, D., Sebille, V., Charpentier, C., Bollaert, P. E., Francois, B., & Korach, J. M. (2002). Effect of treatment with low doses of hydrocortisone and fludrocortisones on mortality in patients with septic shock. *Journal of the American Medical Association*, *288*(7), 862–871. doi:10.1001/jama.288.7.862

Bacon, G. E., Kenny, F. M., Murdaugh, H. V., & Richards, C. (1973). Prolonged serum half-life of cortisol in renal failure. *The Johns Hopkins Medical Journal*, *132*(2), 127–131.

Bocci, V. (1991). Interleukins: Clinical pharmacokinetics and practical implications. *Clinical Pharmacokinetics*, *21*(4), 274–284. doi:10.2165/00003088-199121040-00004

Bone, R., Balk, R., Cerra, F., Dellinger, R., Fein, A., & Knaus, W. (1992). Definitions for sepsis and organ failure and guidelines for the use of innovative therapies in sepsis. *Chest*, *101*(6), 1644–1655. doi:10.1378/chest.101.6.1644

Branger, J., van den Blink, B., Weijer, S., Madwed, J., Bos, C. L., & Gupta, A. (2002). Anti-inflammatory effects of a p38 mitogen-activated protein kinase inhibitor during human endotoxemia. *Journal of Immunology (Baltimore, MD.: 1950)*, *168*(8), 4070–4077.

Branwood, A. (1991). Interleukins: Clinical pharmacokinetics and practical implications. *Clinical Pharmacokinetics*, *21*(4), 274–284.

Coxon, A., Tang, T., & Mayadas, T. N. (1999). Cytokine-activated endothelial cells delay neutrophil apoptosis in vitro and in vivo: A role for granulocyte/macrophage colony-stimulating factor. *The Journal of Experimental Medicine*, *190*(7), 923–934. doi:10.1084/jem.190.7.923

Eiseman, B., Beart, R., & Norton, L. (1977). Multiple organ failure. *Surgery, Gynecology & Obstetrics*, *144*(3), 323–326.

Fuchs, A. C., Granowitz, E. V., Shapiro, L., Vannier, E., Lonnemann, G., & Angel, J. B. (1996). Clinical, hematologic, and immunologic effects of interleukin-10 in humans. *Journal of Clinical Immunology*, *16*(5), 291–303. doi:10.1007/BF01541395

Gogos, C. A., Drosou, E., Bassaris, H. P., & Skoutelis, A. (2000). Pro- versus anti-inflammatory cytokine in patients with severe sepsis: A marker for prognosis and future therapeutic options. *The Journal of Infectious Diseases*, *181*(1), 176–180. doi:10.1086/315214

Huhn, R. D., Radwanski, E., Gallo, J., Affrime, M. B., Sabo, R., & Gonyo, G. (1997). Pharmacodynamics of subcutaneous recombinant human interleukin-10 in healthy volunteers. *Clinical Pharmacology and Therapeutics*, *62*(2), 171–180. doi:10.1016/S0009-9236(97)90065-5

Janeway, C., Travers, P., Walport, M., & Shlomchik, M. (2001). *Immunobiology: The immune system in health and disease*. New York, NY: Garland Publishing.

Kumar, R., Clermont, G., Vodovotz, Y., & Chow, C. C. (2004). The dynamics of acute inflammation. *Journal of Theoretical Biology*, *230*(1), 145–155. doi:10.1016/j.jtbi.2004.04.044

Netlogo. (2005). *Software version 4.0.4*. Retrieved from http://ccl.northwestern.edu/netlogo/

Rello, J., Lisboa, T., Lujan, M., Gallego, M., Kee, C., & Kay, I. (2009). Severity of pneumococcal pneumonia associated with genomic bacterial load. *Chest*, *136*(3), 832–840. doi:10.1378/chest.09-0258

Reyes, W., Brimioulle, S., & Vincent, J. (1999). Septic shock without documented infection: an uncommon entity with a high mortality. *Intensive Care Medicine*, *25*(11), 1267–1270. doi:10.1007/s001340051055

Reynolds, A., Rubin, J., Clermont, G., Day, J., Vodovotz, Y., & Ermentrout, G. B. (2006). A reduced mathematical model of the acute inflammatory response: I. derivation of model and analysis of anti-inflammation. *Journal of Theoretical Biology*, *242*(1), 220–236. doi:10.1016/j.jtbi.2006.02.016

Spector, W. S. (Ed.). (1956). *Handbook of biological data*. London, UK: W. B. Saunders Company.

Tsukaguchi, K., de, Lange B., & Boom, W. H. (1999). Differential regulation of IFN-gamma, TNF-alpha, and IL-10 production by CD4 (+) alphabeta TCR + T cells and vdelta2 (+) gammadelta T cells in response to monocytes infected with Mycobacterium tuberculosis-H37Ra. *Cellular Immunology*, *194*(1), 12–20. doi:10.1006/cimm.1999.1497

Wakeland, W., Macovsky, L., & An, G. (2000). A hybrid simulation model for studying acute inflammatory response. *Toxicological Sciences*, *57*(2), 312–325.

Wang, H., Bloom, O., Zhang, M., Vishnubhakat, J. M., Ombrellino, M., & Che, J. (1999). HMG-1 as a late mediator of endotoxin lethality in mice. *Science*, *285*(5425), 248–251. doi:10.1126/science.285.5425.248

Wang, H., Yang, H., Czura, C. J., Sama, A. E., & Tracey, K. J. (2001). HMGB1 as a late mediator of lethal systemic inflammation. *American Journal of Respiratory and Critical Care Medicine*, *164*(10), 1768–1773.

Zeni, F., Freeman, B., & Natanson, C. (1997). Anti-inflammatory therapies to treat sepsis and septic shock: a reassessment. *Critical Care Medicine*, *25*(7), 1095–1100. doi:10.1097/00003246-199707000-00001

Zouali, M. (2001). Antibodies. In Clarke, A., Ruse, M., Agro, A. F., Bernardini, S., Dotsch, V., & Maccarrone, M., (Eds.), *Encyclopedia of life science*. London, UK: Nature Publishing Group. doi:10.1038/npg.els.0000906

This work was previously published in the International Journal of Artificial Life Research, Volume 2, Issue 2, edited by E. Stanley Lee and Ping-Teng Chang, pp. 105-121, copyright 2011 by IGI Publishing (an imprint of IGI Global).

Chapter 19
Rough Set Based Clustering Using Active Learning Approach

Rekha Kandwal
Ministry of Earth Sciences and Science and Technology, India

Prerna Mahajan
Guru Gobind Singh Indraprastha University, India

Ritu Vijay
Bansthali University, India

ABSTRACT

This paper revisits the problem of active learning and decision making when the cost of labeling incurs cost and unlabeled data is available in abundance. In many real world applications large amounts of data are available but the cost of correctly labeling it prohibits its use. In such cases, active learning can be employed. In this paper the authors propose rough set based clustering using active learning approach. The authors extend the basic notion of Hamming distance to propose a dissimilarity measure which helps in finding the approximations of clusters in the given data set. The underlying theoretical background for this decision is rough set theory. The authors have investigated our algorithm on the benchmark data sets from UCI machine learning repository which have shown promising results.

1. INTRODUCTION

Clustering is one of the major techniques in the data mining as well as an active research topic. It involves identification of clusters (groups) in the data that are similar to each other. The cluster usually corresponds to points that are more similar to each other than they are to points from another cluster (Bagirov, Rubinov, Soukhoroukova, & Yearwood, 2003). Clustering is an unsupervised technique which naturally exploits the subset that has points with greater similarity and it doesn't have a priori information about the structure of cluster. It is basically data driven and it leads to the discovery of the knowledge structures (clusters) which provides information about the data in the cluster. The clustering technique has been extensively studied in many fields such as pattern

DOI: 10.4018/978-1-4666-3890-7.ch019

recognition, image segmentation, data visualization and similarity search, signal processing and trend analysis (Jain, Murthy, & Flynn, 1999; Dy & Brodley, 2004). Many algorithms on clustering exist in literature (Gibson, Kleinberg, & Raghavan, 1998; Zhang, Fu, Cai, & Heng, 2000; Huang, 1997; Wang, Xu, & Liu, 1999; Huang, 1999).

Rough sets have been introduced as a tool to deal with inexact, imprecise or vague knowledge in the real world data (Gibson, Kleinberg, & Raghavan, 1998). It uses the concept of lower and upper approximation of rough sets to handle the uncertainty in information. It has been applied in many fields such as machine learning, document classification and so on (Xu, Xuedong, Sen, & Bin, 2006; Widzlzak & Revett, 2004; Chengdong, Mengxin, Zhonghua, Zhang, & Yong, 2004). Chen et al. have applied rough set theory to clustering analysis and introduced the decision attribute to better define attribute membership matrix. They have used similarity measure based on Euclidean distance (Chen, Cui, Wang, & Wang, 2006). BWang has proposed a clustering technique based on Olary code to transform nominal attributes into integer. It provides a useful way to estimate the number of underlying clusters (Wang, 2010). Chengdong. Wu et al. have given a hierarchical clustering method for attribute discretization. It gives the significant clusters by determining the best classes for discretization picked from scatter plots of several statistics (Chengdong, Mengxin, Zhonghua, Zhang, & Yong, 2004).

Mitra et al have introduced a novel clustering architecture which integrates advantages of both fuzzy sets and rough sets. The algorithm aims to find a common structure revealed at global level which is determined by exchanging prototypes of the subsets of data & by moving prototypes of the corresponding clusters toward each other (Mitra, Banka, & Pedrycz, 2006). Hirano and Tsumoto have proposed a method which secluded the influence of initial equivalence relations by introducing perfect initial equivalence relations defined by the original class of each object and their randomly mutated variants (Hirano & Tsumoto, 2006).

2. BACKGROUND

2.1. Rough Set Theory

Rough set theory proposed by Pawlak (1982), is an extended approach to the classical set theory to represent vagueness or uncertainty. It extends the crisp notion of set by using the concept of approximation. One of the main objectives of Rough set theory is to reduce data size which makes it successful in broad domains such as machine learning, knowledge discovery in databases, pattern recognition. It introduces the notions of indiscernibility, rough set lower and upper approximations and reducts used to approximate inconsistent information and to exclude redundant data (Magnani, 2003).

For a given set of objects U called universe and an indiscernibility relation $RCU \times U$. Let X be a subset of U characterized with respect to R, as shown in Figure 1 (Pawlak, 1982).

The lower approximation of a set X with respect to R is the set of all objects, which can be for certain classified as X with respect to R (certainly)

$$\underline{R}(X) = \cup_{x} \in_{U} \{ R(x): \underline{R}(X) \subseteq X \}$$

The upper approximation of a set X with respect to R is the set of all objects which can be possibly classified as X with respect to R (are possibly X in view of R)

$$\overline{R}(X) = \overline{\cup}_{x} \in_{U} \{ R(x): \overline{R}(X) \cap X \neq \phi \}$$

Figure 1. Rough set model (Pawlak, 1982)

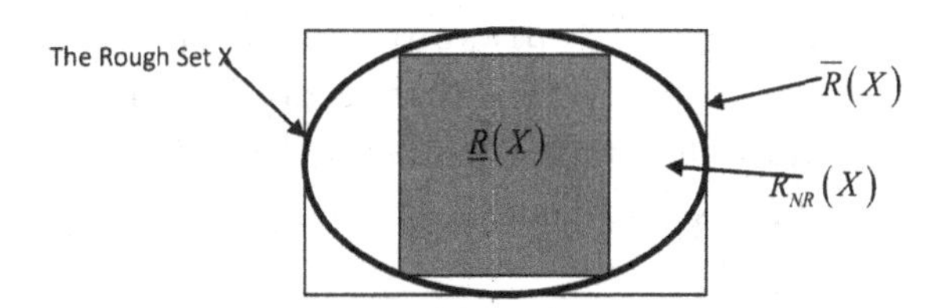

The boundary region of a set X with respect to R is the set of all objects, which can be classified whether as X nor as not-X with respect to R.

$$R_{NR}(X) = \underline{R}\left(X\right) - \overline{R}\left(X\right)$$

2.2. Rough Set Cluster

Rough clusters extends the crisp(precise) notion of cluster, that is in rough clusters some objects are located at the lower approximation of a cluster that is objects that only belong to that cluster implying full membership to it, while others are laid at its upper approximation that is objects which are also members of other clusters. In this way rough clusters manage uncertainty about membership of objects to clusters.

For rough clustering an appropriate distance measure should be used such that the strict requirement of indiscernibility relation used in normal clustering is relaxed (Emilyn & Ramar, 2010).

Some generalizations of rough sets relax the assumption of underlying equivalence relation (Magnani, 2003;Yang & Yang, 2006) and allows for grouping of objects based on a similarity relation rather than based on equivalence relation (Emilyn & Ramar, 2010) .The so-built rough set model is guaranteed to satisfy the following properties.

1. A member of any lower approximation (LC) is also a member of the corresponding upper approximation (UC).
2. A object must belong to at most one lower approximation.
3. If an object is not a member of any lower approximation, then it must belong to two or more upper approximation

Figure 2 represents a Universe $U=\{x_1,x_2,x_3,x_4,x_5\}$, Let R be the equivalence relation which clusters the U and hence $U/R=\{C_1,C_2\}$. As we can see in Figure 2,

$LC_1=\{ x_1\}$, $UC_1=\{ x_1,x_2,x_3\}$,

$LC_2=\{ x_4\}$, $UC_2=\{ x_3,x_4,x_5\}$

And we can easily see from the Figure 2 that the U satisfy all the three properties

$x_1 \overline{\in} LC_1 \, C \, UC_{1,}$ $\{ x_2, x_3, x_5\} \in UCI$ & UC2 both

2.3. Active Learning

The main aim of any cluster based active approach would be to generate or sample the unlabeled instances in such a way that they self- organize into small groups with minimal overlapping. Mahajan et al. proposed an approach which combines the benefits of two approaches namely clustering and best feature selections to generate an active learner (Mahajan, Kandwal, & Vijay, 2011). Cluster approach is also more suitable in real world applications that need automated classification approaches which reduces effort for human annotation in more critical analysis (like fraud detection, disaster management etc.) like the most informative cluster can be picked up for further analysis. The basic idea is to selectively choose most informative features, which are most uncertain and likely candidates to provide more information. The data points will be assigned to clusters based on predefined similarity measure. The active approach works iteratively where in each round the learner actively selects a batch of unlabeled samples for training to improve the internal model, i.e., cluster adjustment as quickly

Figure 2. Rough cluster model

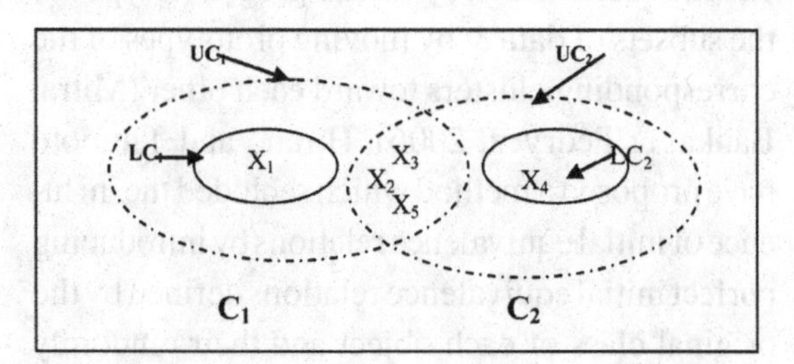

as possible. To incorporate active learning, the learning algorithm will pick up an unlabeled instance, Based on the previous learning, it will compare the feature vector for new instance with feature vectors for clusters centers, if it fit any we will label it, otherwise we will see how much distance exist and label it with the clusters where the similarity is maximum (See Figure 3).

3. ACTIVE LEARNING USING ROUGH SET BASED CLUSTERING

Most of the Active learning methods are supervised, that is, the learning algorithm induces a model that accurately predicts a label for some new instance. However an Active learning algorithm is unsupervised if its task is to simply organize a large amount of unlabeled data in a meaningful way Most importantly the supervised learner try to map instances using a predefined structure, whereas unsupervised learners exploits the inherent structure in the data to explore meaningful patterns. To achieve this we are proposing a rough set based cluster active learning approach and to identify proper clusters we are aiming for the best defining features for sample data set. For cluster definition we have defined a similarity measure based on Hamming Distance as defined next.

3.1. Hamming Distance Metric

Let $A_1,\ldots,A_m$ be a set of categorical attributes with domains $D_1,\ldots,Dm$ respectively. Let the dataset $D=\{X_1,X_2,\ldots,X_n\}$ be a set of objects described by m categorical attributes $A_1,\ldots,A_m$

Let X,Y be two categorical objects with m categorical attributes. The dissimilarity measure between X and Y can be defined by the total Hamming distance of the corresponding attribute values of the two objects. The smaller the distance is, the more similar (closer) the two objects are, Formally

$$d(X,Y)=\sum_{i=1}^{m} w_{i*}d(x_i,y_i)$$

The Hamming distance $d(x, y)$ between two vectors x, y € D is the number of coefficients n which they differ, for example d(00111, 11001) = 4 and d(0122, 1220) = 3.

hd(X,Y)={ 0 if X=Y, 1 if X≠Y }

3.2. Hamming Distance Metric for Rough Clusters

A dissimilarity measure based on Hamming distance is proposed here. The distance between Cluster center and the data point is calculated.

Figure 3. Active learning approach

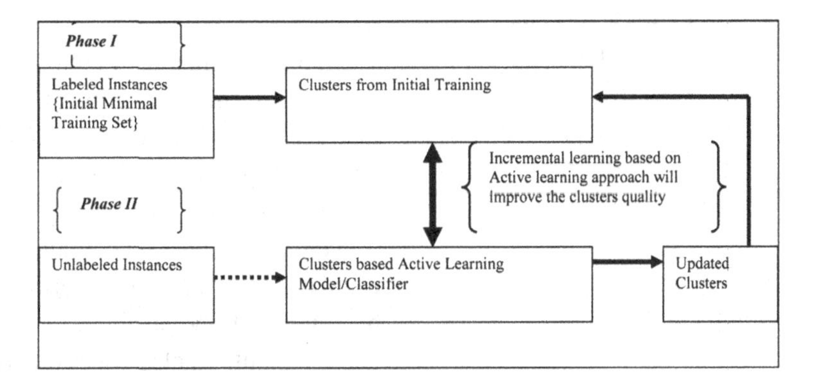

3.2.1.

If the distance is below the threshold (based on domain knowledge and data set) defined, it will certainly belong to that cluster and will belong to lower approximation of that rough cluster.

$hd(x_{cj},x_i) < \alpha\{ \alpha \text{ is the threshold}\} \text{ then } x_i \in x_{cj},$ (1)

3.2.2.

Whereas if the distance is greater than the threshold and is less than the minimum of distance between any two clusters It will be allocated to the upper approximation of that rough cluster. As given in Equation 1.

$\alpha < hd(x_{cj},x_i) \leq \min\{ hd(x_{cj},x_{ck})\} \text{then } x_i \in x_{cj},$ (2)

3.3. Proposed Active Learning Algorithm

Given a dataset of n instances and k be the number of clusters, our clustering algorithm partitions the dataset into k clusters($k \leq n$). The algorithm tries to maximizes the intracluster similarity and minimizes the intercluster similarity.

Algorithm 1

```
Input: The number of clusters, k, and
a dataset containing n objects.
        α is the threshold defined
based on the domain knowledge
Output: Set of Clusters C={c1,.....,ck}
Phase I:
Step1: Let  U be the Universe of set
of n objects located in space. Con-
sider
Let R be equivalence relation R on
U which clusters the U into set of
clusters    U/R={c1,.....,ck}
Step 2
        2a): Randomly choose a data
point.
        2b): Assign it as the cen-
ter of the first cluster.
        2c) Calculate its H
xcj,        // xi belong to lower ap-
proximation
             else if          α <
hd(xcj,xi)≤min{ hd(xcj,xck)}
                           then xi∈
xcj,         // xi belong to upper ap-
proximation
                              end if
        end for
Step 4  for j=1 to k
        4a) Compare (Ucj,Ucj+1)
         4b) If the number of data
points belonging to the upper approx-
imation of these clusters are iden-
tical or similar .
                  { Then merge(Ucj,Ucj+1)
                                  set
k=k-1
                  Go to step 2
                  }
          Else
             Go to Step 5.
        End for
Step 5 Stop
```

4. EXPERIMENTAL RESULTS

We ran experiments on data sets obtained from the UCI Machine Learning Repository. Balloon data set contains 15 instances and each instance has 4 categorical attributes. In this experiment we have assigned equal weight to all attributes.

Hayes Roth data set contains 35 instances and each instance has 5 categorical attributes. We have ignored one attribute i.e., name as it does not contribute to classification and is unique for every instance. Lenses data set contains 24 instances and has 4 categorical attributes. In these experiments we have assigned equal weight to all attributes. Our algorithm has shown promising results in forming the clusters. It has given accurate results and the upper approximation handles the overlapping or vagueness in cluster formation. It is able to satisfy all the three rules given in rough cluster model [21] as it is evident in the cluster model given. The approximations in rough set model given by the proposed algorithm is shown in Tables 1 through 3.

The Figures 4 through 6 shows the Rough set cluster model generated by our algorithm. We have also computed error in classification as given in eq-3 as shown in Figure 7.

Error Percentage=(Number of data points misclassified / total number of points belonging to that approximation). (3)

Table 1. Lenses dataset cluster model

	Distances wrt to clusters			
Data Points	**c1**(x3 center) HD(x1, c1)	**c2**(x13 center) HD(x1,c2)	**c3**(x20 cenetr) HD(x1,c3)	Clusters Allocation
x1	1	2	3	LC1
x2	1	4	3	LC1
x3	0	3	4	LC1
x4	3	2	1	LC3
x5	2	1	2	LC2
x6	2	3	2	UC1,UC3
x7	1	2	3	LC1
x8	4	2	2	UC2,UC3
x9	2	1	3	LC2
x10	2	3	3	LC1
x11	1	2	4	LC1
x12	3	1	1	LC3
x13	3	0	2	LC2
x14	3	2	2	UC2,UC3
x15	2	1	3	LC2
x16	3	3	1	LC3
x17	2	2	2	UC1,UC2,UC3
x18	2	4	2	UC1,UC3
x19	1	3	3	LC1
x20	4	2	0	LC3
x21	3	1	1	UC2,UC3
x22	3	3	1	LC3
x23	2	2	3	UC1,UC2
x24	2	3	2	UC1, UC3

Table 2. Balloon dataset cluster model

Data Points	Distances wrt to clusters		
	c1(x1 center) HD(x1, c1)	c2(x5 center) HD(x1,c2)	Clusters Allocation
x1	0	2	LC1
x2	1	1	UC1,UC2
x3	1	1	UC1,UC2
x4	2	0	LC2
x5	2	0	LC2
x6	1	3	LC1
x7	2	2	UC1,UC2
x8	2	2	UC1,UC2
x9	3	1	LC2
x10	3	1	LC2
x11	2	3	LC1
x12	3	2	LC2
x13	2	2	UC1,UC2
x14	3	1	LC2
x15	3	1	LC2

Table 3. Hayes Roth dataset cluster model

Data Points	Distances wrt to clusters			
	c1(x1 center) HD(x1, c1)	c2(x13 center) HD(x1,c2)	c3(x20 cenetr) HD(x1,c3)	Clusters Allocation
x1	0	4	4	LC1
x2	2	2	3	UC1, UC2
x3	3	3	4	UC1, UC2
x4	0	4	4	LC1
x5	2	2	3	UC1, UC2
x6	3	3	4	UC1, UC2
x7	0	4	4	LC1
x8	3	1	4	LC2
x9	2	2	4	UC1, UC2
x10	2	3	3	UC1, UC2,UC3
x11	3	2	4	UC1, UC2
x12	4	0	4	LC2
x13	4	1	3	LC2
x14	4	1	3	LC2
x15	2	3	4	UC1, UC2
x16	2	2	4	UC1, UC2

continued on following page

Table 3. Continued

	Distances wrt to clusters			
Data Points	**c1(x1 center) HD(x1, c1)**	**c2(x13 center) HD(x1,c2)**	**c3(x20 cenetr) HD(x1,c3)**	**Clusters Allocation**
x17	2	3	3	UC1, UC2,UC3
x18	2	1	4	LC2
x19	2	2	4	UC1, UC2
x20	4	1	4	LC2
x21	2	3	4	UC1, UC2
x22	2	2	4	UC1, UC2
x23	2	2	3	UC1, UC2
x24	4	2	3	UC2,UC3
X25	2	2	4	UC1, UC2
X26	3	3	3	UC1, UC2,UC3
X27	4	3	0	LC3
X28	4	3	2	LC3
X29	4	2	3	UC1, UC2,UC3
X30	3	3	2	LC3
X31	4	3	2	LC3
X32	3	4	2	LC3
X33	4	3	3	UC2,UC3
X34	3	3	1	LC3
X35	3	4	2	LC3

Figure 4. Lenses dataset rough cluster model

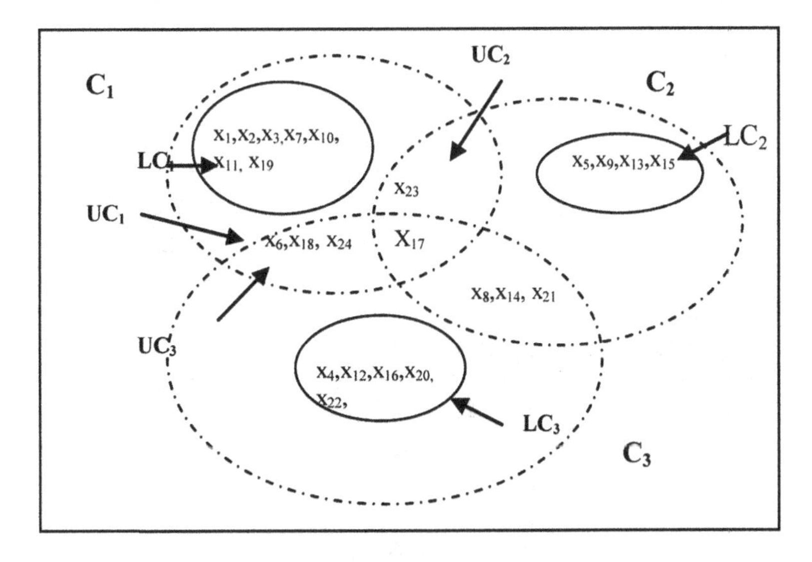

Figure 5. Hayes Roth dataset rough cluster model

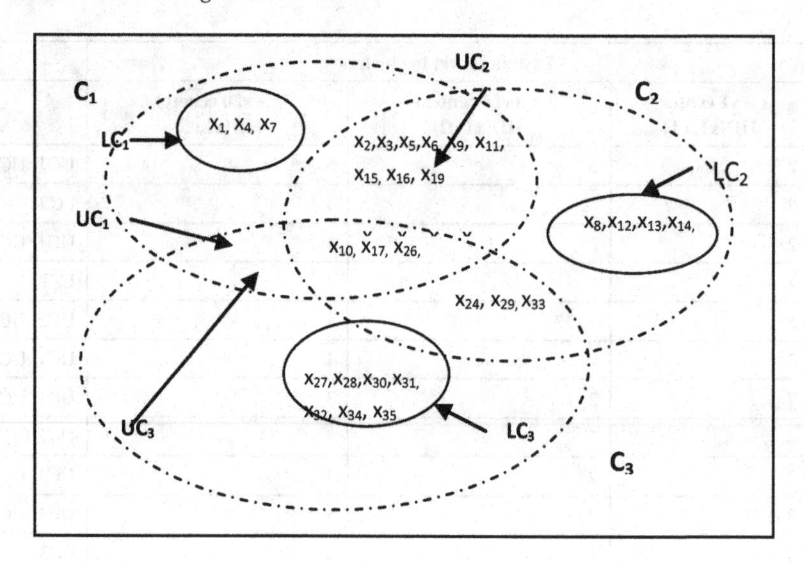

Figure 6. Balloon dataset rough cluster model

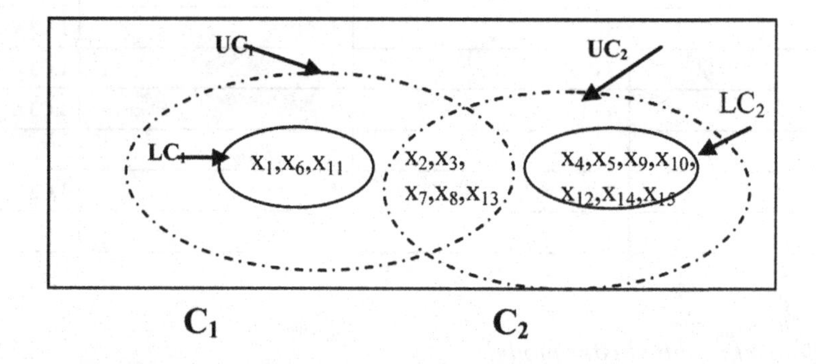

Figure 7. Error rates in prediction

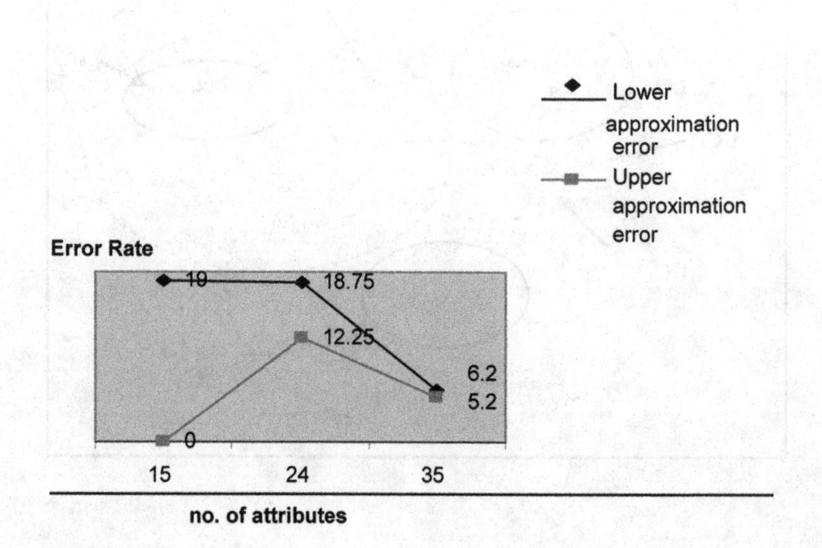

5. CONCLUSION

In this paper, an Active Learning algorithm using rough set based clustering is proposed. The proposed algorithm uses Hamming Distance as the dissimilarity measure and to incorporate uncertainty rough sets have been put forward .The active learning approach explores the inherent structure of the data incrementally and thus helps in merging the clusters based on their upper approximation. The desired cluster model is obtained as the output with minimum labeled data required. The lower approximation of clusters will help in the identification of outliers or exceptions that is being explored.

REFERENCES

Bagirov, A. M., Rubinov, A. M., Soukhoroukova, N. V., & Yearwood, J. (2003). Unsupervised and supervised data classification via nonsmooth and global optimization. *Sociedad da Estadistica e Investigacian Operativa Top*, *11*(1), 1–93.

Chen, D., Cui, D. W., Wang, Ch. X., & Wang, Z. R. (2006). A rough set-based hierarchical clustering algorithm for categorical data. *International Journal of Information Technology*, *12*(3), 149–159.

Chengdong, W., Mengxin, L., Zhonghua, H., Zhang, Y., & Yong, Y. (2004). Discretization algorithms of rough sets using clustering. In *Proceedings of the IEEE International Conference on Robotics and Biomimetics* (pp. 955-960).

Dy, J. G., & Brodley, C. E. (2004). Feature selection for unsupervised learning. *Journal of Machine Learning Research*, *5*, 884–889.

Emilyn, J. J., & Ramar, K. (2010). Rough set based clustering of gene expression data: A survey. *International Journal of Engineering Science & Technology*, *2*(12), 7160–7164.

Gibson, D., Kleinberg, J. M., & Raghavan, P. (1998). Clustering categorical data: An approach based on dynamic systems. In *Proceedings of the International Conference on Very Large Databases* (pp. 311- 323).

Hirano, S., & Tsumoto, S. (2006). On the nature of degree of indiscerniblity for rough clustering. In *Proceedings of the IEEE International Conference on Systems, Man, and Cybernetics* (pp. 3447-3452).

Huang, Z. (1997). A fast clustering algorithm to cluster very large categorical data sets in data mining. In *Proceedings of the SIGMOD Workshop on Research Issues on Data Mining and Knowledge Discovery*.

Huang, Z. (1999). Extensions to the k-means algorithm for clustering large data sets with categorical values. *Data Mining and Knowledge Discovery*, *2*, 283–304. doi:10.1023/A:1009769707641

Jain, A. K., Murthy, M. N., & Flynn, P. J. (1999). Data Clustering: A review. *ACM Computing Surveys*, *31*(3), 264–323. doi:10.1145/331499.331504

Magnani, M. (2003). *Technical report on rough set theory for knowledge discovery in data bases*. Bologna, Italy: University of Bologna.

Mahajan, P., Kandwal, R., & Vijay, R. (2011). General framework for cluster based active learning algorithm. *International Journal on Computer Science and Engineering*, *3*(1), 307–312.

Mitra, S., Banka, H., & Pedrycz, W. (2006). Rough–fuzzy collaborative clustering. *IEEE Transactions on Systems, Man, and Cybernetics. Part B, Cybernetics*, *36*(4), 795–800. doi:10.1109/TSMCB.2005.863371

Pawlak, Z. (1982). Rough sets. *International. Journal of Computer and Information Sciences*, *11*(5), 341–356. doi:10.1007/BF01001956

Wang, B. (2010). A new clustering algorithm on nominal data sets. In *Proceedings of the International MultiConference of Engineers and Computer Scientists* (pp. 605-610).

Wang, K., Xu, C., & Liu, B. (1999). Clustering transactions using large items. In *Proceedings of the ACM International Conference on Information and Knowledge Management* (pp. 483-490).

Widz, S. lzak, D., & Revett, K. (2004). Application of rough set based dynamic parameter optimization to MRI segmentation. In *Proceedings of the 23rd International Conference of the North American Fuzzy Information Processing Society* (pp. 440-445).

Xu, E., Xuedong, G., Sen, W., & Bin, Y. (2006). An clustering algorithm based on rough set. In *Proceedings of the 3rd International IEEE Conference Intelligent Systems* (pp. 475-478).

Yang, L., & Yang, L. (2006). Study of cluster algorithm based on rough sets theory. In *Proceedings of the Sixth International Conference on Intelligent System Design & Application* (pp. 492-496).

Zhang, Y., Fu, A., Cai, C. H., & Heng, P. (2000). Clustering categorical data. In *Proceedings of the IEEE International Conference on Data Engineering* (pp. 305-325).

This work was previously published in the International Journal of Artificial Life Research, Volume 2, Issue 4, edited by E. Stanley Lee and Ping-Teng Chang, pp. 12-23, copyright 2011 by IGI Publishing (an imprint of IGI Global).

Chapter 20
Intuitionistic Fuzzy 2-Metric Space and Some Topological Properties

Q.M. Danish Lohani
South Asian University, India

ABSTRACT

The notion of intuitionistic fuzzy metric space was introduced by Park (2004) and the concept of intuitionistic fuzzy normed space by Saadati and Park (2006). Recently Mursaleen and Lohani introduced the concept of intuitionistic fuzzy 2-metric space (2009) and intuitionistic fuzzy 2-norm space. This paper studies precompactness and metrizability in this new setup of intuitionistic fuzzy 2-metric space.

1. INTRODUCTION AND PRELIMINARIES

Among various developments of the fuzzy set theory (Zadeh, 1965), a progressive development has been made to find the fuzzy analogues of the classical set theory. Infact the fuzzy theory has become an area of active research for the last forty years. It has a wide range of applications in the field of science and engineering, e.g. population dynamics (Barros, Bassanezi, & Tonelli, 2000), chaos control (Fradkov & Evans, 2005), computer programming (Giles, 1980), nonlinear dynamical systems (Hong & Sun, 2006), and medicine (Barro & Marin, 2002).

Fuzzy topology is one of the most important and useful tool studied by various authors, e.g. (Erceg, 1979; Fang, 2002; Felbin, 1992; George & Veeramani, 1994; Kaleva & Seikkala, 1984; Xiao & Zhu, 2002). The most fascinating application of the fuzzy topology in quantum physics arises in $e^{(\infty)}$-theory due to El Naschie (1998, 2000, 2004a, 2004b, 2006a, 2006b, 2007) who presented the relation of fuzzy Kähler interpolation of $e^{(\infty)}$ to

DOI: 10.4018/978-1-4666-3890-7.ch020

the recent work on cosmo-topology and the Poincare dodecahedral conjecture and gave various applications and results of $e^{(\infty)}$ theory from nano technology to brain research.

Atanassov (1986, 1994) introduced the concept of intuitionistic fuzzy sets. For intuitionistic fuzzy topological spaces, we refer to Abbas (2005), Coker (1997), Kaleva and Seikkala (1984), Lupianez (2006), and Park (2004). Recently, Saadati and Park (2006) studied the notion of intuitionistic fuzzy 2-normed spaces. Quite recently, the concept of intuitionistic fuzzy 2-normed spaces and intuitionistic fuzzy 2-normed spaces has been introduced and studied by Mursaleen and Lohani (2009) and Park (2004) respectively. Certainly there are some situations where the ordinary metric does not work and the concept of intuitionistic fuzzy metric seems to be more suitable in such cases, that is, we can deal with such situations by modelling the inexactness of the norm in some situations.

In this paper we study the concept of intuitionistic fuzzy 2-metric space which would provide a more suitable funtional tool to deal with the inexactness of the metric or 2-metric in some situations. We present here analogues of precompactness and metrizability and establish some interesting results in this new setup.

We recall some notations and basic definitions used in this paper.

Definition 1.1: A binary operation $*: [0,1] \times [0,1]$ $[0,1]$ is said to be a *continuous t-norm* if it satisfies the following conditions (Schweizer & Sklar, 1960):

1. $*$ is associative and commutative,
2. $*$ is continuous,
3. $a * 1 = a$ for all $a \in [0,1]$,
4. $a * b$ $c * d$ whenever a c and b d for each $a, b, c, d, \in [0,1]$.

Definition 1.2: A binary operation $\diamond: [0,1] \times [0,1] \rightarrow [0,1]$ is said to be a *continuous t-conorm* if it satisfies the following conditions (Schweizer & Sklar, 1960):

1. $\diamond$ is associative and commutative,
2. $\diamond$ is continuous,
3. $a \diamond 0 = a$ for all $a \in [0,1]$,
4. $a \diamond b \leq c \diamond d$ whenever $a \leq c$ and b d for each $a, b, c, d \in [0,1]$.

Definition 1.3: The five-tuple $(X, \mu, v, *, \diamond)$ is said to be an *intuitionistic fuzzy normed space* (for short, IFNS) if X is a vector space, $*$ is a continuous t-norm, $\diamond$ is a continuous t-conorm, and μ, are fuzzy sets on $X \times (0, \infty)$ satisfying the following conditions. For every $x, y, \in X$ and $s, t > 0$, (Saadati & Park, 2006).

1. $\mu(x, t) + (x, t) \leq 1$,
2. $\mu(x, t) > 0$,
3. $\mu(x, t) = 1$ if and only if $x = 0$,
4. $\mu(a x, t) = \mu(x, \frac{t}{|á|})$ for each $\alpha \neq 0$,
5. $\mu(x, t) * \mu(y, s) \leq \mu(x + y, t + s)$,
6. $\mu(x, \bullet) : (0, \infty) \rightarrow [0, 1]$ is continuous,
7. $\lim_{t \to \infty} \mu(x, t) = 1$ and $\lim_{t \to \infty} \mu(x, t) = 0$,
8. $\nu(x, t) < 1$,
9. $\nu(x, t) = 0$ if any only if $x = 0$,
10. $\nu(ax, t) = v\left(x, \frac{t}{|a|}\right)$ for each $\alpha \neq 0$,
11. $\nu(x, t) \Diamond (y, s) \geq v(x+y, t+s)$
12. $\nu(x, \bullet) : (0, \infty) \rightarrow [0, 1]$ continuous,
13. $\lim_{t \to \infty} \nu(x, t) = 0$ and $\lim_{t \to 0} \nu(x, t) = 1$.

In this case $(\mu,)$ is called an *intuitionistic fuzzy norm*.

Definition 1.4: The five-tuple $(X, M, N, *, \diamond)$ is said to be an *intuitionistic fuzzy metric space* (for short, IFMS) if X is an arbitrary (non-empty) set, $*$ is a continuous t-norm, $\diamond$ is a continuous t-conorm, and M, N fuzzy sets on $X \times X \times (0, \infty)$ satisfying the following conditions. For every $x, y, z \in X$ and $s, t > 0$ (Park, 2004),

1. $M(x, y, t) + N(x, y, t) < 1$,
2. $M(x, y, t) > 0$,
3. $M(x, y, t) = 1$ if and only if $x = y$,
4. $M(x, y, t) = M(y, x, t)$,
5. $M(x, y, t) * M(y, z, s) \leq M(x, z, t + s)$,
6. $M(x, y,.: (0, \infty) \to [0,1]$ is continuous,
7. $N(x, y, t) < 1$,
8. $N(x, y, t) = 0$ if and only if $x = y$,
9. $N(x, y, t) = N(y, x, t)$,
10. $N(x, y, t) \diamond N(y, z, s) \geq N(x, z, t + s)$,
11. $N(x, y,) : (0, \infty) \to [0,1]$ is continuous,

Then (M, N) is called an *intuitionistic fuzzy metric* on X.

Definition 1.5: Let X be a real vector space of dimension d, where $2 \leq d < \infty$. A 2-norm on X is a function $\|.,.\|: X \times X \to$ which satisfies (Gähler, 1965),

1. $\|x, y\| = 0$ if and only if x and y are linearly dependent;
2. $\|x, y\| = \|y, x\|$;
3. $\|\alpha x, y\| = |\alpha| \, \|x, y\|$;
4. $\|x, y + z\| \leq \|x, y\| + \|x, z\|$.

The pair $(X, \|.,.\|)$ is then called a *2-normed space*.

Example 1.1: Take $X = {}^2$ being equipped with the 2-norm $\|x, y\|:=$ the area of the parallelogram spanned by the vectors x and y, which may be given explicitly by the formula,

$$\|x, y\| = |x_1y_2 - x_2y_1|, \ x = (x_1, x_2), \ y = (y_1, y_2).$$

Definition 1.6: Let X be a nonempty set. A real valued function d on $X \times X \times X$ is said to be a *2-metric* on X if (Gähler, 1965)

1. given distinct elements x, y of X, there exists an element z of X such that $d(x, y, z) \neq 0$;
2. $d(x, y, z) = 0$ when at least two of x, y, z are equal;
3. $d(x, y, z) = d(x, z, y) = d(y, z, x)$ for all x, y, z in X;
4. $d(x, y, z) \leq d(x, y, w) + d(x, w, z) + d(w, y, z)$ for all x, y, z, to in X.

The pair (X, d) is then called a *2-metric space*.

Example 1.2: Take $X = {}^3$ being equipped with the 2-metric $d(x, y, z):=$ the area of the triangle spanned by x, y and z, which may be given explicitly by the formula,

$$d(x, y, z) = |x_1(y_2z_3 - z_2y_3) - x_2(y_1z_3 - y_3z_1) + x_3(y_1z_2 - y_2z_1)|.$$

where

$$x = (x_1, x_2, x_3), \ y = (y_1, y_2, y_3), \ z = (z_1, z_2, z_3).$$

2. INTUITIONISTIC FUZZY 2-METRIC SPACE

Quite recently, the concept of intuitionistic fuzzy 2-normed spaces and intuitionistic fuzzy 2-metric space has been introduced by Mursaleen and Lohani (2009), and Park (2004) respectively. In this section we define here some topological concepts in intuitionistic fuzzy 2-metric space.

Definition 2.1 (Mursaleen & Lohani, 2009): The five-tuple $(X \times X, \mu, , *, \diamond)$ is said to be an *intuitionistic fuzzy 2- normed space* (for short, IF-2-NS) if X is a 2-normed space, * is a continuous t-norm, ◇ is a continuous t-conorm, and μ, are fuzzy sets on $X \times X \times (0, \infty)$ satisfying the following conditions for every $x, y, z X$, and $s, t > 0$.

1. $\mu(x, y; t) + (x, y; t) \leq 1$,
2. $\mu(x, y; t) > 0$,
3. $\mu(x, y; t) = 1$ if and only if x and y are linearly dependent,
4. $\mu(\alpha x, y; t) = \mu(x, y; \frac{t}{|\alpha|})$ for each $\alpha \neq 0$,
5. $\mu(x, z; t) * \mu(y, z; s) \leq \mu(x + y, z; t + s)$,
6. $\mu(x, y;) : (0, \infty) \rightarrow [0, 1]$ is continuous,
7. $\lim_{t\rightarrow\infty} \mu(x, y, t) = 1$ and $\lim_{t\rightarrow 0} \mu(x, y, t) = 0$,
8. $\mu(x, y; t) = \mu(y, x; t)$
   a. $v(x, y; t) < 1$,
9. $v(x, y; t) = 0$ if and only if x and y are linearly dependent,
10. $v(\alpha x, y; t) = (x, y; \frac{t}{|\alpha|})$ for each $\alpha \neq 0$,
11. $v(x, z; t) \diamond (y, z; s) \geq (x + y, z; t + s)$,
12. $v(x, y;): (0, \infty) \rightarrow [0, 1]$ is continuous,
13. $\lim_{t\rightarrow\infty} v(x, y, t) = 0$ and $\lim_{t\rightarrow 0} v(x, y, t) = 1$
14. $v(x, y; t) = (y, x; t)$

In this case $(\mu,)$ is called an *intuitionistic fuzzy 2-norm* on X, and we denote it by $(\mu, v)_2$.

Remark 2.1: Let $(X, \|., .\|)$ be a 2-normed space, and let $a * b = ab$ and $a \diamond b = \min\{a + b, 1\}$ for all $a, b \in [0,1]$. For all $x \in X$ and every $t > 0$, consider $\mu(x, y; t) := \frac{t}{t + \|x, y\|}$. and

$$v(x, y; t) := \frac{\|x, y\|}{t + \|x, y\|}$$

Then $(X, \mu, , *, \diamond)$ is an intuitionistic fuzzy 2-normed space.

Definition 2.2: The 5-tuple $(X, M, N, *, 0)$ is said to be an *intuitionistic fuzzy 2- metric space* (for short, IF-2-MS) if X is a 2-metric space, * is a continuous t-norm, ◇ is a continuous t-conorm, and M, N fuzzy sets on $X \times X \times X \times (0,)$, satisfying the following conditions for each $x, y, z, w \in X$ and $s, t > 0$, (Park, 2004)

1. $M(x, y, z; t) + N(x, y, z; t) \leq 1$,
2. given distinct elements x, y of X, there exists an element z of X such that $M(x, y, z; t) > 0$,
3. $M(x, y, z; t) = 1$ if at least two of x, y, z are equal,
4. $M(x, y, z; t) = M(x, z, y; t) = M(y, z, x; t)$ for all x, y, z in X,
5. $M(x, y, w; t) * M(x, w, z; s) * M(w, y, z; r) \leq M(x, y, z; t + s + r)$ for all $x, y, z, w \in X$,
6. $M(x, y, z;): (0,\infty) \rightarrow (0, 1]$ is continuous,
7. $N(x, y, z; t) < 1$,
8. $N(x, y, z; t) = 0$ if at least two of x, y, z are equal,
   a. $N(x, y, z; t) = N(x, z, y; t) = N(y, z, x; t)$ for all x, y, z in X,
9. $N(x, y, w; t) \diamond N(x, w, z; s) \diamond N(w, y, z; r) \geq N(x, y, z, t + s + r)$,
10. $N(x, y, z;): (0, \infty) \rightarrow (0, 1]$ is continuous,

In this case (M, N) is called an *intuitionistic fuzzy 2-metric* on X and we denote it by $(M, N)_2$. The functions $M(x, y, z; t)$ and $N(x, y, z; t)$ denote the degree of nearness and the degree of non nearness between x, y and z with respect to t, respectively.

Remark 2.2: In an intuitionistic fuzzy metric space $(X, M, N, *, \diamond)$, $M(x, y, z; .)$ is non-decreasing and $N(x, y, z; .)$ is non-increasing for all $x, y, z \in X$.

Let (X, d) be a metric space. Denote $a * b = ab$ and $a \diamond b = \min\{1, a + b\}$ for all $a, b \in [0,1]$ and let M_d and N_d be fuzzy sets on $X^3 \times (0, \infty)$ defined by:

$$M_d(x,y,z;t) = \frac{ht^n}{ht^n + md(x,y,z)},$$

$$N_d(x,y,z;t) = \frac{d(x,y,z)}{kt^n + md(x,y,z)}$$

for all h, k, m, n $\in$ R$^+$. Then $(X, M_d, N_d, *, \diamond)$ is an intuitionistic fuzzy-2-metric space.

As in Mursaleen and Lohani (2009) we have the following Remarks

Remark 2.3: Let $(X, \mu, , *, \diamond)$ be an intuitionistic fuzzy 2-normed space. If we define

$M(x, y, z; t) = \mu(x - y, z; t)$ for all $z \in X$, and

$N(x, y, z; t) = v(x - y, z; t)$ for all $z \in X$.

Then $(M,N)_2$ is an intuitionistic fuzzy-2-metric on X, which is induced by the intuitionistic fuzzy 2-norm $(\mu, v)_2$.

Remark 2.4: Let $(X, \mu, v, *, \diamond)$ be an intuitionistic fuzzy 2-normed space. Then, for any $t > 0$, the following hold:

1. $\mu(x, y; t)$ and $(x, y; t)$ are nondecreasing and nonincreasing with respect to t and for all $y \in X$, respectively.
2. $\mu(x - y, z; t) = \mu(y - x, z; t)$ and $(x - y, z; t) = v(y - x, z; t)$ for all $z \in X$.

Definition 2.3: Let $(X, M, N, *,)$ be an intuitionistic fuzzy 2-metric space, and let $r \in (0,1)$, $t > 0$ and $x \in X$. The set $\mathbb{B}(x, r, t) = \{y \in X$: $M(x, y, z; t) 1 - r$, $N(x, y, z; t) < r$, for all $z \in X\}$ is called the *open ball* with center x and radius r with respect to t. (Park, 2004)

Definition 2.4: Let $(X, M, N, *, \diamond)$ be an intuitionistic fuzzy 2-metric space, then a set $U X$ is said to an *open set* if each of its points is the centre of some open ball contained in U. The open set in an intuitionistic fuzzy 2-metric space $(X, M, N, *, \diamond)$ is denoted by $\mathbb{U}$. (Park, 2004)

Definition 2.5: Let $(X, M, N, *, \diamond)$ be an intuitionistic fuzzy 2-metric space. A sequence (x_n) in X is said to be *Cauchy* if for each > 0 and each t > 0, there exists $n_0 \in$ such that $M(x_n, x_m, z; t) > 1 - r$ and $N(x_n, x_m, z; t) < r$ for all $n, m \geq n_0$ and for all $z \in X$. (Park, 2004)

Definition 2.6: Let $(X, M, N, *, \diamond)$ be an intuitionistic fuzzy 2-metric space. A sequence $x = (x_k)$ is said to be *convergent* to $L \in X$, with respect to the intuitionistic fuzzy 2-metric $(M, N)_2$ if, for every $\epsilon > 0$ and $t > 0$, there exists $k_0 \in \mathbb{N}$ such that $M(x_k, L, z; t) > 1 - \epsilon$ and $N(x_k, L, z; t) < \epsilon$ for all $k \geq k_0$ and for all $z \in X$.(Park, 2004)

In this case we write $(M, N)_2$- $\lim x = L$ or $x_k \xrightarrow{(M,N)_2} L$ as $k \to \infty$.

Definition 2.7: Let $(X, M, N, *, \diamond)$ be an intuitionistic fuzzy 2-metric space. Define $\tau_{(M,N)2} = \{A X$: for each $x \in A$, there exists $t > 0$ and $r \in (0,1)$ such that $\mathbb{B}(x, r, t) A\}$. Then $\tau_{(M,N)2}$ is a topology on $(X, M, N, *, \diamond)$. (Park, 2004)

Definition 2.8: Let $(X, M, N, *, \diamond)$ be an intuitionistic fuzzy 2-metric space. Then it is said to be *complete* if every Cauchy sequence is convergent with respect to (Park, 2004)

Now we define the following.

Definition 2.9: Let $(X, M, N, *, \diamond)$ be an intuitionistic fuzzy 2-metric space. A subset A of X is said to be $\mathbf{IF}_2$-*bounded* if there exists $t > 0$ and let $r \in (0,1)$ such that $M(x, y, z; t) > 1 - r$ and $N(x, y, z; t) < r$ for all $x, y \in A$, and for all $z \in X$.

Definition 2.10: Let $(X, M, N, *, \diamond)$ be an intuitionistic fuzzy 2-metric space and $A \subset X$. Then A is *precompact* if for each $r \in (0,1)$ and $t > 0$, there exists a finite subset S of A such that

$$A \subseteq \bigcup_{x \in S} \mathbb{B}(x, r, t)$$

Definition 2.11: Let $(X, M, N, *, \diamond)$ be an intuitionistic fuzzy 2-metric space, $x \in X$ and $\varnothing \neq A \subseteq X$. We define

$D(x, A, t) = \sup\{M(x, y, z, t): y \in A, z \in X\}$ $(t > 0)$

and

$C(x, A, t) = \inf\{N(x, y, z, t): y \in A, z \in X\}$ $(t > 0)$.

Note that $D(x, A, t)$ and $C(x, A, t)$ are a degree of closeness and a degree of non-closeness of x to A at t, respectively.

Definition 2.12: A topological space is called a *topologically complete intuitionistic fuzzy 2-metrizable space* if there exists a complete intuitionistic fuzzy 2-metric inducing the given topology on it.

Example 2.1: Let $X = (0,1]$. The intuitionistic fuzzy 2-metric space $(X, M, N, \min, \max)$, where $M(x, y, z, t) = \dfrac{t}{t + d(x,y,z)}$ and $N(x, y, z, t) = \dfrac{d(x,y,z)}{t + d(x,y,z)}$; where $d(x, y, z) = \min\{|x - y|, |y - z|, |z - x|\}$ is not complete, because the Cauchy sequence $\{\frac{1}{n}\}$ in this space is not convergent. Now consider the 5-tuple $(X, m, n, \min, \max)$, where $m(x, y, z, t) = \dfrac{t}{t + d(x,y,z) + d(\frac{1}{x}, \frac{1}{y}, \frac{1}{z})}$ and $n(x, y, z, t) = \dfrac{d(x,y,z) + d(\frac{1}{x}, \frac{1}{y}, \frac{1}{z})}{t + d(x,y,z) + d(\frac{1}{x}, \frac{1}{y}, \frac{1}{z})}$. It is straight forward to show that $(X, m, n, \min, \max)$ is an intuitionistic fuzzy 2-metric space which is complete. Since x_n tends to x with respect to intuitionistic fuzzy 2-metric $(M, N)_2$, if and only if $|x_n - x| \to 0$, if and only if x_n tends to x with respect to intuitionistic fuzzy 2-metric $(m, n)_2$, hence $(M, N)_2$ and $(m, n)_2$ are equivalent intuitionistic fuzzy 2-metrics. Therefore the intuitionistic fuzzy 2-metric space $(X, m, n, \min, \max)$ is topologically complete intuitionistic fuzzy 2-metrizable.

3. PRECOMPACTNESS IN INTUITIONISTIC FUZZY 2-METRIC SPACES3. PRECOMPACTNESS IN INTUITIONISTIC FUZZY 2-METRIC SPACES

In this section, we establish some results on precompactness in intuitionistic fuzzy 2-metric space.

Lemma 3.1: Let $(X, M, N, *, \diamond)$ be an intuitionistic fuzzy 2-metric space and $A \subset X$. Then A is precompact if and only if for every $r \in (0,1)$ and $t > 0$, there exists a finite subset S of A such that

$$A \subseteq \bigcup_{x \in S} \mathbb{B}(x, r, t) \quad (1)$$

Proof: Let $r \in (0,1)$ and $t > 0$ and condition (1) holds. By the continuity of $*$, $\diamond$, there exists $s \in (0,1)$ such that $(1 - s) * (1 - s) * (1 - s) > 1 - r$ and $s \diamond s \diamond s < r$. Now we apply condition (1) for

s and $\frac{t}{3}$, there exists a subset $S = \{x_1, \ldots, x_n\}$ of X such that $A \subseteq \bigcup_{x_i S} \mathbb{B}\left(x_i, s, \frac{t}{3}\right)$. We assume that $\mathbb{B}\left(x_j, s, \frac{t}{3}\right) \cap A \neq \varnothing$, otherwise we omit x_j from S' and so we have $A \subseteq \bigcup_{x_i \in S-\{x_j\}} \mathbb{B}\left(x_i, s, \frac{t}{3}\right)$.

For every $i = 1, \ldots, n$ we select y_j in $\mathbb{B}\left(x_i, s, \frac{t}{3}\right) \cap A$, therefore $M(x_i, y_i, z; \frac{t}{3}) > 1 - s$ and $N(x_i, y_i, z; \frac{t}{3}) < s$ for all $z \in X$, therefore in particular $M(x_i, y_i, y; \frac{t}{3}) > 1 - s$ and $N(x_i, y_i, y; \frac{t}{3}) < s$ for some $y \in X$ and we put $S = \{y_1, \ldots, y_n\}$. Now for every $y \in A$, there exists $i \in \{1, \ldots, n\}$ such that $M(y, x_i, z; \frac{t}{3}) > 1 - s$ and $N(y, x_i, z; \frac{t}{3}) < s$ for all $z \in X$. Therefore we have $M(y, y_i\, z; t) > M(x_i, y_i\, z; \frac{t}{3}) * M(y, x_i, z; \frac{t}{3}) * M(y, y_i, x_i; \frac{t}{3}) > (1-s)*(1-s)*(1-s) > 1-r$ and

$$N\left(y,\ y_i,\ z;\ t\right) < N(x_i,\ y_i,\ z; \frac{t}{3}) \quad \Diamond$$

$$N(y,\ x_i,\ z; \frac{t}{3})\ N(y,\ y_i,\ x_i; \frac{t}{3}) < s \quad s \quad s < r.$$

which imlies that $A \subseteq \bigcup_{x \in S} \mathbb{B}\left(x, r, t\right)$. The converse is trivial.

Lemma 3.2: Let $(X, M, N, *, \diamond)$ be an intuitionistic fuzzy 2-metric space and $A \subset X$. If A is a precompact set then its closure $\bar{A}$ is also precompact.

Proof: Let $r \in (0,1)$ and $t > 0$. Then by the continuity of $*$, $\diamond$, there exists $s \in (0,1)$ such that $(1 - s) * (1 - s) * (1 - s) > 1 - r$ and $s \diamond s \diamond s < r$, also there exists a finite subset $S' = \{x_1, \ldots, x_n\}$ of X such that $A \subseteq \bigcup_{x_{iS}} \mathbb{B}(x_i, s, \frac{t}{3})$. But for every $y \in \bar{A}$ there exists $x \in A$ such that $M(x, y, z; \frac{t}{3}) > 1 - s$ and $N(x, y, z; t) < s$ for every $z \in X$ and there exists $1 \leq i \leq n$, such that $M(x, x_i, z; \frac{t}{3}) > 1 - s$ and $N(x, x_i, z; \frac{t}{3}) < s$ for all $z \in X$.

Hence in particular $M(x, x_i, y; \frac{t}{3}) > 1 - s$ and $N(x, x_i, y; \frac{t}{3}) < s$ for some $y \in X$. Therefore we have

$$M(y, x_i, z; t) > M(x, x_i\, z; \frac{t}{3}) * M(y, x, z; \frac{t}{3}) * M(y, x_i, x; \frac{t}{3}) > (1-s) * (1-s) * (1-s) > 1-r$$

and

$$N(y, x_i, z; t) < N(x, x_i, z; \frac{t}{3}) \diamond N(y, x, z; \frac{t}{3}) \diamond N(y, x_i, x; \frac{t}{3}) < s \diamond s \diamond s < r.$$

Hence $\bar{A} \subseteq \bigcup_{x_{iS}} \mathbb{B}(x_i, r, t)$. i.e., $\bar{A}$ is precompact set.

Theorem 3.1: Let $(X, M, N, *, \diamond)$ be an intuitionistic fuzzy 2-metric space and $A \subset X$. Then $A>1$ is a precompact set if and only if every sequence has a Cauchy subsequence. $\frac{1}{n_o}\ \frac{1}{n_o} <$

Proof: Let A be a precompact set. Let (p_n) be a sequence in A. For every $k \in \mathbb{N}$. there exists a finite subset S_k of X such that

$A \subseteq \bigcup_{x \in S_k} \mathbb{B}(x, \frac{1}{k}, \frac{1}{k})$. Hence, for $k = 1$, there exists $x_1 \in S_1$ and a subsequence $(p_{1,n})$ of (p_n) such that $p_{1,n} \in \mathbb{B}(x_1, 1, 1)$ for every $n \in \mathbb{N}$. Similarly, there exists $x_2 \in S_2$ and a subsequence $(p_{2,n})$ of $(p_{1,n})$ such that $p_{2,n} \in \mathbb{B}(x_2, \frac{1}{2}, \frac{1}{2})$ for every $n \in \mathbb{N}$. Continuing this process, we get $x_k \in S_k$ and a subsequences $(p_{k,n})$ of $(p_{k-1,n})$ such that $p_{k,n} \in \mathbb{B}(x_k, \frac{1}{k}, \frac{1}{k})$ for every $n \in \mathbb{N}$. Now we consider the subsequence $(p_{n,n})$ of (p_n). For every $r \in (0,1)$ and $t > 0$, by the continuity of $*$, $\diamond$, there exists a $n_o \in \mathbb{N}$ such that $(1 - \frac{1}{n_o}) * (1-) * (1 - \frac{1}{n_o}) > 1 - r$, $\frac{1}{n_o} \diamond \frac{1}{n_o}$ r and $\frac{3}{n_o} < t$. Therefore for every $l, m \geq n_o$, we have

$$M(p_{l,l}, p_{m,m}, z; t) \geq M(p_{l,l}, p_{m,m}, z; \frac{3}{n_o})$$

$$M(p_{l,l}, p_{m,m}, z; \frac{3}{n_o}) \geq M(x_{n_o}, p_{m,m}, z; \frac{1}{n_o}) * M(p_{l,l}, x_{n_o}, z; \frac{1}{n_o}) * M(p_{l,l}, p_{m,m}, z; \frac{1}{n_o})$$

$$> (1 - \frac{1}{n_o}) * (1 - \frac{1}{n_o}) * (1 - \frac{1}{n_o}) > 1 - r$$

and

$$N(p_{l,l}, p_{m,m}, z; t) \leq N(p_{l,l}, p_{m,m}, z; \frac{3}{n_o})$$

$$N(p_{l,l}, p_{m,m}, z; \frac{3}{n_o}) \leq N(x_{n_o}, p_{m,m}, z; \frac{1}{n_o}) \diamond N(p_{l,l}, x_{n_o}, z; \frac{1}{n_o}) \diamond N(p_{l,l}, p_{m,m}, x_{n_o}; \frac{1}{n_o})$$

$$< \frac{1}{n_o} \diamond \frac{1}{n_o} \diamond \frac{1}{n_o} < r.$$

Hence $(p_{n,n})$ is a Cauchy sequence in $(X, M, N, *, \diamond)$.

Conversely, suppose that A is not a precompact set. Then there exists $r \in (0, 1)$ and $t > 0$ such that for every finite subset S of X, A is not a subset of $A \subseteq \bigcup_{xS} \mathbb{B}(x, r, t)$. Fix $p_1 \in A$. Since A is not a subset of $A \subseteq \bigcup_{x\{p_1\}} \mathbb{B}(x, r, t)$, there exists $p_2 \in A$ such that $M(p_1, p_2, z; t) \leq 1 - r$ and $N(p_1, p_2, z; t) \geq r$ for all $z \in X$. Since A is not a subset of $A \subseteq \bigcup_{x\{p_1, p_2\}} \mathbb{B}(x, r, t)$, there exists $p_3 \in A$ such that $M(p_1, p_3, z; t) \leq 1 - r$ and $N(p_1, p_3, z; t) \geq r$ for all $z \in X$. Continuing this process, we construct a sequence (p_n) of distinct points in A such that $M(p_i, p_j, z; t) \leq 1 - r$ and $N(p_i, p_j, z; t) \geq r$ for all $z \in X$ for every $i \neq j$. Therefore (p_n) has not Cauchy subsequence.

Lemma 3.3: Let (x_n) be a Cauchy sequence in an intuitionistic fuzzy 2-metric space $(X, M, N, *, \diamond)$ having a cluster point $x \in X$. Then (x_n) is convergent to x.

Proof: Since (x_n) is a Cauchy sequence in $(X, M, N, *, \diamond)$ having a cluster point $x \in X$. Then, there is a subsequence (x_{n_k}) of (x_n) that converges to x with respect to $\hat{o}_{(M,N)_2}$. Thus, given $r \in (0,1)$ and $t > 0$, there is a $l \in \mathbb{N}$ such that for each $k^3 l$, $M(x, x_{nk}, z; \frac{t}{3}) > 1 - s$ and $N(x, x_{nk}, z; \frac{t}{3}) < s$ for all $z \in X$, therefore in particular $M(x, x_{nk}, x_n; \frac{t}{3}) > 1 - s$ and $N(x, x_{nk}, x_n; \frac{t}{3}) < s$ for some $x_n \in X$, where $s \in (0,1)$ and satisfies $(1 - s) * (1 - s)$

$* (1 - s) > 1 - r$ and $s \diamond s \diamond s < r$. On the other hand, there is $n_t \geq n_l$ such that for each $n, m \geq n_t$, we have $M(x_m, x_n, z; \frac{t}{3}) > 1 - s$ and $N(x_m, x_n, z; \frac{t}{3}) < s$ for all $z \in X$. Therefore for each $n, n_k \geq n_t$, we have

$$M(x, x_n, z; t) \geq M(x_{nk}, x_n, z; \frac{t}{3}) * M(x, x_{nk}, z; \frac{t}{3}) * M(x, x_n, x_{nk}, \frac{t}{3})$$

$$> (1 - s) * (1 - s) * (1 - s) > 1 - r \text{ and}$$

$$N(x, x_n, z; t) \leq N(x_{nk}, x_n, z; \frac{t}{3}) \diamond N(x, x_{nk}, z; \frac{t}{3}) \diamond N(x, x_n, x_{nk}, \frac{t}{3})$$

$$< s \diamond s \diamond s < r.$$

We conclude that the Cauchy sequence converges to x.

4. INTUITIONISTIC FUZZY 2-METRIZABILITY

In this section, we study intuitionistic fuzzy 2-metrizability.

Lemma 4.1: Let $(X, M, N, *, \diamond)$ be an intuitionistic fuzzy 2-metric space. Then $(X, \tau_{(M,N)_2})$ is a metrizable topological space.

Proof: For each $n \in \mathbb{N}$, let

$$U_n = \{(x, y, z) \in X \times X \times X : M\left(x, y, z, \frac{1}{n}\right) > 1 - \frac{1}{n}, N\left(x, y, z, \frac{1}{n}\right) < \text{ for all} z \in X\}.$$

We shall prove that $(U_n : n \in \mathbb{N})$ is a base for a uniformity $\mathcal{U}$ on X whose induced topology coincides with $\tau_{(M,N)_2}$. We first note that for each $n \in \mathbb{N}$, $\{(x, x, x): x \in X\} \subseteq U_n$, $U_{n+1} \subseteq U_n$ and $U_n = U_n^{-1}$ under the operation o defined as $U_1\, o\, U_2 = (U_1 - U_2) \cup (U_2 - U_1)$.

On the other hand, for each $n \in \mathbb{N}$, there is, by the continuity of $*, \diamond$, a $m \in \mathbb{N}$ such that $m > 3n$, $(1 - \frac{1}{m}) * (1 - \frac{1}{m}) * (1 - \frac{1}{m}) > 1 - \frac{1}{n}$ and $\frac{1}{m}\frac{1}{m}\frac{1}{m} < \frac{1}{n}$. Then, $U_m o U_m o U_m \subseteq U_n$. Indeed, let $(w, y, z) \in U_m$, $(x, w, z) \in U_m$ and $(x, y, w) \in U_m$.

Since $M(x, y, z,.)$ is non-decreasing and $N(x, y, z,.)$ is non-increasing for all $x, y, z \in X$, respectively, $M(x, y, z, \frac{1}{n}) \geq M(x, y, z, \frac{3}{n})$ and $N(x, y, z, \frac{1}{n}) \leq N(x, y, z, \frac{3}{n})$. So

$$M(x, y, z, \frac{1}{n}) \geq M(w, y, z, \frac{1}{m}) * M(x, w, z, \frac{1}{m}) * M(x, y, w, \frac{1}{m}) > (1 - \frac{1}{m}) * (1 - \frac{1}{m}) * (1 - \frac{1}{m}) > 1 - \frac{1}{n} \text{ and}$$

$$N(x, w, z, \frac{1}{n}) \leq N(y, w, z, \frac{1}{m}) \diamond N(x, y, z, \frac{1}{m}) \diamond N(x, w, y, \frac{1}{m}) < \frac{1}{m}\frac{1}{m}\frac{1}{m} < \frac{1}{n}$$

Therefore $(x, y, z) \in U_n$. Thus $U_n: n \in \mathbb{N}$ is a base for a uniformity $\mathcal{U}$ on X. Since for each $x \in X$ and each $n \in \mathbb{N}$,

$$U_n(x) = \{y \in X: M(x, y, z, \frac{1}{n}) > 1 - \frac{1}{n}, N(x, y, z, \frac{1}{n}) < \frac{1}{n} \text{ for all } z \in X\} = \mathbb{B}(x, \frac{1}{n}, \frac{1}{n}),$$

we deduce that the topology induced by $\mathcal{U}$ coincides with $\hat{o}_{(M,N)_2}$. Then $\hat{o}_{(M,N)_2}$ is a metrizable topological space.

Note that, in every metrizable space every sequentially compact set is compact.

Corollary 4.1: A subset A of an intuitionistic fuzzy 2-metric space $(X, M,N,*, \diamond)$ is compact if and only if it is precompact and complete.

Lemma 4.2: Let $(X,M,N,*,\diamond)$ be an intuitionistic fuzzy 2-metric space and let (0,1) such that 1. Then there exists an intuitionistic fuzzy 2-metric $(m, n)_2$ on X such that $m(x, y, z, t) \geq$ and $n(x,y,z,t) \leq$ for each $x,y,z \in X$ and $t > 0$ and $(m, n)_2$ and $(M, N)_2$ induce the same topology on X.

Proof: We define $m(x, y, z, t) = \max\{\lambda, M(x, y, z,t)\}$ and $n(x, y, z,t) = \min\{\eta, N(x, y, z, t)\}$. We claim that $(m, n)_2$ is an intuitionistic fuzzy 2-metric on X. The properties of (a), (b),(c),(d),(f),(g),(h),(i),(k) are immediate from the definition. For the inequalities (e) and (j), suppose that $x,y,z,w \in X$ and $t, s, r > 0$. Then $m(x, y, z, t+s+r) \geq \lambda$ and so $m(x, y, z, t+s+r) \geq \lambda * 1 * 1$. Since $0 < m(x, y, w, t), m(x, w, z, s)$ and $m(w, y, z, r) < 1$. Now if either of $m(x, y, w, t) = \lambda$, $m(x, w, z, s) = \lambda$ or $m(w, y, z, r) = \lambda$. Then $m(x, y, z, t+s+r) \geq m(x, y, w, t) * m(x, w, z, s) * m(w, y, z, r)$. The only remaining case is when $m(x, y, z, t) = M(x, y, z, t) > \lambda$, $m(x, w, z, s) = M(x, w, z, s) > \lambda$ and $m(w, y, z, r) = M(w, y, z, r) > \lambda$. But $M(x, y,z,t+s+r) \geq M(x, y, w, t) * M(x, w, z, s) * M(w, y, z, r)$ and $m(x, y, z, t+s+r) \geq M(x, y, z, t+s+r)$ and so $m(x, y, z, t+s+r) \geq m(x, y, w, t) * m(x, w, z, s) * m(w, y, z, r)$. Also, then $n(x, y, z, t+s+r) \leq \eta$ and so $n(x, y, z, t+s+r) \leq n(x, y, w, t) \diamond n(x, w, z, s) \diamond n(w, y, z, r)$ when either $n(x, y, w, t) = \eta$, $n(x, w, z, s) = \eta$ or $n(w, y, z, r) = \eta$. The only remaining case is when $n(x, y, z, t) = N(x, y, z, t) < \eta$, $n(x, w, z, s) = N(x, w, z, s) < \eta$ and $n(w, y, z, r) = N(w, y, z, r) < \eta$. But $N(x, y, z, t+s+r) \leq N(x, y, w, t) \diamond N(x, w, z, s) \lozenge N(w, y, z, r)$ and so $n(x, y, z, t+s+r) \leq n(x, y, w, t) \lozenge n(x, w, z, s) \lozenge n(w, y, z, r)$. Thus $(m, n)_2$ is an intuitionistic fuzzy 2-metric on X. It only remains to show that the topology induced by $(m, n)_2$ is same as that induced by $(M,N)_2$. But we have $M(x_n, x, z, t) \to 1$ and $N(x_n, x, z, t) \to 0$ as $n \to \infty$ for all $z \in X$ and $t > 0$ if and only if $\{\lambda, M(x, y, z, t)\} \to 1$ and $\{\eta, N(x, y, z, t)\} \to 0$ if and only if $M(x_n, x, z, t) \to 1$ and $N(x_n, x, z, t) \to 0$ as $n \to \infty$ for all $z \in X$ and $t > 0$, and we are done.

The intuitionistic fuzzy 2-metric $(m, n)_2$ in above lemma is said to be bounded by (λ,η).

Lemma 4.3: Intuitionistic fuzzy 2-metrizability is preserved under countable Cartesian product.

Proof: Without loss of generality we may assume that the index set is $\mathbb{N}$. Let $((X_n, m_n, n_n, *, \diamond): n \in \mathbb{N})$ be a collection of intuitionistic fuzzy 2-metrizable spaces. Let τ_n be the topology induced by $(m_n, n_n)_2$ on $(X_n, m_n, n_n, *,\diamond)$ for $n \in \mathbb{N}$ and let (X, τ) be the cartesian product of $((X_n, \tau_n) : n \in \mathbb{N})$ with product topology. We have to prove there is an intuitionistic fuzzy 2-metric $(m, n)_2$ on X which induces the topology τ. By Lemma 4.1, we may suppose that $(m_n, n_n)_2$ is bounded by $(1-\epsilon^{(n)}, \epsilon^{[n]})$ Where $\epsilon^{(n)} = \overbrace{\epsilon * \epsilon * \ldots * \epsilon}^{n}$, $\epsilon^{[n]} = \epsilon^{(n)} = \overbrace{\epsilon \lozenge \epsilon \lozenge \ldots \lozenge \epsilon}^{n}$, and $\xi\epsilon$ (0, 1) (Atanassov, 1986), i.e. $m_n(x_n, y_n, z, t) = \max\{1 - \epsilon^{(n)}, M_n(x_n, y_n, z, t)\}$ and $n_n(x_n, y_n, z, t) = \min\{\epsilon^{[n]}, N_n(x_n, y_n, z, t)\}$. Points of $X = \prod_{n\in\mathbb{N}} X_n$ are denoted as sequences as $x = (x_n)$

with $x_n \in X_n$ for $n \in \mathbb{N}$. Define $m(x, y, z, t) = \prod_{n=1}^{\infty} m_n(x_n, y_n, z, t)$ and $n(x, y, z, t) = \coprod_{n=1}^{\infty} n_n(x_n, y_n, z, t)$, for each $x, y, z \in X$ and $t > 0$ where $\prod_{n=1}^{m} a_n = a_1 * a_2 * \ldots * a_m$, $\coprod_{n=1}^{m} a_n = a_1 \diamond a_2 \diamond \ldots \diamond a_m$. First note that $(m, n)_2$ is well defined since $a_j = \prod_{n=1}^{j}(1 - T^{(n)})$ decreasing and bounded then converges to $\alpha \in (0, 1)$ also $b_i = \coprod_{n=1}^{i} T^{[i]}$ is increasing and bounded then converges to $\beta \in (0, 1)$. Also $(m, n)_2$ is an intuitionistic fuzzy 2-metric on X because each $(m_n, n_n)_2$ is an intuitionistic fuzzy 2-metric. Let $\mathcal{U}$ be the topology induced by an intuitionistic fuzzy 2-metric $(m_n, n_n)_2$. We claim that $\mathcal{U}$ coincides with τ.. If $\in$ and $x = (x_n) \in G$, then there exists $r \in (0, 1)$ and $t > 0$ such that $\mathbb{B}(x, r, t) \subset G$. For each $r \in (0, 1)$, we can find a sequence (δ_n) in $(0, 1)$ and a positive integer n_0 such that

$$\prod_{n=1}^{n_o}\left(1-\delta_n\right) * \prod_{n=n_o+1}^{\infty}\left(1-T^{(n)}\right) > 1-r$$

and

$$\coprod_{n=1}^{n_o}\delta_n \diamond \coprod_{n=n_o+1}^{\infty} T^{[n]} < r.$$

For each $n = 1, 2, \ldots, n_o$, let $V_n = \mathbb{B}(x_n, \delta_n, t)$, where the ball is with respect to intuitionistic fuzzy 2-metric $(m_n, n_n)_2$. Let $V_n = X_n$ for n_o. Put $V = \prod_{n \in \mathbb{N}} V_n$, then $x \in V$ and V is an open set in the product topology τ on X denoted by $\mathbb{V}$. Furthermore $\subset \mathbb{B}(x, r, t)$, since for each $y \in \mathbb{V}$.

$m(x,y,z,t) =$

$$\prod_{n=1}^{\infty} m_n\left(x_n, y_n, z, t\right)$$

$$= \prod_{n=1}^{n_o} m_n\left(x_n, y_n, z, t\right) * \prod_{n=n_o+1}^{\infty} m_n\left(x_n, y_n, z, t\right)$$

$$\geq \prod_{n=1}^{n_o}\left(1-\delta_n\right) * \prod_{n=n_o+1}^{\infty}\left(1-T^{(n)}\right) > 1-r$$

$$n\left(x, y, z, t\right) = \coprod_{n=1}^{\infty} n_n\left(x_n, y_n, z, t\right) = \coprod_{n=1}^{n_o} n_n\left(x_n, y_n, z, t\right) \diamond \coprod_{n=n_o+1}^{\infty} n_n\left(x_n, y_n, z, t\right)$$

$$\mathbb{V} - \prod_{n\mathbb{N}} \mathbb{V}_n, \leq \coprod_{n=1}^{n_o} \delta_n \diamond \coprod_{n=n_o+1}^{\infty} T^{[n]} < r.$$

Hence $\mathbb{V} \subset \mathbb{B}(x, r, t) \subset G$. T$\mathbb{V}$...herefore G is open in the product topology.

Conversely, suppose $\mathbb{G}$ is open in the product topology and let $x = (x_n) \in \mathbb{G}$ Choose a standard basic open set such that $x \in \mathbb{V}$ and $\mathbb{V} \subset \mathbb{G}$. Let $\mathbb{V} =$ where each $\mathbb{V}_n$ is open in X_n and $\mathbb{V}_n = X_n$ for all $n > n_o$. For $n = 1, 2,, n_o$, let $1 - r_n = D_n(x_n, X_n - \mathbb{V}_n, t)$ and $q_{n},, = C_n(x_n, X_n \neq \mathbb{V}_n, t)$, if $X_n \neq \mathbb{V}_n$, and $r_n = \epsilon^{(n)}$ and $q_n = \epsilon^{[n]}$, otherwise. Let $r = \min\{r_1, r_2, \ldots, r_o\}$, $r = \min(q_1, q_2, \ldots q_o)$ and $p = \min\{r, q\}$. We claim that $\mathbb{B}(x, p, t) \subset \mathbb{V}$. If $y = (y_n) \in \mathbb{B}(x - \mathbb{G} - \ldots \mathbb{R}, p, t)$, then $m(x,y,z,t) = \prod_{n=1}^{\infty} m_n(x_n, y_n, z, t) > 1 - p$ and so $m_n(x_n, y_n, z, t) > 1\, p \geq 1 - r \geq 1\, r_n$ and $n(x, y, z, t) = \coprod_{n=1}^{\infty} n_n(x_n, y_n, z, t) < p \leq q \leq q_n$ for each $n = 1, 2,, n_o$. Then $y_n \in \mathbb{V}_n$, for $n = 1, 2, \ldots, n_o$. Also for $n > n_o$, $y_n \in \mathbb{V}_n = X_n$. Hence $y \in$ and so $\mathbb{B}(x, p, t) \subset \mathbb{V} \subset \mathbb{G}$ Therefore $\mathbb{G}$ is open with respect to intuitionistic fuzzy 2-metric topology and τ $\mathcal{U}$. Hence τ and $\mathcal{U}$ coincide.

Theorem 4.1: An open subspace of a complete intuitionistic fuzzy 2-metrizable space is a topologically complete intuitionistic fuzzy 2-metrizable space.

Proof: Let $(X, M, N, *, \diamond)$ be a complete intuitionistic fuzzy 2-metric space and $\mathbb{G}$ an open subspace of X. If the restriction of $(M, N)_2$ to $\mathbb{G}$ is not complete we can replace $(M, N)_2$ on $\mathbb{G}$ by other intuitionistic fuzzy 2-metric as follows. Define f: $\mathbb{G} \times (0, \infty) \rightarrow {}^{+}$ by $f(x, t) = \dfrac{1}{1 - D(x, X - \mathbb{G}, t)}$ (f is undefined if X is empty, but then there is nothing to prove.) Fix an arbitrary $s > 0$ and for $x, y \in \mathbb{G}$ and X, define

$$m(x,y,z,t) = \begin{cases} M(x,y,t) * M(f(x,s), f(y,s), f(z,s), t) \; for\, zG \\ M(x,y,z,t) \; for\, zX - G \end{cases}$$

and

$$n(x, y, z, t) = N(x, y, z, t)$$

for each $t > 0$. We claim that $(m, n)_2$ is an is an intuitionistic fuzzy 2-metric on $\mathbb{G}$. The properties of (a), (b), (c), (d), (f), (g), (h), (i), (j) and (k) are immediate from the definition. For inequality (e), suppose that $w, x, y, z \in \mathbb{G}$ and $t, s, u, v > 0$, then

$m(w, y, z, t) * m(x, w, z, u) * m(x, y, w, v) = (M(w, y, z, t) *$

$M(f(w, s), f(y, s), f(z, s), t) * (M(x, w, z, u) * M(f(x, s), f(w, s), f(z, s), u)) *$

$(M(x, y, w, v) * M(f(x, s), f(y, s), f(w, s), v)$

$= (M(w, y, z, t) * M(x, w, z, u) * M(x, y, w, v) * (M(f(w, s), f(y, s), f(z, s), t) * M(f(x, s), f(w, s), f(z, s), u) * M(f(x, s), f(y, s), f(w, s), v))$

$\leq M(x, y, z, t + u + v) * M(f(x, s), f(y, s), f(z, s), t + u + v) = m(x, y, z, t + u + v)$ for $z \in \mathbb{G}$.

Similarly

$m(w, y, z, t) * m(x, w, z, u) * m(x, y, w, v) \leq m(x, y, z, t + u + v)$ for $z \in X - \mathbb{G}$.

Now we show that $(m, n)_2$ and $(M, N)_2$ are equivalent intuitionistic fuzzy 2-metrics on $\mathbb{G}$. We do this by showing that $m(x_n, x, z, t) \rightarrow 1$ if and only if $M(x_n, x, z, t) \rightarrow 1$ and $n(x_n, x, z, t) \rightarrow 0$ if and only if $N(x_n, x, z, t) \rightarrow 0$ of course the second part is trivial. Since $m(x, y, z, t) \leq M(x, y, z, t)$ for all $x, y \in \mathbb{G}$, $z \in X$ and $t > 0$, $M(x_n, x, z, t) \rightarrow 1$ whenever $m(x_n, x, z, t) \rightarrow 1$ for all $z \in X$. To prove the converse, let $M(x_n, x, z, t) \rightarrow 1$ for all $z \in X$, we know from Proposition of Fradkov and Evans (2005), M is continuous funtion on $X \times X \times X \times (0, \infty)$, then since

$$\lim_n D(x_n, X - \mathbb{G}, s) = \lim_n(\sup\{M(x_n, y, z, s) : y\, \mathbb{G} zX\})$$

$$\geq \lim_n M(x_n, y, z, s) = M(x, y, z, s),$$

we have $\lim_n D(x_n, X - \mathbb{G}, s) \geq \lim_n D(x, X - \mathbb{G}, s)$. On the other hand, there exist a $y_o \in X - \mathbb{G}$ and $n_o \in \mathbb{N}$ such that for every $n \geq n_o$ and $z\, X$ we have

$$D(x_n, X - \mathbb{G}, s) * (1 - \frac{1}{n}) \leq M(x_n, y_o, z, s).$$

Then $\lim_n D(x_n, X - \mathbb{G}, s) \leq M(x, y_o, z, s) \leq \sup\{M(x, y, z, s) : y \in X - \mathbb{G}, z \in X\} = D(x, X - \mathbb{G}, s)$. Therefore $\lim_n D(x_n, X - \mathbb{G}, s) = \lim_n D(x, X - \mathbb{G}, s)$. This implies $M(f(x_n, s), f(x, s), f(z, s), t) \rightarrow 1$. Hence $m(x_n, x, z, t) \rightarrow 1$ for all $z \in X$. Therefore $(m, n)_2$ and $(M, N)_2$ are equivalent. Next we show that $(m, n)_2$ is a complete intuitionistic fuzzy 2-metric. Suppose that (x_n) is a Cauchy sequence in $\mathbb{G}$ with respect to $(m,n)_2$. Since, for each m,n

$\in \mathbb{N}$ and $t > 0$, $m(x_m, x_n, z, t) \leq M(x_m, x_n, z, t)$ and $n(x, y, z, t) = N(x, y, z, t)$ for all $z \in X$, the sequence (x_n) is also a Cauchy sequence with respect to $(M,N)_2$. By the completeness of $(X, M, N, *, \diamond)$, (x_n) converges to point p in X. We claim that $p \in \mathbb{G}$. Assume otherwise, then for each $n \in \mathbb{N}$, if $p \in X - \mathbb{G}$ and $M(x_n, p, z, t) \leq D(x_n, X - \mathbb{G}, t)$ for all $z \in X$, then

$$1 - M(x_n, p, z, t) \geq 1 - D(x_n, X - \mathbb{G}, t) > 0.$$

Therefore

$$\frac{1}{1 - D(x_n, X - \mathbb{G}, t)} \geq \frac{1}{1 - M(x_n, p, z, t)},$$

that is

$$f(x_n, t) \geq \frac{1}{1 - M(x_n, p, z, t)},$$

for each $t > 0$ for all $z \in X$. Therefore as $n \to \infty$, for every $t > 0$ we get $f(x_n, s) \to \infty$. In particular, $f(x_n, s) \to \infty$. On the other hand, $M(f(x_n, s), f(x_m, s), f(z, s), t) \geq m(x_m, x_n, z, t)$ for every $m, n \in \mathbb{N}$ and for all $z \in X$, that is $(f(x_n, s))$ is an F-bounded sequence (El Naschie, 1998). This contradiction shows that $p \in \mathbb{G}$. Hence (x_n) converges to p with respect to $(m, n)_2$ and thus $(G, m, n, *, \diamond)$ is a complete intuitionistic fuzzy 2-metrizable space.

Theorem 4.2: Let $(Y, M, N, *, \diamond)$ be an intuitionistic fuzzy 2-metric space and X be a topologically complete intuitionistic fuzzy 2-metrizable subspace of Y. Then X is a G_δ subset of Y.

Proof: Let $(X, M', N', *, \diamond)$ be an intuitionistic fuzzy 2-metric space that induces the same topology for X as does $(M, N)_2$. For each $x \in X$ and each $n \in \mathbb{N}$, , let $r_N(x)$ be a positive real number such tha $\mathcal{G}$ $\frac{1}{n}$ N $\frac{1}{n}$t $r_N(x) <$ and $M'(w, x, z, t) > 1 - \frac{1}{n}$ and $N'(w, x, z, t) < \frac{1}{n}$, whenever $w \in X$ and for all $z \in X$ and $M'(w, x, z, t) > 1 - r_N(x)$ and $N'(w, x, z, t) < r_N(x)$ for each $t > 0$ and for all $z \in X$. Suppose that $_n$ = $\bigcup_n (\mathbb{B}_{(M,N)_2}(x, r_n(x), t) : x \in X, t > 0)$ for each n $\in \mathbb{N}$, and $\Gamma = \bigcup_n (\mathcal{G}_n : n \in \mathbb{N})$. Then Γ is a G_δ subset of Y which clearly contains X. It is enough to shown that $\Gamma \subseteq X$. Let $x_o \in \Gamma$. Then $x_o \in \mathcal{G}_n$ for each $n \in \mathbb{N}$. Hence for each $n \in \mathbb{N}$, there is $x_n \in X$ such that $x_o \in \mathbb{B}_{(M,N)_2}(x, r_n(x), t)$. Therefore $M(x_n, x_o, z, t) > 1 - r_N(x) > 1 - \frac{1}{n}$ and $N(x_n, x_o, z, t) < r_N(x) <$ for each $n \in$ and $t > 0$ and for all $z \in X$. This means that $x_n \overset{(M,N)_2}{\rightarrow} x_o$ in Y.

Now, let $0 < \epsilon < 1$ and $N \in \mathbb{N}$ such that $(1 - \frac{1}{N}) * (1 - \frac{1}{N}) * (1 - \frac{1}{N}) > 1 - \epsilon$ and $(\frac{1}{N})(\frac{1}{N})\Diamond(\frac{1}{N}) < \epsilon$. Let $m \in \mathbb{N}$ be such that

$$(1 - \frac{1}{m}) * M(x_o, x_N, z, t) > 1 - r_N(x_N) \text{ for all } z \in X.$$

Therefore

$$(1 - \frac{1}{m}) * M(x_o, x_N, x_k, t) > 1 - r_N(x_N) \text{ for all } x_k \in X.$$

so

$$(1 - \frac{1}{m}) * M(x_o, x_N, x_k, t) * M(x_o, x_N, z, t) > 1 - r_N(x_N) * M(x_o, x_N, z, t) \text{ for all } x_k \in X.$$

Hence

$$(1-\frac{1}{m}) * M(x_o, x_N, x_k, t) * M(x_o, x_N, z, t) > 1 - r_N(x_N)$$

and

$$\frac{1}{m} \diamond N(x_o, x_N, z, t) < r_N(x_N) \text{ for all } z \in X.$$

so

$$\frac{1}{m} \diamond N(x_o, x_N, x_k, t) \diamond N(x_o, x_N, z, t) < r_N(x_N) \diamond N(x_o, x_N, z, t) \text{ for all } x_k \in X.$$

hence

$$\frac{1}{m} \diamond N(x_o, x_N, x_k, t) \Diamond N(x_o, x_N, z, t) < r_N(x_N)$$

Now for every $k \in \mathbb{N}$ and $k > m$ we have

$$M(x_k, x_N, z, 3t) \geq M(x_k, x_o, z, t) * M(x_o, x_N, x_k, t) * M(x_o, x_N, z, t)$$

$$> (1-\frac{1}{k}) * M(x_o, x_N, x_k, t) * M(x_o, x_N, z, t)$$

$$\geq (1-\frac{1}{m}) * M(x_o, x_N, x_k, t) * M(x_o, x_N, z, t) > 1 - r_N(x_N)$$

and

$$N(x_k, x_N, z, 3t) \leq N(x_k, x_o, z, t) \diamond N(x_o, x_N, x_k, t) \diamond N(x_o, x_N, z, t)$$

$$< \frac{1}{k} \diamond N(x_o, x_N, x_k, t) \diamond N(x_o, x_N, z, t)$$

$$\leq \frac{1}{m} \diamond N(x_o, x_N, x_k, t) \diamond N(x_o, x_N, z, t) < r_N(x_N).$$

Therefore $M'(x_k, x_N, z, 3t) > (1-\frac{1}{N})$ and $N'(x_k, x_N, z, 3t) < \frac{1}{N}$ for all $z \in X$. If $k, l > m$, then

$$M'(x_k, x_p, z, 9t) \geq M'(x_k, x_N, z, 3t) * M'(x_N, x_p, z, 3t) * M'(x_k, x_p, x_N, 3t)$$

$$> > \left(1-\frac{1}{N}\right) * \left(1-\frac{1}{N}\right) * \left(1-\frac{1}{N}\right) > 1 - T$$

and

$$N'(x_k, x_p, z, 9t) \leq N'(x_k, x_N, z, 3t) \diamond N'(x_N, x_p, z, 3t) \diamond N'(x_k, x_p, x_N, 3t)$$

$$< \frac{1}{N} \Diamond \frac{1}{N} \frac{1}{N} < T.$$

Hence the sequence (x_n) is Cauchy in the complete intuitionistic fuzzy 2-metric space $(X, M', N', *, \diamond)$ and then it converges to some element of X. Since $x_n \xrightarrow{(M,N)_2} x_o$ in Y, it follows that $x_o \in X$ and thus $\Gamma \in X$.

ACKNOWLEDGMENT

Research of the author was supported by the Department of Atomic Energy, Government of India under the NBHM-Post Doctoral Fellowship Programme Number 2/40(23)/2009-R&D-II/5152.

REFERENCES

Abbas, S. E. (2005). On intuitionistic fuzzy compactness. *Information Sciences*, *173*, 75–91. doi:10.1016/j.ins.2004.07.004

Atanassov, K. (1986). Intuitionistic fuzzy sets. *Fuzzy Sets and Systems*, *20*, 87–96. doi:10.1016/S0165-0114(86)80034-3

Atanassov, K. (1994). New operations defined over the intuitionistic fuzzy sets. *Fuzzy Sets and Systems*, *61*, 137–142. doi:10.1016/0165-0114(94)90229-1

Barro, S., & Marin, R. (2002). *Fuzzy logic in medicine*. Heidelbrg, Germany: Physica-Verlag.

Barros, L. C., Bassanezi, R. C., & Tonelli, P. A. (2000). Fuzzy modelling in population dynamics. *Ecological Modelling*, *128*, 27–33. doi:10.1016/S0304-3800(99)00223-9

Coker, D. (1997). An introduction to intuitionistic fuzzy topological spaces. *Fuzzy Sets and Systems*, *88*, 81–89. doi:10.1016/S0165-0114(96)00076-0

El Naschie, M. S. (1998). On uncertainty of Cantorian geometry and two-slit experiment. *Chaos, Solitons, and Fractals*, *9*, 517–529. doi:10.1016/S0960-0779(97)00150-1

El Naschie, M. S. (2000). On the unification of heterotic strings theory and $E^{(\infty)}$ theory. *Chaos, Solitons, and Fractals*, *11*(14), 2397–2408. doi:10.1016/S0960-0779(00)00108-9

El Naschie, M. S. (2004a). A review of E-irifinity theory and the mass spectrum of high energy particle physics. *Chaos, Solitons, and Fractals*, *19*, 209–236. doi:10.1016/S0960-0779(03)00278-9

El Naschie, M. S. (2004b). Quantum gravity, Clifford algebras, fuzzy set theory and the fundamental constants of nature. *Chaos, Solitons, and Fractals*, *20*, 437–450. doi:10.1016/j.chaos.2003.09.029

El Naschie, M. S. (2006a). Fuzzy dodecahedron topology and *E*-infinity space time as a model for quantum physics. *Chaos, Solitons, and Fractals*, *30*, 1025–1033. doi:10.1016/j.chaos.2006.05.088

El Naschie, M. S. (2006b). On two new fuzzy Kahler manifolds, Klein modular space and Hooft holographic principles. *Chaos, Solitons, and Fractals*, *29*, 876–881. doi:10.1016/j.chaos.2005.12.027

El Naschie, M. S. (2007). A review of applications and results of *E*-infinity theory. *International Journal of Nonlinear Sciences and Numerical Simulation*, *8*, 11–20. doi:10.1515/IJNSNS.2007.8.1.11

Erceg, M. A. (1979). Metric spaces in fuzzy set theory. *Journal of Mathematical Analysis and Applications*, *69*, 205–230. doi:10.1016/0022-247X(79)90189-6

Fang, J. X. (2002). A note on the completions of fuzzy metric spaces and fuzzy normed spaces. *Fuzzy Sets and Systems*, *131*, 399–407. doi:10.1016/S0165-0114(02)00054-4

Felbin, C. (1992). Finite dimensional fuzzy normed linear space. *Fuzzy Sets and Systems*, *48*, 239–248. doi:10.1016/0165-0114(92)90338-5

Fradkov, A. L., & Evans, R. J. (2005). Control of chaos: Methods and applications in engineering. *Chaos, Solitons, and Fractals*, *29*, 33–56.

Gähler, S. (1963). 2-metrische Räume und ihre topologische Struktur. *Mathematische Nachrichten*, *26*, 115–148. doi:10.1002/mana.19630260109

Gähler, S. (1965). Lineare 2-normietre Räume. *Mathematische Nachrichten*, *28*, 1–43. doi:10.1002/mana.19640280102

George, A., & Veeramani, P. (1994). On some results in fuzzy metric spaces. *Fuzzy Sets and Systems*, *64*, 395–399. doi:10.1016/0165-0114(94)90162-7

Giles, R. (1980). A computer program for fuzzy reasoning. *Fuzzy Sets and Systems*, *4*, 221–234. doi:10.1016/0165-0114(80)90012-3

Hong, L., & Sun, J. Q. (2006). Bifurcations of fuzzy nonlinear dynamical systems. *Communications in Nonlinear Science and Numerical Simulation*, *1*, 1–12. doi:10.1016/j.cnsns.2004.11.001

Kaleva, O., & Seikkala, S. (1984). On fuzzy metric spaces. *Fuzzy Sets and Systems*, *12*, 215–229. doi:10.1016/0165-0114(84)90069-1

Lupianez, F. G. (2006). Nets and filters in intuitionistic fuzzy topological spaces. *Information Sciences*, *176*, 2396–2404. doi:10.1016/j.ins.2005.05.003

Mursaleen, M., & Lohani, Q. M. D. (2009). Intuitionistic fuzzy 2-normed space and some related concepts. *Chaos, Solitons, and Fractals*, *42*, 224–234. doi:10.1016/j.chaos.2008.11.006

Mursaleen, M., Lohani, Q. M. D., & Mohiuddine, S. A. (2009). Intuitionistic fuzzy 2-metric space and its completion. *Chaos, Solitons, and Fractals*, *42*, 1258–1265. doi:10.1016/j.chaos.2009.03.025

Park, J. H. (2004). Intuitionistic fuzzy metric spaces. *Chaos, Solitons, and Fractals*, *22*, 1039–1046. doi:10.1016/j.chaos.2004.02.051

Saadati, R., & Park, J. H. (2006). On the intuitionistic fuzzy topological spaces. *Chaos, Solitons, and Fractals*, *27*, 331–344. doi:10.1016/j.chaos.2005.03.019

Schweizer, B., & Sklar, A. (1960). Statistical metric spaces. *Pacific Journal of Mathematics*, *10*, 313–334.

Xiao, J., & Zhu, X. (2002). On linearly topological structure and property of fuzzy normed linear space. *Fuzzy Sets and Systems*, *121*, 153–161. doi:10.1016/S0165-0114(00)00136-6

Zadeh, L. A. (1965). Fuzzy sets. *Information and Control*, *8*, 338–353. doi:10.1016/S0019-9958(65)90241-X

This work was previously published in the International Journal of Artificial Life Research, Volume 2, Issue 3, edited by E. Stanley Lee and Ping-Teng Chang, pp. 59-73, copyright 2011 by IGI Publishing (an imprint of IGI Global).

Chapter 21
Folding Theory for Fantastic Filters in BL-Algebras

Celestin Lele
University of Dschang, Cameroon

ABSTRACT

In this paper, the author examines the notion of n-fold fantastic and fuzzy n-fold fantastic filters in BL-algebras. Several characterizations of fuzzy n-fold fantastic filters are given. The author shows that every n-fold (fuzzy n-fold) fantastic filter is a filter (fuzzy filter), but the converse is not true. Using a level set of a fuzzy set in a BL-algebra, the author gives a characterization of fuzzy n-fold fantastic filters. Finally, the author establishes the extension property for n-fold and fuzzy n-fold fantastic filters in BL-algebras. The author also constructs some algorithms for folding theory applied to fantastic filters in BL-algebras.

1. INTRODUCTION

Basic logic algebras (BL-algebras for short) introduced by Hájek (1998b) are algebras of Logic BL, their theory is developed in the style of related algebras and logic. The main example of BL-algebras is the unit interval $[0,1]$ endowed with the structure induced by a continuous t-norm. A great deal of literature has been produced on the theory of BCI/BCK/MV/BL-algebras, in particular, emphasis seems to have been put on the ideals and filters theory. From the logical point of view, various ideals and filters correspond to various sets of provable formulas. Zadeh (1965) introduced the notion of fuzzy sets. At present, this concept has been applied to many mathematical branches such as group theory, functional analysis, probability theory, topology and so on. In Lele and Moutari (2008) we have studied the notion of n-fold and fuzzy n-fold various ideals and established many important properties. All the interesting results and the concluding remarks of Jun and Miko (2004) have motivated us to further investigate the foldness of other types of filters in BL-algebras. We find useful to start with the study of foldedness theory of fantastic filters (also called fantastic deductive systems). Thanks to the concept of fuzzy set, we give several characterizations of n-fold

DOI: 10.4018/978-1-4666-3890-7.ch021

and fuzzy n-fold fantastic filters in BL-algebras. Finally, we give the extension property for n-fold and fuzzy n-fold fantastic filters in BL-algebras. Afterwards, we construct some algorithms to determine whether certain finite structures are BL-algebras, n-fold fantastic filters and fuzzy n-fold fantastics filters. All the above results are the natural generalization of the notion of filters and fuzzy filters (namely deductive and fuzzy deductive systems) in BL-algebras (Lele, 2010, in press; Liu & Li, 2005a, 2005b; Motamed & Saied, 2001; Turunen, Tchikapa, & Lele, in press). It is our hope that this work would serve as a foundation of further study of the theory of some types of (fuzzy) filters in BL-algebra.

2. PRELIMINARIES

A BL-algebra is a structure $(X,\wedge,\vee,*,\rightarrow,0,1)$ in which X is a non-empty set with four binary operations $\wedge,\vee,*,\rightarrow$ and two constants 0 and 1 satisfying the following axioms:

BL-1: $(X,\wedge,\vee,0,1)$ is a bounded lattice;
BL-2: $(X,*,1)$ is an abelian monoid; which means that $\ast $ is commutative and associative with $x*1=x$;
BL-3: $x*y\leq z$ iff $x\leq y\rightarrow z$ (residuation);
BL-4: $x\wedge y=x*(x\rightarrow y)$ (divisibility);
BL-5: $(x\rightarrow y)\vee(y\rightarrow x)=1$ (prelinearity);

A BL-algebra X is called a MV-algebra if $\neg(\neg x)=x$, or equivalently $(x\rightarrow y)\rightarrow y=(y\rightarrow x)\rightarrow x$ where $\neg x=x\rightarrow 0$.

Definition 2.1. *A t-norm is a binary operation* T over $[0,1]$., that is commutative, associative, monotone, and has 1 as an identity element. T is a continuous t-norm if it is a t-norm and is a continuous mapping of $[0,1]^2$ into $[0,1]

Example 2.1. The following are important examples of continuous t-norm:

1. **Lukasiewicz T-Norm:** T(x; y) = max(0; x + y - 1).
2. **Godel T-Norm:** T(x; y) = min(x; y).
3. **Product T-Norm:** T(x; y) = x:y

Note that the dual notion of t-norm is a t-conorm: A t-conorm is a binary operation *T* over [0; 1], that is commutative, associative, monotone, and has 0 as an identity element.

Lemma 2.1. (Hájek, 1998b) *Let* T be a continuous t-norm. Then there is a unique operation $x\rightarrow y$ satisfying, for all $x,y,z\in[0,1]$, *the condition* $T(x,z)\leqslant y$ iff $z\leqslant(x\rightarrow y)$, *namely* $x\rightarrow y=\max\{z\,/\,T(x,z)\leqslant y\}$.

Definition 2.2. *The operation* $x\rightarrow y$ from Lemma 2.1 is called the residuum of the t-norm.

The following operations are residual of the three t-norm:

1. **Lukasiewicz Implication:**
$$x\rightarrow y=\begin{cases}1 & ,\ if\ \ x\leqslant y\\ \min(1-x+y;1) & ,\ otherwise.\end{cases}$$
2. **Gödel Implication:**
$$x\rightarrow y=\begin{cases}1 & ,if\ \ x\leqslant y\\ y & ,\ otherwise.\end{cases}$$
3. **Product Implication:**
$$x\rightarrow y=\begin{cases}1 & ,if\ \ x\leqslant y\\ y\,/\,x & ,\ otherwise.\end{cases}$$

The main example of BL-algebras is the unit interval [0; 1] endowed with the structure induced by a continuous t-norm. The following properties also hold in any BL-algebra ((Agliano & Montagna, 2003) (Cignoli & Torrens, 2005) (Dumitru & Dana, 2003) (Haveshki, Saied, & Eslami, 2006) (Hájek, 1998b) (Jun, & Miko, 2004) (Turunen, 1999a) (Turunen, 1999b) (Turunen & Sessa, 2001):

1. $x \leq y$ *iff* $x \to y = 1$;
2. $x \to (y \to z) = (x * y) \to z$;
3. $x * y \leq x \wedge y$;
4. $(x \to y) * (y \to z) \leq x \to z$;
5. $x \vee y = ((x \to y) \to y) \wedge ((y \to x) \to x)$;
6. $x \to (y \to z) = y \to (x \to z)$;
7. $x \to y \leq (y \to z) \to (x \to z)$;
8. $y \to x \leq (z \to y) \to (z \to x)$;
9. If $x \leq y$, then $y \to z \leq x \to z$, and $z \to x \leq z \to y$;
10. $y \leq (y \to x) \to x$;
11. $x \leq y \to (x * y)$;
12. $x * (x \to y) \leq y$;
13. $1 \to x = x$; $x \to x = 1$; $x \to 1 = 1$; $x \leq y \to x$.

We briefly review some fuzzy logic concepts (refer to Jun, Shim, & Lele, 2002; Lele, 2009; Liu, Liu, & Xu, 2006; Liu & Li, 2005a; Zadeh, 1965 for more details).

Definition 2.3. *Let* X be a BL-algebra. A fuzzy subset of X is a function $A : X \leqslant [0,1]$.

We denoted $((x \to (x \to (x \to y))))$ by $x^n \to y$ where x occurs n times for all $x, y \in X$. *Using the fact that in any BL-algebra* $x \to (y \to z) = (x * y) \to z$, *we can prove by induction that* $((x \to (x \to (x \to y))))$ = $(*x * x * x) \to y = x^n \to y$ where x occurs n times for all $x, y \in X$.

Definition 2.4.

- *A filter of a BL-algebra* X is a subset F containing 1 such that

if $x \to y \in F$ and $x \in F$ imply $y \in F$.

- *A fuzzy subset* A of X is a fuzzy filter if

$A(1) \geq A(x)$ and $A(y) \geq \min(A(x \to y), A(x))$, $\forall\, x, y \in X$.

The following theorem gives some characterizations of fuzzy filter.

Theorem 2.1. *Suppose that* A is a fuzzy subset of a BL-algebra X, *then the following conditions are equivalent (Liu & Li, 2005b):*

1. A is a fuzzy filter;
2. $\forall\, t \in [0,1]$, *the t-level subset* $A_t = \{x \in X : A(x) \geq t\}$ is a filter of X if $A_t \neq \varnothing$.

The following properties also hold if A is a fuzzy filter for any BL-algebra X (Liu & Li, 2005b):

1. If $x \leq y$ then $A(y) \geq A(x)$, that is A is order-preserving;
2. If $A(x \to y) = A(1)$, then $A(y) \geq A(x)$;
3. 3. $A(x * y) = A(x) \wedge A(y)$;
4. 4. $A(x \wedge y) = A(x) \wedge A(y)$;
5. 5. $A(0) = A(x) \wedge A(\neg x)$;
6. 6. $A(x \to z) \geq A(x \to y) \wedge A(y \to z)$;
7. 7. $A(x \to y) \leq A((y \to z) \to (x \to z))$;
8. 8. $A(y \to x) \leq A((z \to y) \to (z \to x))$.

3. N-FOLD FANTASTIC FILTER

Definition 3.1. *Let* F be a subset of a BL-algebra X, *if* $1 \in F$ and

$z \to (y \to x) \in F$ and $z \in F$ imply $((x^n \to y) \to y) \to x \in F$,

then F is said to be an n-fold fantastic filter. An 1-fold fantastic filter is called a fantastic filter (Haveshki, Saied, & Eslami, 2006; Kondo & Dudek, 2008).

The following proposition is a characterization of n-fold fantastic filter.

Proposition 3.1. *A filter* F is n-fold fantastic if and only if $y \to x \in F$ implies $((x^n \to y) \to y) \to x \in F$.

We describe the relation between filter and n-fold fantastic filter.

Proposition 3.2. *Every n-fold fantastic filter is a filter.*

The following example shows that a filter may not be an n-fold fantastic filter.

Example 3.1. *Let* $X = \{0, x, y, 1\}$ be a chain with $*$ and $\to$ defined by *Tables 1 and 2.*

It is easy to check that $(X, \wedge, \vee, *, \to, 0, 1)$ is a BL-algebra.

$x \to y = 1 \in \{1\}$, *but* $((y^n \to x) \to x) \to y = y \notin \{1\}$. *So* $\{1\}$ is a filter, but not an n-fold fantastic filter for any $n \geq 1$.

4. FUZZY N-FOLD FANTASTIC FILTERS

Definition 4.1. *A fuzzy subset A of X is a fuzzy n-fold fantastic filter iff* $A(1) \geq A(x)$ and

$$A(((x^n \to y) \to y) \to x) \geq \min(A(z \to (y \to x)), A(z)) \quad \forall\, x, y, z \in X.$$

A fuzzy 1-fold fantastic filter is called a fuzzy fantastic filter (Motamed & Saied, 2011; Turunen, Tchikapa, & Lele, in press).

The following result is a characterization of fuzzy n-fold fantastic filter.

Proposition 4.1. *A fuzzy filter A is fuzzy n-fold fantastic if and only if*

$$A(((x^n \to y) \to y) \leqslant x) \geq A(y \to x) \quad \text{for all} \quad x, y \in X.$$

Table 1. Let $X = \{0, x, y, 1\}$ *be a chain with* $*$

$*$	0	x	y	1
0	0	0	0	0
x	0	0	x	x
y	0	x	y	y
1	0	x	y	1

Table 2. Let $X = \{0, x, y, 1\}$ *be a chain with* $\to$

$\to$	0	x	y	1
0	1	1	1	1
x	x	1	1	1
y	0	x	1	1
1	0	x	y	1

Example 4.1. *Let* $X = \{0, x, y, 1\}$ be a chain with $*$ and $\to$ defined by *Tables 3 and 4.*

With the algorithms at the end of the paper, one can check that $(X, \wedge, \vee, *, \to, 0, 1)$ is an n-fold fantastic BL-algebra for any $n \geq 1$.

Let $t_1, t_2 \in [0,1]$ with $t_1 > t_2$ and A a fuzzy subset on X defined by $t_1 = A(1), t_2 = A(0) = A(x) = A(y)$. It is clear that A is a fuzzy n-fold fantastic filter for any $n \geq 1$.

We analyse the relation between fuzzy filter and fuzzy n-fold fantastic filter.

Proposition 4.2. *Every fuzzy n-fold fantastic filter is a fuzzy filter.*

The next example proves that the converse of the proposition is not always true.

Example 4.2. *Let* $X = \{0, x, y, 1\}$ with $*$ and $\to$ defined as in example 4.1.

Table 3. Let $X = \{0, x, y, 1\}$ be a chain with $*$

$*$	0	x	y	1
0	0	0	0	0
x	0	0	0	x
y	0	0	x	y
1	0	x	y	1

Table 4. Let $X = \{0, x, y, 1\}$ be a chain with $\rightarrow$

$\rightarrow$	0	x	y	1
0	1	1	1	1
x	y	1	1	1
y	x	y	1	1
1	0	x	y	1

Let $t_1, t_2 \in [0,1]$ with $t_1 > t_2$ and A a fuzzy subset on X defined by
$t_1 = A(1), t_2 = A(0) = A(x) = A(y)$. It is easy to see that A is a fuzzy filter, but not a fuzzy n-fold fantastic filter since
$A(((y^n \rightarrow x) \rightarrow x) \leqslant y) < A(x \rightarrow y)$.

Now, we describe the transfer principle (Kondo & Dudek, 2005) for fuzzy n-fold fantastic filter in terms of level subsets as:

Theorem 4.1. *A fuzzy subset A of a BL-algebra X is a fuzzy n-fold fantastic filter if and only if* $A^t = \{x \in X \,/\, A(x) \geq t\}$ is either empty or an n-fold fantastic filter for every $t \in [0,1]$.

Corollary 4.1. *A non empty subset F of X is an n-fold fantastic filter if and only if the characteristic function* χ_F is a fuzzy n-fold fantastic filter.

Now we make a link between fuzzy n-fold fantastic filter and fuzzy (n+1)-fold fantastic filter.

Proposition 4.3. *Every fuzzy n-fold fantastic filter is a fuzzy (n+1)-fold fantastic filter.*

By finite induction, we can prove that every fuzzy n-fold fantastic filter is a fuzzy (n+k)-fold fantastic filter for any interger $k \geq 0$.

Corollary 4.2. *Every n-fold fantastic filter is an (n+k)-fold fantastic filter for any interger* $k \geq 0$.

We close this section with the extension property for n-fold (fuzzy n-fold) fantastic filters.

Proposition 4.4. *Let F and G be two filters of X such that* $F \subseteq G$. If F is n-fold fantastic, then G is also n-fold fantastic.

Theorem 4.2. *(Extention theorem of fuzzy n-fold fantastic filter) Suppose that $A $ and $B $ are two fuzzy filters such that* $A \subseteq B$ and $A(1) = B(1)$. If A is fuzzy n-fold fantastic, then B is also fuzzy n-fold fantastic. .

Proof. We should notice that it cannot be the case that A^t is empty and B^t not since A(1) = B(1): To prove that B is fuzzy n-fold fantastic _lter, it su_ces to show that for any t ∈ [0; 1], the t-level subset $B^t = \{x \in X$: B(x) ≥ t}is an n-fold fantastic filter of *X* when $B^t \neq \varnothing$. Because for any t ∈ [0; 1]: $A^t \subseteq B^t$; we apply the extension theorem of n-fold fantastic filter and obtain the result.

5. CONCLUSION

We have initiated the study of n-fold and fuzzy n-fold fantastic filters in BL-algebras. We have also presented several different characterizations and many important properties of n-fold and fuzzy n-fold fantastic filters in BL-algebras. Our future work will be the investigation of foldedness of other types of filters and fuzzy filters in BL-algebras as well as their relation diagram (Tchikapa & Lele, in press; Tchikapa, Lele, & Turunen, in press). Please see Algorithms in Table 5 of the Appendix

REFERENCES

Agliano, P., & Montagna, F. (2003). Varieties of BL-algebras I: General properties. *Journal of Pure and Applied Algebra*, *181*, 105–129. doi:10.1016/S0022-4049(02)00329-8

Bu§neag, D., &, Piciu, D. (2003). On the lattice of deductive systems of BL-algebras. *Central European Journal of Mathematics*, *1*(2), 221-237.

Cignoli, R., & Torrens, A. (2005). Standard completeness of Hájek basic logic and decompositions of BL-chains. *Soft Computing*, *12*, 862–868. doi:10.1007/s00500-004-0444-x

Hájek, P. (1998a). Basic fuzzy logic and BL-algebras. *Soft Computing*, *2*, 124–128. doi:10.1007/s005000050043

Hájek, P. (1998b). *Metamathematics of fuzzy logic*. Dordrecht, The Netherlands: Kluwer Academic. doi:10.1007/978-94-011-5300-3

Haveshki, M., Saied, A., & Eslami, E. (2006). Some types of filters in BL-algebras. *Soft Computing*, *10*, 657–664. doi:10.1007/s00500-005-0534-4

Jun, Y. B., & Miko, J. (2004). Folding theory applied to BL-algebras. *Central European Journal of Mathematics*, *4*, 584–592. doi:10.2478/BF02475965

Jun, Y. B., Shim, W. H., & Lele, C. (2002). Fuzzy filters /ideals in BCI-algebras. *Journal of Fuzzy Mathematics*, *10*, 469–474.

Kondo, M., & Dudek, W. A. (2005). On the transfer principle in fuzzy theory. *Mathware and Soft Computing*, *12*, 41–55.

Kondo, M., & Dudek, W. A. (2008). Filters theory of BL-algebras. *Soft Computing*, 419–423. doi:10.1007/s00500-007-0178-7

Lele, C. (2009). Folding theory of positive implicative/fuzzy positive implicative filters in BL-algebras. *Journal of Fuzzy Mathematics*, *17*(2), 633–641.

Lele, C. (2010). Algorithms and computations in BL-algebras. *International Journal of Artificial Life Research*, *1*(4), 29–47. doi:10.4018/jalr.2010100103

Lele, C. (in press). Fuzzy n-fold Obstinate filters in BL-algebras. *Afrika Mathematika*.

Lele, C., & Moutari, S. (2008). On some computational methods for study of foldness of H-ideals in BCI-algebras. *Soft Computing*, *12*, 403–407. doi:10.1007/s00500-007-0176-9

Liu, L., & Li, K. (2005a). Fuzzy Boolean and positive implicative filters of BL-algebras. *Fuzzy Sets and Systems*, *152*, 333–348. doi:10.1016/j.fss.2004.10.005

Liu, L., & Li, K. (2005b). Fuzzy filters of BL-algebras. *Information Sciences*, *173*, 141–154. doi:10.1016/j.ins.2004.07.009

Liu, Y. L., Liu, S. Y., & Xu, Y. (2006). An answer to the Jun-Shim-Lele's open problem on the fuzzy filter. *Journal of Applied Mathematics and Computing*, *21*, 325–329. doi:10.1007/BF02896410

Motamed, S., & Saied, A. (2011). n-fold obstinate filters in BL-algebras. *Neural Computing & Applications*, *20*(4), 461–472. doi:10.1007/s00521-011-0548-z

Tchikapa, N., & Lele, C. (in press). Relation diagram between fuzzy n-fold filters in BL-algebras. *Annals of Fuzzy Mathematics and Informatics*.

Tchikapa, N., Lele, C., & Turunen, E. (in press). *n-fold deductive systems in BL-algebras*.

Turunen, E. (1999b). *Mathematics behind fuzzy logic*. Heidelberg, Germany: Physica-Verlag.

Turunen, E. (1999q). BL-algebras and basic fuzzy logic. *Mathware and Soft Computing*, *6*, 49–61.

Turunen, E., & Sessa, S. (2001). Local BL-algebras. *Multi-Valued Logic*, *6*, 229–249.

Turunen, E., Tchikapa, N., & Lele, C. (in press). n-fold implicative logic is Gödel logic. *Soft Computing*.

Zadeh, L. A. (1965). Fuzzy sets. *Information and Control*, *8*, 338–353. doi:10.1016/S0019-9958(65)90241-X

APPENDIX

Table 5. Algorithms

Algorithm for *BL-algebras*
Input(X : set; $\wedge, \vee, *, \rightarrow$: binary operations; \$0, 1\$: constants)
Output(" X is a *BL-algebra* or not")
Begin
If $X = \varnothing$ then
go to (1.);
EndIf
If $(X, \wedge, \vee, 0, 1)$ is not a bounded lattice then
go to (1.);
EndIf
If $(X, *, 1)$ is not an abelian monoid then
go to (1.);
EndIf
Stop:=false;
$i := 1$;
While $i \leq \mid X \mid$ and not(Stop) do
$j := 1$
While $j \leq \mid X \mid$ and not(Stop) do
If $x_i \wedge y_j \neq x_i * (x_i \rightarrow y_j)$ then
Stop:=true;
EndIf
If $(x_i \rightarrow y_j) \vee (y_j \rightarrow x_i) \neq 1$ then
Stop:=true;
EndIf
$k := 1$;

continued on following page

Table 5. Continued

While $k \leq \mid X \mid$ and not($Stop$) do
If $(x_i * y_j \leq z_k)$ and $x_i > (y_j \rightarrow z_k)$ then
Stop:=true;
EndIf
EndWhile
EndWhile
EndWhile
If Stop then
(1.) Output("X is not a *BL-algebra*")
Else
Output("X is a *BL-algebra*")
EndIf
End
Algorithm for n-fold fantastic filters
Input(X: BL - algebra,$ F$: subset of X, $n \in \mathbb{N}$);
Output("F is an n-fold fantastic filter or not");
Begin
If $F = \varnothing$ then
go to (1.);
EndIf
If $1 \notin F$ then
go to (1.);
EndIf
Stop:=false;
$i := 1$;
While $i \leq \mid X \mid$ and not(Stop) do
$j:=1$
While $j \leq \mid X \mid$ and not(Stop) do
$k:=1$
While $k \leq \mid X \mid$ and not(Stop) do
If $(z_k \rightarrow (y_j \rightarrow x_i) \in F)$ and $(z_k \in F)$ then

continued on following page

Table 5. Continued

If $((x_i^n \rightarrow y_j) \rightarrow y_j) \rightarrow x_i \notin F$ then
Stop:=true;
EndIf
EndIf
EndWhile
EndWhile
EndWhile
If Stop then
Output(" F is an n -fold fantastic filter")
Else
(1.) Output(" F is not an n -fold fantastic filter")
EndIf
End
Algorithm for fuzzy subsets
Input($X : BL-algebra, \mathrm{A} : \mathrm{X} \leqslant [0,1]$);
Output(" A is a fuzzy subset of X or not");
Begin
Stop:=false;
$i := 1$;
While $i \leq \mid X \mid$ and not(Stop) do
If ($A(x_i) < 0 \mathrm{n} A(x_i) > 1$) then
Stop:=true;
EndIf
EndWhile
If Stop then
Output(" A is a fuzzy subset of X ")
Else
Output(" A is not a fuzzy subset of X ")
EndIf
End
Algorithm for fuzzy n -fold fantastic filters
Input(X : *BL-algebra*, A : fuzzy subset of X , $n \in \mathbb{N}$);

continued on following page

Table 5. Continued

Output(" A is a fuzzy n -fold fantastic filter or not");
Begin
Stop:=false;
$i := 1$;
While $i \leq \mid X \mid$ and not(Stop) do
If $A(1) < A(x_i)$ then
Stop:=true;
EndIf
$j := 1$
While $j \leq \mid X \mid$ and not(Stop) do
$k := 1$;
While $k \leq \mid X \mid$ and not($Stop$) do
If $A(((x_i^n \to y_j) \to y_j) \to x_i) < \min(A(z_k \to (y_j \to x_i)), A(z_k))$ then
Stop:=true;
EndIf
EndWhile
EndWhile
EndWhile
If Stop then
Output(" A is not a fuzzy n-fold fantastic filter")
Else
Output(" A is a fuzzy n -fold fantastic filter")
EndIf
End

This work was previously published in the International Journal of Artificial Life Research, Volume 2, Issue 4, edited by E. Stanley Lee and Ping-Teng Chang, pp. 32-42, copyright 2011 by IGI Publishing (an imprint of IGI Global).

Chapter 22
An Observer Approach for Deterministic Learning Using Patchy Neural Networks with Applications to Fuzzy Cognitive Networks

H. E. Psillakis
Technological and Educational Institute of Crete, Greece

M. A. Christodoulou
Technical University of Crete, Greece

T. Giotis
Technical University of Crete, Greece

Y. Boutalis
Democritus University of Thrace, Greece

ABSTRACT

In this paper, a new methodology is proposed for deterministic learning with neural networks. Using an observer that employs the integral of the sign of the error term, asymptotic estimation of the respective nonlinear vector field is achieved. Patchy Neural Networks (PNNs) are introduced to identify the unknown nonlinearity from the observer's output and the state measurements. The proposed scheme achieves learning with a single pass from the respective patches and does not need standard persistency of excitation conditions. Furthermore, the PNN weights are updated algebraically, reducing the computational load of learning significantly. Simulation results for a Duffing oscillator and a fuzzy cognitive network illustrate the effectiveness of the proposed approach.

1. INTRODUCTION

Artificial neural networks are well known for their learning ability. They have been efficiently used for deterministic learning in the context of adaptive control (Farrell, 1998; Jiang & Wang, 2000; Rovithakis & Christodoulou, 2000; Spooner, Maggiore, Ordonez, & Passino, 2002; Ge, Hang, Lee, & Zhang, 2002; Farrell & Polycarpou, 2006) and for computational or statistical learning in the context of machine learning (Vapnik, 2000).

DOI: 10.4018/978-1-4666-3890-7.ch022

Recently, a deterministic learning theory (Wang, Hill, & Chen, 2003; Wang & Hill, 2006) was proposed and applied to the dynamical pattern recognition problem (Wang & Hill, 2007, 2010). Wang and Hill (2007) address the dynamical pattern recognition problem of temporal patterns generated from a dynamical system

$$\dot{x}(t) = f(x(t); p) \qquad x(t_0) = x_0 \tag{1}$$

where $x := \begin{bmatrix} x_1 & x_2 & \cdots & x_n \end{bmatrix}^T \in \mathbb{R}^n$ is the state vector, p is a vector with system parameters and $f(x;p) := \begin{bmatrix} f_1(x;p) & f_2(x;p) & \cdots & f_n(x;p) \end{bmatrix}^T$ represents the system dynamics with $f_i(x;p)$ a smooth, unknown, nonlinear function.

Dynamical patterns are defined as general recurrent trajectories generated from (1) and include among others periodic, quasi-periodic or even chaotic trajectories. As described in Wang and Hill (2007) the pattern recognition process involves two main tasks: an initial identification task and a recognition task.

A deterministic learning approach based on localized radial basis function neural networks (RBF NNs) (Sanner & Slotine, 1992) is adopted in Wang, Hill, and Chen (2003) and Wang and Hill (2006) for the initial identification task. Using this learning scheme, information on the dynamical pattern is obtained and stored in the RBF NN weights. After the identification procedure, a set of dynamical models (the so called "test set") is constructed. These models are then employed in the pattern recognition task that involves comparisons between the actual and the test patterns (generated by the test models) based on some suitable similarity measure. A detailed description of the overall methodology can be found in Wang and Hill (2007, 2010).

In this paper, we focus on the initial identification task and propose an alternative approach to deterministic learning. As a first step, an observer is designed based on the robust integral of the sign error (RISE) approach (Xian, Dawson, de Queiroz, & Chen, 2004; Patre, MacKunis, Kaiser, & Dixon, 2008; Patre, MacKunis, Makkar, & Dixon, 2008) that provides an asymptotic time estimate of the smooth vector field $f(x;p)$. A localized neural network can then be employed to extract and store the information of this estimate.

To this end, we introduce a new class of localized neural networks called *patchy neural networks* (PNNs) with basis functions that are "patches" of the state space. We prove their universal approximation capability i.e., it is shown that a PNN with a sufficient number of nodes can approximate with desired accuracy over some compact region a general smooth nonlinear function.

A simple PNN is then employed to extract and store the information obtained from the observer estimate based on an easy to implement algebraic weight update law. The advantages of the proposed methodology with respect to Wang and Hill (2007, 2010) are:

- There is no need for a persistence of excitation condition for the state vector. The PNN is capable of learning the unknown nonlinearity in some region of the state space from a single visit of the state trajectories to the patches of the region. This is in contrast to standard deterministic learning schemes (Wang, Hill, & Chen, 2003; Wang & Hill, 2006, 2007) wherein a recurrence condition for the state trajectories is necessary for learning within the neighborhoods which are periodically excited.
- The PNN weight update laws are defined by an algebraic form and are not given in the form of ordinary differential equations (ODEs) as in Wang and Hill (2006). This results in a significant reduction of the computational cost for learning since we must only solve n ODEs from the observer, opposed to $N_1 \times N_2 \times \cdots \times N_n$ ODEs needed to train the NN weights (for a NN with $N_1 \times \cdots \times N_n$ nodes).

The identification procedure described above can be efficiently applied to the learning problem of the dynamics of Fuzzy Cognitive Networks (Kottas, Boutalis, & Christodoulou, 2007; Boutalis, Kottas, & Christodoulou, 2009). FCNs were proposed in Kottas, Boutalis, and Christodoulou (2007) as generalizations of Fuzzy Cognitive Maps (Kosko, 1986). They are inference networks with cyclic directed graphs that represent the causal relationships between concepts. In Psillakis, Boutalis, Giotis, and Christodoulou (2010), an observer with an adaptation mechanism was effectively applied to the weight identification of a general fully connected FCN under a persistence of excitation condition. As an application, the observer-PNN approach is used in this work to learn the dynamics of an FCN.

The rest of the paper is organized as follows. In the next section, the identification problem is described. The PNN is introduced and its universal approximation property is proved. The proposed learning methodology is then given. The main estimation result is stated and a simulation study is carried out. Finally, some concluding remarks are made in the last section.

1.1 Learning Problem Formulation

In NN pattern recognition of temporal patterns generated by a time-invariant dynamical system in the form of (1), the first step is the NN-based identification of the unknown nonlinearity $f(x)$. Using a test temporal pattern, a localized network $W_i^T P(x)$ is trained to estimate $f(x)$ along the trajectory $x(t; x_0, p_i)$. Then, for each training pattern, N dynamical models are constructed (Wang & Hill, 2007)

$$\dot{\bar{x}}^i(t) = -A(\bar{x}^i - x) + \bar{W}_i^T P(x), (1 \le i \le N) \quad (2)$$

with estimated state

$\bar{x}$, $A := \operatorname{diag}\{a_1, \ldots, a_n\}, a_i > 0$

design constants and

$$\bar{W}_i := \begin{bmatrix} \bar{W}_1^i & \bar{W}_2^i & \cdots & \bar{W}_n^i \end{bmatrix} \in \mathbb{R}^{\ell \times n}$$

the NN weight matrix that corresponds to the i-th test pattern.

In rapid dynamical pattern recognition of a temporal pattern $x(t)$, the most similar from the N test models (2) is identified based on the synchronization error magnitude $\left\| x(t) - \bar{x}^i(t) \right\|$. Thus, from a set of test models, the one describing better a specific temporal pattern is the one with the smallest synchronization error.

In this paper, we focus on the initial identification problem and provide an alternative approach. Particularly, *we propose a novel approach for obtaining an NN estimator $W^T P(x)$ of $f(x)$ of* (1). Rapid dynamical pattern recognition can then be carried out from a set of test models (2) as in Wang and Hill (2007, 2010).

Assumption 1: The state trajectories $x(t; x_0; p)$ are bounded within some compact set $I \subset \mathbb{R}^n$ and there exist positive constants c_f, c_f', c_f'' such that

$$c_f := \max_{1 \le i \le n} \sup_{x \in I} \left| f_i(x) \right| < \infty,$$

$$c_f' := \max_{1 \le i,j \le n} \sup_{x \in I} \left| \partial f_i(x) / \partial x_j \right| < \infty,$$

$$c_f'' := \max_{1 \le i,j,k \le n} \sup_{x \in I} \left| \partial^2 f_i(x) / \partial x_j \partial x_k \right| < \infty.$$

1.2 Patchy Neural Network

A new class of localized NNs is introduced in this paper that is relatively simple and easy to train. Let $I := I_1 \times I_2 \times \cdots \times I_n$ some n-dimensional rectangle and δ-partitions of each interval I_i given by $I_i := \bigcup_{j=1}^{N_i} A_{i,j} := \bigcup_{j=1}^{N_i} [\alpha_{i,j-1}, \alpha_{i,j}]$ with $\alpha_{i,j} := \alpha_{i,0} + j\delta \; (1 \le i \le n)$. On the sets

$A_{1,i_1} \times \cdots \times A_{n,i_n}$ $(1 \le i_j \le N_i; 1 \le i \le n)$ we can now define the "patch" functions

$$p_{i_1,i_2,\ldots,i_n}(x) = \begin{cases} 1\ , & \text{if } x \in A_{1,i_1} \times \cdots \times A_{n,i_n} \\ 0\ , & \text{else} \end{cases} \quad (3)$$

A patchy neural network is a single-hidden layer NN with basis vector consisting of "patch" functions with output given by

$$y = \sum_{i_1=1}^{N_1} \cdots \sum_{i_n=1}^{N_n} w_{i_1,i_2,\ldots,i_n} p_{i_1,i_2,\ldots,i_n}(x) = W^T P(x) \quad (4)$$

where $W := \begin{bmatrix} w_{1,\ldots,1} & \cdots & w_{N_1,\ldots,N_n} \end{bmatrix}^T \in \mathbb{R}^{N_1 \times \cdots \times N_n}$ and $P(x) := \begin{bmatrix} p_{1,\ldots,1}(x) & \cdots & p_{N_1,\ldots,N_n}(x) \end{bmatrix}^T$. A graphical representation of the PNN is shown in Figure 1.

An important property of the PNN is its ability to approximate general nonlinear functions. Particularly, let us consider a nonlinear function $f : I \subset \mathbb{R}^n \to \mathbb{R}$ defined on some n-dimensional rectangle I. Then, for every $\epsilon > 0$ there exists a PNN with $N_1 \times \cdots \times N_n$ nodes (patch functions) and a weight vector $W \in \mathbb{R}^{N_1 \times \cdots \times N_n}$ such that

$$f(x) = W^T P(x) + \epsilon_a(x) \quad (5)$$

with $P(x) := \begin{bmatrix} p_{1,\ldots,1}(x) & \cdots & p_{N_1,\ldots,N_n}(x) \end{bmatrix}^T$ and approximation error $|\epsilon_a(x)| \le \epsilon \ \forall x \in I$. This is a direct consequence of the patch function definition and the Mean Value Theorem (MVT). Particularly, applying the MVT for $f(x)$ in each patch $R_{i_1,\ldots,i_n} := A_{1,i_1} \times \cdots \times A_{n,i_n}$ we have that

$$f(x) = f(x_{i_1,\ldots,i_n}) + \frac{\partial f}{\partial x}\left(\theta_{i_1,\ldots,i_n} x + (1 - \theta_{i_1,\ldots,i_n}) x_{i_1,\ldots,i_n}\right)(x - x_{i_1,\ldots,i_n})$$

$\forall x \in R_{i_1,\ldots,i_n}$ for some $\theta_{i_1,\ldots,i_n} \in (0,1)$ and $x_{i_1,\ldots,i_n} \in R_{i_1,\ldots,i_n}$. Note also that $\left\| x - x_{i_1,\ldots,i_n} \right\| \le \prod_{j=1}^{n} (\alpha_{j,i_j} - \alpha_{j,i_j-1}) = \delta^n$ $\forall x, x_{i_1,\ldots,i_n} \in R_{i_1,\ldots,i_n}$. If we now choose the weight value $w_{i_1,\ldots,i_n} = f(x_{i_1,\ldots,i_n})$ $(1 \le i_j \le N_i; 1 \le i \le n)$ then, $\forall x \in I$ we have that

Figure 1. The patchy neural network (PNN)

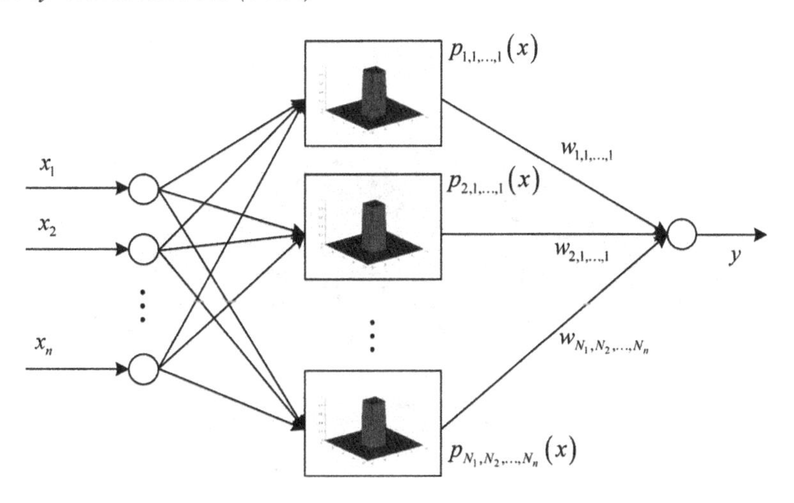

$$\left|f(x) - W^T P(x)\right| \le c' \max_{\substack{1\le i_\ell \le N_\ell \\ 1\le \ell \le n}} \prod_{j=1}^{n} \left(\alpha_{j,i_j} - \alpha_{j,i_j-1}\right) = c'\delta^n$$

where $c' := \max_{x\in I} \left\|\partial f(x)/\partial x\right\|$. Now choosing a sufficiently large number of nodes $N_1,\ldots,N_n$, i.e. selecting a suitably fine δ-partition with $\delta \le \left(\epsilon/c'\right)^{1/n}$ we can achieve the desired approximation property (5).

Example 1: Let for example the nonlinear function $f(x,y)=20-x^2-y^2$ defined on the 2-dimensional rectangle $I := I_1 \times I_2 = [-2,2]\times[-2,2]$ shown in Figure 2a. Consider now the interval partitions $I_i := \cup_{j=1}^{16}\left[0.25(j-1)-2, 0.25j-2\right]$ (i = 1,2) that yield respectively 16 x 16 = 256 patches. Then, the corresponding PNN $\hat{f}(x,y) := W^T P(x,y)$ with weights $w_{i,j} := f\left(0.25(i-1)-2, 0.25(j-1)-2\right)$ $(1 \le i,j \le 16)$ provides a good estimator of $f(x,y)$ as shown in Figure 2b.

1.3 Identification Scheme

In this section, the proposed deterministic learning approach is presented. The overall scheme consists of two distinct estimation tasks. Initially, an asymptotic observer is proposed based on the robust integral of the sign error (RISE) approach

Figure 2. Function f(x,y) and its PNN approximation

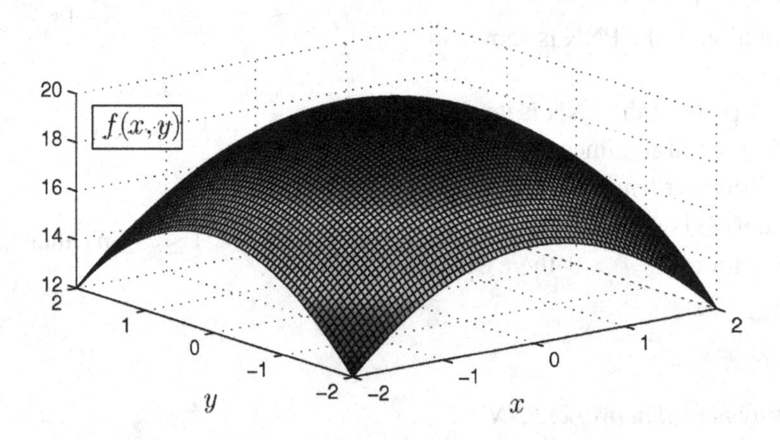

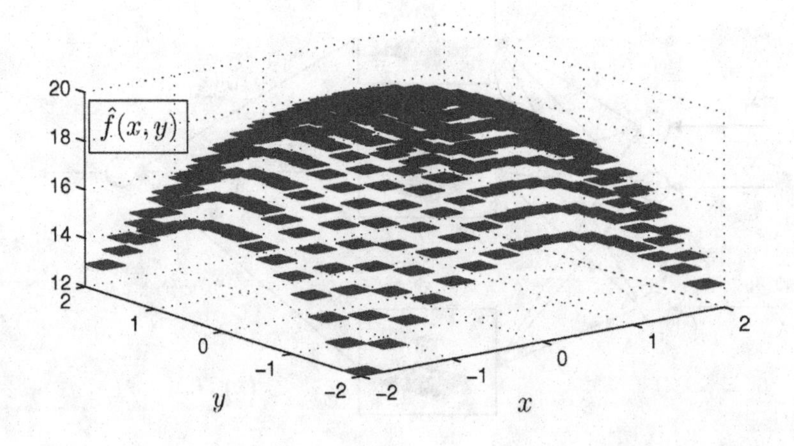

(Xian, Dawson, de Queiroz, & Chen, 2004; Patre, MacKunis, Kaiser, & Dixon, 2008; Patre, MacKunis, Makkar, & Dixon, 2008) that identifies the nonlinear vector field $\xi^*(t) := f(x(t))$. Then, a PNN is employed to extract the general form of $f(x)$ based on the observer's output and the state vector measurements.

1.3.1 RISE Observer

Let the following full state RISE observer

$$\begin{aligned}\dot{\hat{x}}(t) = \xi(t) &:= k\left(x(t) - \hat{x}(t)\right) \\ &+ \lambda k \int_0^t \left(x(s) - \hat{x}(s)\right) ds \\ &+ \beta \int_0^t \operatorname{sgn}\left(x(s) - \hat{x}(s)\right) ds\end{aligned} \tag{6}$$

with constants $k, \lambda, \beta > 0$. For the RISE observer (6) of system (1), the following Lemma can be proved.

Lemma 1: Let the nonlinear dynamical system (1) satisfying Assumption 1 and the RISE observer described by (6), respectively. Then, for the selection $k > \lambda > 0$ and

$$\beta \geq n c_f c_f' + \left(1/\lambda\right) n^2 c_f \left(c_f c_f'' + c_f'^2\right) \tag{7}$$

it holds true that $\lim_{t\to\infty}\left(\hat{x}(t) - x(t)\right) = \lim_{t\to\infty}\left(\xi(t) - f\left(x(t)\right)\right) = 0.$

Proof: The proof is given in the appendix.

Remark 1: The most important property of the RISE observer (6) is that it *provides an asymptotic time estimate $\xi(t)$ of the respective vector field $f(x(t))$* that can be subsequently used for training a localized NN to extract the general form of $f(x)$.

1.3.2 PNN Weight Update

Lemma 1 ensures that the vector signal $\xi(t)$ of the RISE observer converges asymptotically to the vector field $f(x(t))$. The observer's output $\xi(t)$ and the state vector measurements can now be used to train a PNN to recover the general form of $f(x)$ at a neighborhood of a region of the state space $\left\{\nu \in \mathbb{R}^n \mid \nu = x(t), t \in [0, +\infty)\right\}$ defined by the state trajectories. The PNN weights are updated algebraically as follows

$$\begin{aligned}\hat{w}^j_{i_1,\ldots,i_n}(t) &= \left(1 - p_{i_1,\ldots,i_n}\left(x(t)\right)\right) \\ \hat{w}^j_{i_1,\ldots,i_n}(t^-) &+ p_{i_1,\ldots,i_n}\left(x(t)\right)\xi_j(t)\end{aligned} \tag{8}$$

or equivalently

$$\hat{w}^j_{i_1,\ldots,i_n}(t) = \begin{cases}\hat{w}^j_{i_1,\ldots,i_n}(t^-), & \text{if } p_{i_1,\ldots,i_n}\left(x(t)\right) = 0 \\ \xi_j(t), & \text{if } p_{i_1,\ldots,i_n}\left(x(t)\right) = 1\end{cases} \tag{9}$$

$(j = 1, 2, \ldots, n)$ with initial values $\hat{w}^j_{i_1,\ldots,i_n}(0) = 0$ $(1 \leq i_1 \leq N_1, \ldots, 1 \leq i_n \leq N_n; 1 \leq j \leq n)$. Then, the vector $\hat{W}^T P(x)$ with $\hat{W} := \left[\hat{W}_1, \cdots, \hat{W}_n\right]$, $\hat{W}_i := \left[\hat{w}^i_{1,\ldots,1}, \ldots, \hat{w}^i_{N_1,\ldots,N_n}\right]^T$ can be used to estimate $f(x)$.

1.4 PNN Estimation

Now we are ready to state the main result of the paper that the unknown nonlinear vector field elements $f_i(x)$ can be approximated by $\hat{W}_i^T(t) P(x)$ in a region of the state trajectories $R_\delta(T, t) := \bigcup_{\tau \in [T,t]} R\left(x(\tau); \delta\right)$ where $R(s; \delta) := R_{i_1,\ldots,i_n}$ with indexes selected from the set $\left\{i_j \in \mathbb{N} \mid 1 \leq i_j \leq N_i; 1 \leq i \leq n\right\}$ such that

$s \in R_{i_1,\ldots,i_n}$. Thus, $R(s;\delta)$ denotes the δ-patch that includes s among its points and $R(x(\tau);\delta)$ is the δ-patch wherein the state vector belongs at time τ. Aggregating over the time interval $[T, t]$ we obtain $\bar{R}_\delta(T,t)$ that is the union of the patches traversed by the state vector during $[T, t]$.

Theorem 1: Let the nonlinear dynamical system (1) satisfying Assumption 1 and the RISE observer (6), (7). Then, for every $\epsilon > 0$, there exist a time $T(\epsilon) > 0$ and a PNN with patch length $\delta(\epsilon) \leq \frac{\epsilon}{2c_f' n^{1/2}}$ and weight update law (8) such that $\left\|\hat{W}^T(t)P(s) - f(s)\right\| \leq \epsilon$ for all $t \geq T(\epsilon) > 0$ and $s \in \bar{R}_{\delta(\epsilon)}(T(\epsilon),t) := \cup_{\tau \in [T(\epsilon),t]} R(x(\tau);\delta(\epsilon))$.

Remark 2: Practically, the above theorem states, that we can select a sufficiently dense partition of the state space (a small enough δ) so that the PNN can approximate efficiently the nonlinear function f in the region $\bar{R}_{\delta(\epsilon)}(T(\epsilon),t)$ that is the union of individual patches from which the state trajectory passes during $[T(\epsilon),t]$. $T(\epsilon)$ is defined as the time needed for the RISE observer estimate to converge to some region of $f(x(t))$ and remain therein for all time thereafter.

Proof: Since $\lim_{t\to\infty}\left(\xi(t) - f(x(t))\right) = 0$, there exists time $T(\epsilon) \geq 0$ such that $\left\|\xi(t) - f(x(t))\right\| \leq \epsilon/2 \ \forall t \geq T(\epsilon)$. Define now the times $\tau_{[t_1,t_2]}(s;\delta) := \sup\left\{\tau \in [t_1,t_2] \mid s \in R(x(\tau);\delta)\right\}$.

Then, for every $s \in \bar{R}_{\delta(\epsilon)}(T(\epsilon),t)$ it holds true from (9) that $\hat{W}^T(t)P(s) = \xi\left(\tau_{[T(\epsilon),t]}(s;\delta)\right)$. Equivalently, we have that

$$
\begin{aligned}
&\left\|f(s) - \hat{W}^T(t)P(s)\right\| \\
&\leq \left\|f(s) - f\left(x\left(\tau_{[T(\epsilon),t]}(s;\delta)\right)\right)\right\| \\
&+ \left\|f\left(x\left(\tau_{[T(\epsilon),t]}(s;\delta)\right)\right) - \xi\left(\tau_{[T(\epsilon),t]}(s;\delta)\right)\right\| \\
&\leq \sup_{x\in I}\left\|\frac{\partial f}{\partial x}\right\|_F \left\|s - x\left(\tau_{[T(\epsilon),t]}(s;\delta)\right)\right\| \\
&+ \frac{\epsilon}{2} \leq n^{1/2} c_f' \delta + \frac{\epsilon}{2} \leq \epsilon.
\end{aligned}
$$

Remark 3: From Theorem 1 one can see that for a complete learning of some nonlinearity f within some $I \subset \mathbb{R}^n$, the dynamical system (1) should be *mixing*, i.e. the state trajectories should visit all δ-patches of I. This is a characteristic property of chaotic systems (Lasota & Mackey, 1994). If this is not the case, then, the PNN will only learn the form of $f(x)$ at the patches activated by the state trajectories.

Remark 4: There are two main advantages of the proposed methodology w.r.t. existing results (Wang & Hill, 2006, 2007, 2010). First, there is no need for a persistency of excitation condition in order to achieve learning. The PNN can identify the unknown nonlinearity if the state trajectory crosses the respective patch once after the time required for the observer to converge. Second, the new scheme has significantly reduced computational cost since the PNN weights are updated algebraically which means that we have to solve only n ODEs (due to the observer) opposed to $N_1 \times \cdots \times N_n$ ODEs (due to the RBF NN weights) needed in Wang and Hill (2006).

1.5 Simulation Study

Case 1: To verify the effectiveness of the proposed scheme we consider the learning problem for the Duffing oscillator

$$\begin{aligned} \dot{x} &= y \\ \dot{y} &= -p_1 y - p_2 x - p_3 x^3 + q\cos(\omega t) \end{aligned} \tag{10}$$

with initial conditions $x(0) = 0$, $y(0) = -1.8$ and parameters $p_1 = 0.35$, $p_2 = -1.1$, $p_3 = 1$, $q = 1.498$ and $\omega = 1.8$. The nonlinearity to be identified is $f(x,y) := -p_1 y - p_2 x - p_3 x^3$ of (10). The RISE observer

$$\begin{aligned} \dot{\hat{y}} &= q\cos(\omega t) + \xi(t) \\ &= q\cos(\omega t) + k\left(y(t) - \hat{y}(t)\right) \\ &\quad + k\lambda \int_0^t \left(y(s) - \hat{y}(s)\right) ds \\ &\quad + \beta \int_0^t \operatorname{sgn}\left(y(s) - \hat{y}(s)\right) ds \end{aligned}$$

is employed with parameters $k = 2$, $\lambda = 1$ and $\beta = 20$. The square $[-2.15, 2.15] \times [-2.15, 2.15]$ is divided into $44 \times 44 = 1936$ patches with patch length $\delta = 0.1$ for the associated PNN with update law (8) and output $\hat{f}_t(x,y) = \sum_{i_1=1}^{44} \sum_{i_2=1}^{44} \hat{w}_{i_1,i_2}(t)\, p_{i_1,i_2}(x,y)$. Simulation results are shown in Figures 3 through 5. In Figure 3, one can see that $\xi(t)$ converges rapidly to the respective values of f. At $T = 10^2$ sec, the form of f in the attractor region of the Duffing oscillator has been extracted (Figure 4) with some errors due to the initial training at the times before the RISE observer converges. These errors have been corrected at $T=10^3$ sec (Figure 5) since by now, another pass of the state trajectories from the respective patches has occurred. Thus, as can be seen from Figure 5, at that time, the PNN has learned the form of f within the attractor and the NN weights fluctuate among the respective values of f within the patch (Figure 6). For pattern recognition purposes the averaged weight vector $\bar{W} := \left(1/(t_b - t_a)\right) \int_{t_a}^{t_b} \hat{W}(s)\, ds$ can then be employed in (2) to construct the test model where t_a should be chosen after the transient process for the NN weights and $t_b > t_a$ (Wang & Hill, 2007).

Figure 3. Time evolution of $\xi(t)$ and $f(x(t),y(t))$

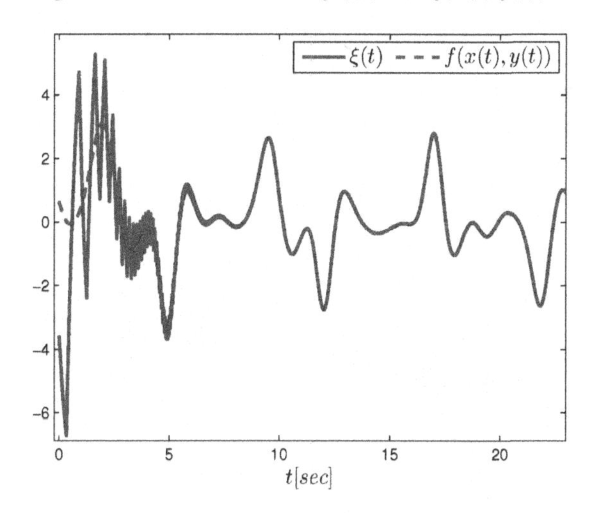

Figure 4. PNN output $\hat{f}_t(x,y) = \sum_{i_1=1}^{44} \sum_{i_2=1}^{44} \hat{w}_{i_1,i_2}(t)\, p_{i_1,i_2}(x,y)$ graph for $t = 10^2$ sec

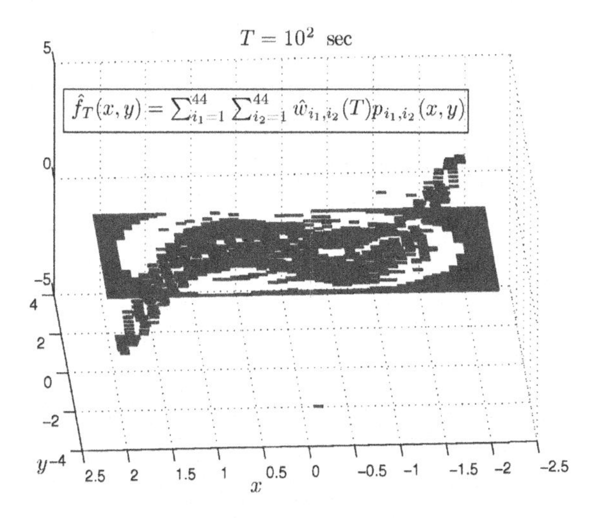

Case 2 (FCN Learning): Let now a second order FCN.

Figure 5. PNN output $\hat{f}_t(x,y) = \sum_{i_1=1}^{44} \sum_{i_2=1}^{44} \hat{w}_{i_1,i_2}(t) p_{i_1,i_2}(x,y)$ *graph for* $t = 10^3$ sec

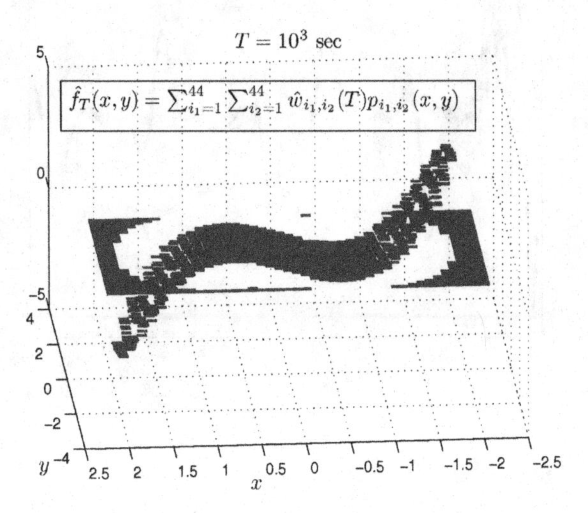

Figure 6. The PNN weight norm $\left\| \hat{W}(t) \right\|$ *time evolution*

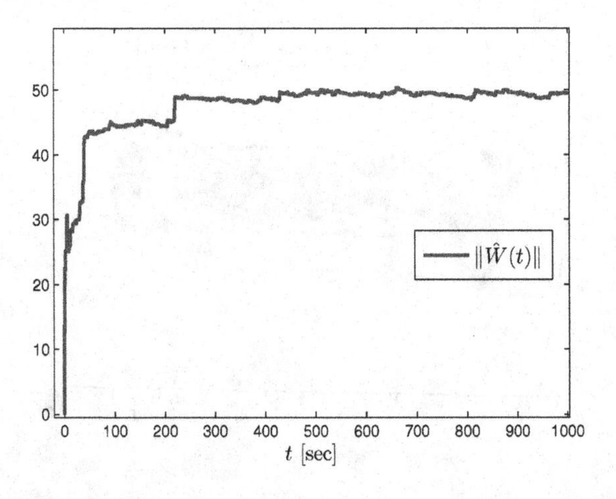

$$\begin{bmatrix} \dot{A}_1(t) \\ \dot{A}_2(t) \end{bmatrix} = \begin{bmatrix} f_1(A_1, A_2, u) \\ f_2(A_1, A_2, u) \end{bmatrix} = f\left(\begin{bmatrix} -0.2 & 0.1 & 0.1 \\ -0.1 & -0.3 & 0.2 \end{bmatrix} \cdot \begin{bmatrix} A_1(t) \\ A_2(t) \\ u(t) \end{bmatrix} \right)$$

with initial conditions $A_1(0) = -0.15, A_2(0) = 0.25$, $f(x) = \tanh(x)$ and external input $u(t) = 0.1 + \sin(t)$. The above system converges towards a limit cycle shown in Figure 7a. The observer-PNN scheme is employed to identify the unknown vector fields f_1, f_2. Even though the input appears in the dynamics, a similar analysis to the proof of Lemma 1 shows that, for bounded $u, \dot{u}, \ddot{u}$ and a suitable selection of k, λ, β, $\lim_{t\to\infty}\left(\xi_i(t) - f_i\left(A_1(t), A_2(t), u(t)\right)\right) = 0$ with ξ defined by (6). Now, the (u, A_1, A_2) state space is decomposed into 3-dimensional rectangles by decomposing the intervals $I_u := [-1, 1.38] = \cup_{i=1}^{14} [-1 + 0.17(i-1), -1 + 0.17i]$. $I_{A_1} := [0.02, 0.14] = \cup_{j=1}^{12} [0.02 + 0.01(j-1), 0.02 + 0.01j]$, $I_{A_2} := [-0.061, 0.167]$. $= \cup_{k=1}^{12} [-0.061 + 0.019(k-1), -0.061 + 0.019k]$ and considering the resulting $14 \times 12 \times 12 = 2016$ patches. For k, λ, β the values of Case 1 are used. Similarly, the PNN weights reach a steady state and two averaged vectors $\bar{W}_i$ $(i = 1,2)$ can be obtained after the

Figure 7. State trajectories $(u(t), A_1(t), A_2(t))$ and the PNN approximation $\bar{W}_i^T P(A_1, A_2, u)$ *of* $f_i(u(t), A_1(t), A_2(t))$ *in the periodic steady state*

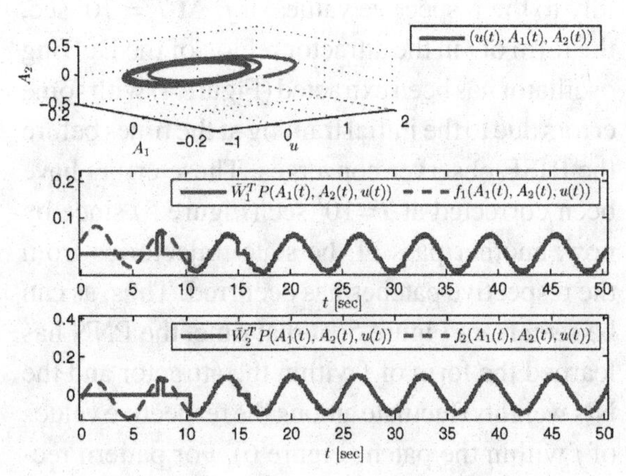

transient phase. From Figure 7b-c one can see that the PNNs $\bar{W}_i^T P(A_1, A_2, u)$ provide good estimators for $f_i(A_1,A_2,u)$ $(i = 1,2)$ at the patches excited after the initial training phase.

2. CONCLUSION

An observer-based deterministic learning approach is presented in this paper. The proposed scheme consists of an observer providing asymptotic estimates of the unknown nonlinearly parametrized vector field. The observer's output and the state vector measurements are used subsequently by a localized PNN to extract the general form of the nonlinearity in the regions wherein the state trajectories remain. Detailed analysis demonstrates the capability of learning with limited assumptions. A simulation study was also carried out that verifies our theoretical results.

REFERENCES

Boutalis, Y., Kottas, T., & Christodoulou, M. A. (2009). Adaptive estimation of fuzzy cognitive maps with proven stability and parameter convergence. *IEEE Transactions on Fuzzy Systems*, *17*, 874–889. doi:10.1109/TFUZZ.2009.2017519

Farrell, J. A. (1998). Stability and approximator convergence in nonparametric nonlinear adaptive control. *IEEE Transactions on Neural Networks*, *9*, 1008–1020. doi:10.1109/72.712182

Farrell, J. A., & Polycarpou, M. M. (2006). *Adaptive approximation based control: Unifying, neural, fuzzy and traditional adaptive approximation approaches*. Hoboken, NJ: John Wiley & Sons.

Ge, S. S., Hang, C. C., Lee, T. H., & Zhang, T. (2002). *Stable adaptive neural network control*. Dordrecht, The Netherlands: Kluwer Academic Publishers.

Ioannou, P. A., & Sun, J. (1996). *Robust adaptive control*. Upper Saddle River, NJ: Prentice Hall.

Jiang, D., & Wang, J. (2000). On-line learning of dynamical systems in the presence of model mismatch and disturbances. *IEEE Transactions on Neural Networks*, *11*, 1272–1283. doi:10.1109/72.883420

Kosko, B. (1986). Fuzzy cognitive maps. *International Journal of Man-Machine Studies*, 65–75. doi:10.1016/S0020-7373(86)80040-2

Kottas, T. L., Boutalis, Y. S., & Christodoulou, M. A. (2007). Fuzzy cognitive networks: A general framework. *Intelligent Decision Technologies*, *1*(4), 183–196.

Lasota, A., & Mackey, M. C. (1994). *Chaos, fractals, and noise: Stochastic aspects of dynamics*. New York, NY: Springer.

Patre, P. M., MacKunis, W., Kaiser, K., & Dixon, W. E. (2008). Asymptotic tracking for uncertain dynamic systems via a multilayer neural network feedforward and RISE feedback control structure. *IEEE Transactions on Automatic Control*, *53*, 2180–2185. doi:10.1109/TAC.2008.930200

Patre, P. M., MacKunis, W., Makkar, C., & Dixon, W. E. (2008). Asymptotic tracking for systems with structured and unstructured uncertainties. *IEEE Transactions on Control Systems Technology*, *16*, 373–379. doi:10.1109/TCST.2007.908227

Psillakis, H. E., Boutalis, Y., Giotis, T., & Christodoulou, M. A. (2010). Adaptive identification of fuzzy cognitive networks. In *Proceedings of the 4th International Symposium on Communications, Control and Signal Processing,* Limassol, Cyprus (pp. 1-6).

Rovithakis, G. A., & Christodoulou, M. A. (2000). *Adaptive control with recurrent high-order neural networks: Theory and industrial applications*. New York, NY: Springer.

Sanner, R. M., & Slotine, J.-J. E. (1992). Gaussian networks for direct adaptive control. *IEEE Transactions on Neural Networks, 3*, 837–863. doi:10.1109/72.165588

Spooner, J. T., Maggiore, M., Ordonez, R., & Passino, K. M. (2002). *Stable adaptive control and estimation for nonlinear systems- neural and fuzzy approximator techniques*. Hoboken, NJ: John Wiley & Sons. doi:10.1002/0471221139

Vapnik, V. P. (2000). *The nature of statistical learning theory* (2nd ed.). New York, NY: Springer.

Wang, C., & Hill, D. J. (2006). Learning from neural control. *IEEE Transactions on Neural Networks, 17*, 130–146. doi:10.1109/TNN.2005.860843

Wang, C., & Hill, D. J. (2007). Deterministic learning and rapid dynamical pattern recognition. *Transactions on Neural Networks, 18*, 617–630. doi:10.1109/TNN.2006.889496

Wang, C., & Hill, D. J. (2010). *Deterministic learning theory for identification, recognition and control*. Boca Raton, FL: CRC Press.

Wang, C., Hill, D. J., & Chen, G. (2003). Deterministic learning of nonlinear dynamical systems. In *Proceedings of the 18th IEEE International Symposium on Intelligent Control*, Houston, TX (pp. 87-92).

Xian, B., Dawson, D. M., de Queiroz, M. S., & Chen, J. (2004). A continuous asymptotic tracking control strategy for uncertain nonlinear systems. *IEEE Transactions on Automatic Control, 49*, 1206–1211. doi:10.1109/TAC.2004.831148

APPENDIX

Proof of Lemma 1

Let us define the observation error $\tilde{x}(t) := \hat{x}(t) - x(t)$ and the filtered observation error $r(t) := \dot{\tilde{x}}(t) + \lambda\tilde{x}(t)$. Then for the nonnegative function

$$V(t) := \frac{\lambda^2}{2}\tilde{x}^T(t)\tilde{x}(t) + \frac{1}{2}r^T(t)r(t) \tag{11}$$

we have that

$$\dot{V} = -\lambda^3\tilde{x}^T(t)\tilde{x}(t) + r^T(t)\left[\ddot{\hat{x}}(t) - \ddot{x}(t) + \lambda r(t)\right]. \tag{12}$$

Differentiating the observer form (6) we obtain

$$\ddot{\hat{x}}(t) = -kr(t) - \beta\operatorname{sgn}(\tilde{x}(t)). \tag{13}$$

Using the above identity in (12) we result in

$$\dot{V} = -\lambda^3\tilde{x}^T(t)\tilde{x}(t) - (k-\lambda)r^T(t)r(t) + L(t) \tag{14}$$

with $L(t) := -r^T(t)\left[\ddot{x}(t) + \beta\operatorname{sgn}(\tilde{x}(t))\right]$. Following a similar analysis to Xian, Dawson, de Queiroz, and Chen (2004) we obtain

$$\begin{aligned}
\int_0^t L(s)ds &= -\lambda\int_0^t \tilde{x}^T(s)\left(\ddot{x}(s) + \beta\operatorname{sgn}(\tilde{x}(s))\right)ds \\
&\quad - \int_0^t \dot{\tilde{x}}^T(s)\ddot{x}(s)ds - \beta\int_0^t \dot{\tilde{x}}^T(s)\operatorname{sgn}(\tilde{x}(s))ds \\
&= -\lambda\int_0^t \tilde{x}^T(s)\left(\ddot{x}(s) + \beta\operatorname{sgn}(\tilde{x}(s))\right)ds \\
&\quad - \tilde{x}^T(s)\ddot{x}(s)\Big|_0^t + \int_0^t \tilde{x}^T(s)\dddot{x}(s)ds - \beta\sum_{i=1}^n\left|\tilde{x}_i(s)\right|\Big|_0^t \\
&\le \lambda\sum_{i=1}^n\int_0^t\left|\tilde{x}_i(s)\right|\left(\left|\ddot{x}_i(s)\right| + \frac{1}{\lambda}\left|\dddot{x}_i(s)\right| - \beta\right)ds + \beta\sum_{i=1}^n\left|\tilde{x}_i(0)\right| \\
&\quad + \sum_{i=1}^n\left|\tilde{x}_i(s)\right|\left(\left|\ddot{x}_i(s)\right| - \beta\right) + \tilde{x}^T(0)\ddot{x}(0).
\end{aligned} \tag{15}$$

Since $\dot{x}_i(t) = f_i(x)$ we have that

$$\ddot{x}_i(t) = \sum_{j=1}^{n} \frac{\partial f_i(x)}{\partial x_j} f_j(x) \tag{16}$$

from which we result in

$$\max_{1\leq i\leq n} \sup_{0\leq t<\infty} \left|\ddot{x}_i(t)\right| \leq n c_f c_f'. \tag{17}$$

Further differentiation yields

$$\dddot{x}_i(t) = \sum_{k=1}^{n}\sum_{j=1}^{n}\left[\frac{\partial^2 f_i}{\partial x_j \partial x_k}(x) f_j(x) f_k(x) + \frac{\partial f_i}{\partial x_j}(x)\frac{\partial f_j}{\partial x_k}(x) f_k(x)\right] \tag{18}$$

and the following bound can be obtained directly

$$\max_{1\leq i\leq n} \sup_{0\leq t<\infty} \left|\dddot{x}_i(t)\right| \leq n^2 c_f \left(c_f c_f'' + c_f'^2\right). \tag{19}$$

Thus, for β satisfying (7) we have from (18), (19) and (15) that

$$\int_0^t L(s)\,ds \leq \beta \sum_{i=1}^{n} \left|\tilde{x}_i(0)\right| + \tilde{x}^T(0)\ddot{x}(0). \tag{20}$$

The above inequality ensures the nonnegativity of $P(t)$ defined by

$$P(t) := \beta \sum_{i=1}^{n} \left|\tilde{x}_i(0)\right| + \tilde{x}^T(0)\ddot{x}(0) - \int_0^t L(s)\,ds \geq 0. \tag{21}$$

Let us now consider the nonnegative function $\bar{V}(t) := V(t) + P(t)$. Then, from (14), (21) the dynamics of $\bar{V}$ take the following form

$$\dot{\bar{V}} \leq -\lambda^3 \tilde{x}^T(t)\tilde{x}(t) - (k-\lambda) r^T(t) r(t) \leq 0 \tag{22}$$

as long as $k > \lambda > 0$. The fact that function $\bar{V}(t)$ is decreasing and bounded from below ensures convergence (i.e. $\lim_{t\to\infty} \bar{V}(t) = \bar{V}_\infty < \infty$) and boundedness of $\bar{V}(t)$. Thus, it holds true that $\left\|\tilde{x}(t)\right\|, \left\|r(t)\right\| \in \mathcal{L}_\infty$. Integrating (22) we obtain

$$\lambda^3 \int_0^\infty \tilde{x}^T(s)\tilde{x}(s)\,ds + (k-\lambda)\int_0^\infty r^T(s)r(s)\,ds \le \bar{V}(0) - \bar{V}_\infty < \infty \tag{23}$$

i.e. $\|\tilde{x}(t)\|, \|r(t)\| \in \mathcal{L}_\infty \cap \mathcal{L}_2$. Furthermore, from (13) it holds true that

$$\dot{\tilde{x}}(t) = -\lambda\tilde{x}(t) + r(t) \tag{24}$$

$$\dot{r}(t) = -(k-\lambda)r(t) - \lambda^2\tilde{x}(t) - \beta\operatorname{sgn}(\tilde{x}(t)) - \ddot{x}(t) \tag{25}$$

from which we obtain $\|\dot{\tilde{x}}(t)\|, \|\dot{r}(t)\| \in \mathcal{L}_\infty$. Since $\|\tilde{x}(t)\|, \|r(t)\| \in \mathcal{L}_\infty \cap \mathcal{L}_2$ and $\|\dot{\tilde{x}}(t)\|, \|\dot{r}(t)\| \in \mathcal{L}_\infty$ we can now invoke Barbalat's Lemma (Ioannou & Sun, 1996) that yields $\lim_{t\to\infty} \tilde{x}(t) = \lim_{t\to\infty} r(t) = 0$

Finally, we have $\lim_{t\to\infty}\left(\xi(t) - f(x(t))\right) \begin{aligned} &= \lim_{t\to\infty}\left(\dot{\hat{x}}(t) - \dot{x}(t)\right) \\ &= \lim_{t\to\infty} \dot{\tilde{x}}(t) = \lim_{t\to\infty}\left(r(t) - \lambda\tilde{x}(t)\right) = 0 \end{aligned}$ that concludes the proof of Lemma 1.

This work was previously published in the International Journal of Artificial Life Research, Volume 2, Issue 1, edited by E. Stanley Lee and Ping-Teng Chang, pp. 1-16, copyright 2011 by IGI Publishing (an imprint of IGI Global).

Compilation of References

Abbas, S. E. (2005). On intuitionistic fuzzy compactness. *Information Sciences*, *173*, 75–91. doi:10.1016/j.ins.2004.07.004

Abeku, T., Hay, S., Ochola, S., Langi, P., Beard, B., & De Vlas, S. (2004). Malaria epidemic early warning and detection in African highlands. *Trends in Parasitology*, *20*, 400–405. doi:10.1016/j.pt.2004.07.005

Abidi1, S. S. R., & Goh, A. (2006). *Applying knowledge discovery to predict infectious disease epidemics*. Berlin, Germany: Springer-Verlag.

Addison, P. (1997). *Fractals and Chaos: An Illustrated Course*. Bristol, UK: Institute of Physics. doi:10.1887/0750304006

Agarwal, R. P., Meehan, M., & Regan, D. O. (2001). *Fixed point theory and applications*. Cambridge, UK: Cambridge University Press. doi:10.1017/CBO9780511543005

Agliano, P., & Montagna, F. (2003). Varieties of BL-algebras I: General properties. *Journal of Pure and Applied Algebra*, *181*, 105–129. doi:10.1016/S0022-4049(02)00329-8

Alam, J. B., Alam, M. J. B., Rahman, M. M., Dikshit, A. K., & Khan, S. K. (2006). Study on traffic noise level of sylhet by multiple regression analysis associated with health hazards. *Iranian Journal of Environmental Health Sciences & Engineering*, *3*(2), 71–78.

Alexander, R. D. (1971). The search for an evolutionary philosophy of man. *Proceedings of the Royal Society of Victoria*, *84*, 99–120.

Alexander, R. D. (1990). *How did humans evolve?: Reflections on the uniquely unique species*. Ann Arbor, MI: University of Michigan Museum of Zoology.

Alexandersen, S., Zhang, Z., Donaldson, A. I., & Garland, A. J. M. (1997). The pathogenesis and diagnosis of foot-and-mouth disease. *Transactions of the ASAE. American Society of Agricultural Engineers*, *40*, 247–252.

Al-Refaie, A., Ghnaimat, O., & Ko, J.-H. (2011). The effects of quality management practices on customer satisfaction and innovation: a perspective from Jordan. *International Journal of Productivity and Quality Management*, *8*(4).

Anderson, I. (2002). Foot and mouth disease 2001: Lessons to be learned inquiry. *The Stationary Office, 187.*

Anderson, J. (2009). *Simulating epidemics in rural areas and optimizing preplanned quarantine areas using a clustering heuristic*. Unpublished master's thesis, Kansas State University, Manhattan, KS.

Anderson, P. (1996). *Use of a PC printer port for control and data acquisition.* The Electronic Journal for Engineering Technology.

Andreyev, Y. V., Belsky, Y. L., Dmitriev, A. S., & Kuminov, D. A. (1996). Information processing using dynamical chaos: Neural networks implantation. *IEEE Transactions on Neural Networks*, *7*, 290–299. doi:10.1109/72.485632

An, G. (2000). Agent-based computer simulation and SIRS: Building a bridge between basic science and clinical trials. *Shock (Augusta, Ga.)*, *16*(4), 266–273. doi:10.1097/00024382-200116040-00006

An, G. (2004). In silico experiments of existing and hypothetical cytokine-directed clinical trials using agent-based modeling. *Critical Care Medicine*, *32*(10), 2050–2060. doi:10.1097/01.CCM.0000139707.13729.7D

Angus, D. C., Linde-Zwirble, W. T., Lidicker, J., Clermont, G., Carcillo, J., & Pinsky, M. R. (2001). Epidemiology of severe sepsis in the United States: Analysis of incidence, outcome, and associated costs of care. *Critical Care Medicine*, *29*(7), 1303–1310. doi:10.1097/00003246-200107000-00002

Annane, D., Sebille, V., Charpentier, C., Bollaert, P. E., Francois, B., & Korach, J. M. (2002). Effect of treatment with low doses of hydrocortisone and fludrocortisones on mortality in patients with septic shock. *Journal of the American Medical Association*, *288*(7), 862–871. doi:10.1001/jama.288.7.862

Argoul, F., Huth, J., Merzeau, P., Arnrodo, A., & Swinney, H. (1993). Experimental evidence for homoclinic chaos in an electrochemical growth process. *Physica D. Nonlinear Phenomena*, *62*, 170. doi:10.1016/0167-2789(93)90279-A

Atanassov, K. (1986). Intuitionistic fuzzy sets. *Fuzzy Sets and Systems*, *20*, 87–96. doi:10.1016/S0165-0114(86)80034-3

Atanassov, K. (1994). New operations defined over the intuitionistic fuzzy sets. *Fuzzy Sets and Systems*, *61*, 137–142. doi:10.1016/0165-0114(94)90229-1

Atchison, S. N., Burford, R. P., & Hibbert, D. B. (1994). Chemical effects on the morphology of supported electrodeposited metals. *International Journal of Electroanalytical Chemistry*, *371*, 137. doi:10.1016/0022-0728(94)03245-9

Atkinson, M. P., & Wein, L. M. (2008). Quantifying the routes of transmission for pandemic influenza. *Bulletin of Mathematical Biology*, *70*(3), 820–867. doi:10.1007/s11538-007-9281-2

Aunins, J., Lee, A., & Volkin, D. (1995). Vaccine production. In Bronzino, J. D. (Ed.), *The biomedical engineering handbook* (pp. 1502–1517). Boca Raton, FL: CRC Press.

Awaaz Foundation. (2010). *Noise Free Mumbai*. Retrieved from http://www.awaaz.org/downloads/noise-free-mumbai.pdf

Bacon, G. E., Kenny, F. M., Murdaugh, H. V., & Richards, C. (1973). Prolonged serum half-life of cortisol in renal failure. *The Johns Hopkins Medical Journal*, *132*(2), 127–131.

Baddeley, A. (2000). The episodic buffer: A new component of working memory? *Trends in Cognitive Sciences*, *4*(11), 417–423. doi:10.1016/S1364-6613(00)01538-2

Bagirov, A. M., Rubinov, A. M., Soukhoroukova, N. V., & Yearwood, J. (2003). Unsupervised and supervised data classification via nonsmooth and global optimization. *Sociedad da Estadistica e Investigacian Operativa Top*, *11*(1), 1–93.

Balkenius, C., Zlatev, J., Brezeal, C., Dautenhahn, K., & Kozima, H. (Eds.). (2001). *Proceedings of the First International Workshop on Epigenetic Robotics: Modeling Cognitive Development in Robotic Systems*, Lund, Sweden (Vol. 85).

Ball, F. G., & Lyne, O. D. (2002). Optimal vaccination policies for stochastic epidemics among a population of households. *Mathematical Biosciences*, *177*, 333–354. doi:10.1016/S0025-5564(01)00095-5

Barea, R., Boquete, L., Mazo, M., & López, E. (2002). System for Assisted Mobility Using Eye Movements Based on Electrooculography. *Neural Systems and Rehabilitation Engineering. IEEE Transactions on*, *10*(4), 209–218.

Barker, M., & Bostock, H. (1992). Ectopic activity in demyelinated spiral root axons of the rat. *The Journal of Physiology*, *451*, 539–552.

Barnsley, M. F. (1993). *Fractals everywhere* (2nd ed.). Boston, MA: Academic Press Professional.

Barnsley, M. F. (2006). *Superfractals*. Cambridge, UK: Cambridge University Press.

Barnsley, M. F. (2009). Transformations between Self-Referential Sets. *The American Mathematical Monthly*, *116*(4), 291–304. doi:10.4169/193009709X470155

Barnsley, M. F., Hutchinson, J. E., & Stenflo, Ö. (2005). A fractal valued random iteration algorithm and fractal hierarchy. *Fractals*, *13*(2), 111–146. doi:10.1142/S0218348X05002799

Barnsley, M. F., Hutchinson, J. E., & Stenflo, Ö. (2008). Fractals with partial self-similarity. *Advances in Mathematics*, *218*(6), 2051–2088. doi:10.1016/j.aim.2008.04.011

Barrett, C. L., Eubank, S. G., & Smith, J. (2005). If smallpox strikes Portland (simulation of the spread of disease in social networks). *Scientific American*, *292*(3), 54–61. doi:10.1038/scientificamerican0305-54

Barro, S., & Marin, R. (2002). *Fuzzy logic in medicine*. Heidelbrg, Germany: Physica-Verlag.

Barros, L. C., Bassanezi, R. C., & Tonelli, P. A. (2000). Fuzzy modelling in population dynamics. *Ecological Modelling*, *128*, 27–33. doi:10.1016/S0304-3800(99)00223-9

Bates, T. W., Carpenter, T. E., & Thurmond, M. C. (2003). Description of an epidemic simulation model for use in evaluating strategies to control an outbreak of foot-and-mouth disease. *American Journal of Veterinary Research*, *64*, 195–204. doi:10.2460/ajvr.2003.64.195

Bates, T. W., Carpenter, T. E., & Thurmond, M. C. (2003). Results of epidemic simulation modeling to evaluate strategies to control an outbreak of foot-and-mouth disease. *American Journal of Veterinary Research*, *64*, 205–210. doi:10.2460/ajvr.2003.64.205

Becker, N. G., & Starczak, D. N. (1997). Optimal vaccination strategies for a community of households. *Mathematical Biosciences*, *139*(2), 117–132. doi:10.1016/S0025-5564(96)00139-3

Bedau, M. A. (1998). Four puzzles about life. *Artificial Life*, *4*, 125–140. doi:10.1162/106454698568486

Bedau, M. A., McCaskill, J., Packard, P., Rasmussen, S., Green, D., & Ikegami, T. (2000). Open problems in artificial life. *Artificial Life*, *6*(4), 363–376. doi:10.1162/106454600300103683

Beigel, J., Farrar, J., Han, A., Hayden, F., Hyer, R., & De Jong, M. (2005). Avian influenza A (H5N1) infection in humans. *The New England Journal of Medicine*, *353*(13), 1374. doi:10.1056/NEJMra052211

Bell, D. M., & World Health Organization Writing Group. (2006). Non-pharmaceutical interventions for pandemic influenza, national and community measures. *Emerging Infectious Diseases*, *12*(1), 88–94.

Benedicks, M., & Carleson, L. (1991). The dynamics of the Hénon maps. *The Annals of Mathematics*, *133*, 1–25. doi:10.2307/2944326

Ben-Jacob, E. (1993). From snowflake formation to the growth of bacterial colonies, Part I: Diffusive patterning in non-living systems. *Contemporary Physics*, *34*, 247–273. doi:10.1080/00107519308222085

Berglund, B., Lindvall, T., & Schwela, D. H. (1999). *Guidelines for community noise*. Paper presented at the World Health Organization Expert Task Force Meeting, London, UK.

Berglund, B., & Lindvall, T. (1995). Community noise. In Berglund, B., & Lindvall, T. (Eds.), *Archives of the Center for Sensory Research* (*Vol. 2*, pp. 1–195).

Bhandari, A., Khare, V., Santhosh, J., & Anand, S. (2007). Wavelet based compression technique of Electro-oculogram signals. *3rd Kuala Lumpur International Conference on Biomedical Engineering 2006. 15*, pp. 440-443. Kuala Lumpur: Springer.

Bickerton, D. (2002). Foraging versus social intelligence in the evolution of protolanguage. In Wray, A. (Ed.), *The transition to language* (pp. 207–225). Oxford, UK: Oxford University Press.

Blendon, R. J., DesRoches, C. M., Cetron, M. S., Benson, J. M., Meinhardt, T., & Pollard, W. (2006). Attitudes toward the use of quarantine in a public health emergency in four countries. *Health Affairs*, *25*(2), 15. doi:10.1377/hlthaff.25.w15

Blendon, R. J., Koonin, L. M., Benson, J. M., Cetron, M. S., Pollard, W. E., & Mitchell, E. W. (2008). Public response to community mitigation measures for pandemic influenza. *Emerging Infectious Diseases*, *14*(5), 778. doi:10.3201/eid1405.071437

Bocci, V. (1991). Interleukins: Clinical pharmacokinetics and practical implications. *Clinical Pharmacokinetics*, *21*(4), 274–284. doi:10.2165/00003088-199121040-00004

Bone, R., Balk, R., Cerra, F., Dellinger, R., Fein, A., & Knaus, W. (1992). Definitions for sepsis and organ failure and guidelines for the use of innovative therapies in sepsis. *Chest*, *101*(6), 1644–1655. doi:10.1378/chest.101.6.1644

Bose, P., & Morin, P. (2004). Competitive online routing in geometric graphs. *Online Algorithms in Memoriam. Steve Seiden*, *324*(2-3), 273–288.

Boutalis, Y., Kottas, T., & Christodoulou, M. A. (2009). Adaptive estimation of fuzzy cognitive maps with proven stability and parameter convergence. *IEEE Transactions on Fuzzy Systems, 17*, 874–889. doi:10.1109/TFUZZ.2009.2017519

Branger, J., van den Blink, B., Weijer, S., Madwed, J., Bos, C. L., & Gupta, A. (2002). Anti-inflammatory effects of a p38 mitogen-activated protein kinase inhibitor during human endotoxemia. *Journal of Immunology (Baltimore, MD.: 1950), 168*(8), 4070–4077.

Branwood, A. (1991). Interleukins: Clinical pharmacokinetics and practical implications. *Clinical Pharmacokinetics, 21*(4), 274–284.

Brinck, I., & Gärdenfors, P. (2003). Co-operation and communication in apes and humans. *Mind & Language, 18*, 484–501. doi:10.1111/1468-0017.00239

Brinkhoff, T. (2002). A framework for generating network-based moving objects. *GeoInformatica, 6*(2), 153–180. doi:10.1023/A:1015231126594

Brit, O., Vounatsou, P., Gunawardena, D., Galappaththy, G., & Amerasinghe, P. (2008). Models for short term malaria prediction in Sri Lanka. *Malaria Journal, 7*, 1475–2875.

Britton, T., Janson, S., & Martin-Lof, A. (2007). Graphs with specified degree distributions, simple epidemics, and local vaccination strategies. *Advances in Applied Probability, 39*, 922–948. doi:10.1239/aap/1198177233

Brothers, L. (1990). The social brain: A project for integrating primate behavior and neurophysiology in a new domain. *Concepts in Neuroscience, 1*, 27–51.

Brundage, J. F., & Shanks, G. D. (2008). Deaths from bacterial pneumonia during 1918–19 influenza pandemic. *Emerging Infectious Diseases, 14*(8), 1193. doi:10.3201/eid1408.071313

Bu§neag, D., &, Piciu, D. (2003). On the lattice of deductive systems of BL-algebras. *Central European Journal of Mathematics, 1*(2), 221-237.

Bulling, A., Roggen, D., & Tröster, G. (2008). It's in Your Eyes - Towards Context-Awareness and Mobile HCI Using Wearable EOG Goggles. *ACM Proceedings of the 10th International Conference on Ubiquitous Computing (UbiComp 2008)*, (pp. 84-93). Seoul, Korea.

Bulling, A., Roggen, D., & Tröster, G. (2008). Robust Recognition of Reading Activity in Transit Using Wearable Electrooculography. *Proceedings of the 6th International Conference on Pervasive Computing (Pervasive 2008)* (pp. 19-37). Sydney, Australia: Springer.

Bureau of Transportation Statistics. (2002). *2001 national household travel survey (NTHS)*. Retrieved from http://www.bts.gov/programs/national_household_travel_survey/

Burg, J. P. (1975). *Maximal Entropy Spectral Analysis, PhD Thesis*. Stanford University, Department of Geophysics.

Byrne, R. W. (1997). The technical intelligence hypothesis: An additional evolutionary stimulus to intelligence? In A. Whiten & R. W. Byrne (Eds.), *Machiavellian intelligence, vol. II: Extensions and evaluations* (pp. 289-211). Cambridge, UK: Cambridge University Press.

Byrne, B. (2001). *Structural Equation Modeling with Amos: Basic Concepts, Applications, and Programming*. Mahwah, NJ: Lawrence Erlbaum.

Carden, R., & DelliFraine, J. (2004). An examination of hospital satisfaction with blood suppliers. *Transfusion, 44*, 1648–1655. doi:10.1111/j.0041-1132.2004.04184.x

Carlyle, K. (2009). *Optimizing quarantine regions through graph theory and simulation*. Unpublished master's thesis, Kansas State University, Manhattan, KS.

Carpenter, T., Thurmond, M., & Bates, T. (2004). A simulation model of intraherd trans- mission of foot and mouth disease with reference to disease spread before and after clinical diagnosis. *Journal of Veterinarian Diagnostic Investigation, 16*, 11–16. doi:10.1177/104063870401600103

Carpio, A. (2005). Asymptotic construction of pulses in the discrete Hodgkin-Huxley model for myelinated nerves. *Physical Review Letters E, 72.*

Carrat, F., Lavenu, A., Cauchemez, S., & Deleger, S. (2006). Repeated influenza vaccination of healthy children and adults: Borrow now, pay later? *Epidemiology and Infection, 134*(1), 63–70. doi:10.1017/S0950268805005479

Cauchemez, S., Carrat, F., Viboud, C., Valleron, A., & Boelle, P. (2004). A Bayesian MCMC approach to study transmission of influenza: Application to household longitudinal data. *Statistics in Medicine, 23*(22), 3469–3487. doi:10.1002/sim.1912

CDC. (2010). *The Center for Disease Control and Prevention's flu website.* Retrieved from http://flu.gov/

Center for Polymer Studies. (2010). *Growing rough patterns: Electrodeposition.* Retrieved from http://polymer.bu.edu/ogaf/html/chp41.htm

Centers for Disease Control and Prevention (CDC). (1998). Rift Valley Fever-East Africa, 1997-1998. *Morbidity and Mortality Weekly Report, 47*(13), 261–264.

Centers for Disease Control and Prevention. (2006). *CDC influenza operational plan.* Retrieved from http://www.cdc.gov/flu/pandemic/cdcplan.htm

Centers for Disease Control and Prevention. (2007). *Interim pre-pandemic planning guidance: Community strategy for pandemic influenza mitigation in the united states.* Retrieved from http://www.pandemicflu.gov/plan/community/community_mitigation.pdf

Centers for Disease Control and Prevention. (2008). *Avian influenza: Current H5N1 situation.* Retrievedfrom http://www.cdc.gov/flu/avian/outbreaks/current.htm

Centers for Disease Control and Prevention. (2009). *Preparing for pandemic influenza.* Retrieved from http://www.cdc.gov/flu/pandemic/preparedness

Chakrabarty, R. K., Moosmüller, H., Arnott, W. P., Garro, M. A., Tian, G., & Slowik, J. G. (2009). Low fractal dimension cluster-dilute soot aggregates from a premixed flame. *Physical Review Letters, 102*(23). doi:10.1103/PhysRevLett.102.235504

Chang, H., & Chang, C. (2008). An assessment of technology-based service encounters & network security on the e-health care systems of medical centers in Taiwan. *BMC Health Services Research, 8*(87).

Chang, W. K., Wei, C., & Huang, N. T. (2006). An approach to maximize hospital service quality under budget constraints. *Total Quality Management, 17*(6), 757–774. doi:10.1080/14783360600725040

Chao, D. L., Halloran, M. E., Obenchain, V. J., & Longini, I. M. (2010). FluTE, a publicly available stochastic influenza epidemic simulation model. *PLoS Computational Biology, 6*(1). doi:10.1371/journal.pcbi.1000656

Chen, D., Cui, D. W., Wang, Ch. X., & Wang, Z. R. (2006). A rough set-based hierarchical clustering algorithm for categorical data. *International Journal of Information Technology, 12*(3), 149–159.

Chengdong, W., Mengxin, L., Zhonghua, H., Zhang, Y., & Yong, Y. (2004). Discretization algorithms of rough sets using clustering. In *Proceedings of the IEEE International Conference on Robotics and Biomimetics* (pp. 955-960).

Chi, C. G., & Gursoy, D. (2009). Employee Satisfaction, Customer Satisfaction and Financial Performance: An Empirical Examination. *International Journal of Hospitality Management, 28*(2), 245–253. doi:10.1016/j.ijhm.2008.08.003

Chowdhury, R. S. (2010). Mathematical models for prediction and optimal mitigation of epidemics (Doctoral dissertation, University of Minnesota). *K-State Electronic Theses, Dissertations, and Reports.*

Chowdhury, R. S., Scoglio, C., & Hsu, W. (2009). Evolution and control strategies of the foot and mouth disease epidemic on a weighted contact network. In *Proceedings of the Second International Conference on Infectious Diseases Dynamics*, Athens, Greece (p. 2.01).

Chowdhury, S. R., Scoglio, C., & Hsu, W. (2010). Simulative modeling to control the foot and mouth disease epidemic. *Procedia Computer Science, 1*(1), 2261–2270. doi:10.1016/j.procs.2010.04.253

Chtioui, Y., Panigrahi, S., & Francl, L. (1999). A generalized regression neural network and its application for leaf wetness prediction to forecast plant disease. *Chemometrics and Intelligent Laboratory Systems, 48*, 47–58. doi:10.1016/S0169-7439(99)00006-4

Cignoli, R., & Torrens, A. (2005). Standard completeness of Hájek basic logic and decompositions of BL-chains. *Soft Computing, 12*, 862–868. doi:10.1007/s00500-004-0444-x

Clark, A. (1997). *Being there: Putting brain, body, and world together again.* Cambridge, MA: MIT Press.

Coker, D. (1997). An introduction to intuitionistic fuzzy topological spaces. *Fuzzy Sets and Systems, 88*, 81–89. doi:10.1016/S0165-0114(96)00076-0

Colardo Department of Human Services Division of Mental Health. (2009). *Pandemic influenza: Quarantine, isolation and social distancing.* Retrieved from http://www.flu.gov/news/colorado_toolbox.pdf

Colizza, V., Barrat, A., Barthelemy, M., Valleron, A., & Vespignani, A. (2007). Modeling the worldwide spread of pandemic influenza: Baseline case and containment interventions. *PLoS Medicine, 4*(1), 95. doi:10.1371/journal.pmed.0040013

Colizza, V., Barrat, A., Barthélemy, M., & Vespignani, A. (2006). The role of the airline transportation network in the prediction and predictability of global epidemics. *Proceedings of the National Academy of Sciences of the United States of America, 103*(7), 2015. doi:10.1073/pnas.0510525103

Committee on Modeling Community Containment for Pandemic Influenza. (2006). *Modeling community containment for pandemic influenza: A letter report.* Retrieved from http://www.nap.edu/catalog/11800.html

Cooley, P., Ganapathi, L., Ghneim, G., Holmberg, S., Wheaton, W., & Hollingsworth, C. R. (2008). Using influenza-like illness data to reconstruct an influenza outbreak. *Mathematical and Computer Modelling, 48*(5-6), 929–939. doi:10.1016/j.mcm.2007.11.016

Costa, J. M., Sagues, F., & Vilarrasa, M. (1991). Growth rate of copper electrodeposits: Potential and Concentration effects. *Physical Review Letters, 43*(12), 7057–7060.

Coxon, A., Tang, T., & Mayadas, T. N. (1999). Cytokine-activated endothelial cells delay neutrophil apoptosis in vitro and in vivo: A role for granulocyte/macrophage colony-stimulating factor. *The Journal of Experimental Medicine, 190*(7), 923–934. doi:10.1084/jem.190.7.923

Cristea, A., Zaharia, C., Deutsch, I., Bunescu, E., & Blujdescu, M. (1992). Mathematical modeling of measles epidemics and the optimization of corresponding antiepidemic programs. In *Proceedings of the 4th International Symposium on Systems Analysis and Simulation* (pp. 623-628).

Cronemberger, C. M., & Sampaio, L. C. (2006). Growth of fractal electrodeposited aggregates under action of electric and magnetic fields using a modified diffusion-limited aggregation algorithm. *Physical Review Letters, 73*(4).

Cudina, M., & Prezelj, J. (2005). Noise due to firecracker explosions. *Journal of Mechanical Engineering Science, 219*(6), 523–537.

Cummings, K. M., Jette, A. M., Brock, B. M., & Haefner, D. P. (1979). Psychosocial determinants of immunization behavior in a swine influenza campaign. *Medical Care, 17*(6), 639. doi:10.1097/00005650-197906000-00008

Curtis, C. E., & D'Esposito, M. (2003). Persistent activity in the prefrontal cortex during working memory. *Trends in Cognitive Sciences, 7*(9), 415–423. doi:10.1016/S1364-6613(03)00197-9

Darwin, C. (1871). *The descent of man, and selection in relation to sex.* London, UK: John Murray.

Das, T., & Savachkin, A. (2008). A large scale simulation model for assessment of societal risk and development of dynamic mitigation strategies. *IIE Transactions, 40*(9), 893–905. doi:10.1080/07408170802165856

Daubney, R., Hudson, J. R., & Garnham, P. C. (1931). Enzootic hepatitis of Rift Valley Fever, an undescriptible virus disease of sheep, cattle and man from East Africa. *The Journal of Pathology and Bacteriology, 34*, 543–579. doi:10.1002/path.1700340418

Davies, D. L., & Bouldin, D. W. (1979). A cluster separation measure. *Pattern Analysis and Machine Intelligence. IEEE Transactions on, PAMI-1*(2), 224–227.

De Duve, C. (2003). *Live evolving: Molecules, mind, and meaning.* New York, NY: Oxford University Press.

Deacon, T. (1997). *The symbolic species: The co-evolution of language and the brain.* New York, NY: W. W. Norton.

deCastro, M., Hofer, E., Munuzurri, A. P., Gomez-Gesteira, M., Plamck, G., & Schafferhofer, I. (1999). Comparison between the role of discontinuities in cardiac conduction and in a one-dimensional hardware model. *Physical. Review Letters E, 59*, 5962–5969.

Denney, D., & Denney, C. (1984). The eye blink electro-oculogram. *The British Journal of Ophthalmology, 64*, 225–228. doi:10.1136/bjo.68.4.225

Deshpande, A. (2005). *Noise pollution scenario in Maharashtra during Diwali festival.* Mumbai, India: Maharashtra Pollution Control Board. Retrieved from http://mpcb.gov.in/images/pdf/noisereport2005.pdf

Devaney, R. L. (1986). *An introduction to chaotic dynamical systems*. Menlo Park, CA: Benjamin/Cummings.

Devaney, R. L. (1992). *A first course in chaotic dynamical systems, theory and experiment*. Reading, MA: Addison-Wesley.

Dhillon, H. S., Singla, R., Rekhi, N. S., & Jha, R. (2009). EOG and EMG Based Virtual Keyboard: A Brain-Computer Interface. *Computer Science and Information Technology, 2nd International Conference on* (pp. 259-262). Beijing, China: IEEE.

Di Paolo, E. A., Noble, J., & Bullock, S. (2000). Simulation models as opaque thought experiments. In *Proceedings of the Seventh International Conference on Artificial Life* (pp. 497-506).

Diekmann, O., & Heesterbeek, J. A. P. (2000). *Mathematical epidemiology of infectious diseases: Model building, analysis, and interpretation*. New York, NY: John Wiley & Sons.

Donini, L. M., Castellaneta, E., & Guglielmi, S. (2008). Improvement in the quality of the catering service of a rehabilitation hospital. *Clinical Nutrition (Edinburgh, Lothian), 27*, 105–114. doi:10.1016/j.clnu.2007.10.004

Dorin, A. (2005). Artificial life, death and epidemics in evolutionary, generative electronic art. In F. Rothlauf, J. Branke, S. Cagnoni, D. Wolfe Corne, R. Drechsler, Y. Jin et al. (Eds.), *Proceedings of the Evo Workshops on Applications of Evolutionary Computing* (LNCS 3449, pp. 448-457).

Douady, A., & Hubbard, J. H. (1985). On the dynamics of polynomial-like mappings. *Annales Scientifiques de l'E.N.S, 18*(2), 287-343.

Dovgoshey, O., Martio, O., Ryazanov, V., & Vuorinen, M. (2006). The Cantor function. *Expositiones Mathematicae, 24*(1), 1–37. doi:10.1016/j.exmath.2005.05.002

Dowell, L. J., & Bruno, M. L. (2000). Connectivity of random graphs and mobile networks: Validation of Monte Carlo simulation results. In *Proceedings of the ACM Symposium on Applied Computing* (pp. 77-81).

Dunbar, R. I. M. (1998). The social brain hypothesis. *Evolutionary Anthropology, 6*(5), 178–190. doi:10.1002/(SICI)1520-6505(1998)6:5<178::AID-EVAN5>3.0.CO;2-8

Dutta, P., & Horn, P. M. (1981). Low-frequency fluctuations in solids: 1/*f* noise. *Reviews of Modern Physics, 53*, 497. doi:10.1103/RevModPhys.53.497

Dy, J. G., & Brodley, C. E. (2004). Feature selection for unsupervised learning. *Journal of Machine Learning Research, 5*, 884–889.

Eiseman, B., Beart, R., & Norton, L. (1977). Multiple organ failure. *Surgery, Gynecology & Obstetrics, 144*(3), 323–326.

El Naschie, M. S. (1998). On uncertainty of Cantorian geometry and two-slit experiment. *Chaos, Solitons, and Fractals, 9*, 517–529. doi:10.1016/S0960-0779(97)00150-1

El Naschie, M. S. (2000). On the unification of heterotic strings theory and $E^{(\infty)}$ theory. *Chaos, Solitons, and Fractals, 11*(14), 2397–2408. doi:10.1016/S0960-0779(00)00108-9

El Naschie, M. S. (2004). A review of E-irifinity theory and the mass spectrum of high energy particle physics. *Chaos, Solitons, and Fractals, 19*, 209–236. doi:10.1016/S0960-0779(03)00278-9

El Naschie, M. S. (2004). Quantum gravity, Clifford algebras, fuzzy set theory and the fundamental constants of nature. *Chaos, Solitons, and Fractals, 20*, 437–450. doi:10.1016/j.chaos.2003.09.029

El Naschie, M. S. (2006). Fuzzy dodecahedron topology and *E*-infinity space time as a model for quantum physics. *Chaos, Solitons, and Fractals, 30*, 1025–1033. doi:10.1016/j.chaos.2006.05.088

El Naschie, M. S. (2006). On two new fuzzy Kahler manifolds, Klein modular space and Hooft holographic principles. *Chaos, Solitons, and Fractals, 29*, 876–881. doi:10.1016/j.chaos.2005.12.027

El Naschie, M. S. (2007). A review of applications and results of *E*-infinity theory. *International Journal of Nonlinear Sciences and Numerical Simulation, 8*, 11–20. doi:10.1515/IJNSNS.2007.8.1.11

Emilyn, J. J., & Ramar, K. (2010). Rough set based clustering of gene expression data: A survey. *International Journal of Engineering Science & Technology, 2*(12), 7160–7164.

Encarnacao, J. L., Peitgen, H. O., Sakas, G., & Englert, G. (1992). *Fractal geometry and computer graphics*. Berlin, German: Springer-Verlag.

Epstein, J. M. (2009). Modeling to contain pandemics. *Nature*, 406.

Epstein, J. M., Parker, J., Cummings, D., & Hammond, R. A. (2008). Coupled contagion dynamics of fear and disease: Mathematical and computational explorations. *PLoS ONE*, *3*(12), 3955. doi:10.1371/journal.pone.0003955

Erceg, M. A. (1979). Metric spaces in fuzzy set theory. *Journal of Mathematical Analysis and Applications*, *69*, 205–230. doi:10.1016/0022-247X(79)90189-6

Estrany, B., Fuster, P., Garcia, A., & Luo, Y. (2009). EOG signal processing and analysis for controlling computer by eye movements. *Pervasive Technologies Related to Assistive Environments, Proceedings of the 2nd International Conference on*. Corfu, Greece: ACM.

Eubank, S., Guclu, H., Kumar, A. V. S., Marathe, M., Srinivasan, A., & Toroczkai, Z. (2004). Modeling disease outbreaks in realistic urban social networks. *Nature*, *429*, 180–184. doi:10.1038/nature02541

Falconer, K. (1990). *Fractal Geometry: Mathematical Foundation and Applications*. Chichester, UK: John Wiley & Sons.

Fang, J. X. (2002). A note on the completions of fuzzy metric spaces and fuzzy normed spaces. *Fuzzy Sets and Systems*, *131*, 399–407. doi:10.1016/S0165-0114(02)00054-4

Fargues, M. P., & Bennett, R. (1995). Comparing Wavelet Transforms and AR Modeling as Feature Extraction Tools for Underwater Signal Classification. *Proceedings of the Conference Record of the Twenty-Ninth Asilomar Conference on Signals, Systems and Computers (ASILOMAR '95)* (pp. 915 - 919). Pacific Grove, CA, USA: IEEE.

Farrell, J. A. (1998). Stability and approximator convergence in nonparametric nonlinear adaptive control. *IEEE Transactions on Neural Networks*, *9*, 1008–1020. doi:10.1109/72.712182

Farrell, J. A., & Polycarpou, M. M. (2006). *Adaptive approximation based control: Unifying, neural, fuzzy and traditional adaptive approximation approaches*. Hoboken, NJ: John Wiley & Sons.

Faust, O., Acharyaa, R., Allen, A., & Lin, C. (2008). Analysis of EEG signals during epileptic and alcoholic states using AR modeling techniques. *IRBM*, *29*(1), 44–52. doi:10.1016/j.rbmret.2007.11.003

Fedson, D. S. (2003). Pandemic influenza and the global vaccine supply. *Clinical Infectious Diseases*, *36*(12), 1552–1561. doi:10.1086/375056

Feigenbaum, M. (1979). The universal metric properties of nonlinear transformation. *Journal of Statistical Physics*, *21*, 669–706. doi:10.1007/BF01107909

Felbin, C. (1992). Finite dimensional fuzzy normed linear space. *Fuzzy Sets and Systems*, *48*, 239–248. doi:10.1016/0165-0114(92)90338-5

Feng, W. (1997). On m-point nonlinear boundary value problem. *Nonlinear Analysis*, *30*(6), 5369–5374. doi:10.1016/S0362-546X(97)00360-X

Feng, W., & Webb, J. R. L. (1997). Solvability of m-point boundary value problems with nonlinear growth. *International Journal of Mathematical Analysis and Applications*, *212*, 467–480. doi:10.1006/jmaa.1997.5520

Ferguson, N. M., Cummings, D. A. T., Cauchemez, S., Fraser, C., Riley, S., & Meeyai, A. (2005). Strategies for containing an emerging influenza pandemic in southeast Asia. *Nature*, *437*(7056), 209–214. doi:10.1038/nature04017

Ferguson, N. M., Cummings, D. A. T., Fraser, C., Cajka, J. C., Cooley, P. C., & Burke, D. S. (2006). Strategies for mitigating an influenza pandemic. *Nature*, *442*(7101), 448–452. doi:10.1038/nature04795

Ferguson, N. M., Mallett, S., Jackson, H., Roberts, N., & Ward, P. (2003). A population-dynamic model for evaluating the potential spread of drug-resistant influenza virus infections during community-based use of antivirals. *The Journal of Antimicrobial Chemotherapy*, *51*(4), 977. doi:10.1093/jac/dkg136

Ferguson, N., Donnelly, C., & Anderson, R. (2001). The foot-and-mouth epidemic in Great Britain: pattern of spread and impact of interventions. *Science*, *292*, 1155–1160. doi:10.1126/science.1061020

Fleury, V., Chazalviel, J.-N., Rosso, M., & Sapoval, B. (1990). The growth speed of electrochemical deposits. *International Journal of Electroanalytical Chemistry*, *290*, 249. doi:10.1016/0022-0728(90)87434-L

Flinn, M. V., Geary, D. C., & Ward, C. V. (2005). Ecological dominance, social competition, and coalitionary arms races: Why humans evolved extraordinary intelligence. *Evolution and Human Behavior*, *26*, 10–46. doi:10.1016/j.evolhumbehav.2004.08.005

Fortunato, S. (2005). Damage spreading and opinion dynamics on scale-free networks. *Physica A*, *348*, 683–690. doi:10.1016/j.physa.2004.09.007

Fradkov, A. L., & Evans, R. J. (2005). Control of chaos: Methods and applications in engineering. *Chaos, Solitons, and Fractals*, *29*, 33–56.

Francl, L., & Panigrahi, S. (1997). Artificial neural network models of wheat leaf wetness. *Agricultural and Forest Meteorology*, *88*, 57–65. doi:10.1016/S0168-1923(97)00051-8

Fraser, C., Donnelly, C. A., Cauchemez, C., Hanage, W. P., Van Kerkhove, M. D., & Hollingsworth, T. D. (2009). Pandemic potential of a strain of influenza A (H1N1): Early findings. *Science*, *324*(5934), 1557–1561. doi:10.1126/science.1176062

Fuchs, A. C., Granowitz, E. V., Shapiro, L., Vannier, E., Lonnemann, G., & Angel, J. B. (1996). Clinical, hematologic, and immunologic effects of interleukin-10 in humans. *Journal of Clinical Immunology*, *16*(5), 291–303. doi:10.1007/BF01541395

Fuks, H., Lawniczak, A., & Duchesne, R. (2006). Effects of population mixing on the spread of SIR epidemics. *The European Physical Journal B*, *50*(1-2), 209–214. doi:10.1140/epjb/e2006-00136-7

Gaff, H. D., Hartley, D. M., & Leahy, N. P. (2007). An epidemiological model of Rift Valley Fever. *Electronic Journal of Differential Equations*, *115*, 1–12.

Gähler, S. (1963). 2-metrische Räume und ihre topologische Struktur. *Mathematische Nachrichten*, *26*, 115–148. doi:10.1002/mana.19630260109

Gähler, S. (1965). Lineare 2-normietre Räume. *Mathematische Nachrichten*, *28*, 1–43. doi:10.1002/mana.19640280102

Gambart, R., Myncke, H., & Cops, A. (1976). Study of annoyance by traffic noise in Leuven (Belgium). *Applied Acoustics*, *9*(3), 193. doi:10.1016/0003-682X(76)90017-7

Gardner, M. (1978). Mathematical games -- white and brown music, fractal curves and one-over-f fluctuations. *Scientific American*, *238*(4), 16. doi:10.1038/scientificamerican0478-16

George, A., & Veeramani, P. (1994). On some results in fuzzy metric spaces. *Fuzzy Sets and Systems*, *64*, 395–399. doi:10.1016/0165-0114(94)90162-7

Gerdes, G. H. (2004). Rift Valley Fever. *Revue Scientifique et Technique (International Office of Epizootics)*, *23*(2), 613–623.

Germann, T. C., Kadau, K., Longini, I. M., & Macken, C. A. (2006). *Mitigation strategies for pandemic influenza in the United States*. Washington, DC: The National Academy of Sciences.

Gershenson, C. (2007). The world as evolving information. In *Proceedings of the International Conference on Complex Systems*.

Gershenson, C. (2004). Cognitive paradigms: Which one is the best? *Cognitive Systems Research*, *5*(2), 135–156. doi:10.1016/j.cogsys.2003.10.002

Ge, S. S., Hang, C. C., Lee, T. H., & Zhang, T. (2002). *Stable adaptive neural network control*. Dordrecht, The Netherlands: Kluwer Academic Publishers.

Gianvitorio, J., & Rahmat, Y. (2002). Fractal Antennas: A Novel Antenna Miniaturization Technique and Applications. *IEEE Antennas and Propagation Magazine*, *44*(1), 20–36. doi:10.1109/74.997888

Gibson, D., Kleinberg, J. M., & Raghavan, P. (1998). Clustering categorical data: An approach based on dynamic systems. In *Proceedings of the International Conference on Very Large Databases* (pp. 311- 323).

Gilden, D. L., Thornton, T., & Mallon, M. W. (1995). 1/f noise in human cognition. *Science*, *267*, 1837–1839. doi:10.1126/science.7892611

Giles, R. (1980). A computer program for fuzzy reasoning. *Fuzzy Sets and Systems*, *4*, 221–234. doi:10.1016/0165-0114(80)90012-3

Glass, R. J., Glass, L. M., Beyeler, W. E., & Min, H. J. (2006). Targeted social distancing design for pandemic influenza. *Emerging Infectious Diseases, 12*(11), 1671–1681.

Goebel, K., & Kirk, W. A. (1990). *Topics in metric fixed point theory* (pp. 27–35). Cambridge, UK: Cambridge University Press. doi:10.1017/CBO9780511526152.004

Gogos, C. A., Drosou, E., Bassaris, H. P., & Skoutelis, A. (2000). Pro- versus anti-inflammatory cytokine in patients with severe sepsis: A marker for prognosis and future therapeutic options. *The Journal of Infectious Diseases, 181*(1), 176–180. doi:10.1086/315214

Goldman, L., & Albus, J. S. (1968). Computation of impulse conduction in myelinated fibres; theoretical basis of the velocity-diameter relation. *Biophysical Journal, 8*, 596–607. doi:10.1016/S0006-3495(68)86510-5

Goldstein, M. (1994). *Low-frequency components in complex noise and their perceived loudness and annoyance.* Solna, Sweden: National Institute of Occupational Health.

Gordis, L., & Saunders, W. B. (2008). *Epidemiology* (4th ed.). Amsterdam, The Netherlands: Elsevier.

Gosavi, A., Das, T. K., & Sarkar, S. (2004). A simulation-based learning automata framework for solving semi-markov decision problems under long-run average reward. *IIE Transactions, 36*(6), 557–567. doi:10.1080/07408170490438672

Grand, S. (2003). *Creation: Life and how to make it.* Cambridge, MA: Harvard University Press.

Gui, M. (2009). Advanced methods for prediction of animal-related outages in overhead distribution systems (Doctoral dissertation, Kansas State University). *K-State Electronic Theses, Dissertations, and Reports.*

Guimaraes, P., de Menezes, M., Baird, R., Lusseau, D., Guimaraes, P., & dos Reis, S. (2007). Vulnerability of a killer whale social network to disease outbreaks. *Physical Review E: Statistical, Nonlinear, and Soft Matter Physics, 76*, 1–4. doi:10.1103/PhysRevE.76.042901

Gulz, A. (1991). *The planning of action as a cognitive and biological phenomenon.* Lund, Sweden: Lund University Cognitive Studies.

Guo, D., & Lakshmikantham, V. (1988). *Nonlinear Problems in Abstract Cones.* San Diego, CA: Academic Press.

Gupta, G. P. (1992). Solvability of a three point nonlinear boundary value problem for a second order ordinary differential equation. *International Journal of Mathematical Analysis and Applications, 168*, 5450–5551.

Gupta, G. P., Ntouyas, S. K., & Tsamatos, P. C. (1994). On m-point boundary value problem for second order differential equations. *Nonlinear Analysis, 23*(11), 1427–1436. doi:10.1016/0362-546X(94)90137-6

Gupta, P., & Kumar, P. R. (1998). Critical power for asymptotic connectivity in wireless networks. In McEneany, W. M., Yin, G. G., & Zhang, Q. (Eds.), *Stochastic analysis, control, optimization and applications* (pp. 547–566). Boston, MA: Birkhausen.

Hair, J. F., Anderson, R. E., Tatham, R. L., & Black, W. C. (1995). *Multivariate data analysis.* Upper Saddle River, NJ: Prentice Hall.

Hájek, P. (1998). Basic fuzzy logic and BL-algebras. *Soft Computing, 2*, 124–128. doi:10.1007/s005000050043

Hájek, P. (1998). *Metamathematics of fuzzy logic.* Dordrecht, The Netherlands: Kluwer Academic. doi:10.1007/978-94-011-5300-3

Halder, N., Kelso, J. K., & Milne, G. J. (2010). Analysis of the effectiveness of interventions used during the 2009 A/H1N1 influenza pandemic. *BMC Public Health, 10*, 168. doi:10.1186/1471-2458-10-168

Halfhill, T. (2009). *Inflation calculator.* Retrieved from http://www.halfhill.com/inflatation.html

Halloran, M. E. (2006). Invited commentary: Challenges of using contact data to understand acute respiratory disease transmission. *American Journal of Epidemiology, 164*(10), 945. doi:10.1093/aje/kwj318

Halloran, M. E., Ferguson, N. M., Eubank, S., Longini, I. M., Cummings, D. A. T., & Lewis, B. (2008). Modeling targeted layered containment of an influenza pandemic in the United States. *Proceedings of the National Academy of Sciences of the United States of America, 105*(12), 4639. doi:10.1073/pnas.0706849105

Handel, A., Longini, I. M., & Antia, R. (2010). Towards a quantitative understanding of the within-host dynamics of influenza A infections. *Journal of the Royal Society, Interface*, *7*(42), 35. doi:10.1098/rsif.2009.0067

Hanson, C., Towers, D., & Meister, L. (2006). *Transit noise and vibration impact assessment* (Tech. Rep. No. FTA-VA-90-1003-06). Washington, DC: Department of Transportation Federal Transit Administration Office of Planning and Environment.

Haq, R. U., Sulaiman, N., & Rani, M. (2010, June 28-29). Superior fractal antennas. In *Proceedings of the Malaysian Technical University Conference on Engineering and Technology*, Melaka, Malaysia (pp. 23-26).

Hardin, G. (1968). The tragedy of the commons. *Science*, *162*(3859), 1243–1248. doi:10.1126/science.162.3859.1243

Haveshki, M., Saied, A., & Eslami, E. (2006). Some types of filters in BL-algebras. *Soft Computing*, *10*, 657–664. doi:10.1007/s00500-005-0534-4

Hawksworth, A., Hansen, C., Good, P., Ryan, M., Russell, K., & Kelley, P. (2003). Using autoregressive epidemic modeling to augment the existing department of defense (dod) febrile respiratory illness surveillance system at military training centers. *Journal of Urban Health*, *80*, 38. doi:10.1007/BF02416933

Hénon, M. (1976). A two dimensional mapping with a strange attractor. *Communications in Mathematical Physics*, *50*, 69–77. doi:10.1007/BF01608556

Hibbert, D. B. (1991). Fractals in chemistry. *Chemometrics and Intelligent Laboratory Systems*, *11*, 1–11. doi:10.1016/0169-7439(91)80001-7

Hill, K. (1982). Hunting and human evolution. *Journal of Human Evolution*, *11*, 521–544. doi:10.1016/S0047-2484(82)80107-3

Hirano, S., & Tsumoto, S. (2006). On the nature of degree of indiscernibility for rough clustering. In *Proceedings of the IEEE International Conference on Systems, Man, and Cybernetics* (pp. 3447-3452).

Hirota, R., & Suzuky, K. (1970). Studies on lattice solitons by using electrical networks. *Journal of the Physical Society of Japan*, *28*(5), 1366. doi:10.1143/JPSJ.28.1366

Hobbie, R. K. (1982). *Intermediate physics for medicine and biology* (3rd ed.). New York, NY: Springer.

Hong, L., & Sun, J. Q. (2006). Bifurcations of fuzzy nonlinear dynamical systems. *Communications in Nonlinear Science and Numerical Simulation*, *1*, 1–12. doi:10.1016/j.cnsns.2004.11.001

Hooge, F. N., Kleinpenning, T. G. M., & Vandamme, L. K. J. (1981). Experimental studies on 1/f noise. *Reports on Progress in Physics*, *44*(5), 479. doi:10.1088/0034-4885/44/5/001

Hoon, M. J., Hagen, T. H., Schoonewelle, H., & Dam, H. v. (1996). Why Yule-Walker should not be used for autoregressive modelling. *Annals of Nuclear Energy*, *23*(15), 1219–1228. doi:10.1016/0306-4549(95)00126-3

Hopfield, J. J. (1994). Physics, computation, and why biology looks so different. *Journal of Theoretical Biology*, *171*, 53–60. doi:10.1006/jtbi.1994.1211

Horiguchi, T., & Morita, T. (1984). Fractal dimension related to devil's staircase for a family of piecewise linear mappings. *Physica A*, *128*(1-2), 289–295. doi:10.1016/0378-4371(84)90092-X

Hori, J., Sakano, K., Miyakawa, M., & Saitoh, Y. (2006). Eye Movement Communication Control System Based on EOG and Voluntary Eye Blink. *Computers Helping People with Special Needs. Lecture Notes in Computer Science*, *4061*, 950–953. doi:10.1007/11788713_138

Hsiao, C., Pai, J., & Chiu, H. (2009). The study on the outsourcing of Taiwan's hospitals: a questionnaire survey research. *BMC Health Services Research*, *9*, 78. doi:10.1186/1472-6963-9-78

Huang, Z. (1997). A fast clustering algorithm to cluster very large categorical data sets in data mining. In *Proceedings of the SIGMOD Workshop on Research Issues on Data Mining and Knowledge Discovery*.

Huang, C.-Y., Hsieh, J.-L., Sun, C.-T., & Cheng, C.-Y. (2006). Teaching epidemic and public health policies through simulation. *WSEAS Transactions on Information Science and Applications*, *3*, 899–904.

Huang, Z. (1999). Extensions to the k-means algorithm for clustering large data sets with categorical values. *Data Mining and Knowledge Discovery*, *2*, 283–304. doi:10.1023/A:1009769707641

Huhn, R. D., Radwanski, E., Gallo, J., Affrime, M. B., Sabo, R., & Gonyo, G. (1997). Pharmacodynamics of subcutaneous recombinant human interleukin-10 in healthy volunteers. *Clinical Pharmacology and Therapeutics*, *62*(2), 171–180. doi:10.1016/S0009-9236(97)90065-5

Humphrey, N. K. (1976). The social function of intellect. In Bateson, P. P. G., & Hinde, R. A. (Eds.), *Growing points in ethology* (pp. 303–317). Cambridge, UK: Cambridge University Press.

Hutchinson, J. E. (1981). Fractals and self-similarity. *Indiana University Mathematics Journal*, *30*(5), 713–747. doi:10.1512/iumj.1981.30.30055

Huttenlocher, D. P., Klanderman, G. A., & Rucklidge, W. J. (1993). Comparing Images using the Hausdorff Distance. *IEEE Transactions on Pattern Analysis and Machine Intelligence*, *15*(9), 850–863. doi:10.1109/34.232073

Institute of Medicine (Ed.). (2008). *Antivirals for pandemic influenza: Guidance on developing a distribution and dispensing program*. Washington, DC: The National Academies Press.

Ioannou, P. A., & Sun, J. (1996). *Robust adaptive control*. Upper Saddle River, NJ: Prentice Hall.

Jacksonville Aviation Authority. (2010). *Daily traffic volume data*. Retrieved from http://www.jaa.aero/General/Default.aspx

Jain, A. K., Murthy, M. N., & Flynn, P. J. (1999). Data Clustering: A review. *ACM Computing Surveys*, *31*(3), 264–323. doi:10.1145/331499.331504

Janeway, C., Travers, P., Walport, M., & Shlomchik, M. (2001). *Immunobiology: The immune system in health and disease*. New York, NY: Garland Publishing.

Jiang, X., & Wallstrom, G. (2006). A bayesian network for outbreak detection and prediction. In *Proceedings of the 21st National Conference on Artificial Intelligence* (Vol. 2, pp. 1155-1160).

Jiang, D., & Wang, J. (2000). On-line learning of dynamical systems in the presence of model mismatch and disturbances. *IEEE Transactions on Neural Networks*, *11*, 1272–1283. doi:10.1109/72.883420

Jibrael, F. J. (2008). Miniature Dipole Antenna Based on Fractal Square Koch Curve. *European Journal of Scientific Research*, *21*(4), 700–706.

Jolly, A. (1999). *Lucy's legacy: Sex and intelligence in human evolution*. Cambridge, MA: Harvard University Press.

Jun, Y. B., & Miko, J. (2004). Folding theory applied to BL-algebras. *Central European Journal of Mathematics*, *4*, 584–592. doi:10.2478/BF02475965

Jun, Y. B., Shim, W. H., & Lele, C. (2002). Fuzzy filters /ideals in BCI-algebras. *Journal of Fuzzy Mathematics*, *10*, 469–474.

Jupp, P. G., Kemp, A., Grobbelaar, A., Leman, P., Burt, F. J., & Alahmed, A. M. (2002). The 2000 epidemic of Rift Valley Fever in Saudi Arabia: Mosquito vector studies. *Medical and Veterinary Entomology*, *15*(3), 245–252. doi:10.1046/j.1365-2915.2002.00371.x

Kaleva, O., & Seikkala, S. (1984). On fuzzy metric spaces. *Fuzzy Sets and Systems*, *12*, 215–229. doi:10.1016/0165-0114(84)90069-1

Kauffman, S. A. (2000). *Investigations*. New York, NY: Oxford University Press.

Keane, M. T., Walter, M. V., Patel, B. I., Moorthy, S., Stevens, R. B., & Bradley, K. M. (2005). Confidence in vaccination: A parent model. *Vaccine*, *23*(19), 2486–2493. doi:10.1016/j.vaccine.2004.10.026

Keeling, M. J., Woolhouse, M. E. J., Shaw, D. J., Matthews, L., Chase-Topping, M., & Haydon, D. T. (2001). Dynamics of the 2001 UK foot and mouth epidemic: Stochastic dispersal in a heterogeneous landscape. *Science*, *294*, 813–817. doi:10.1126/science.1065973

Kelso, J. K., Milne, G. J., & Kelly, H. (2009). Simulation suggests that rapid activation of social distancing can arrest epidemic development due to a novel strain of influenza. *BMC Public Health*, *9*, 117. doi:10.1186/1471-2458-9-117

Khopkar, S. M. (1993). *Environmental pollution analysis*. New Delhi, India: New Age International.

Kiamini, M., Alirezaee, S., Perseh, B., & Ahmadi, M. (2008). A wavelet based algorithm for Ocular Artifact detection In the EEG signals. *Multitopic Conference, 2008. INMIC 2008. IEEE International*, (pp. 165-168). Karachi.

Kim, T., Hwang, W., Zhang, A., Sen, S., & Ramanathan, M. (2008). Multi-agent model analysis of the containment strategy for avian influenza (AI) in South Korea. In *Proceedings of the IEEE International Conference on Bioinformatics and Biomedicine* (pp. 353-356).

Kim, T., Hwang, W., Zhang, A., Ramanathan, M., & Sen, S. (2009). Damage isolation via strategic self-destruction: A case study in 2d random networks. *Europhysics Letters*, *86*(2), 24002. doi:10.1209/0295-5075/86/24002

Kim, T., Hwang, W., Zhang, A., Sen, S., & Ramanathan, M. (2010). Multi-agent modeling of the South Korean avian influenza epidemic. *BMC Infectious Diseases*, *10*(236).

Kim, Y., Cho, C., & Ahn, S. (2008). A study on medical services quality and its influence upon value of care and patient satisfaction. *Total Quality Management*, *19*(11), 1155–1171. doi:10.1080/14783360802323594

Kobayashi, M., Dickey, B., Carpenter, T., & Howitt, R. E. (2007). A dynamic optimal disease control model for foot-and-mouth disease: Model results and policy implications. *Preventive Veterinary Medicine*, *79*, 274–286. doi:10.1016/j.prevetmed.2007.01.001

Kogan, S. M. (1985). Low-frequency current 1/f-noise in solids. *Uspekhi Fizicheskikh Nauk*, *145*, 285–328. doi:10.3367/UFNr.0145.198502d.0285

Kolesin, I. D. (2007). Mathematical model of the development of an epidemic process with aerosol transmission. *Biophysics*, *52*(1), 92–94. doi:10.1134/S0006350907010150

Koles, Z. J., & Rasminsky, M. (1972). A computer simulation of conduction in demyelinated nerve fibres. *The Journal of Physiology*, *227*, 351–364.

Kondo, M., & Dudek, W. A. (2005). On the transfer principle in fuzzy theory. *Mathware and Soft Computing*, *12*, 41–55.

Kondo, M., & Dudek, W. A. (2008). Filters theory of BL-algebras. *Soft Computing*, 419–423. doi:10.1007/s00500-007-0178-7

Korzeniewski, B. (2001). Cybernetic formulation of the definition of life. *Journal of Theoretical Biology*, *209*(3), 275–286. doi:10.1006/jtbi.2001.2262

Kosko, B. (1986). Fuzzy cognitive maps. *International Journal of Man-Machine Studies*, 65–75. doi:10.1016/S0020-7373(86)80040-2

Kottas, T. L., Boutalis, Y. S., & Christodoulou, M. A. (2007). Fuzzy cognitive networks: A general framework. *Intelligent Decision Technologies*, *1*(4), 183–196.

Krasnoselkii, M. A. (1964). *Positive Solutions of Operator Equations*. Groningen, The Netherlands: P. Noordhoff.

Krishnamachari, B., Wicker, S. B., & Bajar, R. (2001). Phase transition phenomena in wireless ad-hoc networks. In *Proceedings of the Global Telecommunications Conference*.

Krishnamurti, T., & Chakraborti, V. A.and Mehta, & A.V., M. (2007). *Experimental prediction of climate-related malaria incidence*. Retrieved from www.pitt.edu/~super7/29011-30001/29021.ppt

Krolak, A., & Strumillo, P. (2009). Eye-Blink Controlled Human-Computer Interface for the Disabled. *Human-Computer Systems Interaction. Advances in Soft Computing*, *60*, 123–133.

Krowinski, W. J., & Steiber, S. R. (1996). *Measuring and managing patient satisfaction* (2nd ed.). Chicago, IL: Hospital Publishing.

Kucukarslan, S. (2008). Evaluating medication-related services in a hospital. *Research in Social & Administrative Pharmacy*, *4*, 12–22. doi:10.1016/j.sapharm.2007.01.001

Kumar, M., & Rani, M. (2005). A new approach to superior Julia sets. *Journal of Natural and Physical Science*, *19*(2), 148–155.

Kumar, R. (2002). Models of competing host-parasite pairs. *Indian Journal of Pure and Applied Mathematics*, *33*(10), 1515–1528.

Kumar, R., Clermont, G., Vodovotz, Y., & Chow, C. C. (2004). The dynamics of acute inflammation. *Journal of Theoretical Biology*, *230*(1), 145–155. doi:10.1016/j.jtbi.2004.04.044

Kurland, N. B., & Pelled, L. H. (2000). Passing the word: Toward a model of gossip and power in the workplace. *Academy of Management Review*, *25*(2), 428–438.

Lagazio, C., Dreassi, E., & Biggeri, A. (2001). A hierarchical Bayesian model for spacetime variation of disease risk. *Statistical Modelling*, *1*, 1729. doi:10.1191/147108201128069

Lai, D. (2005). Monitoring the sars epidemic in China: Time series analysis. *Journal of Data Science*, *3*, 279–293.

Langton, C. (1989). Artificial life. In Langton, C. (Ed.), *Artificial life: Santa Fe Institute studies in the sciences of complexity* (pp. 1–47). Reading, MA: Addison-Wesley.

Lasota, A., & Mackey, M. C. (1994). *Chaos, fractals, and noise: Stochastic aspects of dynamics*. New York, NY: Springer.

Lawless, J. F., & Lawless, J. (Eds.). (1982). *Statistical models and methods for lifetime data*. New York, NY: John Wiley & Sons.

Lee, P. Y., Matchar, D. B., Clements, D. A., Huber, J., Hamilton, J. D., & Peterson, E. D. (2002). Economic analysis of influenza vaccination and antiviral treatment for healthy working adults. *Annals of Internal Medicine*, *137*(4), 225.

Lee, S. M., Rho, B. H., & Lee, S. G. (2003). Impact of Malcolm Baldrige National Quality Award criteria on organizational quality performance. *International Journal of Production Research*, *41*(9), 2003–2020. doi:10.1080/00207540310000077329

Lee, T. H. (2006). Device physics: Electrical solitons come of age. *Nature*, *440*, 36–37. doi:10.1038/440036a

Lele, C. (2009). Folding theory of positive implicative/fuzzy positive implicative filters in BL-algebras. *Journal of Fuzzy Mathematics*, *17*(2), 633–641.

Lele, C. (2010). Algorithms and computations in BL-algebras. *International Journal of Artificial Life Research*, *1*(4), 29–47. doi:10.4018/jalr.2010100103

Lele, C. (in press). Fuzzy n-fold Obstinate filters in BL-algebras. *Afrika Mathematika*.

Lele, C., & Moutari, S. (2008). On some computational methods for study of foldness of H-ideals in BCI-algebras. *Soft Computing*, *12*, 403–407. doi:10.1007/s00500-007-0176-9

Leventhall, H. G. (1988). Low frequency noise in buildings–internal and external sources. *Journal of Low Frequency Noise and Vibration*, *7*, 74.

Li, W. (2009). *A bibliography on 1/f noise*. Retrieved from http://www.nslij-genetics.org/wli/1fnoise/index.html

Liagkou, V., Makri, E., Spirakis, P., & Stamatiou, Y. C. (2006). The threshold behaviour of the fixed radius random graph model and applications to the key management problem of sensor networks. In S. E. Nikoletseas & J. D. P. Rolim (Eds.), *Proceedings of the Second International Workshop on Algorithmic Aspects of Wireless Sensor Networks* (LNCS 4240, pp. 130-139).

Lian, W. L., Wong, F. H., & Yeh, C. C. (1996). On the existence of positive solutions of nonlinear second order differential equations. In *Proceedings of the American Mathematical Society* (Vol. 124).

Liben-Nowell, D., Novak, J., Kumar, R., Raghavan, P., & Tomkins, A. (2005). Geographic routing in social networks. *Proceedings of the National Academy of Sciences of the United States of America*, *102*, 11623–11628. doi:10.1073/pnas.0503018102

Li, C., & Chen, G. (2004). Estimating the Lyapunov exponents of discrete systems. *Chaos (Woodbury, N.Y.)*, *14*(2), 343–346. doi:10.1063/1.1741751

Lipsitch, M., Cohen, T., Murray, M., & Levin, B. R. (2007). Antiviral resistance and the control of pandemic influenza. *PLoS Medicine*, *4*(1), 111. doi:10.1371/journal.pmed.0040015

Liu, L., & Li, K. (2005). Fuzzy Boolean and positive implicative filters of BL-algebras. *Fuzzy Sets and Systems*, *152*, 333–348. doi:10.1016/j.fss.2004.10.005

Liu, L., & Li, K. (2005). Fuzzy filters of BL-algebras. *Information Sciences*, *173*, 141–154. doi:10.1016/j.ins.2004.07.009

Liu, Y. L., Liu, S. Y., & Xu, Y. (2006). An answer to the Jun-Shim-Lele's open problem on the fuzzy filter. *Journal of Applied Mathematics and Computing*, *21*, 325–329. doi:10.1007/BF02896410

Llyin, V. A., & Moiseev, E. I. (1987). Nonlocal boundary value problem of the first kind for a Sturm-Liouville operator on its differential and finite difference aspects. *Differential Equations*, *23*(7), 803–810.

Longini, I. M. Jr, Halloran, M. E., Nizam, A., & Yang, Y. (2004). Containing pandemic influenza with antiviral agents. *American Journal of Epidemiology*, *159*(7), 623. doi:10.1093/aje/kwh092

Longini, I. M. Jr, Nizam, A., Xu, S., Ungchusak, K., Hanshaoworakul, W., & Cummings, D. A. (2005). Containing pandemic influenza at the source. *Science*, *309*, 1083–1087. doi:10.1126/science.1115717

Luisi, P. L. (1998). About various definitions of life. *Origins of Life and Evolution of the Biosphere*, *28*(4-6), 613–622. doi:10.1023/A:1006517315105

Luper, S. (2009). Death. In Zalta, E. N. (Ed.), *The Stanford encyclopedia of philosophy*. Stanford, CA: Stanford University Press.

Luper, S. (2009). *The philosophy of death*. Cambridge, UK: Cambridge University Press.

Lupianez, F. G. (2006). Nets and filters in intuitionistic fuzzy topological spaces. *Information Sciences*, *176*, 2396–2404. doi:10.1016/j.ins.2005.05.003

Magnani, M. (2003). *Technical report on rough set theory for knowledge discovery in data bases*. Bologna, Italy: University of Bologna.

Magosso, E., Ursino, M., Zaniboni, A., & Gardella, E. (2006). A wavelet based method for automatic detection of slow eye movements: A pilot study. *Medical Engineering & Physics*, *28*(9), 860–875. doi:10.1016/j.medengphy.2006.01.002

Magosso, E., Ursino, M., Zaniboni, A., & Gardella, E. (2009). A wavelet-based energetic approach for the analysis of biomedical signals: Application to the electroencephalogram and electro-oculogram. *Applied Mathematics and Computation*, *207*(1), 42–62. doi:10.1016/j.amc.2007.10.069

Mahajan, P., Kandwal, R., & Vijay, R. (2011). General framework for cluster based active learning algorithm. *International Journal on Computer Science and Engineering*, *3*(1), 307–312.

Malarz, K., Szvetelszky, Z., Szekfu, B., & Kulakowski, K. (2007). *Gossip in random networks*. Retrieved from http://arxiv.org/pdf/physics/0601158.pdf

Mallat, S. G. (1989). A Theory for Multiresolution Signal Decomposition: The Wavelet Representation. *IEEE Transactions on Pattern Analysis and Machine Intelligence*, *11*(7), 674–693. doi:10.1109/34.192463

Malomed, B. A. (1992). Propagation of solitons in damped ac-driven chains. *Physical Review Letters A*, *45*, 4097–4101.

Mandelbrot, B. B. (1982). *The fractal geometry of nature*. San Francisco, CA: W. H. Freeman.

Manier, J. (2007). *How staph became drug-resistant threat 94,000 infections a year, many occurring outside of hospitals*. Chicago Tribune.

Mann, W. R. (1953). Mean value methods in iteration. *Proceedings of the American Mathematical Society*, *4*, 506–510. doi:10.1090/S0002-9939-1953-0054846-3

Manor, Y., Koch, C., & Segev, I. (1991). Effect of geometrical irregularities on propagation delay in axonal trees. *Biophysical Journal*, *60*, 1424–1437. doi:10.1016/S0006-3495(91)82179-8

Ma, R. (1998). Positive solutions of a nonlinear three point boundary value problem. *Electronic Journal of Differential Equations*, *34*, 1–8.

Ma, R. (2001). Positive solutions of a nonlinear m-point boundary value problem. *Computers & Mathematics with Applications (Oxford, England)*, *42*(6-7), 755–765. doi:10.1016/S0898-1221(01)00195-X

Martin, S. J. (2002). Numerical modelling of median road traffic noise barrier. *Journal of Sound and Vibration*, *251*(4), 671. doi:10.1006/jsvi.2001.3955

Maturana, H. R., & Varela, F. J. (1980). *Autopoiesis and cognition: The realization of the living* (2nd ed.). Dordrecht, The Netherlands: D. Reidel Publishing.

Maturana, H. R., & Varela, F. J. (1987). *The tree of knowledge: The biological roots of human understanding*. Boston, MA: Shambhala.

Maunder, R., Hunter, J., Vincent, L., Bennett, J., Peladeau, N., & Leszcz, M. (2003). The immediate psychological and occupational impact of the 2003 SARS outbreak in a teaching hospital. *Canadian Medical Association Journal*, *168*(10), 1245.

Mboussi, N. A., & Woafo, P. (2010). Effects of imperfection of ionic channels and exposure to electromagnetic fields on the generation and propagation of front waves in nervous fibre. *Communications in Nonlinear Science and Numerical Simulation*, *15*, 2350–2360. doi:10.1016/j.cnsns.2009.09.040

McFadden, K., Henagan, S., & Gowen, C. (2009). The patient safety chain: Transformational leadership's effect on patient safety culture, initiatives, and outcomes. *Journal of Operations Management, 27*, 390–404. doi:10.1016/j.jom.2009.01.001

Meltzer, M. I., Cox, N. J., & Fukuda, K. (1999). The economic impact of pandemic influenza in the united states: Priorities for intervention. *Emerging Infectious Diseases, 5*(5), 659. doi:10.3201/eid0505.990507

Menach, A. L. (2006). Key strategies for reducing spread of avian influenza among commercial poultry holdings: lessons for transmission to humans. *Proceedings. Biological Sciences, 273*(1600), 2467–2475. doi:10.1098/rspb.2006.3609

Miami International Airport. (2010). *Daily traffic volume data*. Retrieved from http://www.miami-airport.com

Migliore, M. (1996). Modeling the attenuation and failure of action potentials in the dendrites of hippocampal neurons protoplasmic. *Biophysical Journal, 71*, 2394–2403. doi:10.1016/S0006-3495(96)79433-X

Miller, G., Randolph, S., & Patterson, J. E. (2008). Responding to simulated pandemic influenza in San Antonio, Texas. *Infection Control and Hospital Epidemiology, 29*(4), 320–326. doi:10.1086/529212

Mills, C. E., Robins, J. M., & Lipsitch, M. (2004). Transmissibility of 1918 pandemic influenza. *Nature, 432*(7019), 904–906. doi:10.1038/nature03063

Milne, G. J., Kelso, J. K., Kelly, H. A., Huband, S. T., & McVernon, J. (2008). A small community model for the transmission of infectious diseases: Comparison of school closure as an intervention in individual-based models of an influenza pandemic. *PLoS ONE, 3*(12), 4005. doi:10.1371/journal.pone.0004005

Milotti, E. (2002). *1/f noise: a pedagogical review*. Retrieved from http://arxiv.org/abs/physics/0204033

Mitra, S., Banka, H., & Pedrycz, W. (2006). Rough–fuzzy collaborative clustering. *IEEE Transactions on Systems, Man, and Cybernetics. Part B, Cybernetics, 36*(4), 795–800. doi:10.1109/TSMCB.2005.863371

Miyara, F. (2010). *A note on Firecracker's noise*. Retrieved from http://www.eie.fceia.unr.edu.ar/~acustica/biblio/firecr1.htm

Models of Infectious Disease Agent Study (MIDAS). (2004). *Report from the models of infectious disease agent study (MIDAS) steering committee*. Retrieved from http://www.nigms.nih.gov/News/Reports/midas_steering_050404.htm

Mohawald, M., & Douglas, R. (1991). A silicon neuron. *Nature, 354*, 515–518. doi:10.1038/354515a0

Montgomery, D. C. (2008). *Design and analysis of experiments*. New York, NY: John Wiley & Sons.

Moore, J. W., & Westerfield, M. (1983). Action potential propagation and threshold parameters in inhomogeneous regions of squid axons. *The Journal of Physiology, 336*, 285–300.

Moreno, A. (2002). Artificial life and philosophy. *Leonardo, 35*(4), 401–405. doi:10.1162/002409402760181204

Morris, R., Wilesmith, J., Stern, M., Sanson, R., & Stevenson, M. A. (2001). Predictive spatial modeling of alternative control strategies for the foot-and-mouth disease epidemic in Great Britain. *The Veterinary Record*, 137–144. doi:10.1136/vr.149.5.137

Motamed, S., & Saied, A. (2011). n-fold obstinate filters in BL-algebras. *Neural Computing & Applications, 20*(4), 461–472. doi:10.1007/s00521-011-0548-z

Munster, V. J., de Wit, E., van den Brand, J. M. A., Herfst, S., Schrauwen, E. J. A., & Bestebroer, T. M. (2009). Pathogenesis and transmission of swine-origin 2009 A(H1N1) influenza virus in ferrets. *Science, 325*(5939), 481–483.

Mursaleen, M., & Lohani, Q. M. D. (2009). Intuitionistic fuzzy 2-normed space and some related concepts. *Chaos, Solitons, and Fractals, 42*, 224–234. doi:10.1016/j.chaos.2008.11.006

Mursaleen, M., Lohani, Q. M. D., & Mohiuddine, S. A. (2009). Intuitionistic fuzzy 2-metric space and its completion. *Chaos, Solitons, and Fractals, 42*, 1258–1265. doi:10.1016/j.chaos.2009.03.025

National Institute of General Medical Sciences. (2008). *Models of infectious disease agent study*. Retrieved from http://www.nigms.nih.gov/Initiatives/MIDAS/

Neal, P. (2008). The SIS great circle epidemic model. *Journal of Applied Probability, 45*(2), 513–530. doi:10.1239/jap/1214950364

Negi, A., & Rani, M. (2008). A new approach to dynamic noise on superior Mandelbrot set. *Chaos, Solitons, and Fractals, 36*(4), 1089–1096. doi:10.1016/j.chaos.2006.07.026

Netlogo. (2005). *Software version 4.0.4.* Retrieved from http://ccl.northwestern.edu/netlogo/

Neumann, G., Noda, T., & Kawaoka, Y. (2009). Emergence and pandemic potential of swine-origin H1N1 influenza virus. *Nature, 459*(7249), 931–939. doi:10.1038/nature08157

Newcomb, R. W., & Sathyan, S. (1983). An RC op amp chaos generator. *IEEE Transactions on Circuits and Systems, 30*, 54–56. doi:10.1109/TCS.1983.1085277

Newman, M. E. J. (2002). Spread of epidemic disease on networkds. *Physical Review E: Statistical, Nonlinear, and Soft Matter Physics, 66*(1).

Newman, M. E. J., Forrest, S., & Balthrop, J. (2002). Email networks and the spread of computer viruses. *Physical Review E: Statistical, Nonlinear, and Soft Matter Physics, 66*, 035101. doi:10.1103/PhysRevE.66.035101

Newton, E. A. C., & Reiter, P. (1992). A model of the transmission of dengue fever with an evaluation of the impact of ultra-low volume (ULV) insecticide applications on dengue epidemics. *The American Journal of Tropical Medicine and Hygiene, 47*, 709–720.

Ney, H., Mergel, D., Noll, A., & Paeseler, A. (1992). Data driven organization of the dynamic programming beam search for continuous speech recognition. *IEEE Transactions on Signal Processing, 40*(2), 272–281. doi:10.1109/78.124938

Nguimdo, R. M., Nubissie, S., & Woafo, P. (2008). Waves amplification in discrete nonlinear electrical lines: Direct numerical simulation. *Journal of the Physical Society of Japan*, 77.

Niederhauser, V. P., Baruffi, G., & Heck, R. (2001). Parental decision-making for the varicella vaccine. *Journal of Pediatric Health Care, 15*(5), 236–243.

Nigmatulina, K. R., & Larson, R. C. (2009). Living with influenza: Impacts of government imposed and voluntarily selected interventions. *European Journal of Operational Research, 195*(2), 613–627. doi:10.1016/j.ejor.2008.02.016

Nuno, M., Castillo-Chaves, C., Feng, Z., & Marcheva, M. (2008). Mathematical models of influenza: The role of cross-immunity, quarantine and age structure. *Mathematical Epidemiology,* 349-364.

Ogren, P., & Martin, C. F. (2000). Optimal vaccination strategies for the control of epidemics in highly mobile populations. In *Proceedings of the 39th IEEE Conference on Decision and Control* (Vol. 2, pp. 1782-1787).

Olsen, M., Siegelmann-Danieli, N., & Siegelmann, H. (2008). Robust artificial life via artificial programmed death. *Artificial Intelligence, 172*(6-7), 884–898. doi:10.1016/j.artint.2007.10.015

Olsson, B. (1996). Optimization using a host-parasite model with variable-size distributed populations. In *Proceedings of the IEEE Conference on Evolutionary Computation* pp. 295-299).

Onuu, M. U. (2000). Road traffic noise in Nigeria: Measurements, analysis and evaluation of nuisance. *Journal of Sound and Vibration, 233*(3), 391. doi:10.1006/jsvi.1999.2832

Osvath, M., & Gärdenfors, P. (2005). Oldwan culture and the evolution of anticipatory cognition. *Lund University Cognitive Science, 122.*

Park, J. H. (2004). Intuitionistic fuzzy metric spaces. *Chaos, Solitons, and Fractals, 22*, 1039–1046. doi:10.1016/j.chaos.2004.02.051

Pasteur, S. (2009). *Influenza A(H1N1) 2009 monovalent vaccine.* Retrieved from http://www.fda.gov/downloads/biologicsbloodvaccines/vaccines/approvedproducts/ucm182404.pfd

Patel, R., & Longini, I. M. (2005). Finding optimal vaccination strategies for pandemic influenza using genetic algorithms. *Journal of Theoretical Biology, 234*(2), 201–212. doi:10.1016/j.jtbi.2004.11.032

Patil, A. G., Chisty, S. Q., Khan, A. R., Basit, M. A., & Behere, S. H. (2001). Fractal growth in copper sulphate solution. In *Proceedings of the Indian Science Congress Association,* Delhi, India.

Patre, P. M., MacKunis, W., Kaiser, K., & Dixon, W. E. (2008). Asymptotic tracking for uncertain dynamic systems via a multilayer neural network feedforward and RISE feedback control structure. *IEEE Transactions on Automatic Control, 53*, 2180–2185. doi:10.1109/TAC.2008.930200

Pawlak, Z. (1982). Rough sets. *International. Journal of Computer and Information Sciences, 11*(5), 341–356. doi:10.1007/BF01001956

Peacock, C. (1994). *Interfacing the standard parallel port*. Retrieved from http://www.beyondlogic.org/spp/parallel.pdf

Pearson, M. L., Bridges, C. B., & Harper, S. A. (2006). Influenza vaccination of health-care personnel. Recommendations of the Healthcare Infection Control Practices Advisory Committee (HICPAC) and the Advisory Committee on Immunization Practices (ACIP). *Morbid and Mortality Weekly Report, 55*(2).

Peitgen, H. O., Jürgens, H., & Saupe, D. (2004). *Chaos and fractals: New frontiers of science* (2nd ed.). New York, NY: Springer.

Peters, C. J., & Linthicum, K. J. (1994). Rift Valley Fever. In G. W. Beran (Ed.), *Handbook of zoonoses, section B: Viral zoonoses* (2nd ed.) (pp. 125-138). Boca Raton, FL: CRC Press.

Pickover, C. A., & McCarty, K. (1990). Visualizing Cantor cheese construction. *Computers & Graphics, 14*(2), 337–341. doi:10.1016/0097-8493(90)90046-Z

Piqueira, J., Navarro, B., & Monteiro, L. (2005). Epidemiological models applied to viruses in computer networks. *Journal of Computer Sciences, 1*(1), 31–34. doi:10.3844/jcssp.2005.31.34

Pitzer, V. E., Leung, G. M., & Lipsitch, M. (2007). Estimating variability in the transmission of severe acute respiratory syndrome to household contacts in Hong Kong, China. *American Journal of Epidemiology, 166*(3), 355. doi:10.1093/aje/kwm082

Pitzer, V. E., Olsen, S. J., Bergstrom, C. T., Dowell, S. F., & Lipsitch, M. (2007). Little evidence for genetic susceptibility to influenza A (H5N1) from family clustering data. *Emerging Infectious Diseases, 13*(7), 1074.

Pitzer, V. E., Viboud, C., Simonsen, L., Steiner, C., Panozzo, C. A., & Alonso, W. J. (2009). Demographic variability, vaccination, and the spatiotemporal dynamics of rotavirus epidemics. *Science, 325*(5938), 290–294. doi:10.1126/science.1172330

Pompilia, D., Scoglio, C., & Lopez, L. (2008). Multicast algorithms in service overlay networks. *Computer Communications, 31*(3).

Potts, R. (1998). Variability selection in hominid evolution. *Evolutionary Anthropology, 7*, 81–96. doi:10.1002/(SICI)1520-6505(1998)7:3<81::AID-EVAN3>3.0.CO;2-A

Pourbohloul, B., Ahued, A., Davoudi, B., Meza, R., Meyers, L. A., & Skowronski, D. M. (2009). Initial human transmission dynamics of the pandemic (H1N1) 2009 virus in north America. *Influenza and Other Respiratory Viruses, 3*(5), 215–222. doi:10.1111/j.1750-2659.2009.00100.x

Press, W. H. (1978). Flicker noises in astronomy and elsewhere. *Comments on Astrophysics, 7*, 103.

Psillakis, H. E., Boutalis, Y., Giotis, T., & Christodoulou, M. A. (2010). Adaptive identification of fuzzy cognitive networks. In *Proceedings of the 4th International Symposium on Communications, Control and Signal Processing*, Limassol, Cyprus (pp. 1-6).

Ram, B. (1991). *Fundamentals of microprocessors and microcomputers* (5th ed.). Retrieved from http://www.national.com

Rani, M. (2010a, February 26-28). Superior Antifractals. In *Proceedings of the IEEE International Conference on Computer and Automation Engineering*, Singapore (Vol. 1, pp. 798-802).

Rani, M. (2010b, March 23-25). Superior tricorns and multicorns. In *Proceedings of the 9th WSEAS International Conference on Application of Computer Engineering*, Penang, Malaysia (pp. 58-61).

Rani, M., & Agarwal, R. (2010a, February 26-28). Effect of noise on Julia sets generated by logistic map. In *Proceedings of the 2nd IEEE International Conference on Computer and Automation Engineering*, Singapore (Vol. 2, pp. 55-59).

Rani, M., & Kumar, M. (2009, December 14-16). Circular saw Mandelbrot sets. In *Proceedings of the WSEAS 14th International Conference on Recent Advances in Applied Mathematics* (pp. 131-136).

Rani, M., & Kumar, V. (2003, Oct 18-20). Fractals in Vedic heritage and fractal carpets. In *Proceedings of the National Seminar on History, Heritage and Development of Mathematical Science*, Allahabad, India (pp. 110-121).

Rani, M., & Negi, A. (2008). New Julia sets for complex Carotid-Kundalini function. *Chaos, Solitons & Fractals: The Interdisciplinary Journal of Nonlinear Science – Nano and Quantum Technology, 36*(2), 226-236.

Rani, M., & Agarwal, R. (2008). A new experimental approach to study the stability of logistic map. *Chaos, Solitons, and Fractals, 41*(4), 2062–2066. doi:10.1016/j.chaos.2008.08.022

Rani, M., & Agarwal, R. (2009). Generation of fractals from complex logistic map. *Chaos, Solitons, and Fractals, 42*(1), 447–452. doi:10.1016/j.chaos.2009.01.011

Rani, M., & Agarwal, R. (2010). Effect of stochastic noise on superior Julia sets. *Journal of Mathematical Imaging and Vision, 36*, 63–68. doi:10.1007/s10851-009-0171-0

Rani, M., & Chandra, M. (2009). Categorization of fractal plants. *Chaos, Solitons, and Fractals, 41*(3), 1442–1447. doi:10.1016/j.chaos.2008.05.024

Rani, M., & Goel, S. (2010). A new approach to pattern recognition in fractal ferns. *International Journal of Artificial Life Research, 1*(2). doi:10.4018/jalr.2010040102

Rani, M., & Kumar Prasad, S. (2010). Superior Cantor sets and superior devil staircases. In Koetsier, T., & Bergmans, L. (Eds.), *Mathematics and the divine: A historical study* (pp. 532–547). Amsterdam, The Netherlands: Elsevier. doi:10.4018/jalr.2010102106

Rani, M., & Kumar, M. (2002). A Fractal hedgehog theorem. *Journal of the Korean Society of Mathematical Education Series B: Pure & Applied Mathematics, 9*(2), 91–105.

Rani, M., & Kumar, M. (2005). A new approach to superior Julia sets. *Journal of Natural and Physical Science, 19*(2), 148–155.

Rani, M., & Kumar, V. (2004). Superior Julia set. *Journal of the Korean Society of Mathematical Education Series D: Research in Mathematical Education, 8*(4), 261–277.

Rani, M., & Kumar, V. (2004). New fractal carpets. *Arabian Journal for Science and Engineering: Section B. Engineering, 29*(2), 125–134.

Rani, M., & Kumar, V. (2005). A new experiment with the logistic function. *Journal of the Indian Academy of Mathematics, 27*(1), 143–156.

Rani, M., & Negi, A. (2008). Midgets of superior Mandelbrot set. *Chaos, Solitons, and Fractals, 36*(2), 237–245. doi:10.1016/j.chaos.2006.06.059

Rani, M., & Negi, A. (2008). New Julia sets for complex CarotidKundalini function. *Chaos, Solitons, and Fractals, 36*(2), 226–236. doi:10.1016/j.chaos.2006.06.058

Rani, M., & Negi, A. (2008). A new approach to dynamic noise on superior Mandelbrot set. *Chaos, Solitons, and Fractals, 36*(4), 1089–1096. doi:10.1016/j.chaos.2006.07.026

Rani, M., Negi, A., & Mahanti, P. K. (2008). Computer simulation of the behavior of Julia sets using switching processes. *Chaos, Solitons, and Fractals, 37*(4), 1187–1192. doi:10.1016/j.chaos.2006.10.061

Rani, M., & Prasad, S. (2010). Superior Cantor sets and superior Devils Staircases. *International Journal of Artificial Life Research, 1*(1), 78–84. doi:10.4018/jalr.2010102106

Rasmussen, S., Bedau, M. A., Chen, L., Deamer, D., Krakauer, D. C., Packard, N. H., & Stadler, P. F. (Eds.). (2008). *Protocells: Bridging nonliving and living matter bridging nonliving and living matter*. Cambridge, MA: MIT Press.

Ray, T. S. (1994). An evolutionary approach to synthetic biology: Zen and the art of creating life. *Artificial Life, 1*(1-2), 195–226.

Rello, J., Lisboa, T., Lujan, M., Gallego, M., Kee, C., & Kay, I. (2009). Severity of pneumococcal pneumonia associated with genomic bacterial load. *Chest, 136*(3), 832–840. doi:10.1378/chest.09-0258

Reutsky, S., Rossoni, E., & Tirozzi, B. (2003). Conduction in bundles of demyelinated nerve fibres: competer simulation. *Biological Cybernetics*, *89*, 439–448. doi:10.1007/s00422-003-0430-x

Reyes, W., Brimioulle, S., & Vincent, J. (1999). Septic shock without documented infection: an uncommon entity with a high mortality. *Intensive Care Medicine*, *25*(11), 1267–1270. doi:10.1007/s001340051055

Reynolds, A., Rubin, J., Clermont, G., Day, J., Vodovotz, Y., & Ermentrout, G. B. (2006). A reduced mathematical model of the acute inflammatory response: I. derivation of model and analysis of anti-inflammation. *Journal of Theoretical Biology*, *242*(1), 220–236. doi:10.1016/j.jtbi.2006.02.016

Rhodes, S. D., & Hergenrather, K. C. (2002). Exploring hepatitis B vaccination acceptance among young men who have sex with men: Facilitators and barriers* 1. *Preventive Medicine*, *35*(2), 128–134. doi:10.1006/pmed.2002.1047

Ricketts, D. S., Li, X., & Ham, D. (2006). Electrical soliton oscillator. *IEEE Transactions on Microwave Theory and Techniques*, *54*(1), 373–382. doi:10.1109/TMTT.2005.861652

Robertson, E., Hershenfield, K., Grace, S. L., & Stewart, D. E. (2004). The psychosocial effects of being quarantined following exposure to SARS: A qualitative study of Toronto health care workers. *Canadian Journal of Psychiatry*, *49*, 403–407.

Rosenberg, R. (1997). *Typhoid Mary: The sad story of a woman responsible for several typhoid outbreaks*. Retrieved from http://history1900s.about.com/od/1900s/a/typhoidmary.htm

Rosenthal, S. L., Kottenhahn, R. K., Biro, F. M., & Succop, P. A. (1995). Hepatitis B vaccine acceptance among adolescents and their parents. *The Journal of Adolescent Health*, *17*(4), 248–254. doi:10.1016/1054-139X(95)00164-N

Rothman, K. (2002). *Epidemiology: An introduction*. Naperville, IL: Sourcebooks.

Rovithakis, G. A., & Christodoulou, M. A. (2000). *Adaptive control with recurrent high-order neural networks: Theory and industrial applications*. New York, NY: Springer.

Rvachev, L. A. (1968). Simulation of large-scale epidemics on a digital computer. *Soviet Physics, Doklady*, *13*(5), 384–386.

Saadati, R., & Park, J. H. (2006). On the intuitionistic fuzzy topological spaces. *Chaos, Solitons, and Fractals*, *27*, 331–344. doi:10.1016/j.chaos.2005.03.019

Sadique, M. Z., Edmunds, W. J., Smith, R. D., Meerding, W. J., De Zwart, O., & Brug, J. (2007). Precautionary behavior in response to perceived threat of pandemic influenza. *Emerging Infectious Diseases*, *13*(9), 1307.

Safranek, T. J., Lawrence, D. N., Kuriand, L. T., Culver, D. H., Wiederholt, W. C., & Hayner, N. S. (1991). Reassessment of the association between guillain-barré syndrome and receipt of swine influenza vaccine in 1976-1977: Results of a two-state study. *American Journal of Epidemiology*, *133*(9), 940.

Sanner, R. M., & Slotine, J.-J. E. (1992). Gaussian networks for direct adaptive control. *IEEE Transactions on Neural Networks*, *3*, 837–863. doi:10.1109/72.165588

Savachkin, A., Uribe-Sanchez, A., Das, T., Prieto, D., Santana, A., & Martinez, D. (2010). *Supplemental data and model parameter values for cross-regional simulation-based optimization testbed*. Retrieved from http://imse.eng.usf.edu/pandemic/supplement.pdf

Scharfstein, D. O., Halloran, M. E., Chu, H., & Daniels, M. J. (2006). On estimation of vaccine efficacy using validation samples with selection bias. *Biostatistics (Oxford, England)*, *7*(4), 615. doi:10.1093/biostatistics/kxj031

Schoenstadt, A. (2010). *Spanish flu*. Retrieved from http://flu.emedtv.com/spanish-flu/spanish-flu.html

Schroder, M. (1990). *Fractals, chaos, power laws: Minutes from an infinite paradise*. New York, NY: FreeMan.

Schweizer, B., & Sklar, A. (1960). Statistical metric spaces. *Pacific Journal of Mathematics*, *10*, 313–334.

Scoglio, C., Schumm, W., Schumm, P., Sydney, A., Chowdhury, S., & Youssef, M. (2010). Efficient mitigation strategies to control epidemics. *PLoS ONE*, *5*(7), 11569. doi:10.1371/journal.pone.0011569

Scott, A. C. (1999). *Nonlinear science: Emergence and dynamics of coherent structure*. Oxford, UK: Oxford University Press.

Shaikh, Y. H., Khan, A. R., Iqbal, M. I., Behere, S. H., & Bagare, S. P. (2008). Sunspot Data Analysis using time series. *Fractal*, *16*(3), 259. doi:10.1142/S0218348X08004009

Shaikh, Y. H., Khan, A. R., Pathan, J. M., Patil, A., & Behere, S. H. (2009). Fractal pattern growth simulation in electrodeposition and study of the shifting of center of mass. *Chaos, Solitons, and Fractals*, *42*(5), 2796–2803. doi:10.1016/j.chaos.2009.03.192

Shonkwiler, R. (1989). An image algorithm for computing the Hausdorff distance efficiently in linear time. *Information Processing Letters*, *30*(2), 87–89. doi:10.1016/0020-0190(89)90114-2

Shonkwiler, R. (1991). Computing the Hausdorff set distance in linear time for any Lp point distance. *Information Processing Letters*, *38*(4), 201–207. doi:10.1016/0020-0190(91)90101-M

Singh, S. L., Jain, S., & Mishra, S. N. (2009). A new approach to Superfractals. *Chaos, Solitons, and Fractals*, *42*, 3110–3120. doi:10.1016/j.chaos.2009.04.052

Singh, S. L., Jain, S., & Mishra, S. N. (2011). Orbit of an Image under iterated system. *Communications in Nonlinear Science and Numerical Simulation*, *16*, 1469–1482. doi:10.1016/j.cnsns.2010.07.012

Singh, S. L., Mishra, S. N., & Sinkala, W. (2012). A new iterative approach to fractal models. *Communications in Nonlinear Science and Numerical Simulation*, *17*, 521–529. doi:10.1016/j.cnsns.2011.06.014

Sinha, S. (1996). Transient 1/f noise. *Physics Review Letters E*, *53*, 5.

Smailbegovic, M., Laing, G., & Bedford, H. (2003). Why do parents decide against immunization? The effect of health beliefs and health professionals. *Child: Care, Health and Development*, *29*(4), 303–311. doi:10.1046/j.1365-2214.2003.00347.x

Smith, D. J. (2006). Predictability and preparedness in influenza control. *Science*, *312*(5772), 392–394. doi:10.1126/science.1122665

Smith, G. J., Vijaykrishna, D., Bahl, J., Lycett, S. J., Worobey, M., & Pybus, O. G. (2009). Origins and evolutionary genomics of the 2009 swine-origin H1N1 influenza A epidemic. *Nature*, *459*(7250), 1122–1125. doi:10.1038/nature08182

Smith, K. J., & McDonald, W. I. (1999). The pathophysiology of multiple sclerosis: The mechanism underlying the production of symptoms and natural history of disease. *Philosophical Transactions of the Royal Society of London*, *354*, 1649–1673. doi:10.1098/rstb.1999.0510

Solomyak, B. (1997). On the measure of arithmetic sums of Cantor sets. *Indagationes Mathematicae*, *8*(1), 133–141. doi:10.1016/S0019-3577(97)83357-5

Spector, W. S. (Ed.). (1956). *Handbook of biological data*. London, UK: W. B. Saunders Company.

Spooner, J. T., Maggiore, M., Ordonez, R., & Passino, K. M. (2002). *Stable adaptive control and estimation for nonlinear systems- neural and fuzzy approximator techniques*. Hoboken, NJ: John Wiley & Sons. doi:10.1002/0471221139

Sprott, J. C. (1993). *Strange attractors: Creating patterns in chaos*. New York, NY: M & T Books.

Squire, L. R., Ojemann, J. G., Miezin, F. M., Petersen, S. E., Videen, T. O., & Raichle, M. E. (1992). Activation of the hippocampus in normalhumans: A functional anatomical study of memory. *Proceedings of the National Academy of Sciences of the United States of America*, *89*, 1837–1841. doi:10.1073/pnas.89.5.1837

Steels, L. (1993). Building agents out of autonomous behavior systems. In Steels, L., & Brooks, R. A. (Eds.), *The artificial life route to artificial intelligence: Building embodied situated agents*. Mahwah, NJ: Lawrence Erlbaum.

Sterelny, K., & Griffiths, P. E. (1999). *Sex and death*. Chicago, IL: University of Chicago Press.

Stewart, J. (1996). Cognition = life: Implications for higher-level cognition. *Behavioural Processes*, *35*(1-3), 311–326. doi:10.1016/0376-6357(95)00046-1

Suddendorf, T., & Corballis, M. C. (1997). Mental time travel and the evolution of human mind. *Genetic, Social, and General Psychology Monographs*, *123*, 133–167.

Sun, Y. (2005). Positive solutions of nonlinear second-order m-point boundary value problem. *Nonlinear Analysis*, *61*, 1283–1294. doi:10.1016/j.na.2005.01.105

Tallahassee Regional Airport. (2010). *Daily traffic volume data*. Retrieved from http://www.talgov.com/airport/index.cfm

Tampa International Airport. (2010). *Daily traffic volume data*. Retrieved from http://www.tampaairport.com

Tandon, N. (2003). Firecrackers noise. *Noise and Vibration Worldwide*, *34*, 5.

Taubenberger, J., & Morens, D. (2006). *1918 influenza: The mother of all pandemics*. Retrieved from http://www.cdc.gov/ncidod/EID/vol12no01/05-0979.htm

Tchikapa, N., Lele, C., & Turunen, E. (in press). *n-fold deductive systems in BL-algebras*.

Tchikapa, N., & Lele, C. (in press). Relation diagram between fuzzy n-fold filters in BL-algebras. *Annals of Fuzzy Mathematics and Informatics*.

Teccea, J. J., Gips, J., Olivieri, C. P., Pok, L. J., & Consiglio, M. R. (1998). Eye movement control of computer functions. *International Journal of Psychophysiology*, *29*(3), 319–325. doi:10.1016/S0167-8760(98)00020-8

Tempest, W. (Ed.). (1985). *The noise handbook*. London, UK: Academic Press.

The New York Times. (2009). *Doctors swamped by swine flu vaccine fears*. Retrieved from http://www.msnbc.msn.com/id/33179695/ns/health-swine_flu/

The New Yorker. (2009). *The fear factor*. Retrieved from http://www.newyorker.com/talk/comment/2009/10/12/091012taco_talk_specter

Thiele, R. (2005). Georg Cantor and the divine. In Koetsier, T., & Bergmans, L. (Eds.), *Mathematics and the divine: A historical study*. Amsterdam, The Netherlands: Elsevier.

Tiehong, T., & Zheng, Z. (2003). A Novel Multiband Antenna: Fractal Antenna. In *Proceedings of the International Conference on Communication Technology* (pp. 1907-1910).

Tildesley, M. J., Savill, N. J., Shaw, D. J., Deardon, R., Brooks, S. P., & Woolhouse, M. E. J. (2005). Optimal reactive vaccination strategies for a foot-and-mouth outbreak in the UK. *Nature*, *440*, 83–86. doi:10.1038/nature04324

Tocci, R., & Widmer, N. (1998). *Digital systems: Principles and applications* (8th ed.). Upper Saddle River, NJ: Prentice Hall.

Tooby, J., & DeVore, I. (1987). The reconstruction of hominid behavioral evolution through strategic modelling. In Kinzey, W. (Ed.), *The evolution of human behavior: Primate models* (pp. 183–238). Albany, NY: State University of New York Press.

Treanor, J. J., Campbell, J. D., Zangwill, K. M., Rowe, T., & Wolff, M. (2006). Safety and immunogenicity of an inactivated subvirion influenza A (H5N1) vaccine. *The New England Journal of Medicine*, *354*(13), 1343. doi:10.1056/NEJMoa055778

Troll, G. (1991). A devil's staircase into chaotic scattering. *Physica D. Nonlinear Phenomena*, *50*(2), 276–296. doi:10.1016/0167-2789(91)90180-H

Tseng, S.-Y., Chen, R.-C., Chong, F.-C., & Kuo, T.-S. (1995). Evaluation of parametric methods in EEG signal analysis. *Medical Engineering & Physics*, *17*(1), 71–78. doi:10.1016/1350-4533(95)90380-T

Tsonis, A. A. (1992). *Chaos: From theory to applications*. New York, NY: Plenum Press.

Tsukaguchi, K., de, Lange B., & Boom, W. H. (1999). Differential regulation of IFN-gamma, TNF-alpha, and IL-10 production by CD4 (+) alphabeta TCR + T cells and vdelta2 (+) gammadelta T cells in response to monocytes infected with Mycobacterium tuberculosis-H37Ra. *Cellular Immunology*, *194*(1), 12–20. doi:10.1006/cimm.1999.1497

Tufan, M. (2003). Report of the foot-and-mouth situation in Turkey from 1990 to 2002. In *Consultant's report to the food and agriculture organisation of the United Nations* (p. 44). New York, NY: United Nations.

Tulving, E. (1993). What is episodic memory? *Current Directions in Psychological Science*, *2*, 67–70. doi:10.1111/1467-8721.ep10770899

Turell, M. J., Dohm, D. J., Mores, C. N., Terracina, L., Wallette, D. L., & Hribar, L. J. (2008). Potential for North American mosquitoes to transmit Rift Valley Fever virus. *Journal of the American Mosquito Control Association*, *24*(4), 502–507. doi:10.2987/08-5791.1

Turunen, E. (1999). *Mathematics behind fuzzy logic*. Heidelberg, Germany: Physica-Verlag.

Turunen, E. (1999q). BL-algebras and basic fuzzy logic. *Mathware and Soft Computing*, *6*, 49–61.

Turunen, E., & Sessa, S. (2001). Local BL-algebras. *Multi-Valued Logic*, *6*, 229–249.

Turunen, E., Tchikapa, N., & Lele, C. (in press). n-fold implicative logic is Gödel logic. *Soft Computing*.

U.S Census Bureau. (2000). *2001 American community survey*. Retrieved from http://www.census.gov/prod/2001pubs/statab/sec01.pdf

U.S. Department of Health & Human Services. (2007). *HHS pandemic influenza plan*. Retrieved from http://www.hhs.gov/pandemicflu/plan/

Unser, M. (1996). Wavelets, Statistics, and Biomedical Applications. *Proceedings of the 8th IEEE Signal Processing Workshop on Statistical Signal and Array Processing (SSAP '96)*, (pp. 244 - 249). Corfu, Greece.

Usakli, A. B., & Gurkan, S. (2010). Design of a Novel Efficient Human–Computer Interface: An Electrooculagram Based Virtual Keyboard. *Instrumentation and Measurement. IEEE Transactions on*, *59*(8), 2099–2108.

Van Vliet, C. M. (1991). A survey of results and future prospects on quantum 1/f noise and 1/f noise in general. *Solid-State Electronics*, *34*, 1. doi:10.1016/0038-1101(91)90195-5

Vapnik, V. P. (2000). *The nature of statistical learning theory* (2nd ed.). New York, NY: Springer.

Varela, F. J., Maturana, H. R., & Uribe, R. (1974). Autopoiesis: The organization of living systems, its characterization and a model. *Bio Systems*, *5*, 187–196. doi:10.1016/0303-2647(74)90031-8

Viboud, C., Pierre-Yves, B., Alain-Jacques, F. C. V., & Flahault, A. (2003). Prediction of the spread of influenza epidemics by the method of analogues. *American Journal of Epidemiology*, *158*, 996–1006. doi:10.1093/aje/kwg239

Vicsek, T. (1992). *Fractal growth phenomena*. Singapore: World Scientific.

Von Gierke, H. E., & Nixon, C. W. (1976). Effects of intense infrasound on man. In Tempest, W. (Ed.), *Infrasound and low frequency noise vibration* (p. 115). London, UK: Academic Press.

Wakeland, W., Macovsky, L., & An, G. (2000). A hybrid simulation model for studying acute inflammatory response. *Toxicological Sciences*, *57*(2), 312–325.

Wallace, S., Maki-Petaja, K., Cheriyan, J., Davidson, E., McEniery, C., & Wilkinson, I. (2006). Simvastatin prevents acute inflammation-induced aortic stiffening and endothelial dysfunction in healthy volunteers. *British Journal of Clinical Pharmacology*, *70*(6), 799–806. doi:10.1111/j.1365-2125.2010.03745.x

Wang, B. (2010). A new clustering algorithm on nominal data sets. In *Proceedings of the International MultiConference of Engineers and Computer Scientists* (pp. 605-610).

Wang, C., Hill, D. J., & Chen, G. (2003). Deterministic learning of nonlinear dynamical systems. In *Proceedings of the 18th IEEE International Symposium on Intelligent Control*, Houston, TX (pp. 87-92).

Wang, K., Xu, C., & Liu, B. (1999). Clustering transactions using large items. In *Proceedings of the ACM International Conference on Information and Knowledge Management* (pp. 483-490).

Wang, C., & Hill, D. J. (2006). Learning from neural control. *IEEE Transactions on Neural Networks*, *17*, 130–146. doi:10.1109/TNN.2005.860843

Wang, C., & Hill, D. J. (2007). Deterministic learning and rapid dynamical pattern recognition. *Transactions on Neural Networks*, *18*, 617–630. doi:10.1109/TNN.2006.889496

Wang, C., & Hill, D. J. (2010). *Deterministic learning theory for identification, recognition and control*. Boca Raton, FL: CRC Press.

Wang, H., Bloom, O., Zhang, M., Vishnubhakat, J. M., Ombrellino, M., & Che, J. (1999). HMG-1 as a late mediator of endotoxin lethality in mice. *Science*, *285*(5425), 248–251. doi:10.1126/science.285.5425.248

Wang, H., Yang, H., Czura, C. J., Sama, A. E., & Tracey, K. J. (2001). HMGB1 as a late mediator of lethal systemic inflammation. *American Journal of Respiratory and Critical Care Medicine*, *164*(10), 1768–1773.

Ward, L. M., & Greenwood, P. E. (2007). 1/f noise. *Scholarpedia*, *2*(12), 1537. doi:10.4249/scholarpedia.1537

Weissman, M. B. (1988). 1/f noise and other slow, non exponential kinetics in condensed matter. *Reviews of Modern Physics, 60*(2), 537. doi:10.1103/RevModPhys.60.537

West Bengal Pollution Control Board. (2005). *Report of assessment of noise pollution survey in Kolkata during Kalipuja and Diwali festivals*. Retrieved from http://www.wbpcb.gov.in/html/downloads/kalipuja_diwali_06.pdf

West, B. J., & Shlesinger, M. F. (1990). The noise in natural phenomena. *American Scientist, 78*, 40.

Wetterling, F. L. (2001). The Internet and the spy business. *International Journal of Intelligence and CounterIntelligence, 14*(3), 342–365. doi:10.1080/08850600152386846

Widz, S. lzak, D., & Revett, K. (2004). Application of rough set based dynamic parameter optimization to MRI segmentation. In *Proceedings of the 23rd International Conference of the North American Fuzzy Information Processing Society* (pp. 440-445).

Wikipedia. (n.d.). *Colors of noise*. Retrieved from http://en.wikipedia.org/wiki/Colors_of_noise

Wilson, S. W. (1985). Knowledge growth in an artificial animal. In *Proceedings of the First International Conference on Genetic Algorithms and Their Applications* (pp. 16-23).

Witten Jr., T. A. & Sander, L. M. (2000). Diffusion-limited aggregation, a kinetic critical phenomenon. *Critical Review Letters, 47*(19).

Wong, H. (2003). Low-frequency noise study in electron devices: review and update. *Microelectronics and Reliability, 43*(4), 585–589. doi:10.1016/S0026-2714(02)00347-5

World Health Organization (WHO). (2007). *Rift Valley Fever.* Retrieved from http://www.who.int/mediacentre/factsheets/fs207/en/

World Health Organization. (2004). *WHO guidelines on the use of vaccine and antivirals during influenza pandemics*. Retrieved from http://www.who.int/csr/resources/publications/influenza/11_29_01_A.pdf

World Health Organization. (2007). *Typhoid fever.* Retrieved from http://www.who.int/vaccine_research/diseases/diarrhoeal/en/index7.html

World Health Organization. (2009). *Pandemic (h1n1) 2009 vaccine deployment update - 17 December 2009.* Retrieved from http://www.who.int/csr/disease/swineflu/vaccines/h1n1_vaccination_deployment_update_20091217.pdf

World Health Organization. (2009). *Pandemic influenza preparedness and response.* Retrieved from http://www.who.int/csr/disease/influenza/pipguidance2009/en/index.html

World Health Organization. (2010). *Cumulative number of confirmed human cases of avian InfluenzaA(H5N1) reported to WHO.* Retrieved from http://www.who.int/csr/disease/avian_influenza/country/cases_ table_2010_08_31/en/index.html

Wu, J. T., Riley, S., Fraser, C., & Leung, G. M. (2006). Reducing the impact of the next influenza pandemic using household-based public health interventions. *PLoS Medicine, 3*(9), 361. doi:10.1371/journal.pmed.0030361

Wu, J. T., Riley, S., & Leung, G. M. (2007). Spatial considerations for the allocation of pre-pandemic influenza vaccination in the united states. *Proceedings. Biological Sciences, 274*(1627), 2811. doi:10.1098/rspb.2007.0893

Xian, B., Dawson, D. M., de Queiroz, M. S., & Chen, J. (2004). A continuous asymptotic tracking control strategy for uncertain nonlinear systems. *IEEE Transactions on Automatic Control, 49*, 1206–1211. doi:10.1109/TAC.2004.831148

Xiao, J., & Zhu, X. (2002). On linearly topological structure and property of fuzzy normed linear space. *Fuzzy Sets and Systems, 121*, 153–161. doi:10.1016/S0165-0114(00)00136-6

Xu, E., Xuedong, G., Sen, W., & Bin, Y. (2006). An clustering algorithm based on rough set. In *Proceedings of the 3rd International IEEE Conference Intelligent Systems* (pp. 475-478).

Yale University. (n. d.). *White noise.* Retrieved from http://classes.yale.edu/fractals/CA/OneOverF/WhiteNoise.html

Yang, L., & Yang, L. (2006). Study of cluster algorithm based on rough sets theory. In *Proceedings of the Sixth International Conference on Intelligent System Design & Application* (pp. 492-496).

Yang, Y., Halloran, M. E., Sugimoto, J. D., & Longini, I. M. Jr. (2007). Detecting human-to-human transmission of avian influenza A (H5N1). *Emerging Infectious Diseases*, *13*(9), 1348.

Yang, Y., Sugimoto, J. D., Halloran, M. E., Basta, N. E., Chao, D. L., & Matrajt, L. (2009). The transmissibility and control of pandemic influenza A (H1N1) virus. *Science*, *326*(5953), 729. doi:10.1126/science.1177373

Yaroslavsky, L. P. (2007). Stochastic nonlinear dynamics pattern formation and growth models. *Nonlinear Biomedical Physics*, *1*(4).

Yasuda, H., & Suzuki, K. (2009). Measures against transmission of pandemic H1N1 influenza in Japan in 2009: Simulation model. *Euro Surveillance: European Communicable Disease Bulletin*, *14*, 44.

Yu, Y., Luo, X., Gao, G., & Ai, S. (2006, November 1-4). Research of a potential worm propagation model based on pure P2P principle. In *Proceedings of the International Conference on Communication Technology*.

Zadeh, L. A. (1965). Fuzzy sets. *Information and Control*, *8*, 338–353. doi:10.1016/S0019-9958(65)90241-X

Zeni, F., Freeman, B., & Natanson, C. (1997). Anti-inflammatory therapies to treat sepsis and septic shock: a reassessment. *Critical Care Medicine*, *25*(7), 1095–1100. doi:10.1097/00003246-199707000-00001

Zeraoulia, E., & Sprott, J. C. (2008). A minimal 2-D quadratic map with quasi-periodic route to chaos. *International Journal of Bifurcation and Chaos in Applied Sciences and Engineering*, *18*(5), 1567–1577. doi:10.1142/S021812740802118X

Zeraoulia, E., & Sprott, J. C. (2008). A two-dimensional discrete mapping with C^{∞}-multifold chaotic attractors. *Electronic Journal of Theoretical Physics*, *5*(17), 111–124.

Zhang, Y., Fu, A., Cai, C. H., & Heng, P. (2000). Clustering categorical data. In *Proceedings of the IEEE International Conference on Data Engineering* (pp. 305-325).

Zhigalskii, G. P. (1997). 1/f noise and nonlinear effects in thin metal films. *Uspekhi Fizicheskikh Nauk*, *167*, 623–648. doi:10.3367/UFNr.0167.199706c.0623

Zouali, M. (2001). Antibodies. In Clarke, A., Ruse, M., Agro, A. F., Bernardini, S., Dotsch, V., & Maccarrone, M., (Eds.), *Encyclopedia of life science*. London, UK: Nature Publishing Group. doi:10.1038/npg.els.0000906

About the Contributors

George Magoulas is Professor of Computer Science, Director of Postgraduate Studies at the Department of Computer Science and Information Systems, Birkbeck College, University of London, and a member of the London Knowledge Lab (www.lkl.ac.uk). He was educated at the University of Patras, Greece, in Electrical and Computer Engineering (BEng/MEng, PhD), and holds a Post Graduate Certificate in Teaching and Learning in Higher Education from Brunel University, UK. He has been teaching and researching in the area of Computational Intelligence and Learning Systems since 1993. His research focuses on the development of intelligent systems that exhibit different levels of learning, seamlessly combine explicit knowledge representation with significant learning capabilities, and can handle uncertainty which is inherent in complex environments. Prof. Magoulas serves on the board of several international journals and his research portfolio includes research grants from the EPSRC, the ESRC, the AHRC, and the JISC, UK.

* * *

Abbas Al-Refaie is currently the chairman of the department of industrial engineering at University of Jordan. Dr Al-Refaie has several papers accepted and published in reputable international journals, including IEEE Transactions; IIE; PPC; JIM; QREI; JSCS; JEM; AIEDAM, JOS; IJALR and QTQM. His research interests include: data envelopment analysis, statistical quality control, design and analysis of experiments, Taguchi methods, and quality management.

Joe Anderson received a B.S. in Operations Research with a minor in Systems Engineering from the United States Military Academy (1999), West Point, New York and a M.S. in Operations Research from Kansas State University (2009), Manhattan, Kansas. He is currently working for the U.S. Army at the Pentagon.

T. Arita received his B.E., M.E., and Ph.D. degrees from the University of Tokyo in 1983, 1985 and 1988. He was a research associate and then an assistant professor at Nagoya Institute of Technology, and a visiting researcher at University of California, Los Angeles. Since 2003, he has been a professor in the graduate school of information science at Nagoya University. His research interests are in the area of artificial life, in particular in the following fields: evolution of language, evolution of cooperation, interaction between evolution and learning, and swarm intelligence. He is the author of several books on artificial life or artificial intelligence.

Subhash Behere retired on 30th Oct 2010 as a professor and former head of the department of physics at the Dr. Babasaheb Ambedkar Marathwada University, Aurangabad, Maharashtra after 37 years of teaching experience to post graduate courses. He was also working as a director of UGC-Academic Staff College of the university since Jan 2009 where he conducted 35 orientation and refresher courses each of nearly three weeks duration. He obtained MS degree in 1971 from the Marathwada University with first rank and PhD degree in 1978. His special interests are in Molecular spectroscopy and he has carried out elaborate work in the field of high resolution rotational spectroscopy, vibrational spectroscopy, potential energy functions, Franck Condon factors, intensity measurements and work related to other molecular interactions. During his research work he was the sole person behind the construction and fabrication of a 35ft Concave grating spectrograph. He has also made substantial contributions in the field of Fractal growth and Simulation studies. Recently he has also completed the study of noise pollution in the Aurangabad city especially due to vehicular traffic. He has published around 50 research papers in reputed international journals and has guided 16 PhD students and 9 are working. He has attended several conferences and chaired many sessions. He was a UGC visiting fellow at Sant Gadgebaba University Amaravati and Saurashtra University Rajkot. He has collaborative research work with the HPPD of BARC Trombay Mumbai and was also an associate of IUCAA during 1992 to 1995. He is a life member of ISCA, IPA, ILA, IAPT. He has presented a paper at the International symposium on molecular spectroscopy held at Columbus Ohio, (USA) in June 2006. He is a referee for national and international journals as well as for PhD thesis of various universities. Besides his regular teaching and research, he is also interested in popularization of science for which he lectured at many colleges and universities, delivered 60 radio talks on All India Radio Aurangabad and wrote several articles in newspapers on current sciences topics in English and Indian Languages. He was sectional recorder of the physical science section of the Indian Science Congress for 2007 & 2008. And also president for the ISCA 2010 held at Trivandrum. His hobbies include philately, bird watching and Indian classical music.

Y. Boutalis received the diploma of Electrical Engineer in 1983 from Democritus University of Thrace (DUTH), Greece and the PhD degree in Electrical and Computer Engineering (topic Image Processing) in 1988 from the Computer Science Division of National Technical University of Athens, Greece. Since 1996, he serves as a faculty member, at the Department of Electrical and Computer Engineering, DUTH, Greece, where he is currently an Associate Professor and director of the Automatic Control Systems lab. Currently, he is also a Visiting Professor for research cooperation at Friedrich-Alexander University of Erlangen-Nuremberg, Germany, chair of Automatic Control. He served as an assistant visiting professor at University of Thessaly, Greece, and as a visiting professor in Air Defence Academy of General Staff of airforces of Greece. He also served as a researcher in the Institute of Language and Speech Processing (ILSP), Greece, and as a managing director of the R&D SME Ideatech S.A, Greece, specializing in pattern recognition and signal processing applications. His current research interests are focused in the development of Computational Intelligence techniques with applications in Control, Pattern Recognition, Signal and Image Processing Problems.

Colleen R. Burgess, MS is the managing partner of MathEcology, LLC, a technical and mathematical services firm located in Phoenix, and a woman-owned small business. Her area of expertise is the application of modeling and simulation to biological systems, and she specializes in the development of mathematical models to help solve policy and national security issues in epidemiological, biological, ecological, environmental and agricultural sciences, with special emphasis on the development of simulation models to assess the threat of infectious diseases to military and civilian populations within the United States and globally. She has developed epidemiological models for international non-profit public health organizations and the US Federal Government agencies and Department of Defense, as well as a number of academic institutions and private industry groups. In addition to her role at MathEcology, Ms. Burgess also holds a Research Faculty Associate appointment in the School of Life Sciences at Arizona State University.

Kyle Carlyle received a B.S. and M.S. in Industrial Engineering from Kansas State University in 2009. His research focused on using graph theory and simulation to optimize quarantine regions. He currently works as a Logistics Engineer for J.B. Hunt Transportation in Lowell, AR.

S. Roy Chowdhury obtained her B.E. in Electronics and Communication Engineering at the Birla Institute of Technology, India in May 2007. Subsequently she earned her MS in Electrical and Computer Engineering at Kansas State University in 2010. Currently, she is pursuing her PhD in Electrical Engineering as a member of the SPINCOM group at the University of Minnesota. Her research interests include complex networks, machine intelligence, network modeling and analysis, signal processing and wireless communication.

M. A. Christodoulou was born in Kifissia, Greece, in 1955. He received the diploma degree (EE'78) from the National Technical University of Athens, Greece, the M.S. degree (EE'79) from the University of Maryland, College Park the engineer degree (EE'82) from the University of Southern California, Los Angeles, and the Ph.D. degree (EE'84) from the Democritus University, Thrace, Greece. He joined The Technical University of Crete, Greece in 1988, where he is currently a Professor of Control. He has been a Visiting Professor at Georgia Tech, Syracuse University, the University of Southern California, Tufts University, Victoria University and the Massachusetts Institute of Technology. He has authored and co-authored more than 200 journal articles, book chapters, books, and conference publications in the areas of control theory and applications, robotics, factory automation, computer integrated manufacturing in engineering, neural networks for dynamic system identification and control, in the use of robots for minimally invasive surgeries and recently in systems biology. Dr. Christodoulou is the organizer of various conferences and sessions of IEEE and IFAC and guest editor in various special issues of International Journals. He is managing and cooperating on various research projects in Greece, in the European Union and in collaboration with the United States. He has held many administrative positions such as the Vice Presidency of the Technical University of Crete, as Chairman of the office of Sponsored research and as a member of the board of governors of the University of Peloponnese. He is a member of the Technical Chamber of Greece. He has been active in the IEEE CS society as the founder and first Chairman of the IEEE Control Systems Society Greek Chapter, which received the 1997 Best Chapter of the Year Award and as the founder of the IEEE Mediterranean Conference on Control and Automation, which became an annual event. Dr Christodoulou received the MCA Founders award in 2005. He is a member of the board of governors of the Mediterranean Control Association since 1993.

C. J. Chyan, born in 1952, received the Master degree from National Taiwan Normal University, PhD degree from Vanderbilt University. He is currently servicing as an Professor in the Department of Mathematics of Tamkang University. His major research interests include Ordinary Differential Equations, Difference Equation, Measure Chain, Boundary Value Problem, Mathematical Biology. He has authored and coauthored about 40 papers.

Todd Easton received a B.S. in Mathematics with a minor in Statistics from Brigham Young University (1993), Provo, Utah, an M.S. in Operations Research from Stanford University (1994), Stanford, California and a Ph.D. in Industrial and Systems Engineering from Georgia Institute of Technology (1999), Atlanta GA. He worked as a post-doctoral fellow at Georgia Institute of Technology and then moved to Manhattan, Kansas where he is currently an Associate Professor in the Industrial and Manufacturing Systems Engineering Department at Kansas State University. His research interests are in combinatorial optimization with an emphasis in integer programming and graph theory.

Holly Gaff is an Assistant Professor in the Department of Biological Sciences at Old Dominion University and is affiliated with the Virginia Modeling, Analysis and Simulation Center. Dr. Gaff earned her Ph.D. in Mathematics at the University of Tennessee, Knoxville, in 1999. Dr. Gaff's research interests have focused mainly on studying the dynamics and control of infectious diseases using mathematical modeling and computer simulation. Most of her research has focused on developing mathematical models for exploring the ecology of vector-borne diseases including Rift Valley fever and tick-borne diseases in the Hampton Roads area. She has had funding for these and other projects from NIH, NSF, DHS, CDC and the VA.

T. Giotis received the Diploma and the MSc degree both in electrical and computer engineering from the Technical University of Crete, Chania, Greece in 2005 and 2008, respectively. Currently, he is a PhD student in the same department. He has also been an assistant in the Automation Laboratory of the Technical University of Crete. His research interests include linear and nonlinear systems, neural networks, systems biology and adaptive and neuro-fuzzy control.

David Hartley is an Associate Professor at the Georgetown University Medical Center. He earned a BS (1990), MS (1992), and PhD (1996) in physics and an MPH with an emphasis in epidemiology (2006). His current research interests include the ecology of infectious disease, hospital infection control, and biosurveillance. His research has focused on studying the dynamics and control of infection using mathematical modeling and computer simulation as well as more traditional epidemiologic methods. He has studied Rift Valley fever, West Nile virus, cholera, Staphylococcus aureus, and food-borne infections among others.

W. Hsu is currently an Associate Professor in the Department of Computing and Information Sciences at the Kansas State University. He earned his PhD in Computer Science from the University of Illinois at Urbana-Champaign in 1998. Previously, he earned his MSE and BS degrees in Computer Science with a double major in Math Sciences and Computer Science at The John Hopkins University in 1993. As the Director of the Laboratory for Knowledge Discovery in Databases (KDD) at Kansas Stet University, his research interests include knowledge discovery in databases, information extraction, and knowledge-based systems, computational neuroscience, medical informatics, bioinformatics (especially computational genomics and proteomics), crisis management, multisensor integration, atomic-scale computational physics, ecological genomics, and sensor networks.

Jacqueline Jackson is a researcher for Sentara Healthcare and an adjunct professor at Regent University. Dr. Jackson earned her Ph.D. in Health Services Research from Old Dominion University with a concentration in modeling and simulation in 2010. Dr. Jackson's research interest has focused on using geographic information sciences to study infectious diseases. Most of her research has focused on understanding the environmental factors related to vector-borne diseases, such as Rift Valley fever and West Nile virus.

Sarika Jain, assistant professor, school of engineering and technology, Amity University, Noida (U.P.) has carried out her research work under the supervision of Prof. S. L. Singh and has submitted her doctoral thesis to H. N. B. Garhwal (Central) University, Srinagar-Garhwal. She has an M. Phil. degree in mathematics and master's degree in both mathematics as well as computer science from M. D. University, Rohtak. Her primary research interest is in computer graphics and stability theory of iterations.

Matthew James is a senior at Kansas State University where he is pursuing a concurrent B.S./M.S. in Industrial Engineering and a B.S. in Economics (anticipated 2012). He is highly involved in student government, having served as Vice-President of the College of Engineering and chairing the University's student fee allocation committee.

Rekha Kandwal is currently working in India Meteorological Department, Ministry of Ministry of Earth Sciences and Science and Technology, India. She received her PhD in computer science from Jawahar Lal Nehru University and MS degree in mathematics from University of Delhi, India. Her current research interests include knowledge engineering, knowledge representation, machine learning, active learning, clustering and rough sets analysis.

A. R. Khan is Head, Department of Bioinformatics at Maulana Azad College, Dr. Rafique Zakeria Campus, Aurangabad, Maharashtra since last 4 years. He did his M.Sc.(Physics) in 1975 and Ph.D (Physics) in 1985 from the Marathwada University, Aurangabad (MS) India. He has experience of over thirty years of teaching different subject at different levels. He has published around several research papers in reputed international journals. He has attended several conferences and workshops and presented research papers. Besides his regular teaching and research, he has contribution to the field of pharmacokinetics in pharmacy and worked as consultant to multinational industries. He has interest in microcontrollers and embedded systems.

Taehyong Kim is currently a PhD candidate in the Department of Computer Science and Engineering at the State University of New York at Buffalo. He is a research assistant in bioinformatics and data mining laboratory at the department of Computer Science and Engineering of the State University of New York at Buffalo. He was an advisory software engineer at the advanced software engineering center at Samsung SDS before he joined PhD program. His research interests include bioinformatics, data mining, stochastic simulation, dynamics modeling and software engineering.

Sanjeev Kumar Prasad was born in 1977 in India. He received master's of technology degree from Utter Pradesh Technical, Lucknow, India. He is presently assistant professor in Ajay Kumar Garg Engineering College, Ghaziabad, an affiliated college of Uttar Pradesh Technical University, Lucknow, India. Before joining this college, he has served many other colleges affiliated to Uttar Pradesh Technical University, Lucknow, India. He has published 1 paper in International Journal and 5 papers presented and published in national and international conference.

Celestin Lele obtained his master degree at Yaound university (Cameroon) and his PhD in mathematics at Harbin Institute of Technology (China). He did a postdoctorate at the university of Nice (France), a year visiting fellow at the university of Cambridge (UK). He is also a Fulbrith research scholar at the university of Oregon (USA). His current research interests are in the area of Algebra, fuzzy algebras and logic. He is a permanant professor at the University of Dschang.

Kang Li is currently a Ph.D. student in the Department of Computer Science and Engineering of the State University of New York at Buffalo, under the supervision of Prof. Aidong Zhang. He received his B.S. in Electronic Information Science and Technology from University of Science and Technology of China in 2009. His research focuses on network modeling and data mining.

S. W. Lin, born in 1978, received the Master degree from National Taipei University of Education, supervised by Dr. Fu-Hsiang Wong, now is studying in the Department of Mathematics of Tamkang University for PhD degree, supervised by Dr. Chuan Jen Chyan. His study is about Ordinary Differential Equations and Dynamic Equations on Time Scales.

Q. M. Danish Lohani is working as a Post Doctoral Fallow in the Department of Mathematics, Aligarh Muslim University, India. It is a prestigious fellowship awarded by the National Board of Higher Mathematics and Department of Atomic Energy, Government of India, after evaluating his research work through the experts in that field. A Ph. D. in Mathematics with thirteen international publication in various specialization such as Functional Analysis, Summability theory, Sequence Space, Fuzzy Topology with Applications. Beside research he is actively involved in teaching to undergraduate classe in the Department of Mathematics, Aligarh Muslim University, Aligarh from 2006 till now. During these years he taught Co-ordinate geometry and Advance calculus.

Prerna Mahajan is currently working in computer science department, Institute of Information Technology and Management, Indraprastha University, Delhi, India and also working towards a doctoral degree in computer science from Banasthali University, India. She received her MCA degree from Gurukul kangri University Hardwar, India. Her current research interests include active learning, clustering and rough sets analysis, data mining/KDD.

Kenichi Minoya was born in 1983 in Japan. He graduated from Department of Information Systems Science in the Faculty of Engineering at Soka University in 2006, and received Master's degree from Department of Information Systems Science, Graduate School of Engineering in 2009 from the same university. He is now a phD student of Department of Complex Systems Science, Graduate School of Information Science at Nagoya University. His research interests include planning, symbolic communication, and relation between planning and symbolic communication. Also, he is interested in the information processing procedure in brain and a computational model for the theory of minds.

S. N. Mishra obtained his doctorate degree in mathematics from the University of Allahabad, India in 1977. Since then he has served in different capacities in several countries. Currently, he is a professor of mathematics and the director of the school of mathematical and computational sciences at the Walter Sisulu University in South Africa. His current research interests are Nonlinear Functional Analysis and General Topology, mainly, fixed point theory and its applications. He has published a number of papers in this area.

Tianchan Niu works as a Project Researcher in the Imaging Science and Information Systems (ISIS) Center at Georgetown University Medical Center. Dr. Niu earned her Ph.D. in Computational Analysis and Modeling at Louisiana Tech University, Ruston, in 2007. Dr. Niu's research interests have focused mainly on applying mathematical and computational methods to understand the dynamics of infectious disease in agricultural populations and human as well as to discover and analyze strategies to detect and control such diseases. Most of her research has focused on developing epidemiological models to explore the ecology of vector-borne diseases, such as Rift Valley fever and West Nile virus.

Aïssatou Mboussi Nkomidio is born on 16 April 1980 at Gueboba (Cameroon). She is assistant at the Department of Physics at the University of Yaoundé I. She is holder of a Ph. D. Doctorat obtained in 2010 in the field of Mechanics (Nonlinear Dynamics of Biological systems). She is associate researcher at Laboratory of Modeling and Simulation in Engineering and Biological Physics of Department of Physics with strong interests in applied Nonlinear Dynamics, biological physics and bio-engineering and appropriate technologies for development (www.lamsebp.org). She is presently member of various scientific organizations at the national and international levels.

Ramaswamy Palaniappan received his first degree and MEngSc degree in electrical engineering and Phd degree in microelectronics/biomedical engineering in 1997, 1999 and 2002, respectively from University of Malaya, Kuala Lumpur, Malaysia. He is currently with the School of Computing and Electronic Systems, University of Essex, United Kingdom. Prior to this, he was the Associate Dean and Senior Lecturer at Multimedia University, Malaysia and Research Fellow in the Biomedical Engineering Research Centre-University of Washington Alliance, Nanyang Technological University, Singapore. His current research interests include biological signal processing, brain-computer interfaces, biometrics, artificial neural networks, genetic algorithms, and image processing. To date, he has published a book and written over 120 papers in peer-reviewed journals, book chapters, and conference proceedings. Dr. Palaniappan is a senior member of the Institute of Electrical and Electronics Engineers (Engineering in Medicine and Biology Society), member in Institution of Engineering and Technology, and Biomedical Engineering Society. He also serves as editorial board member for several international journals. His pioneering work on using brain signals as biometrics has received international recognition.

Yiannis Papelis is a Research Associate Professor at Old Dominion University's Virginia Modeling Analysis & Simulation Center (VMASC). Dr. Papelis earned a BSEE (with honors) from Southern Illinois University in 1988, a MSEE from Purdue University in 1989 and a Ph.D. degree in Electrical & Computer Engineering from the University of Iowa in 1993. Dr. Papelis' research interests include modeling and simulation of autonomous agents in constructive and immersive virtual environments, and human behavior modeling and assessment. He is currently conducting research on autonomous agent modeling issues as applied to a wide range of topics, including use of serious games for education, disease modeling and propagation, realistic simulation of crowds and control of autonomous ground and aerial vehicles.

K. B. Patange is working as a lecturer in physics at Deogiri College, Aurangabad, Maharashtra since last 7 years. He had teaching experience to bachelor of education and also to the engineering diploma. He obtained master of science degree in 1995 and master of education degree in 1997 from the Marathwada University, Aurangabad. He received PhD degree in physics from Dr. Babasaheb Ambedkar Marathwada

University, Aurangabad for the topic entitled as "Study of Noise Pollution in Aurangabad City" in 2010. His special interests are in study of noise, fractals, and electronics. He has attended several conferences and workshops. He has published around several research papers in reputed international journals. Also he has organized workshops on topics such as combating aids, etc. He presented paper entitled "Impact of noise pollution on students in educational institutions" in national conference of academy of environmental biology in 1997 and also presented paper entitled as "Fractal geometry of lakes" at 97th Indian Science Congress Association held at University of Kerala, Thiruvananthapuram, January 3 to7, (2010). He is a life member of IAPT and organizing physics olympiad exams at his college. Besides his regular teaching and research, he is also interested in delivering lectures on value education. Till date he had delivered many lectures on the said topic. He wrote articles in newspapers on noise pollution. His interest is in meditation to realize the truth of life.

J. M. Pathan is working as a Vice-Principal at Maulana Azad College, Dr. Rafique Zakeria Campus, Aurangabad, Maharashtra since last 25 years. He obtained M.Sc. degree in 1985 from the Marathwada University with first division. Also he obtained D.H.E. degree in 1988 and recently submitted thesis for award of Ph.D. degree. His special interests are in electronics and he has carried out elaborate work in the field of fractal growth studies. He has published around 5 research papers in reputed international journals. He has attended several conferences and workshops and presented research paper at 97th Indian Science congress, Thiruvanantpuram held during 3-7 Jan 2010. He is a member of ISCA, IAPT. Besides his regular teaching and research, he is also interested in popularization of science amongst students of junior college. He arranged many workshops for poor and weak students to enhance their knowledge of Physics. He is the organizer of free allied coaching classes which is run by central government for minority students in his college since last four years. His hobbies include playing and watching cricket and listening Gazals.

H. E. Psillakis received the Diploma and the PhD degree both in electrical and computer engineering from the University of Patras, Patras, Greece in 2000 and 2006, respectively. Since 2007, he has been with the Technological and Educational Institute (TEI) of Crete, Heraklion, Crete, Greece teaching courses in the Applied Informatics and Multimedia and at the Electrical Engineering Department. He has also been an Adjunct Lecturer for the Electronic & Computer Engineering Department, Technical University of Crete, Chania, Crete, Greece and the Financial & Management Engineering Department, University of the Aegean, Chios, Greece. His research interests include intelligent control of uncertain and stochastic systems, passive and nonlinear systems and advanced control and stability applications on power, drive and robotic systems.

Murali Ramanathan received his PhD in Bioengineering from the University of California-San Francisco and University of California-Berkeley Joint Program in Bioengineering. He is a full professor of Pharmaceutical Sciences at the State University of New York at Buffalo. His research focus is the treatment of multiple sclerosis (MS), an inflammatory-demyelinating disease of the central nervous system that affects over 1 million patients worldwide. MS is a complex, variable disease that causes physical and cognitive disability and nearly 50% if patients diagnosed with MS are unable to walk after 15 years. The etiology and pathogenesis of MS remains poorly understood. The focus of the research is to identify the molecular mechanisms by which the autoimmunity of MS is translated into neurological damage in the CNS.

Alex Savachkin is an assistant professor of the Department of Industrial and Management Systems Engineering at the University of South Florida at Tampa. His research interests include analytical support of enterprise risk analysis, public health disaster mitigation and emergency planning, and (re)engineering healthcare systems. His research has been funded by the National Science Foundation and Department of Defense. Dr. Savachkin teaches courses in applied probability, stochastic processes, and risk analysis. Dr. Savachkin received a Ph.D. in industrial engineering from Texas A&M University in 2005.

C. Scoglio is currently an Associate Professor in the Department of Electrical and Computer Engineering at Kansas State University. She obtained her Dr. Eng. in Electronics Engineering from 'Sapienza' University of Rome, Italy in 1987. She had been with Fondazione Ugo Bordoni – Italy, and with the 'Broadband and Wireless Networking Laboratory' of the Georgia Institute of Technology. She is the Co-Director of the 'Sunflower Networking Group' and K-State EPICENTER at Kansas State University. Her research interests include network science, complex networks, modeling and control of epidemics, overlay and virtual networks, smart grids and renewable energy.

Surajit Sen received his PhD in theoretical statistical physics from the University of Georgia in 1990. He is a full professor in the department of Physics at the State University of New York at Buffalo. He research is currently supported by the National Science Foundation and the Solid Mechanics and Mathematics divisions of the US Army Research Office. He is serving the second term as the President of the American Chapter of the Indian Physics Association; he is an editor of International Journal of Modern Physics B and Modern Physics Letters B and an associate editor of the physics collection of the open digital library at merlot.org. He is currently also an elected member of the Committee on International Scientific Affairs of the American Physical Society and serve in several of American Physical Society's awards committees. He is currently working on nonlinear dynamics, statistical physics, and complex systems in biology and conflicts.

Yusuf H. Shaikh is working as a assistant professor in Shivaji Arts, Commerce and Science College Kannad; Dist Aurangabad., Maharashtra State since last 16 years. He did his M.Sc.(Physics) in 1993 and PhD (Physics) in 2007 from the Marathwada University, Aurangabad (MS) India. He has experience of over sixteen years of teaching at degree levels and several talks at Teachers Training i.e., orientations and refresher courses. He has published around several research papers in reputed international journals. He has attended several conferences and workshops and presented research papers. He is a member of ISCA. Besides his regular teaching and research, he is also interested in popularization of science amongst students of college.. His hobbies include playing and watching cricket and listening Gazals.

S. L. Singh was born in 1942 in village Chowkhara of district Mirzapur (U.P., India). He obtained his BA and MA degrees from the University of Allahabad in 1964 & 1966 respectively. He received his PhD (mathematics) from Kumaun Univeristy, Nainital in 1979. He retired from Gurukula Kangri University, Haridwar in 2004 as professor and principal. Besides publishing more than 200 research papers, he supervised 29 doctoral theses in mathematics and computer graphics. He is Fellow of *Allahabad Mathematical Society* and Vijñāna Parisada (Science Academy of India) of India. Currently, he is UGC Emeritus Fellow at Pt. L. M. S. Govt. Autonomous Postgraduate College, Rishikesh.

R. Suzuki received his Ph.D. degree from the graduate school of human informatics, Nagoya University in 2003. He is now an associate professor in the graduate school of information science, Nagoya University. His main research fields are artificial life and evolutionary computation. By using individual-based evolutionary models, he is investigating how evolutionary processes can mutually interact with various ecological processes such as lifetime learning (phenotypic plasticity), niche construction, and dynamically changing network structures of interactions. He is also interested in an application of biological concepts to artistic works such as interactive evolutionary computation.

Tatsuo Unemi was born in Kanazawa, Japan in 1956. He has graduated at Tokyo Institute of Technology in 1978. He received a Master's degree from the Department of System Sciences in 1980, and a Doctor's degree in 1994 from the same university. He worked as a research associate from 1981 to 1987 at the Tokyo Institute of Technology, as an assistant professor at the Nagaoka University of Technology from 1987 to 1992, and from 1992 to 1995 as a visiting scholar at the Laboratory for International Fuzzy Engineering. Since 1992, he is employed as an assistant professor at the Soka University. As a visiting professor, he stayed at the Artificial Intelligence Laboratory of the University of Zurich from April to September 2000. He conducted research in the fields of Artificial Intelligence, Soft Computing, and Robotics. Current interests include Sociological and Artistic applications of techniques derived from Artificial Life.

Andres Uribe-Sanchez is a doctoral candidate of the Department of Industrial and Management Systems Engineering at the University of South Florida at Tampa. His areas of research interest include engineering risk analysis, support of healthcare enterprise capacity management, and decision support for mitigation of large-scale public health disasters.

Ritu Vijay is currently working in department of computer science and electronics, Banasthali University. She received her PhD in computer science and electronics from Banasthali University, MS degree in electronics from Banasthali University, India. Her research interests include neural networks, wavelets compression, artificial intelligence and digital signal processing.

Tobias Wissel received his MSc degree in Embedded Systems in 2010 from the University of Essex, United Kingdom. In the context of a one-year scholarship awarded by the German Academic Exchange Service, DAAD, his thesis was aligned to the research of the Brain-Computer Interface group at Essex. For his diploma thesis, he is currently involved in a research project at the University of Magdeburg, Germany, where he has been a student of electrical engineering/information electronics since 2005. He is also working as a research assistant for the Chair for Healthcare Telematics and Medical Engineering. Prior to this he was a research assistant in the fields of digital signal processing and embedded systems/integrated automation, which are his two major research interests. Tobias Wissel is a student member of the Institute of Electrical and Electronics Engineers and IEEE Signal Processing Society.

Paul Woafo is professor of Physics at the University of Yaoundé I, Cameroon. He is holder of a Doctorat d'Etat obtained in 1997 in the field of Mechanics (Nonlinear Dynamics). He is the director of the Laboratory of Modeling and Simulation in Engineering and Biological Physics with strong interests in applied Nonlinear Dynamics, Electromechanical devices, control of vibrations, dynamics of semiconductor lasers, chaos cryptography, biological physics and bio-engineering, and appropriate technologies for development (www.lamsebp.org). He is presently co-author of more than 100 papers published in peer-reviewed journals and member of various scientific organizations at the national and international levels.

F. H. Wong is an professor of Mathematics at National Taipei University of Education, Taiwan. He received his Ph.D degree from National Central University, in 1992. Some of his interests are boundary value problems for ordinary differentail equations, dynamic equations on time scale and oscillation theory. He has authored and coauthored about 90 papers.

Aidong Zhang received her PhD in Computer Science from Purdue University. She is a full professor and chair in the department of Computer Science and Engineering at the State University of New York at Buffalo and the principal investigator and program director of the Buffalo Center for Biomedical Computing (BCBC). She is the author of over one hundred research publications pertaining to her research in such areas as bioinformatics, geographic information systems, content-based image retrieval, distributed database systems, multimedia database systems, digital libraries, and database mining. In March of 2003, she received a $1.6 million NSF grant for bioinformatics research and in August 2003, she also received a National Institutes of Health (NIH) grant to establish a pre-Center for Biomedical Computing. Her research interests include bioinformatics, data mining, database systems, content-based retrieval, and multimedia systems.

Index

F

G

H

I

K

L

M

N

O

P

Q

R

S

T

U

V

W